Behavioral Neurobiology

BEHAVIORAL NEUROBIOLOGY

AN INTEGRATIVE APPROACH

THIRD EDITION

GÜNTHER K. H. ZUPANC
Northeastern University, Boston, Massachusetts

FOREWORD BY

THEODORE H. BULLOCK

Great Clarendon Street, Oxford, OX2 6DP,
United Kingdom

Oxford University Press is a department of the University of Oxford.
It furthers the University's objective of excellence in research, scholarship,
and education by publishing worldwide. Oxford is a registered trade mark of
Oxford University Press in the UK and in certain other countries

First edition 2004
Second edition 2010

Published in the United States of America by Oxford University Press
198 Madison Avenue, New York, NY 10016, United States of America

British Library Cataloguing in Publication Data
Data available

Library of Congress Control Number: 2018932382

ISBN 978–0–19–873872–5

Printed and bound by
CPI Group (UK) Ltd, Croydon, CR0 4YY

To Cecilia

ABOUT THE AUTHOR

Günther K. H. Zupanc is Professor, Department of Biology of Northeastern University in Boston, Massachusetts. After graduating in biology (1985) and physics (1987) from the University of Regensburg, Germany, he did his graduate studies in neurosciences at the University of California, San Diego, where he received his Ph.D. in 1990. Subsequently, he worked as a Research Biologist at the Scripps Institution of Oceanography in La Jolla, California (1990–1992), headed a Junior Research Group at the Max Planck Institute for Developmental Biology in Tübingen, Germany (1992–1997), and was on the faculty of the University of Manchester, U.K. (1997–2002) and of Jacobs University Bremen (formerly International University Bremen), Germany (2002–2009). Zupanc was also Visiting Professor at the University of Ottawa, Canada; Visiting Scientist at the Salk Institute for Biological Studies and Scripps Research Institute, both in San Diego, the University of Chicago, Tufts University in Boston, and the Max Planck Institute for Behavioral Physiology, Seewiesen; and adjunct faculty member of the University of Tübingen. He has taught lecture and laboratory courses in behavioral neurobiology at both undergraduate and graduate level to numerous students in the U.S.A. and Europe. In recognition of his research and his contributions to the public understanding of science, he has received many awards. His research focuses on the exploration of cellular mechanisms underlying behavioral and neuronal plasticity in teleost fish. In addition to a large number of articles, his book publications include: *Fish and Their Behavior* (1982); *Praktische Verhaltensbiologie* (editor; 1988); *Fische im Biologieunterricht* (1990); *Adult Neurogenesis: a Comparative Approach* (editor; 2002); *Electric Fish: Model Systems for Neurobiology* (editor; 2006); *Integrative and Comparative Neurobiology: Papers in Memoriam of Theodore H. Bullock* (editor; 2008); and *Towards a Comparative Understanding of Adult Neurogenesis* (editor, jointly with Luca Bonfanti and Ferdinando Rossi; 2011). Zupanc was senior editor of the *Journal of Zoology* (2007–2011). Since 2008, he has served as editor of the *Journal of Comparative Physiology A*.

FOREWORD

How profoundly human it is, how deeply characteristic of our species—ethologically speaking—to wonder at, study, and investigate behavior of our own and other species toward understanding, in the sense of accounting for, the actions, the appetites, the drives, alternative modes, and sensory guidance that we observe. How rewarding we find it to think of hypotheses, to test and discard them, and to achieve a degree of understanding—at one or more levels—and report it to our colleagues.

In all its angles and aspects, this human urge to embrace, comprehend, and explain what we and other species do—our behavior—is now recognized as a field of endeavor called neuroethology, rivaling those other distinctively human traits such as cooking, dancing in innovative ways, recounting the past and imagining the future, making each other laugh, and making music.

It strikes me as some kind of a pinnacle to bring together between book covers what has been learnt and how it was done so that we can exercise another very human trait—wonder at it!

One reason for this somewhat obtuse opening is that it underlines a strongly felt view of Günther Zupanc that the *raison d'etre* of such a book need not depend on its potential relevance to practical human concerns, such as medicine or the psychology of human aggression, but, in today's world, upon simple curiosity and knowing for its own sake. This latest summing up follows worthily the path pioneered by Jörg-Peter Ewert (*Neuro-Ethologie: Einführung in die neurophysiologischen Grundlagen des Verhaltens*, Springer-Verlag, Berlin, 1976), Jeff Camhi (*Neuroethology: Nerve Cells and the Natural Behavior of Animals*, Sinauer, Sunderland/Massachusetts, 1984), and most recently Tom Carew (*Behavioral Neurobiology: The Cellular Organization of Natural Behavior*, Sinauer, Sunderland/Massachusetts, 2000), each in its own style. Exploiting the freedom of a foreword writer, I would like to underline some of the features of special importance in Günther Zupanc's treatment, without implying that they are neglected in previous books.

First and foremost, it is a feature of the subject that it deals with biodiversity in the extreme. The consequence of evolution is an accumulation of diversity, particularly in behavior—more conspicuously and significantly, and more inviting to analysis and explanation, than form, color, or pattern. The usual distillate of biology—that 'life is genes propagating genes'—is a serious misrepresentation because evolution has created diversity in what animals do between generations; life consists of diverse states and actions—of course including keeping alive and reproducing. But the big picture is missed if we do not hold up for scrutiny the different ways of doing these things in beetles that burrow, butterflies that migrate, corals that luminesce, and lions that sleep away most of the day.

Zupanc has recognized the importance of organizing the wealth of detail in a reader-friendly way—with Leitmotifs in special categories and sidebars. He has also recognized the importance of descriptive natural history preceding and leading into reductionist analysis—neuroethology begins with adequate ethology.

Particularly important in neuroethology, and well represented in this treatment, is comparison—comparison of sensory stimuli that trigger, of pathways and central structures involved, of background state dependence, and alternative tricks for canceling self-generated signals. Comparison is the essence and applies to different taxa, ontogenetic stages, and readiness states. Nature appears rarely to use a single mechanism for all animals that exhibit a similar behavior—nor does she use a large number of alternate mechanisms; several is the norm. The limitation in our knowledge is a limitation of research endeavor, perhaps because there is less glamour in looking at the same behavior in other taxa than in looking at hitherto unstudied behavior.

Another feature that stands out in the present treatment, although frequently underplayed in the primary literature, is that every case is identified with a broad or basic issue. This is complemented by showing how some species are particularly favorable for the given study and for closer analysis.

A distinctive feature of this book is its concern with the history of each strand of the fabric of neuroethology, the drama and dependence on serendipity and mindset. The results of this historical research are distributed throughout the text, often in sidebars or boxes. May this book illumine the science, influence the way investigators proceed, and draw new ones into the field.

Theodore Holmes Bullock
La Jolla
June 2003

PREFACE

About the book

This book is based on two courses that I designed and taught on joining the faculty of the School of Biological Sciences of the University of Manchester in 1997. One of these behavioral neurobiology courses was targeted at a beginner's level, the other at an advanced level. Over the five years that followed, they were taken by several hundred students coming from a wide range of degree programs. It was this positive interaction, as well as the students' request to make available a text covering these courses, that encouraged me to write this book. These students also provided me with invaluable feedback on how to stimulate interest in the subject, without compromising the quality of the science taught.

Like the Manchester courses, this book is designed primarily for undergraduate and graduate students, but postdoctoral scientists and instructors of such courses are also likely to benefit from it. I have assumed that the students have some basic knowledge in biology, physics, and mathematics. However, courses in neurobiology are not a prerequisite. Concepts and approaches from this discipline that are important for understanding behavioral neurobiology are introduced in Chapters 2.

Any scientific discovery can be fully understood only within its historical context. I have therefore also included a chapter (Chapter 3) on the historical development of behavioral neurobiology, and added to several chapters a description of the work and life of those who have pioneered this development.

The approach used in the book has been to focus on a few selected systems that, in my opinion, best illuminate the key principles. These examples are then discussed in depth, while relating them to the general principles. Unavoidably, such an approach leads to the exclusion of a number of other excellent studies. I apologize to those whose work has been neglected.

I have also placed particular emphasis on presenting not just the findings of research, but also describing the experimental approaches employed to obtain these results, and discussing the limitations of the experimental design and the scientific methodology. I consider such a discussion to be of vital importance for the learning process of students and their ability to critically evaluate the literature.

ACKNOWLEDGEMENTS

The foundation to writing this book was laid during my own graduate education. Four teachers have been particularly influential. The late Walter Heiligenberg, in whose laboratory at the Scripps Institution of Oceanography of the University of California, San Diego (UCSD) I had the privilege to work both as a Ph.D. student and as a postdoctoral fellow. The late Ted Bullock, also of UCSD, has been a source of inspiration since my graduate student days. He made many suggestions on the manuscript of the first edition, and also kindly contributed the Foreword. Larry Swanson, then at the Salk Institute for Biological Sciences in La Jolla, California, now at the University of Southern California at Los Angeles, taught me how to use neuroanatomy as a tool to analyze the structural basis of behavioral control mechanisms. Equally importantly, he also showed me that even the teaching of a difficult subject can be joy for both instructor and student. Len Maler of the University of Ottawa, Canada, taught me, through many joint projects, how to integrate behavioral, anatomical, and physiological data to gain an appreciation of how the central nervous system generates behavior.

In the course of writing the three editions of this book, I greatly benefited from the advice of the following colleagues: Jon Banks (University of Manchester, U.K.); Jon Barnes (University of Glasgow, U.K.); Bob Beason (University of Louisiana, Monroe, U.S.A.); Rob Bell (Queen's University Belfast, U.K.); Gary Boyd (University of the West of Scotland, Paisley, U.K.); Kenneth H. Britten (University of California, Davis, U.S.A.); Toby Carter (Anglia Ruskin University, U.K.); Bruce Carlson (Washington University, St. Louis, U.S.A.); Chris Elliott (University of York, U.K.); Jack Gray (University of Saskatchewan, Canada); Natalie Hempel de Ibarra (University of Exeter, U.K.); Jane Hoyle (Sheffield, U.K.); Franz Huber (Starnberg, Germany); Jason Jones (Vassar College, U.S.A.); Mark Konishi (California Institute of Technology, Pasadena, California, U.S.A.); Bill Kristan (UCSD, San Diego, U.S.A.); Jürg Lamprecht (Max-Planck-Institut für Verhaltensphysiologie, Seewiesen, Germany); Ken Lohmann (University of North Carolina, Chapel Hill, U.S.A.); Anne Lyons (Oxford, U.K.); Eve Marder (Brandeis University, Waltham, Massachusetts, U.S.A.); Tom Matheson (University of Leicester, U.K.); Roswitha M. Marx (University of Victoria, Canada); Tom Matheson (University of Leicester, U.K.); James Mazer (Yale University, U.S.A.); Jo Ostwald (Eberhard Karls Universität Tübingen, Germany); Sarah A. Parker (Stratford-upon-Avon, U.K.); Alan Roberts (University of Bristol, U.K.); David Skingsley (Staffordshire University, Stoke-on-Trent, U.K.); Julian Thomas (Sterna Word Services, Somerset, U.K.); Wim van de Grind (Universiteit Utrecht, The Netherlands); Gerhard von der Emde (University of Bonn, Germany); Neil Watson (Simon Fraser University, Burnaby, British Columbia, Canada); David Wilcockson (Aberystwyth University, U.K.); Janine Wotton (Gustavus Adolphus College, U.S.A.); and Jayne E. Yack (Carleton University, Ottawa, Canada).

My friend Cecilia Ubilla (UCSD) spent numerous hours on the manuscript of the three editions to optimize my style of writing. Her comments from the perspective of a naive reader on the one hand, and an experienced professional writer and teacher on the other, were instrumental in polishing up the text.

Another factor that proved to be indispensable in writing this book was the excellent collaboration with the staff of Oxford University Press. Esther Browning proposed this project and got me started. Five publishing editors oversaw the writing and production of the three editions—Jon Crowe, Ross Bowmaker, Dewi Jackson, Jess White, and Lucy Wells. They have been reliable sources of advice and encouragement, and I am grateful to each of them.

Major parts of the manuscript for the book were written while I spent summers at UCSD. I am grateful to the Conrad Naber Foundation (Germany) for partially funding these visits. I would also like to thank the Royal Society in the U.K. for the award of a History of Science grant that enabled me to collect the information necessary to write Chapter 3.

Finally, I am indebted to my wife Marianne, and our children Frederick, Christina, and Daniel for their (almost) never-ending forbearance. Without their support, this book would not have been possible.

Günther K. H. Zupanc
Boston, Massachusetts
November 2018

TABLE OF CONTENTS

HOW TO USE THIS BOOK

Each of the 13 chapters is organized in a similar way and contains a number of valuable tools which aim to help you to understand and learn.

The **introduction** and **key concepts** sections outline the major topics and ideas covered in each chapter.

Key concepts

The orienting responses of animals are extremely diverse. A first major attempt to classify the different types of these behaviors was undertaken by the German zoologist Alfred Kühn (1885–1968) and is

The **take-home messages** that are most important to remember in each section are highlighted.

➤ Orienting behavior does not depend on the existence of a nervous system. The unicellular protist *Paramecium caudatum*, for example, exhibits several taxis reactions, including chemotaxis and galvanotaxis.

Text boxes explore key concepts, such as the structure and function of central pattern generators, and key people in the field of neuroethology.

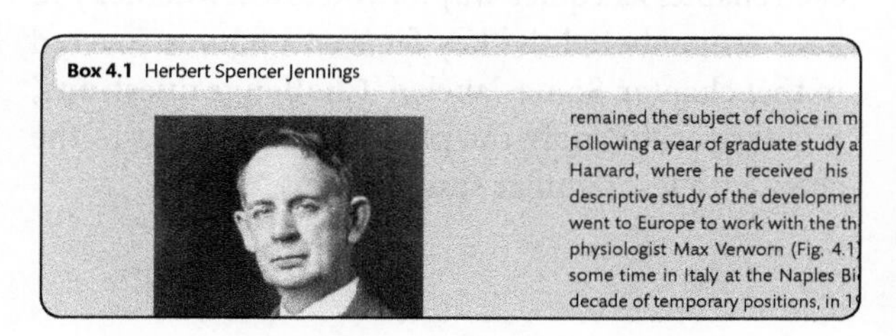
Box 4.1 Herbert Spencer Jennings

remained the subject of choice in m
Following a year of graduate study a
Harvard, where he received his
descriptive study of the developme
went to Europe to work with the th
physiologist Max Verworn (Fig. 4.1
some time in Italy at the Naples Bi
decade of temporary positions, in 1

Key terms are highlighted in bold. **Glossary term** boxes provide on-the-page explanations of the most important of these terms, or ones with which the reader may not be familiar; in addition, there is a glossary at the end of the book, in which these terms are collated.

Labyrinth Otolith organ plus semicircular canals plus cochlea.

The otolith organ, whose structure and function will be examined in more detail below, consists of a patch (called the **macula**) of sensory cells and a covering

There is a **summary** at the end of each chapter listing the important points which have been covered in that chapter.

Summary

- Among orienting responses, taxes constitute an important category. They involve an orienting reaction or movement in freely moving organisms directed in relation to a stimulus.

Each chapter includes a section at the end entitled '**The bigger picture**' which connects the chapter-specific content with the general principles, conclusions, and perspectives in the wider field of research and application.

The bigger picture

As we have seen in Chapter 3, the beginning of the twentieth century was a critical time in the development of the behavioral sciences. Scientists in different countries and from different disciplines

There is a **recommended reading** list at the end of each chapter for those wishing to read further around the topic.

Recommended reading

Adler, J. (1987). How motile bacteria are attracted and repelled by chemicals: An approach to neurobiology. *Biological Chemistry Hoppe-Seyler* **368:**163–173.

Short-answer questions are provided at the end of each chapter as a quick way for you to test whether you have remembered the key facts and figures covered in that chapter. Some, labeled '**Challenge questions**', require you to apply the principles explained in the chapter to less familiar situations.

Short-answer questions

4.1 Define, in one sentence, the term 'positive phototaxis'.

4.2 Attracted by the calling song of an invisible conspecific male, a female cricket walks to the source of sound.

Essay questions are also provided at the end of each chapter. These broader questions could be used by instructors to set assessments, or by students for exam practice. These questions can typically be answered by two to three written pages.

Essay questions

4.1 Describe how mechanical stimulation of paramecia leads, through proper activation of the cilia, to a phobotactic response.

Advanced topics at the end of each chapter offer a more extended assignment with starter references provided. These will be most suitable for senior undergraduates and graduate students. It is envisaged that the assignment would take the form of a review article of around five to ten written pages.

Advanced topic Physical modeling of galvanot

Background information

Galvanotaxis involves orientation in a DC electric field of *Paramecium* toward the cathode. This orienting movement is mediated by specific changes in the beat pattern of the

A list of **references** accompanies each chapter. They are gathered into a single **bibliography** at the end of the book, organized by chapter. These references, both those referred to in the chapter and those listed under further reading, include original research articles, review papers, book chapters, and books, from publications that are of historical importance as well as those that report the latest developments of current research.

1 Introduction

Baerends, G. P. and Baerends-van Roon, J. M. (1950). An introduction to the study of the ethology of cichlid fishes. *Behaviour* 1 (Suppl.):1–242.

Jørgensen, C. B. (2001). August Krogh and Claude Bernard on basic principles in experimental physiology. *BioScience* 51:59–61.

There are also a number of resources online to supplement the book and help you to test and extend your understanding:

For students:

- Multiple choice questions for you to test yourself
- Useful weblinks
- A biography of the author and interview about key questions in behavioral neurobiology

For lecturers:

- Answers to the questions provided at the end of each chapter
- Figures from the book in a downloadable format
- Journal Club material for each chapter

www.oup.com/uk/zupanc3e

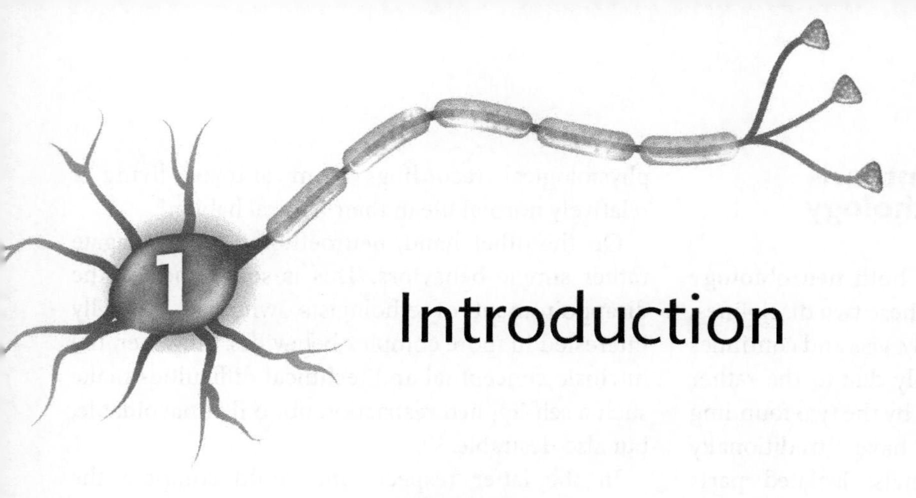

Introduction

Introduction

This book introduces the reader to the fascinating field of **neuroethology**. As with other disciplines studying animal behavior, a major task of neuroethology is to understand the causal factors that lead to the production of behavior. There are two principal approaches to achieve this goal. One approach aims at a 'software' explanation of behavior. As shown in Fig. 1.1, the animal is treated as a black box, which, in response to a biologically relevant stimulus, generates a behavior. Such an approach is used by all behavioral sciences, including ethology, one of the founder disciplines of neuroethology. The second approach, which is employed by neuroethology, aims at understanding how the central nervous system translates the stimulus into behavioral activity. In other words, neuroethology seeks a 'hardware' explanation of behavior by elucidating the structure and the function of the black box.

Neuroethology The biological discipline that attempts to understand how the nervous system controls the natural behavior of animals.

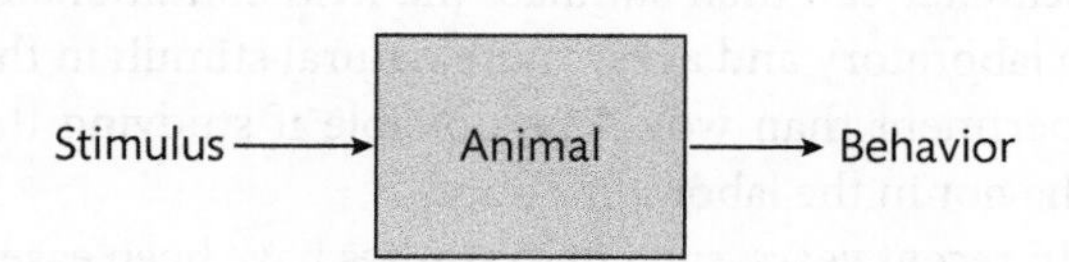

Figure 1.1 The black-box approach. Scientific disciplines restricting their research to the behavioral level treat the animal as a black box that, upon stimulation with a biologically relevant stimulus, produces a behavioral pattern. Such disciplines thus provide 'software' explanations of behavior. In contrast, neuroethology attempts to give 'hardware' explanations by exploring the structure and function of the black box in relation to the production of behavior. (Courtesy: G. K. H. Zupanc.)

Neuroethology: The synthesis of neurobiology and ethology

Neuroethology has its roots in both **neurobiology** and **ethology**. The synthesis of these two disciplines, which created a new area of study, was and continues to be challenging. This is mainly due to the rather diametric approaches employed by the two founding disciplines. Neurobiologists have traditionally worked on anesthetized animals, isolated parts of tissue, or even single cells. They are primarily interested in the structure and function of such particular cells or tissues. The species is often chosen based on technical considerations, such as the presence of large nerve cells, and the ease by which the animal preparation can be obtained. Ethologists, on the other hand, employ a **whole-animal approach**, with the animal kept under conditions as natural as possible. Preferably, at least part of their observations should be conducted in the field. If this is not possible, then the animal is transferred to or bred in the laboratory, where it is kept under semi-natural conditions to minimize the occurrence of unnatural behavior.

Despite the obvious differences between neurobiology and ethology, the success of neuroethology is based on the incorporation of a blend of neurobiological and ethological approaches into its own scientific armory. Particularly, its focus on 'natural' and biologically relevant behavioral patterns makes neuroethology distinct from other disciplines studying the neural basis of behavior. As part of the overall strategy, this should include investigations of the animal in its natural habitat. The researcher can then simulate the field conditions in the laboratory and apply more natural stimuli in the experiment than would be possible if studying the behavior in the laboratory only.

In recent years, such **field studies** have been eased by many technological developments, such as the availability of battery-powered laptop computers, which allow the researcher to characterize the animal's natural behavior with an unprecedented degree of precision. With the enormous advances made in the miniaturization of instruments, it is no longer unthinkable that neuroethologists will, at some point in the future, be able to obtain physiological recordings from animals living a relatively normal life in their natural habitat!

On the other hand, neuroethologists investigate rather simple behaviors. This is sometimes to the disappointment of ethologists, who are typically interested in more complex behaviors. However, the intrinsic conceptual and technical difficulties make such a self-applied restriction not only unavoidable, but also desirable.

In the latter respect, one could compare the situation of today's neuroethology with that of physics in the seventeenth century. The initial restriction to simple models to analyze the motion of objects (neglecting, for example, air resistance when examining falling objects) led to the discovery and establishment of many fundamental principles, such as Newton's laws of motion. An attempt in the early days of mechanics to analyze more complex systems, although closer to reality, would almost certainly have failed and tremendously delayed the further development of the physical sciences.

➤ Neuroethological research combines both neurobiological and ethological approaches.

Choosing the right level of simplicity

Progress in neuroethology is crucially dependent upon choosing the right level of simplicity. Thus, although the ultimate goal of neuroethology is to understand the neural mechanisms underlying behavior, it would at present not be sensible to examine the entire behavior of an animal.

In any animal, the behavior consists of many individual elements. It involves not only what is generally associated with behavior, movements of the body in particular, but may also include specific body postures, production of sound, color changes, electric discharges, and even glandular activity, for example the secretion of pheromones. The entire behavioral repertoire of an animal is called an **ethogram**. As immediately evident, this entire set of behavior is too complex to be analyzed by the neuroethologist, or even by the ethologist. Therefore, the behavior of an animal has to be split into its individual components, referred to as the individual behaviors or behavioral patterns. As a first step, the total behavioral repertoire is often divided into major groups representing functional categories, for example 'sleep,' 'feeding,' 'courtship,' or 'aggression.' Yet, these categories are still too large to be quantified and analyzed in a meaningful way. This makes a further subdivision necessary.

Ethogram The entire behavioral repertoire of an animal species.

Example: *Cichlids are a family of more than 2000 teleost fishes. They are well known for their highly developed aggressive, courtship and parental-care behavior. As illustrated in Fig. 1.2, the behavior subsumed under the term 'aggression' actually consists of a number of individual behavioral patterns, including chasing, butting, frontal display, lateral display, and mouth wrestling.*

Obviously, by dividing the behavior of an animal further and further down, a hierarchical arrangement results. Niko Tinbergen, one of the founders of ethology (see Chapter 2), called the different levels within this hierarchical system **levels of integration**. The lateral display of cichlids, for example, involves alignment of the fish beside its opponent, typically within a few centimeters, and a rapid, powerful sideways thrust of the tail. During this display, the dorsal, anal, and pelvic fins are erect and the opercula extended. The erection of the fins, on the other hand, is defined by the action of the individual fin rays, whose movement is the result of the contraction of muscle fibers controlled by neural motor units. Fig. 1.3 illustrates this stepwise subdivision of behavior into smaller and smaller elements.

While it hardly makes sense to undertake a study aimed at elucidating the neural basis of 'aggression' in a cichlid fish, one would likely succeed, with the techniques available, in identifying the structures within the central nervous system that control the

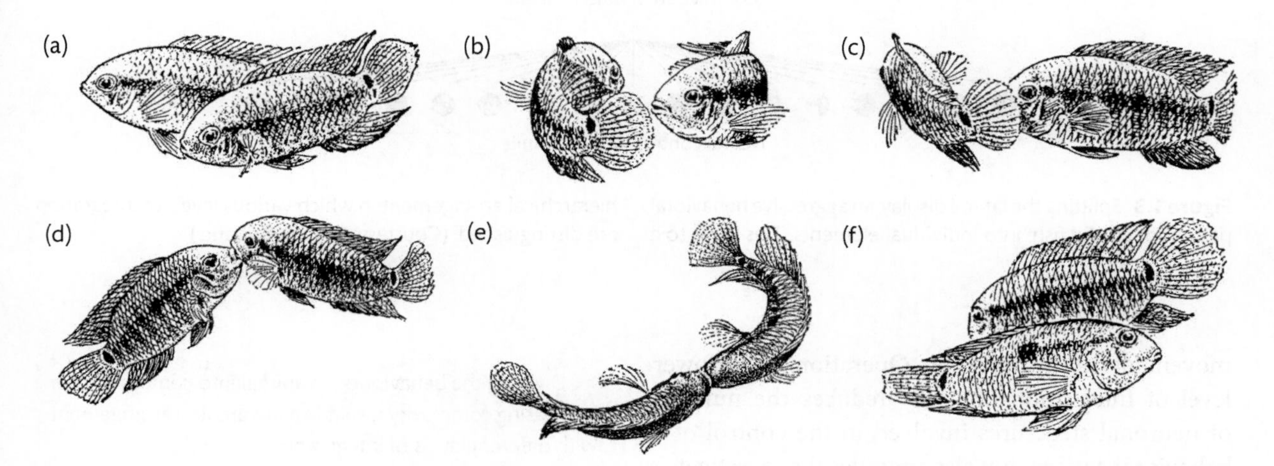

Figure 1.2 Aggressive behavior of the blue acara (*Aequidens pulcher*), a teleost fish of the cichlid family. (a) Lateral display. The fish align beside each other, spread the dorsal, anal, and pelvic fins, and intensify the coloration of their bodies. These threats are accompanied by light tail beats. (b) Circling. While circling each other, the fish have the ventral part of the mouth lowered. (c) Tail beating. One fish beats its tail against the head of the opponent. (d) Mouth grasping. The fish grasp each other at the lower or upper mandible. (e) Mouth pulling. Each fish tries to pull the opponent, after having grasped each other's mouth. (f) Defeat. The fish at the front gives up. It folds its fins, adopts a pale coloration, and swims off. (After: Wicker, W. (1968).)

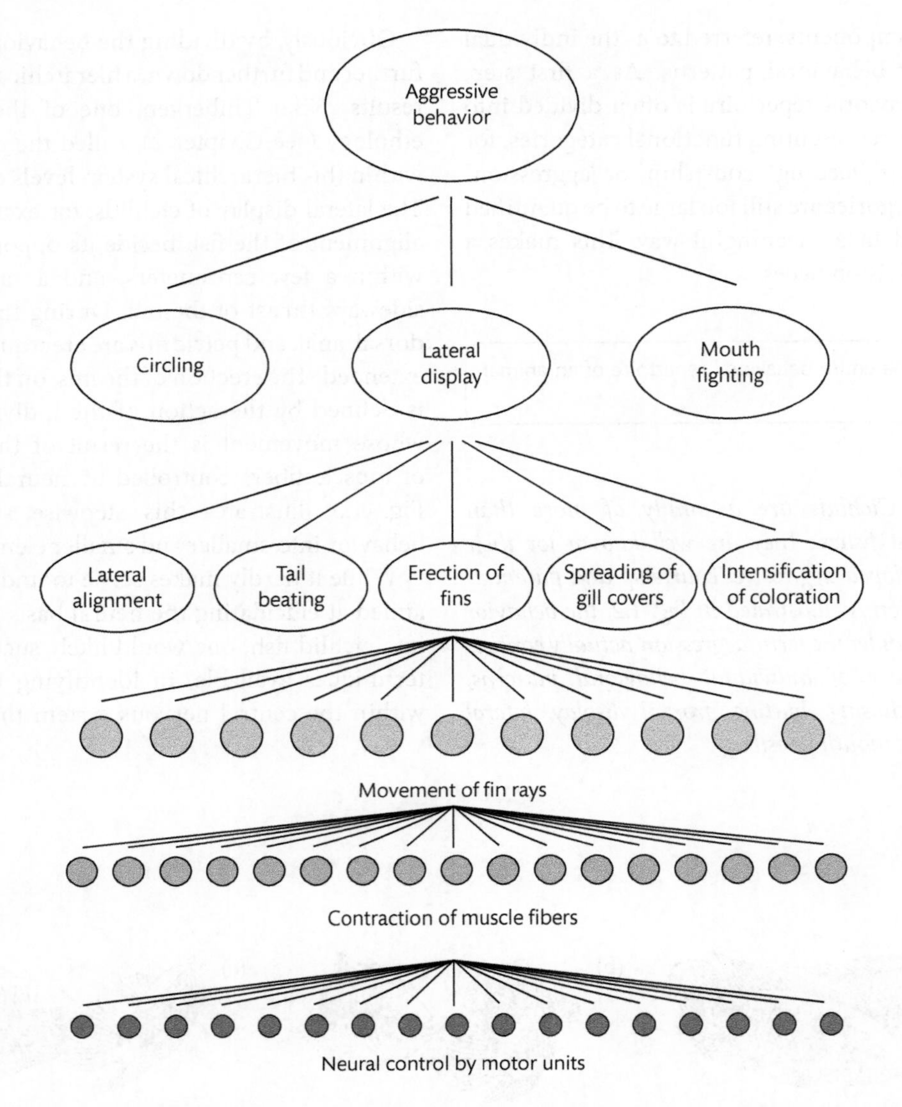

Figure 1.3 Splitting the lateral display, an aggressive behavioral pattern of cichlid fish, into individual elements. This leads to a hierarchical arrangement in which various levels of integration are distinguished. (Courtesy: G. K. H. Zupanc.)

movements of the dorsal fin. Operation at this lower level of integration not only reduces the number of neuronal structures involved in the control of a behavioral pattern, but also provides the investigator with behavioral elements typically much better defined than those encountered at a higher level of integration. This makes it markedly easier to quantify behavioral patterns—a prerequisite for many types of analysis at both the behavioral and the neurobiological level.

➤ Dividing the behavior of an animal into components of decreasing complexity results in a hierarchical arrangement with different levels of integration.

Quantifying behavior: A prerequisite for neuroethological research

In general, behavioral patterns can be quantified using their rate of occurrence ('How often is the dorsal fin erected?'), their duration ('How long is the dorsal fin kept in an erect position?'), and/or their intensity ('Is the dorsal fin erected maximally, with the fin rays almost perpendicular to the dorsal edge of the fish's body, or do the fin rays adopt positions intermediate between the maximal and the minimal angle?'). A similar attempt to quantify the lateral display would be very difficult, particularly because of the complexity of the different actions involved in the execution of this behavior. As not all individual actions are necessarily executed simultaneously during lateral display, a main problem would be to identify the end points of this behavior. Lack of such information makes it virtually impossible to determine the parameters' 'rate of occurrence' and 'duration.' Also, measurement of the intensity would be difficult: Is the behavior more intense when a larger number of individual actions are displayed, or when the degree of execution of individual patterns is maximized?

> A behavioral pattern can be quantified using its rate of occurrence, duration, and/or intensity.

Finding the right model system

The above considerations underline the importance of choosing the right **model system**. Such systems are *not* primarily studied to provide insights into the neural mechanisms underlying the behavior of the respective species. Rather, their characterization enables the neuroethologist to extract principles applicable to many, if not all, animals. This is possible because there are, probably in any case, only a finite number of solutions to a given behavioral problem. Keeping the body oriented, for example, requires the analysis of geophysical invariants, but the number of options available is limited to a few, such as the measurement of the direction of the incident light or of the animal's angle relative to gravity (see Chapter 4).

Many solutions to such problems were invented very early in evolution, so that frequently the neural implementations of these solutions are **homologous** in different species. Such homologous developments are a major reason why many fundamental cellular mechanisms underlying learning and memory are very similar among animals, including both vertebrates and invertebrates. On the other hand, these universals of life make it possible to establish principles of learning and memory processes by studying the rather simple neural network of the sea slug *Aplysia*, although ultimately most researchers would like to understand more complex systems, including those of humans (see Chapter 13).

Despite the existence of similar solutions to a given behavioral problem in many species, not each of them is equally suited for exploring the neural implementations of these solutions. Recognizing the dilemma of the right choice of experimental animals, the Danish physiologist August Krogh published in 1929 an essay in which he wrote: 'For a large number of problems there will be some animal of choice or a few such animals in which it can be most conveniently studied.' **Krogh's Principle**, as it was dubbed later, has been widely used by comparative physiologists and neuroethologists as a guide in the experimental-design process. For example, the success in the identification and characterization of the neural circuitry that controls rhythmic motor activity was largely based on the choice of *Xenopus* tadpoles, with their relatively simple organization of the spinal cord, and the possibility of evoking neural activity associated with the motor behavior from physiological preparations (see Chapter 6). Barn owls as nocturnal hunters rely heavily on acoustic cues for localization of prey. They are, therefore, particularly well suited to elucidating the neural mechanisms that underlie processing of auditory information (see Chapter 7).

Besides selecting a model system based on a good match of the research question and the relevant biological properties of the organism, other important considerations are that the behavior under scrutiny should be simple and robust, readily accessible, and ethologically relevant; and the animal displaying this behavioral pattern should be inexpensive, suitable for examination in the laboratory, easy to maintain, and possible to breed. These requirements are far from trivial to meet, and the right choice of a suitable model system always demands profound knowledge in both animal biology and husbandry.

Example: *Two orders of teleost fish produce, by means of a specialized organ, electric discharges of low voltage. As will be demonstrated in detail in Chapter 8, these so-called electric organ discharges are quite simple in terms of their biophysical properties, highly robust, and they can readily be monitored by placing recording electrodes near the fish. Their rate of occurrence, duration, and intensity can easily be measured, thus allowing the researcher to quantify this behavior. This is illustrated by Fig. 1.4, which shows the discharge pattern of the elephant nose* (Gnathonemus petersii). *This mormyriform fish produces very brief electric pulses, which are, even over hours and days, highly stable in terms of their physical appearance (Fig. 1.4a). However, the fish are able to modulate the discharge pattern, for example by altering the pulse repetition rate or by changing the mode of regularity (Fig. 1.4b, c). These modulations are used, for example, to encode information in the context of intraspecific communication, such as during aggressive encounters (Fig. 1.5). In addition, the fish employ their discharges for object detection. The electric behavior, therefore, meets the above requirement of ethological relevance. Moreover, many weakly electric fishes can be kept in aquaria under semi-natural conditions; several species have even been bred successfully in the laboratory. Taken together, these properties make them ideal subjects for neuroethological research.*

This and other examples of good neuroethological model systems are discussed in detail in the following chapters of this book. Their exploration over the last decades has greatly advanced our understanding of how the brain controls behavior. Moreover—and equally important—this research has also deepened our appreciation for the biology of the whole animal.

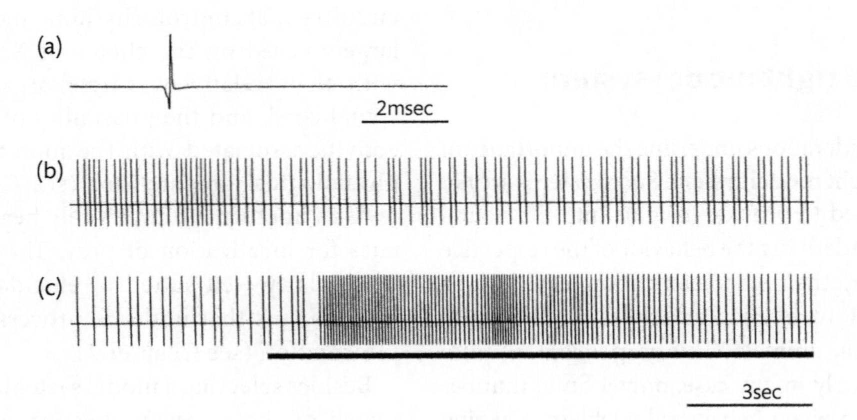

Figure 1.4 Electric organ discharge of the elephant nose, *Gnathonemus petersii.* (a) Each discharge results in a brief electric pulse, which is highly constant in terms of duration, amplitude, and waveform. (b) Resting discharge pattern over a slower timescale. As the individual pulses are highly constant, this behavior can readily be quantified by counting the number of pulses (here represented as vertical bars) produced per second and by analyzing the pattern of discharge. In the example shown, approximately five pulses per second are generated, and the intervals between the individual pulses are somewhat irregular. (c) After stimulating an isolated fish with the discharges of a second elephant nose (indicated by the horizontal bar below the trace), the discharge pattern changes significantly. In the example shown, the fish discharges at almost three times the resting rate, and the pattern of pulse production becomes more regular. The result of this experiment underlines the ethological relevance of the electric organ discharge. (Courtesy: G. K. H. Zupanc and J. R. Banks.)

Figure 1.5 The elephant nose shown in various behavioral situations. (a) At rest, the fish likes to stay in caves or, in this case, in a clay pipe. In such a situation, the fish typically emits only a few pulses, separated by rather irregular intervals. (b) If a second fish is introduced into the tank, both fish initially stay at a short distance from each other. (c) They then frequently adopt a head-to-tail stance alongside each other, with their 'chin' appendage (sometimes incorrectly called a 'nose') projecting rigidly forward. (d) If the intruder fails to move off, the territory-holder attacks and rams the intruder. (e) Finally, the intruder is beaten by the territory-holder. The defeated fish turns light brown. Now, both fish have curled in their chin appendages. The social interactions shown in (b)–(e) are accompanied by specific patterns of electric organ discharges, which clearly differ from the resting discharge. (f) Elephant nose fish, like other electric fish, can be stimulated by mimics of the discharges of a neighboring fish played back via an electric fish model. This model consists of plexiglass (Perspex) rods with built-in electrodes. (Courtesy: G. K. H. Zupanc.)

Summary

- Neuroethology attempts to understand the neural mechanisms governing animal behavior by employing approaches derived from ethology and neurobiology.
- A crucial requirement for neuroethological research is the choice of suitable model systems. These systems are characterized by behavioral patterns that, although simple, robust, and readily accessible, are also ethologically relevant.

The bigger picture

A defining feature of neuroethology is its focus on neural correlates of natural animal behavior. As a consequence, the design of experiments and the interpretation of results are based on the integration of information about the animal's behavior in its natural habitat. Another distinguishing feature of neuroethology is the frequent incorporation of a comparative approach into its research strategy. While such a strategy does not exclude restriction in individual investigations to a specific organism, nor imply random 'collection' of species, it transpires as a philosophical framework to many aspects of study. Although 'comparative thinking' demands from an investigator broad biological knowledge and experience, it offers a reward impossible to get through a single-model-system approach—insights into how evolution has shaped nervous systems so that they can generate biologically meaningful behavior.

Recommended reading

Martin, P. and Bateson, P. (1993). *Measuring Behaviour: An Introductory Guide*, 2nd edition. Cambridge University Press, Cambridge.

An excellent guide to the principles and methods of quantitative studies of behavior, with emphasis on techniques of observation, recording, and analysis.

Tinbergen, N. (1951). *The Study of Instinct*. Oxford University Press, London.

A classic: Even more than half-a-century after publication of the first edition, it is still a source of inspiration.

Short-answer questions

1.1 What are the two founder disciplines of neuroethology?

1.2 Define, in one sentence, the term 'ethogram'.

1.3 Why is it not possible to specify the sum of all behavioral patterns of an animal by a single fixed number?

1.4 List three fundamental properties of behavioral patterns that can be used for their quantification.

1.5 Name three features that qualify an animal species as a good model organism for neuroethological research.

1.6 **Challenge question** Over the last few decades, scientists have faced increasing pressure from funding agencies and institutions to carry out either applied research or basic research with high translational potential. Given such pressure, how do you justify neuroethological research?

Essay questions

1.1 Why is it important for an ethologist or psychologist not just to focus on 'software' explanations of behavior, but also to consider 'hardware' explanations?

1.2 Certain behavioral disciplines focus on one or a few model systems, instead of performing research on a variety of systems and comparing the results across species. Discuss the advantages and disadvantages of such an approach.

1.3 You plan to submit a research grant proposal to develop a novel neuroethological model system. What animal species and which behavior would you propose to examine? Justify your choice.

1.4 Songbirds, particularly during the breeding season, produce characteristic songs that subserve a variety of behavioral functions. How would you split the songs into individual behavioral patterns? How would you quantify these behavioral patterns?

1.5 Any model system is a simplification of reality. Using one of the model systems presented in this book, discuss aspects that, in your opinion, were neglected for the sake of simplicity. How would inclusion of one or several of these aspects have impeded neuroethological analysis? On the other hand, how could consideration of such aspects provide more realistic explanations of the neural mechanisms' underlying behavior?

Advanced topic *Drosophila* as a model system of behavioral genetics

Background information

In 1967, Seymour Benzer (see Chapter 10) of the California Institute of Technology in Pasadena, California, published a seminal paper in which he proposed the use of mutants to analyze the genetic basis of behavior. This event marked the beginning of the genetic dissection of behavior and the associated nervous system. The success achieved since then has been made possible by using the fruit fly *Drosophila* as a major model system in such studies.

Essay topic

In an extended essay, describe how classic genetic analysis and modern molecular genetic technology applied to *Drosophila* can contribute to a better understanding of the molecular and cellular factors involved in control of specific behaviors. Illustrate your description by discussing the approaches used, and the results obtained in investigations by examining the following two aspects: First, the role of the *sevenless* gene in the development of the compound eye and in visual orientation; secondly, the significance of sex-specific translational modification of the *fruitless* gene in the regulation of courtship behavior. Why is *Drosophila* particularly well suited as a model system to perform such studies?

Starter references

Benzer, S. (1967). Behavioral mutants of *Drosophila* isolated by countercurrent distribution. *Proceedings of the National Academy of Sciences U.S.A.* **58:**1112–1119.

Heisenberg, M. and Buchner, E. (1977). The role of retinula cell types in visual behavior of *Drosophila melanogaster*. *Journal of Comparative Physiology A* **117:**127–162.

Heisenberg, M. (1997). Genetic approaches to neuroethology. *BioEssays* **19:**1065–1973.

Usui-Aoki, K., Ito, H., Ui-Tei, K., Takahashi, K., Lukacsovich, T., Awano, W., Nakata, H., Piao, Z. F., Nilsson, E. E., Tomida, J.-Y., and Yamamoto, D. (2000). Formation of the male-specific muscle in female *Drosophila* by ectopic fruitless expression. *Nature Cell Biology* **2:**500–506.

Wasserman, S. A. (2000). Multi-layered regulation of courtship behaviour. *Nature Cell Biology* **2:**E145–E146.

To find answers to the short-answer questions and the essay questions, as well as interactive multiple choice questions and an accompanying Journal Club for this chapter, visit **www.oup.com/uk/zupanc3e.**

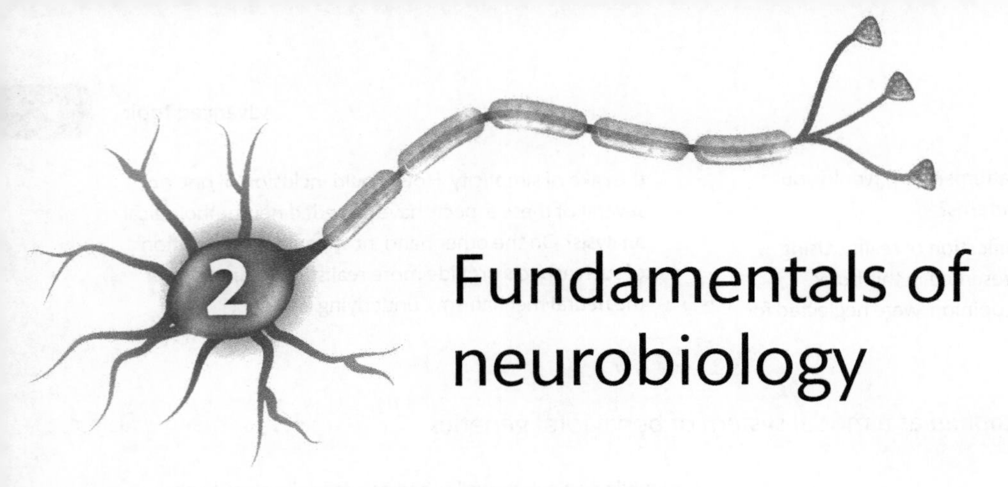

Fundamentals of neurobiology

Introduction

The last chapter demonstrated that behavioral neurobiology has its roots in both the behavioral sciences and neurobiology. As a consequence, its own approaches are often derived from these two disciplines. The aim of this chapter is to introduce the reader to the fundamentals of neurobiology. Special emphasis will be placed on those concepts and methodologies that are of immediate relevance to behavioral neurobiology.

Cellular and subcellular composition of nervous systems

Cell theory versus reticular theory

While nowadays there is no doubt that the brain is composed of individual cells, it took a rather long time to firmly establish the **cell theory**. In the nineteenth century, a common hypothesis was that the brain is made up of nerve cells forming long, thin processes fused together to form a continuous network—just as arteries and veins are linked by capillaries. This has become known as the **reticular theory** of nervous organization. By contrast, the opposing cell theory proposes the existence of entirely separate entities ('cells'). According to the latter hypothesis, the branches of the nerve cells terminate in **free nerve endings**.

Cell theory The nervous system is composed of individual cells.

Reticular theory The nervous system forms a continuous network ('reticulum') of fused processes.

The cell theory received an enormous boost after Camillo Golgi, an Italian physician, invented, in

1873, a new histological staining technique, the **Golgi method.** Its essential feature is silver impregnation of all of the components of a nerve cell, thus permitting visualization of a black neuron against a transparent background of unstained tissue. Another remarkable feature is the capricious staining behavior of this method: In a section, only a tiny fraction of the cells present are stained by the Golgi method (Fig. 2.1). The reasons for this staining selectivity are unknown. However, what may at first sight be considered a disadvantage turns out to be the major strength of the Golgi method. Due to the enormous structural complexity of the nervous system, a more complete staining of the tissue would make a study of individual cells utterly impossible.

It was the ingenious Spanish neurohistologist Santiago Ramón y Cajal (see Box 2.1) who realized the potential of the Golgi method to provide strong evidence in support of the cell theory. He further improved the staining technique and applied it to many parts of the nervous system. Reconstruction of the structures stained with the Golgi method clearly

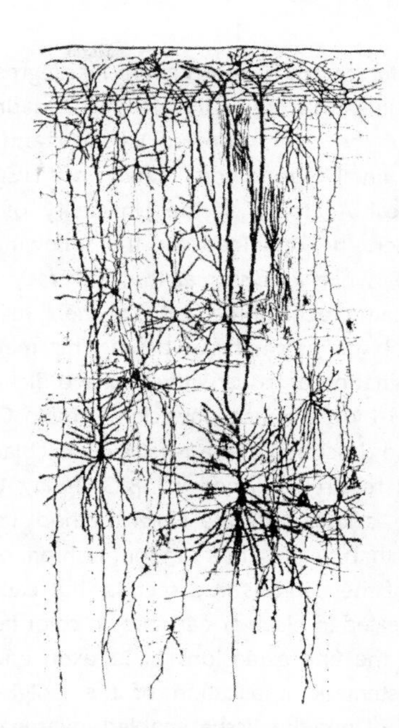

Figure 2.1 The principal neuronal types in the different layers of the motor cortex of a newborn human infant. The composite figure is based on drawings of individual neurons stained with the Golgi method. (After: Marin-Padilla, M. (1987).)

Box 2.1 Santiago Ramón y Cajal

Santiago Ramón y Cajal. (Photographer unknown.)

When Santiago Ramón y Cajal died in 1934, the British medical journal *The Lancet* wrote in an obituary: '*With Santiago Ramón y Cajal there passes one of the great pioneers of science. To most medical men in this country his name means little more than an association with methods for staining the nervous system with silver and gold. But in Spain all men, even the peasants, who understand nothing of his scientific achievements, revered the name of Don Santiago. For Cajal in addition to being a great scientist was a great man.*'

Today, Ramón y Cajal is regarded by many as the greatest neuroscientist of all times. Born in Petilla de Aragón, in the province of Zaragoza, Spain, in 1852, he developed an early fondness for drawing and painting, and wished to become an artist. However, his father, then Professor of Applied Anatomy at the University of Zaragoza, persuaded him to study medicine, specializing in anatomy. Later, Cajal combined his profession with his passion to create numerous anatomical drawings of unsurpassed artistic and scientific quality, and to illustrate all of his books and papers himself.

After he received his medical license in 1873, Cajal served as a medical lieutenant in the army and joined an expeditionary force that landed in La Habana, Cuba, to fight nationalist guerrillas. There, however, he contracted malaria and tuberculosis and had to be repatriated after a few months, suffering from jaundice

... continued

and anemia. In 1877, he obtained his degree of Doctor of Medicine from the University of Madrid. Passing by way of the Universities of Valencia and Barcelona, he was finally appointed Chair of Histology and Pathological Anatomy at the University of Madrid in 1892, where he remained for the following 30 years.

In 1887, while Professor at the University of Valencia, Cajal became acquainted with a then relatively new method of studying nervous tissue. This method, based on impregnation of tissue with silver and dichromate, was developed by the Italian physician Camillo Golgi (1843–1926), who described it as a *reazione nera* ('black reaction') in a brief report published in 1873. Today, this staining technique is known as the Golgi method. Its distinctive feature is that it stains, in a random fashion, only a minor fraction of the neurons in a section, but each individual cell is revealed in black or dark brown color in its entirety, including the fine extensions of its axon and dendrites. Cajal's systematic application of the Golgi method to the study of nervous tissue enabled investigators for the first time to resolve the elementary unit of the nervous system—the neuron. This approach provided strong arguments against the then prevailing theory that the nervous system is organized in the form of a continuous reticular network. Instead, Cajal's histological observations clearly favored the hypothesis that the nervous system is, like other organs, composed of individual cells. This theory is often referred to as the 'neuron doctrine.'

The demonstration that the neuron doctrine adequately reflects the structural organization of the nervous system meant a quantum leap in the historical development of neuroscience, enabling investigators to advance from a macroanatomical description of the brain and spinal cord to a microanatomical analysis at the level of neurons and associated structures. In addition, Cajal revolutionized the study of the nervous system in many other ways. He formulated the law of functional polarity, stating that electrical activity is conducted in one direction from the cell body or the dendrites to the axonal terminals. He proposed that this activity is transmitted from '*neuron to neuron by contact rather than continuity*,' thus paving the way for the recognition of synapses as specialized points of intercellular contact. He discovered the growth cone and the dendritic spine. He suggested that growth cones are guided toward the targets by gradients of chemoattractive substances. He produced the first diagrams of reflex pathways. He hypothesized that learning is based on a strengthening of specific synapses. And he laid the foundations for systematically studying nerve degeneration and regeneration. His observations and theories appeared in the form of numerous, brilliant books, monographs, and papers, including his masterpiece, *Textura del Sistema Nervioso del Hombre y de los Vertebrados* (Fine Structure of the Nervous System of Man and Vertebrates), which appeared in three volumes between 1897 and 1904 and comprised 1800 pages and 807 original illustrations.

Last but not least, Cajal also pioneered other areas of science and education. Prompted by the outbreak of cholera in Spain, in 1885 he carried out a comprehensive study of the epidemic and presented the then novel idea of stimulating antibody production and thus protecting animals from bacterial infection by injecting a bacterial culture that had been killed by heat. An avid photographer himself, he was one of the early pioneers of studying the physical and chemical principles of photography; his activities in this area culminated in the publication of a comprehensive monograph on color photography. In addition to research, he played a central role in reforming science and education in Spain.

In 1906, Santiago Ramón y Cajal and Camillo Golgi shared the Nobel Prize in Physiology and Medicine. Although Cajal had always acknowledged the importance of Golgi's staining method for his success in the elucidation of the cellular structure of the nervous system, and although the neuron doctrine was widely accepted by that time, Golgi behaved towards Cajal in a hostile fashion and, in his Nobel lecture, he vigorously defended the reticular theory. Later, in his autobiography, Cajal commented on his sharing of the Nobel prize with Golgi: '*What a cruel irony of fate to pair, like Siamese twins united by the shoulders, scientific adversaries of such contrasting character!*'

showed individual cells. Moreover, the existence of 'discontinuities' in the nervous system was also strongly suggested by physiological findings, such as the presence of **synaptic delay**, i.e., delay of conduction of neural signals across chemical synapses. Nevertheless, the cell theory received its final confirmation only in the 1950s when electron microscopy revealed the structure of gaps at chemical synapses.

➤ Application of the Golgi method to histological sections of the nervous system provided strong support for the cell theory.

Neurons and nervous systems

Neurons constitute the elementary units of nervous systems. Based on their size, shape, and the branching pattern of their processes, a large and sometimes confusing variety of different neuronal types is defined. It is not the purpose of this section to consider this terminology further; rather, we will try to extract some of the salient features typical of neurons.

➤ 1μm (micrometer) = 10^{-6}m.

As in other cells, the **cell body** or **soma** (plural **somata**) is the cytoplasmic region of the cell body, including the nucleus. The cytoplasmic region of the cell body without the nucleus is called the **perikaryon.** Depending on the animal and the type of neuron, the size of the soma ranges between approximately 5μm and more than 1000μm in diameter. The **nucleus,** as in other eukaryotic cells, contains the DNA and its associated proteins. During development, the nucleus provides DNA for mitotic replication. Upon exiting the mitotic cycle and becoming a fully differentiated neuron, this function ceases. A second important function of the nucleus, not restricted to developmental stages, is to transcribe DNA into RNA in order to synthesize proteins.

➤ The major parts of the neuron are the soma, dendrites, and axon.

In most neurons, the cell body gives rise to several extensions called **dendrites.** They vary greatly in number, length, and branching pattern according to the type of neuron. Together with the cell body, dendrites constitute the major domain of the neuron that receives synaptic input. As a consequence, the arborization of the dendrites largely defines the receptive field of the neuron.

In addition, and contrary to the classical view of chemical signaling at synapses, dendrites may also release neurotransmitters and modulatory neuropeptides. This mode of transmission is thought to play a role in graded tuning of synaptic input, thereby empowering dendrites to filter input and adjust its strength.

An important specialized structure found on dendrites of certain neurons is **dendritic spines.** These are short dendritic extensions that exhibit a variety of morphologies, including thin, mushroom, branched, and stubby (Fig. 2.2). The main function of spines is to increase the area of a neuron available for synaptic input. Typically, one spine is contacted by one chemical synapse (for more information on chemical synapses, see section 'Chemical synapse: Structure and function,' later). On the largest **pyramidal neurons,** as many as 40 000 spines, and thus a similar number of synapses, may be present.

Pyramidal neurons Neurons with a pyramidal-shaped cell body and two distinct dendritic trees. They are found in a number of areas within the forebrain. They represent the most common excitatory cell type in the mammalian cortex.

The second type of process found in neurons is the **axon.** Together with the cell body, it constitutes the major conducting unit of the neuron. Unlike dendrites, the axon may travel over considerable distances (up to several meters in some large animals). Typically, an axon has a number of branches; they are referred to as **collaterals.**

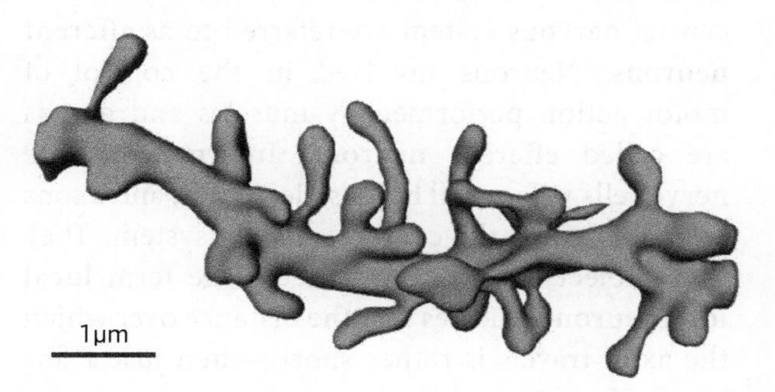

Figure 2.2 Three-dimensional reconstruction of part of a dendrite of a pyramidal neuron in the CA1 region of the hippocampus. The short dendritic extensions are spines, which display a variety of morphologies. (After: McKinney, R. A. (2005).)

The axon arises from a cone-shaped extension (called **axon hillock**) either of the soma or of a major dendrite. The axon hillock gives rise to a specialized, unmyelinated part of the axon, the initial segment, also referred to as the **axon initial segment**. This region is 10–60μm long and distinguished by the presence of **voltage-gated Na^+, K^+, and Ca^{2+} channels**. These channels play a critical role in initiating action potentials, one of the main functions of the axon initial segment.

The **projection of a neuron** is a term to describe the route of an axon from the area of origin to its target region.

> **Voltage-gated ion channels** A class of transmembrane proteins that form ion channels, which open and close in response to changes in the membrane potential.

> **Projection of a neuron** The route of its axon from the site of origin to the target region.

Example: *In fishes, amphibians, reptiles, and birds, ganglion cells of the vertebrate retina send axons to the* ***optic tectum*** *(the homolog of the* ***superior colliculus*** *in mammals). The ganglion cells are therefore said to project to the optic tectum. This axonal pathway is also referred to as a* ***retinotectal projection****. In an equivalent expression, retinal ganglion cells are said to* ***innervate*** *the optic tectum.*

In general, neurons that convey information from sensory organs in the periphery to the central nervous system are referred to as **afferent neurons**. Neurons involved in the control of motor action performed by muscles and glands are called **efferent neurons**. **Interneurons** are nerve cells whose cell bodies, dendrites, and axons are confined to the central nervous system. They may project over long distances. The term **local interneuron** indicates that the distance over which the axon travels is rather short—often just a few hundreds of micrometers.

Axon ensheathment by myelin

In vertebrates, many axons are coated by multiple, tightly wound layers of **myelin**. The main function of myelination is to provide electrical insulation of the axon (see section 'Propagation of action potentials: Myelinated versus unmyelinated axons,' below). These sheaths, or **lamellae**, of myelin do not envelop the axon over its entire length. Instead, they are interrupted at regular intervals (approximately 1mm in peripheral nerves) by zones free of myelin. These zones are called **nodes of Ranvier**. The myelinated segment between two nodes is referred to as the **internode**.

The myelin sheaths originate from non-neuronal cells, the so-called **glia**. In the peripheral nervous system, these glial cells belong to the class of **Schwann cells**. As shown in Fig. 2.3(a), each Schwann cell closely lines up along the axon to define a single internode. In the central nervous system, the myelin-producing cells are the **oligodendrocytes**. One oligodendrocyte extends several processes, each of which envelopes distinct internodes (Fig. 2.3b). These internodes may belong to different axons.

Myelin produced by both Schwann cells and oligodendrocytes is characterized by a high concentration of **lipids**. The exact composition of these lipids, and of the myelin proteins, differs between these two types of myelin. Quite unusually, the myelin proteins in both Schwann cells and oligodendrocytes are synthesized in the growing myelin process, rather than in the cell body. This is achieved by transport of mRNA in the form of ribonucleoprotein granules and of ribosomes from the cell nucleus to distant sites along microtubules (Fig. 2.3b).

Myelinated axons have been found in all vertebrates, except hagfish and lampreys, which are considered primitive members of the vertebrate line. This observation has led to the hypothesis that vertebrate myelin first evolved in placoderms, the antecedents of contemporary chondrichthyans (cartilaginous fishes) and osteichthyans (bony fishes). Myelination of axons also occurs in several taxa of invertebrates (where it most likely emerged through convergent evolution), such as crustaceans and annelids. So far, myelin has not been found in molluscs and insects.

Interestingly, in malacostracans and copepods, the membranous layers of myelin are arranged concentrically around the axon—in contrast to

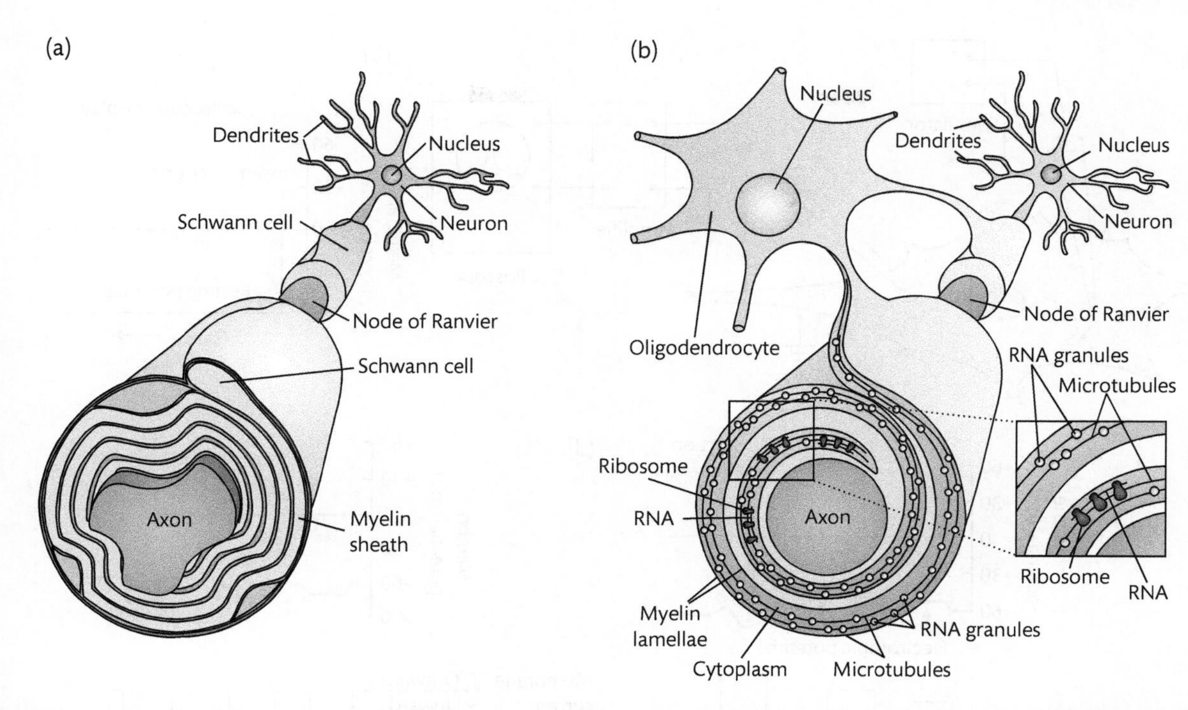

Figure 2.3 Axon ensheathment and myelin growth. (a) Myelination of an axon in the peripheral nervous system by Schwann cells. Each of the two internodes is formed by one closely apposed Schwann cell. (b) Myelination of an axon in the central nervous system by a single oligodendrocyte. The two internodes are formed by the processes of a single oligodendrocyte. Synthesis of myelin proteins takes place distant to the cell body, after microtubule-based transport of mRNA (in the form of RNA granules) and ribosomes from the cell nucleus. (After: Sherman, D. L. and Brophy, P. J. (2005).)

vertebrates, in which the myelin sheaths coat the axon in a spiral-like fashion. Furthermore, in oligochaetes, penaeid shrimps, and copepods, **focal nodes** occur along the axon, forming small openings in the myelin sheath, rather than extending completely around the nerve fiber as they do in vertebrates and palaemonid shrimps. It has been shown that the focal nodes perform the same function as the circumferential nodes: To mediate fast impulse conduction along axons (see section 'Propagation of action potentials: Myelinated versus unmyelinated axons,' later).

Considering the important role of myelin in signal propagation, it is not surprising that damage or loss of this substance has devastating effects. This is evident in patients suffering from **multiple sclerosis**, an autoimmune condition in which the immune system attacks the central nervous system, leading to demyelination and thus to a slowing down of signal conduction in nerve fibers.

Physiology and ionic basis of the electrical properties of neurons

Measuring membrane potential

Investigations of the electrical properties of neurons depend on the availability of proper techniques to monitor the electric signals produced by the cells. To measure the so-called membrane potential, **intracellular recordings** have been intensively used since the 1950s.

> **Intracellular recording** Insertion of the tip of a glass microelectrode inside a cell and measurement of the potential difference between the intracellular space and an extracellular reference point.

To record from the inside of neurons, special glass micropipettes, called **microelectrodes**, are prepared (Fig. 2.4a). These electrodes are made from glass capillaries heated in the middle until the glass

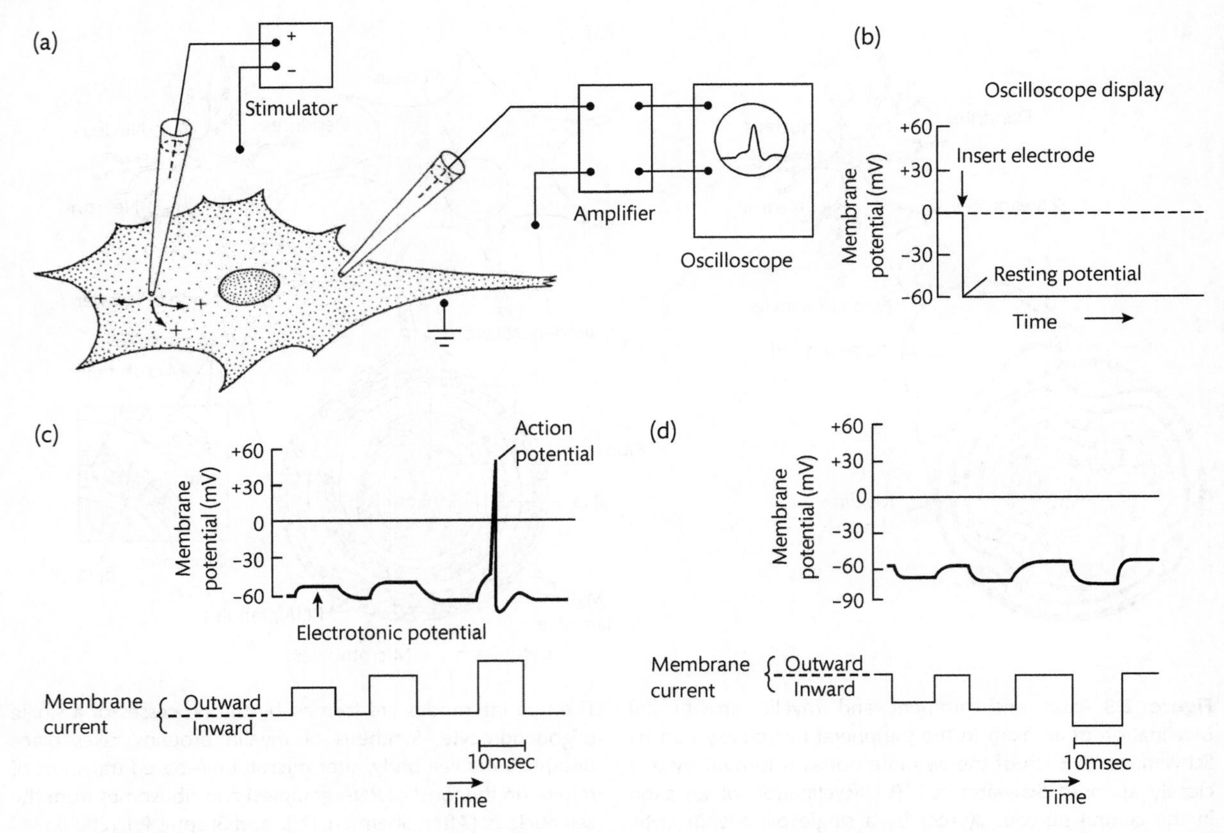

Figure 2.4 Illustration of the intracellular recording and stimulation technique. (a) Arrangement of the stimulation and recording electrodes. A glass microelectrode, with a very fine tip, is placed inside the cell. A second electrode is positioned outside the cell. After amplification, the potential measured by the two electrodes across the cell membrane is fed into an oscilloscope. For stimulation, a second pair of electrodes is used. One of these electrodes is inserted into the cell, while the other is left outside in the surrounding fluid. These two electrodes are connected to a stimulator that can pass current into or out of the cell. (b) No current is applied. Initially, both recording electrodes are left outside the cell; thus, no potential difference between the two electrodes is measured. As soon as the microelectrode impales the cell, a potential difference of about –60mV, reflecting the resting membrane potential, is recorded. (c) Injection of positive charge into the cell results in a decrease in the membrane potential. This is also referred to as depolarization. The degree of this depolarization varies proportionally with the amount of positive current passed into the cell. However, as soon as a threshold of approximately 15mV of depolarization is reached, a graded potential is no longer recorded. Rather, a new, active type of electrical response is generated. This type of signal is called an action potential. (d) Reversal of the direction of current flow by withdrawing positive charge from the inside of the cell leads to an increase in the potential difference measured across the cell membrane, thereby hyperpolarizing the cell. (After: Camhi, J. M. (1984) and Kandel, E. R. and Schwartz, J. (1985).)

becomes soft. Then, the two ends of the glass capillary are quickly pulled apart, so that the capillary breaks at the center and two fine capillary tubes result. The tips of these tubes are open and have a diameter of less than 1μm. This allows the investigator to impale the neuron's cell body without destroying it and to place the tip of the microelectrode inside the cell. The capillary is filled with an electrolyte solution, potassium chloride for example, to conduct electricity. At the wide end, a wire is inserted such that it is partially immersed in the electrolyte solution. The other end of the wire is connected, via an amplifier, to a recording device. A second electrode is placed in the physiological buffer in which the tissue, or the cell from which the recordings are made, is bathed. This electrode is connected to the other lead of the amplifier. By convention, the potential of the extracellular fluid is defined as the zero value. This arrangement enables the investigator to measure the intracellular potential with respect to this reference potential. The difference between the intracellular and extracellular potential values is referred to as the **membrane potential.**

An alternate approach to measure the membrane potential involves the use of **voltage-sensitive fluorescent dyes.** This method is particularly useful when studying electric events in organelles and cells that are too small to allow the use of microelectrodes, for example in the case of dendritic spines or when mapping variations in membrane potentials across excitable cells. One class of these probes operates based on changes in their electronic structure, and thus in their fluorescent properties, in response to alterations in the surrounding electric field. Probes of a second class exhibit voltage-dependent changes in their transmembrane distribution, which are accompanied by changes in their fluorescent properties.

Resting potential

When the membrane potential of a nerve cell is recorded by employing the intracellular recording technique, and initially both electrodes are placed outside a cell, the potential difference is zero. However, the moment the intracellular electrode impales the cell, a potential difference of approximately –60 to –80mV, with the inside being negative, occurs (Fig. 2.4b). This indicates that the membrane of a neuron is polarized at rest. The resulting potential difference is called the **resting potential.**

What causes the resting potential? In 1902, the German physiologist Julius Bernstein (1839–1917; Fig. 2.5) proposed that the electrical potential is generated by a concentration gradient of electrolytes between the inside and the outside of the cell. The gradient is maintained by the cell membrane, which acts as a semipermeable membrane. The ion selectivity of this membrane leads to a build-up of negative charge at the interior surface of the cell membrane, and of positive charge at its exterior surface. The membrane theory, as Bernstein called it, provided the first plausible physico-chemical explanation of a bioelectric event, and laid the foundation for the biophysical modeling of other electrical phenomena of living cells.

Figure 2.5 Julius Bernstein. (Courtesy of Universität Halle-Wittenberg.)

Bernstein's concept of explaining the resting potential by the differential distribution of electrolytes across the plasma membrane has remained valid to this day, but now many more details are known. The major ion species that contribute to the establishment of this potential are K^+, Na^+, and Cl^-. The unequal distribution of these ions between the inside and the outside of the cell is actively maintained. **Sodium–potassium exchange pumps,** for example, transport three molecules of Na^+ out of the neuron, while pumping two molecules of K^+ into the cell. This process is driven by the breakdown of ATP. As a result of such active transport mechanisms, K^+ is found at a much higher concentration inside the cell than outside the cell. Na^+ and Cl^- are more concentrated outside the cell. Due to these unequal concentrations, the ions tend to diffuse down their concentration gradients through specialized channels, called **ion channels** or **ionophores,** which bridge the plasma membrane.

In the case of K^+, this movement causes the inside of the cell to become more negative, as the membrane is impermeable to the large anions inside the cell, preventing them from following the K^+ ions across the membrane. This leads to an increase in negative force inside the cell, so that at one point the electrostatic attraction of the positively charged K^+ ions by the negative membrane potential inside the cell will balance the thermodynamic forces, causing the K^+ ions to move down the concentration gradient. At this point, there is no net flow of K^+ ions. The membrane potential at which this equilibrium is reached is referred to as the **equilibrium potential.** This situation is summarized in Fig. 2.6.

Among other factors, the equilibrium potential depends on the relative concentrations of the respective ions inside and outside the cell. For a single ion species, the membrane potential E_{ion} can be calculated using the Nernst equation, named after the German physico-chemist Walther Hermann Nernst (1864–1941):

$$E_{ion} = \frac{RT}{FZ} \ln \frac{[Ion]_e}{[Ion]_i}$$

E_{ion}: membrane potential at which the ionic species is at equilibrium
R: gas constant [8.315 joules per Kelvin per mole ($JK^{-1}mol^{-1}$)]
T: temperature in Kelvin ($T_{Kelvin} = 273.15 + T_{Celsius}$)
F: Faraday's constant [96 485 coulombs per mole ($C\ mol^{-1}$)]
Z: valence of the ion
$[Ion]_e$: concentration of the ion outside the cell
$[Ion]_i$: concentration of the ion inside the cell

For a monovalent, positively charged ion at a temperature of 20°C, the equation can be simplified into the following form:

$$E_{ion} = 58.2 \log ([ion]_{outside}/[ion]_{inside})$$

***Exercise:** Calculate the equilibrium potential of the K^+ ions for a classic physiological preparation—the squid giant axon. The concentrations of K^+ are 400mM inside and 20mM outside the axon.*

***Solution:** Using the full or simplified Nernst equation, an equilibrium potential of −76mV, with the inside negative, results.*

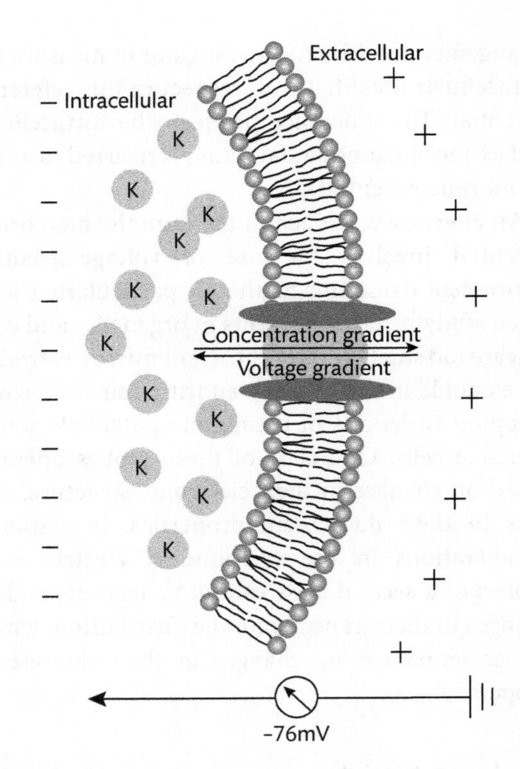

Figure 2.6 The equilibrium potential as the result of a concentration gradient and voltage gradient across the membrane. In neurons, K^+ ions are actively concentrated inside the cell. Due to a diffusion process, they flow down the concentration gradient from the inside to the outside of the cell. However, the negative potential inside the cell counteracts this diffusion by attracting the positively charged K^+ ions. At the equilibrium potential, these two forces balance one another. In the squid giant axon, this potential is −76mV. (After: Zigmond, M. J., Bloom, F. E., Landis, S. C., Roberts, J. L., and Squire, L. R. (eds) (1999).)

Similar considerations apply to other ions. In particular, Na^+ and Cl^- are important due to their relatively high concentrations. Therefore, the resting potential is somewhat in between the equilibrium potentials of these three ions. In the case of the squid giant axon, typically a resting potential of approximately −65mV is measured.

Generation of action potentials

If, during intracellular recordings, positive current pulses are passed through the membrane of the cell with the aid of an additional intracellular (**stimulation**) **electrode,** a reduction in the potential difference across the membrane is observed (Fig. 2.4c).

This reduction, called **depolarization,** brings the membrane potential closer to zero. An increase in membrane potential, on the other hand, is referred to as **hyperpolarization** (Fig. 2.4d).

➤ 1 mV (millivolt) = 10^{-3} V (volt).

For small current pulses, the membrane potential V increases roughly linearly with the increase in current flow I in accordance with Ohm's law:

$$V = R \cdot I$$

where R denotes the membrane resistance. Within a certain range, any increase in current flow is truly reflected by a proportional increase in membrane potential; therefore, the resulting voltage differences are termed **graded potentials.**

If, however, depolarization reaches a critical **threshold** level, the potential across the membrane does not continue to increase linearly with current. Rather, a completely new type of signal called an **action potential** (often also referred to as an **impulse** or **spike**) is generated (Fig. 2.4c). The action potential is characterized by a rapid increase in voltage, crossing the zero mark, and becoming transiently positive (**overshooting potential**). The membrane potential then rapidly returns toward the resting value, often accompanied by a brief undershooting.

Action potential A transient depolarization of the membrane potential generated in an all-or-nothing fashion.

Action potentials are distinguished in that they are not graded. Rather, they have a constant amplitude and are of constant duration. As long as a depolarizing stimulus causes the membrane potential to exceed the threshold voltage, an action potential is produced. If, however, the threshold value is not reached, no action potential at all is generated (**all-or-nothing property**).

New action potentials cannot be generated before the previous one is completed. There is also a brief **refractory period** following each action potential. As a consequence, the rate of impulses produced has an upper limit of about 1000 spikes/sec.

The changes in membrane potential measured during generation of an action potential are caused by transient alterations in the permeability of specific ion channels (Fig. 2.7). Depolarization of the membrane induces an opening of voltage-gated Na^+ channels (= increase in Na^+ conductivity), resulting in an inward flux of Na^+ ions. This increase in the concentration of Na^+ ions inside the cell causes further depolarization of the membrane, resulting in the opening of even more voltage-gated Na^+ channels, thus driving the membrane potential rapidly into the positive range.

The reversal of the depolarizing potential is mediated by two mechanisms: First, by inactivation of the Na^+ channels (for further details of the inactivated state of Na^+ channels, see section 'Voltage-gated sodium channels,' below), and secondly, with some delay, by the opening of voltage-gated K^+ channels and thus an increase in K^+ conductance (Fig. 2.7). The latter is accompanied by an efflux of K^+ ions, thereby causing a loss of positive charge inside the cell. Both the inactivation of the Na^+ channels and the activation of the K^+ channels lead to a repolarization of the membrane.

The cellular cause of the refractory period is the inactivation of the Na^+ channels. Upon removal of this inactivation, the channels are ready for reopening, as soon as a depolarizing stimulus is applied.

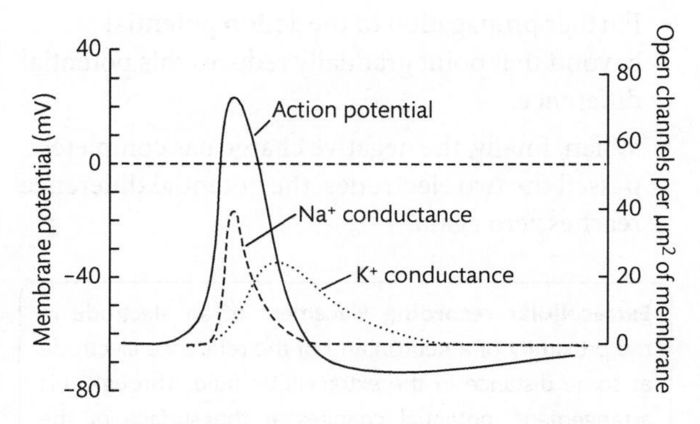

Figure 2.7 Contribution of transient changes in conductivity of ion channels to the generation of the action potential. Depolarization of the membrane causes opening of voltage-gated Na^+ channels. This increase in Na^+ conductance leads to the depolarizing phase of the action potential. The major factor attributable to the repolarizing phase is the increase in K^+ conductance by opening of the voltage-gated K^+ channels. (After: Kandel, E. R., Schwartz, J. H., and Jessell, T. M. (2000).)

Recording of propagating action potentials

One of the major functions of the axon is to conduct action potentials from the cell body to the terminal region(s). How can the propagation of the action potential along the axon be monitored? Most commonly, an **extracellular recording** technique is used. Two wires are hooked, typically about 1cm apart, on the outside of the axon or a nerve bundle. These wires are connected via an amplifier to an oscilloscope. The following can then be observed (Fig. 2.8):

- Before the action potential has arrived at the electrodes, no potential difference between the two wires is observed.
- As the negative charge on the outside of the cell passes the first electrode, this electrode becomes negative, in relation to the second electrode.
- As the negative charge passes on, the second electrode also becomes more and more negative, thus diminishing the potential difference between the two electrodes. This becomes visible as a reversal of polarity.
- As the negative charge propagates further, the negative charge recorded by the second electrode exceeds more and more the negative charge encountered at the first electrode. At this point, the first electrode 'sees' a positive charge with reference to the second electrode.
- Further propagation of the action potential beyond that point gradually reduces this potential difference.
- When, finally, the negative charge has completely passed the two electrodes, the potential difference reaches zero again.

> **Extracellular recording** Placement of an electrode in the proximity of a neuron and of the reference electrode at some distance in the extracellular fluid. Through this arrangement, potential changes at the surface of the cell membrane are recorded either from single neurons ('single-unit recording') or, in the form of field potentials, from many neurons.

The overall result of this recording is displayed on the oscilloscope as the typical waveform associated with an action potential, although the actual shape depends on several factors, including the exact placement of the electrodes. On a slower timescale, the entire width of an action potential is reduced to a thin stick visible as a rapid deflection from the base line. This slow display mode allows the experimenter to visualize many spikes simultaneously and to analyze both their rate of production and their inter-spike interval pattern.

Propagation of action potentials: Myelinated versus unmyelinated axons

Rapid signal conduction by axons is crucial in many pathways of the nervous system. Often, life-or-death decisions made by the animal, such as those required during escape responses or predatory behaviors, depend on fast and efficient neural processing of the relevant signals.

A critical factor that determines how fast action potentials are propagated after the influx of Na^+ ions in one segment of the membrane is the speed by which depolarization passively spreads to the next segment. Both theoretical considerations and experimental evidence have shown that the **resistance of the axoplasm** and the **conductance of the membrane** play a central role in this process. According to Ohm's law, the larger the resistance, the smaller the current flow, and thus the lower the speed of the depolarizing current. Similarly, the larger the membrane capacitance, the more current flow is required to change the potential across the membrane; again, this slows down the speed of current spread. The product of axial resistance and membrane capacitance defines the **time constant** in this circuit—the smaller the product, the higher the speed of passive spread, and thus the larger the distance over which the action potential can be propagated.

> **Axoplasm** The cytoplasm of an axon.

One strategy to reduce the value of the product of resistance and capacitance is to **increase the diameter of the axon.** As in any conducting wire, the axial resistance is inversely proportional to the square of the axon diameter. In contrast, the

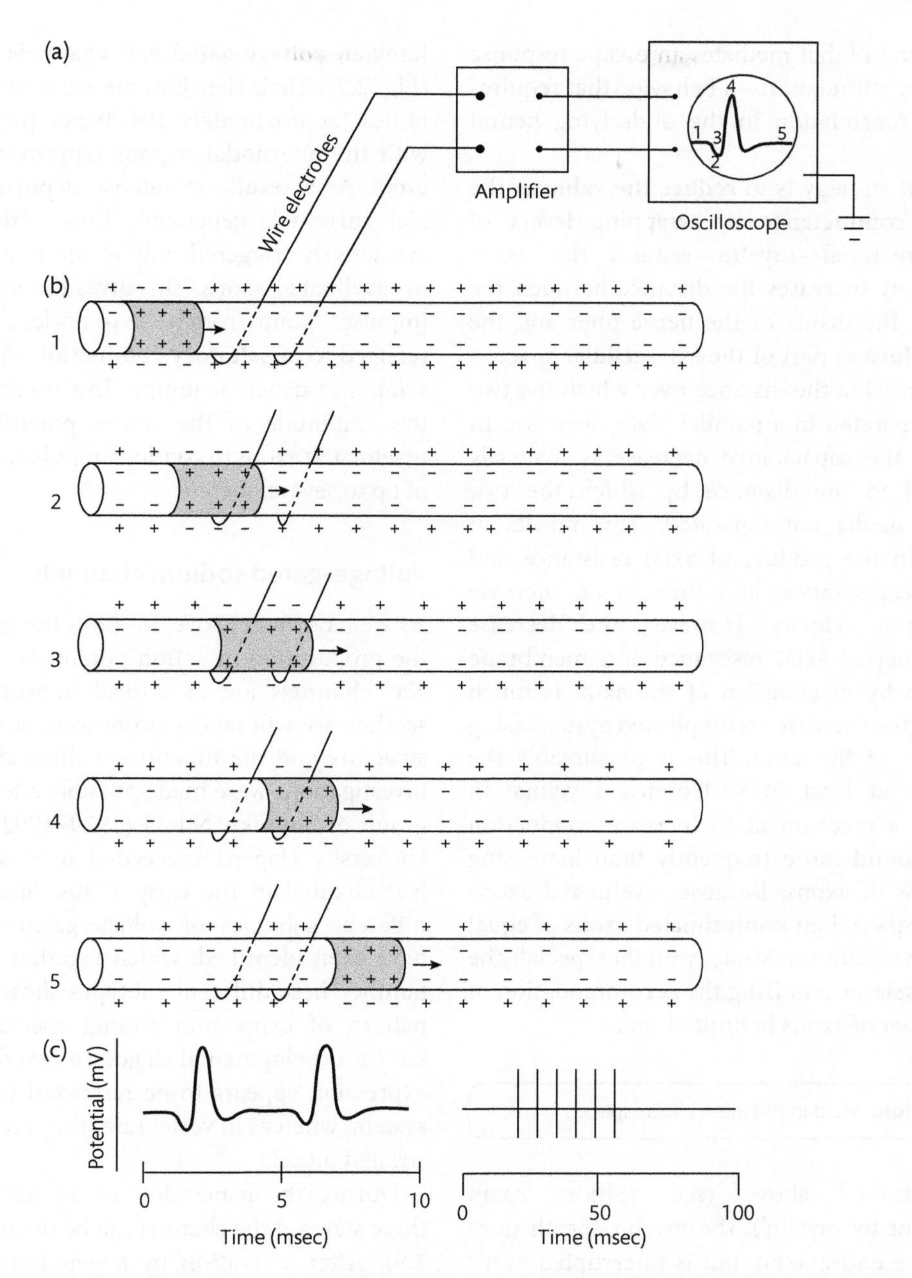

Figure 2.8 Principle of the extracellular recording technique and conduction of an action potential. (a) Two wire electrodes are hooked on the outside of an axon. The signal obtained is amplified and fed into an oscilloscope. (b) The spreading of an action potential alongside the axon (its location is indicated by the gray-shaded portion) is shown at five consecutive moments. The resulting five phases of the action potential are indicated by the corresponding numbers in the oscilloscope trace. (c) Oscilloscope traces of action potentials at faster (left) and slower (right) timescales. (Courtesy: G. K. H. Zupanc.)

membrane conductance increases only linearly with the axon diameter. As a consequence, any increase in the thickness of an axon will lead to a decrease in the time constant defined by axial resistance and membrane conductance, and this, in turn, will result in further spread of depolarization. An example of such a strategy is the squid giant axon, which assumes diameters as large as 1mm. This axon is part of a pathway that mediates an escape response upon sensory stimulation—a behavior that requires fast signal transmission in the underlying neural network.

A different strategy is to **reduce the value of the membrane conductance** by wrapping layers of insulating material—**myelin**—around the axon. This effectively increases the distance between the axoplasm at the inside of the nerve fiber and the **interstitial fluid** as part of the extracellular space—just like increasing the distance over which the two plates are separated in a parallel plate capacitor. In either case, the capacitance decrease is inversely proportional to the distance by which the two conducting media are separated. This results in a decrease in the product of axial resistance and membrane capacitance, and thus in an increase in conduction velocity. However, the decrease in the product of axial resistance and membrane conductance by myelination of the axon is much greater than the decrease accomplished by increasing the diameter of the axon. This is presumably the reason why, at least in vertebrates, myelination of axons as a mechanism to increase conduction velocity is found more frequently than increasing the thickness of axons. Because myelinated axons occupy less space than unmyelinated axons of equal conduction velocity, this strategy might especially be favored in systems requiring the accommodation of a large number of axons in limited space.

Interstitial fluid Fluid in the intercellular spaces.

As mentioned above (see section 'Axon ensheathment by myelin'), the myelin sheath does not cover the entire axon, but is interrupted every millimeter or so by small, bare regions of membrane, the **nodes of Ranvier**. These nodal regions play an important role in the propagation of action potentials in myelinated axons. When an action potential is triggered at the **axon hillock**, the depolarizing inward current spreads passively down the axon by discharging the capacitance along its way. Although the capacitance at the internodal regions is quite small due to ensheathing of the axon by myelin, the current is nevertheless continuously reduced. However, as soon as the depolarizing current reaches a node of Ranvier, **voltage-gated Na^+ channels** are activated (Fig. 2.9). Their densities are extremely high at the nodes (approximately 10 000 per μm^2), compared with the internodal regions (approximately 20 per μm^2). As a result, an intense depolarizing inward Na^+ current is generated. Thus, action potentials are actively triggered only at the nodes of Ranvier in myelinated axons. This gives the impression that impulses 'jump' from node to node, a phenomenon referred to as **saltatory conduction** (from the Latin *saltare*, to dance or jump). This mechanism boosts the amplitude of the action potentials, enabling myelinated axons to conduct impulses over distances of up to several meters.

Voltage-gated sodium channels

As we have seen above, for both the generation and the propagation of action potentials, **voltage-gated Na^+ channels** are of critical importance. In this section, we will take a closer look at the molecular structure and the function of these channels. Such investigations were made possible after the research group of Shosaku Numa (1929–1992) from Kyoto University (Japan) succeeded in cloning the first Na^+ channel in the early 1980s. Since then, nine different subtypes of voltage-gated Na^+ channels have been identified, which together form a single family. These different subtypes show a differential pattern of expression among species, cell types, and/or developmental stages. In invertebrates, their expression appears to be restricted to the nervous system, whereas in vertebrates they are also found in striated muscle.

During the generation of an action potential, three states of the channel can be distinguished (Fig. 2.9). After **activation** by a depolarizing stimulus, the channel opens, leading to a rapid influx of Na^+ ions and thus to the depolarizing phase of the action potential. However, shortly after activation and while still in the depolarizing phase, **inactivation** of the channel takes place. During this state, which lasts for a few milliseconds, the channel is not just closed but also unavailable for reopening. This causes the plasma membrane in this region to become **refractory**. Only after repolarization of the cell membrane can the channel adopt the third state, known as **deactivated**—the channel remains closed,

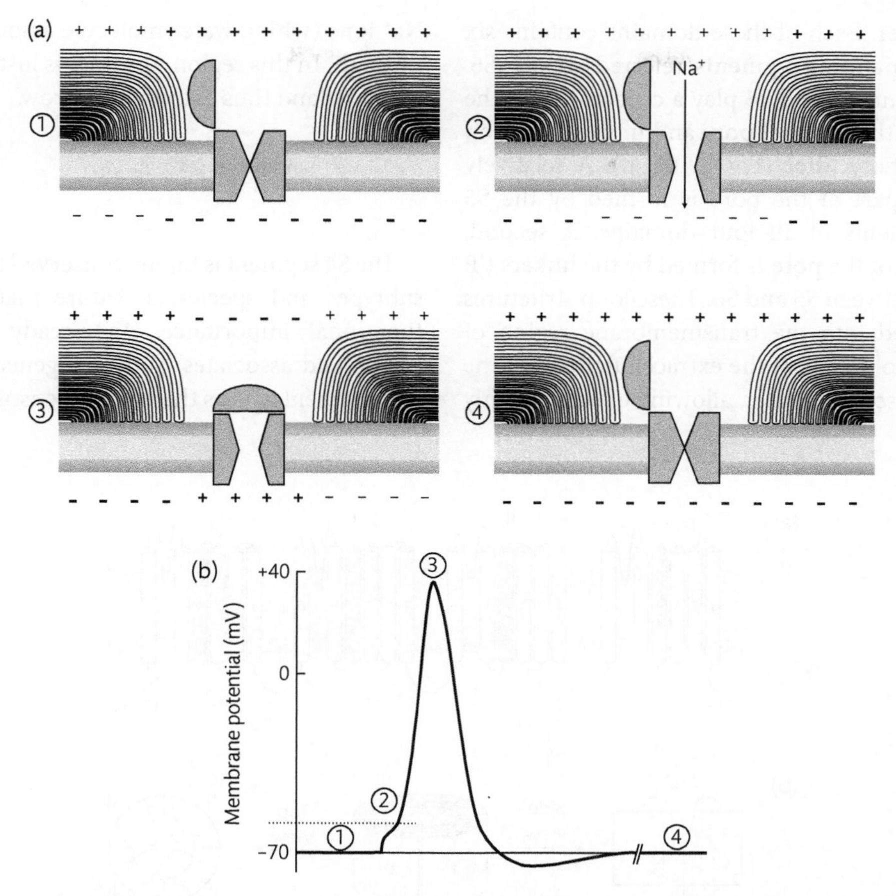

Figure 2.9 Role of voltage-gated Na^+ channels in the generation of action potentials at the node of Ranvier. (a) Different states of the Na^+ channel (labeled 1–4). Although the density of voltage-gated Na^+ channels is extremely high at the nodal region, for the sake of simplicity only one channel is shown. The resting potential is symbolized by large (–) signs on the cytoplasmic side and large (+) signs on the extracellular side of the axon. Depolarization of the membrane is indicated by small (–) and (+) signs on the cytoplasmic and extracellular sides, respectively. (b) Membrane potential during generation of an action potential at the node of Ranvier. Before arrival of the depolarizing stimulus, the region around the node of Ranvier displays the resting membrane potential, and the Na^+ channel is in a deactivated state (1). As soon as depolarization arrives and reaches the threshold, the generation of an action potential is initiated by activation of the channel, leading to an influx of Na^+ ions into the cell and thus to further depolarization of the plasma membrane (2). Shortly thereafter, while the cytoplasmic face of the membrane becomes positive, the channel enters an inactivated state, being unable to reopen in response to depolarization (3). Some time after repolarization of the membrane, the channel switches back to its deactivated state (4). (Courtesy: G. K. H. Zupanc.)

but is now ready for activation during the next action potential.

Structural analysis has revealed that all subtypes of voltage-gated Na^+ channels are **integral membrane proteins** composed of one α-**subunit** and one or more auxiliary β-**subunits**. In the mammalian brain, the α-subunit has a molecular weight of approximately 260kDa, whereas the β-subunits are much smaller molecules. Studies have shown that expression of the α-subunit alone is sufficient to form a functional channel, and that the role of the β-subunits appears to be largely restricted to modification of certain channel properties. In addition, the β-subunits may act like cell-adhesion molecules.

As shown in Fig. 2.10(a), the α-subunit consists of **four domains**, labeled I–IV, which are similar to one another. Each of these domains contains **six membrane-spanning segments**, referred to as S1–S6.

The segments S5 and S6 play a crucial role in the formation of the **channel pore** and in the definition of the **selectivity filter** (Fig. 2.10a, b). A relatively wide, inner part of the pore is formed by the S5 and S6 segments of all four domains. A second, external part of the pore is formed by the linkers ('P segments') between S5 and S6. These loop structures are embedded into the transmembrane region of the channel to define, at the extracellular end of the pore, an ion-selective filter, allowing effectively only Na^+ ions (with a water molecule associated) to pass through. In this region, the pore is just 0.3–0.5nm in diameter and thus extremely narrow.

➤ 1nm (nanometer) = 10^{-9}m.

The S4 segment is highly conserved across channel subtypes and species, a feature that indicates its functional importance. As already proposed by Numa and associates, and now generally accepted, this segment acts as the **voltage sensor**. It appears to

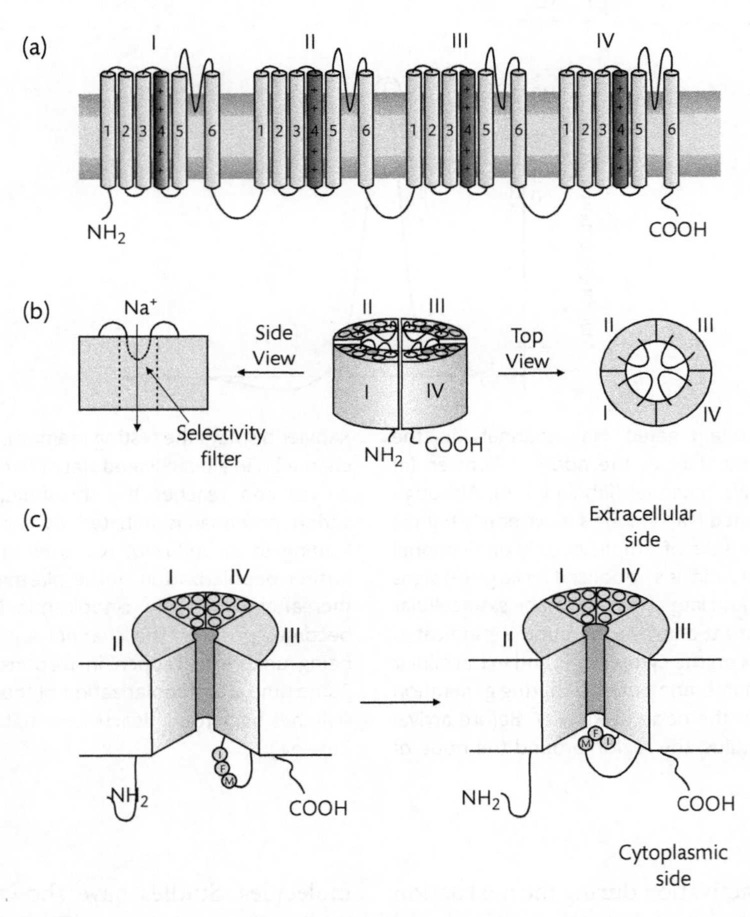

Figure 2.10 Molecular structure of the α-subunit of the voltage-gated Na^+ channel. (a) Putative transmembrane folding. Each of the four domains (labeled I-IV) consists of six membrane-spanning segments (labeled 1-6). Each of the four S4 segments has a positively charged amino acid in every third position. These segments act as voltage sensors. The inner part of the pore is formed by segments 5 and 6 of all four domains. The P segments, linking segments 5 and 6, form an ion-selective filter at the extracellular mouth of the pore. The intracellular loop between domains III and IV is thought to act as the inactivation gate. (b) Details of the structure of the selectivity filter. (c) Proposed mechanism of the inactivation gate. Within the loop, the IFM motif is shown. (After: Marban, E., Yamagishi, T., and Tomaselli, G. F. (1998), and Yu, F. H. and Catterall, W. A. (2003).)

exert this function through the positively charged amino acids arginine and lysine located at every third position. In the voltage-gated Na^+ channel expressed in the rat brain, there are five such residues in the S4 segments of the first two domains, six in the S4 of the third domain, and eight in S4 of the fourth domain. Shifts in transmembrane voltage lead to a translocation of the S4 segment. Although experimental evidence has shown that all four S4 segments are involved in activation and inactivation of the Na^+ channel, the exact mechanism that leads to the conformational changes associated with these two processes is still unclear.

An intracellular loop between domains III and IV has been proposed to function as the **inactivation gate**. Within this loop, a key motif consisting of three hydrophobic amino acid residues (isoleucine, phenylalanine, and methionine) is positioned such that it can block the inner mouth of the pore (Fig. 2.10c). Using the **one-letter code** for these three amino acid residues, this cluster is referred to as the IFM motif.

> **One-letter code of amino acids** For each of the 20 amino acids, a unique letter of the alphabet is assigned: A, alanine; C, cysteine; D, aspartic acid; E, glutamic acid; F, phenylalanine; G, glycine; H, histidine; I, isoleucine; K, lysine; L, leucine; M, methionine; N, asparagine; P, proline; Q, glutamine; R, arginine; S, serine; T, threonine; V, valine; W, tryptophan; Y, tyrosine.

Synapses

Chemical synapse: Structure and function

Based on the proposal of Santiago Ramón y Cajal that the nervous system is composed of individual nerve cells (see Box 2.1), the English neurophysiologist Charles Sherrington (see Box 6.1) introduced the term **synapse** in 1897 to describe the structure that mediates transmission of nerve impulses from one cell to the next. However, it still remained unknown how information was transmitted across the synaptic cleft between two cells. Observations, especially by a group of English physiologists headed by John Newport Langley (1852–1925), suggested, at the beginning of the 1920s, the involvement of chemical substances in this process. This was confirmed in 1921, when the Austrian pharmacologist Otto Loewi (1873–1961) published the results of his classic experiments. He demonstrated that a non-stimulated frog heart could be stimulated by applying ventricular fluid onto it from a stimulated heart. Since then, a large number of such chemical substances, called **neurotransmitters**, have been identified and characterized. They mediate **synaptic transmission** at **chemical synapses**. A second type of synapse involving transmission of electrical signals is referred to as **electrical synapse**; this type will be described in a separate section below.

> **Synapses** Specialized contact zones in the nervous system where a neuron communicates with another cell.

A chemical synapse is comprised of the following three major structural components (Fig. 2.11):

- A presynaptic element consisting of a swelling of the axon. Most frequently, this swelling is called a **varicosity** when it occurs along the axon and a **bouton** when it is found at the tip of the axonal terminal. A neuron with all of its terminal branches may form hundreds of such presynaptic swellings and thus make potentially synaptic contact with a large number of other neurons. An important feature of varicosities and boutons is that they contain **transmitter molecules** packaged in a large number of membrane-bound structures with a diameter of approximately 50nm; these structures are referred to as **synaptic vesicles**.
- A **postsynaptic element** of the recipient cells, which is often on dendrites, including dendritic spines. On the postsynaptic membrane are specific **receptors** for binding of the transmitter molecules.
- A space 20–40nm wide between the two cells (the **synaptic cleft** or **synaptic gap**).

Synaptic transmission is initiated following the arrival of an action potential at the presynaptic swelling. The depolarization of the presynaptic membrane leads to a cascade of molecular events that culminates in (i) the **release of neurotransmitter** from vesicles fusing with the presynaptic membrane,

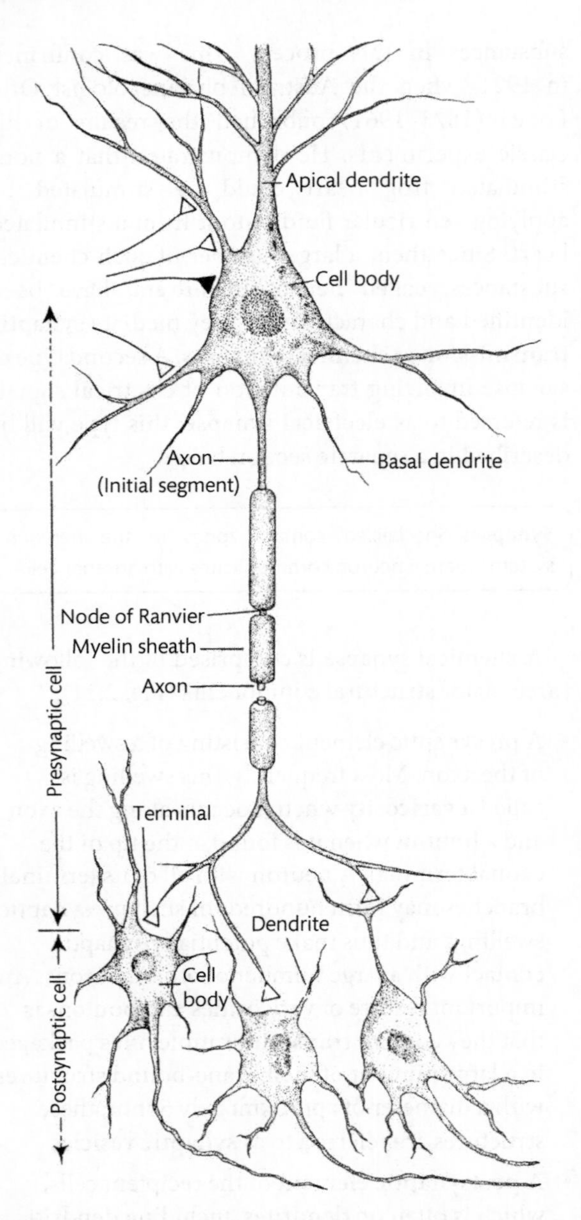

Figure 2.11 Illustration of a neuron with its various points of contact. Both the cell body and the dendrites receive synaptic input from excitatory synapses (open triangles) and inhibitory synapses (filled triangles). The corresponding somatic and dendritic regions serve as postsynaptic elements. At its terminal region, the axon of the neuron forms synaptic contact with the cell bodies and dendrites of three other cells. There, the neuron is considered presynaptic, whereas the three other cells are postsynaptic. Also, note the insulation of the axon by myelin sheaths. These are interrupted at the nodes of Ranvier. For diagrammatic purposes, the relative dimensions of the axon and the synapses are considerably distorted. In reality, the axon is very thin and often extremely long. Moreover, the terminal branches of the axon may form synapses with as many as a thousand other neurons. (After: Kandel, E. R. and Schwartz, J. H. (1985).)

and (ii) the generation of a specific electrical response, the **postsynaptic potential**, on the postsynaptic side (Fig. 2.12).

Analysis of the sub-cellular events that mediate synaptic transmission has shown that a rather minor portion of the synaptic vesicles within the presynaptic swelling participates in the release of transmitter at a given time. These vesicles are located near the presynaptic membrane in an area distinguished by its electron-dense structure (hence appearing dark in electron micrographs). This area is commonly referred to as the **active zone**.

The fusion of the membranes of the synaptic vesicles and of the presynaptic membrane is driven by a process that involves the interaction of specific membrane-associated proteins called **SNAREs** (soluble *N*-ethylmaleimide-sensitive factor attachment protein receptors). SNAREs associated with the vesicles are referred to as v-SNAREs, whereas those of the target (= presynaptic) membrane are called t-SNAREs. The exact molecular mechanism of this reaction is not fully understood. However, it is well established that fusion of the vesicles with the presynaptic membrane is triggered by a local rise in the concentration of **Ca^{2+} ions**, caused by the opening of voltage-sensitive Ca^{2+} channels in the active zone and followed by an influx of Ca^{2+} ions into the presynaptic cell. A potential candidate for sensing the increase in the Ca^{2+} concentration is **synaptotagmin**, an integral membrane protein of synaptic vesicles. Upon fusion of one vesicle, approximately 5000 transmitter molecules are released into the synaptic gap. The transmitter molecules of one vesicle diffuse across the synaptic gap and bind rapidly to the postsynaptic receptors. The entire process from arrival of an action potential at the presynaptic bouton to the generation of a postsynaptic potential takes about 1msec.

A characteristic feature of the chemical synapse is that it is **polarized**: communication is primarily **unidirectional**; it occurs from the presynaptic neuron to the postsynaptic cell.

Neurotransmitters

Commonly, synapses are grouped into various types according to the transmitter released by the presynaptic element. They are further divided into subtypes based on the effect of various **agonists** that

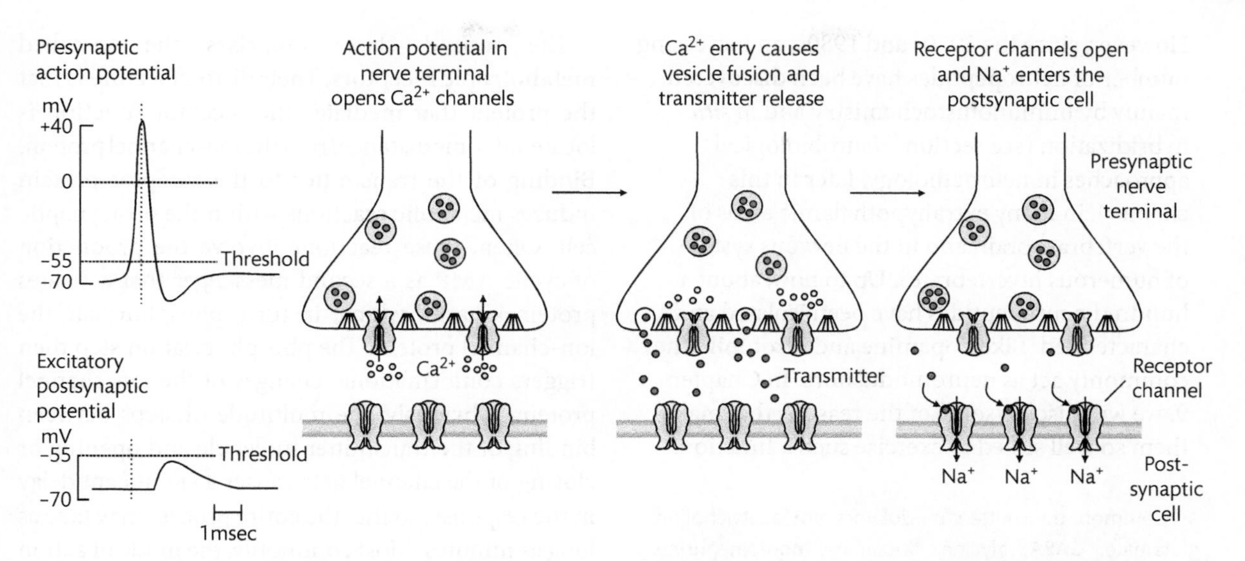

Figure 2.12 Synaptic transmission at the chemical synapse. Upon arrival of an action potential at the presynaptic terminal, voltage-gated Ca^{2+} channels open, leading to an influx of Ca^{2+} ions. The local rise in the concentration of intracellular Ca^{2+} causes fusion of synaptic vesicles with the presynaptic membrane at the active zone and release of the neurotransmitter into the synaptic gap. The neurotransmitter molecules diffuse across the gap and bind to specific receptors on the postsynaptic membrane. In the case illustrated here, binding of the transmitter results in opening of the receptor channel and influx of Na^+ ions into the cell. The rise in intracellular Na^+ concentration results in the generation of an excitatory postsynaptic potential. (After: Kandel, E. R., Schwartz, J. H., and Jessell, T. M. (2000).)

mimic the action of the transmitter molecule, or of various **antagonists** that inhibit the action of the transmitter molecule. Agonists and antagonists have proven extremely powerful tools to study details of the processes involved in synaptic transmission.

Example: *Synapses using the transmitter* ***acetylcholine*** *are called* ***cholinergic.*** *Cholinergic synapses are divided into two subtypes,* ***nicotinic*** *and* ***muscarinic,*** *named after the agonists nicotine and muscarine, respectively. The latter two compounds simulate the action of acetylcholine, but bind to different types of cholinergic receptors.*

According to their chemical nature, various types of transmitters are distinguished:

- **Acetylcholine** is the major transmitter at the neuromuscular junction. It is also used by some neurons in various regions of the central nervous system.
- **Amino acid transmitters** include substances like **glutamate, glycine,** and **γ-aminobutyric acid (GABA)**. Glutamate is one of the major excitatory transmitters in the central nervous system. The receptors of the glutamatergic synapses fall into two main categories. The first category is distinguished by binding of the agonist ***N*-methyl-D-aspartate (NMDA) receptors**; the second category is characterized by binding of quisqualate and kainate (**non-NMDA receptors**). Application of glutamate, or one of these agonists, to brain regions with glutamatergic synapses causes depolarization of neurons. GABA and glycine are common inhibitory transmitters in the brain.
- **Biogenic amines** are molecules with an amine group. They form three subgroups: **catecholamines (dopamine; norepinephrine,** also referred to as **noradrenaline;** and **epinephrine,** also known as **adrenaline);** **serotonin,** also called **5-hydroxytryptamine;** and **histamine.** Dopamine and serotonin often act as **modulators;** thus, they exert a modulatory action on other neurotransmitters, rather than having a direct excitatory or inhibitory effect of their own.
- **Neuropeptides** have only been recognized in the last few decades to be involved in synaptic transmission. Originally, their distribution was thought to be restricted to the hypothalamus.

However, since the 1970s and 1980s, an increasing number of neuropeptides have been discovered, mainly by immunohistochemistry and *in situ* hybridization (see section 'Neurobiological approaches in neuroethology,' later in this chapter), in many extrahypothalamic areas of the vertebrate brain and in the nervous system of numerous invertebrates. Up to now, about a hundred neuropeptides have been isolated and characterized. Like dopamine and serotonin, they commonly act as neuromodulators. In Chapter 9, we will discuss some of the reasons that make them so well suited to exercise such a function.

> Common transmitters/modulators are acetylcholine, glutamate, GABA, glycine, dopamine, norepinephrine, epinephrine, serotonin, histamine, and a large number of neuropeptides.

Ionotropic versus metabotropic receptors

The effect of a transmitter is defined by the properties of the receptor to which it binds, rather than by the properties of the transmitter molecule itself. This is readily demonstrated by the fact that many transmitters can exert either an excitatory or an inhibitory effect, depending on the specific receptor (sub)type they activate.

The following two features are common to all neurotransmitter receptors:

1. They bind the transmitter molecule with part of their extracellular domain.
2. This activation of the receptor protein site induces conformational changes in an associated ion-channel protein, leading to opening or closing of this channel.

The nature of the link between the receptor site and the channel protein defines two major classes of neurotransmitter receptors (Fig. 2.13). The first class consists of so-called **ionotropic receptors**. Their distinct property is that the receptor domain forms an integral part of the ion channel, thus resulting in a single macromolecule. The response of ionotropic receptors is fast (typically in the order of milliseconds). Their mode of action is often referred to as **direct gating**.

The second class comprises the so-called **metabotropic receptors**. Their distinct feature is that the protein that mediates the receptor function is located at some distance from the ion-channel protein. Binding of the transmitter to the receptor protein induces metabolic reactions within the postsynaptic cell. Often, these reactions involve the production of cyclic AMP as a second messenger that activates protein kinases, which, in turn, phosphorylate the ion-channel protein. The phosphorylation step then triggers conformational changes of the ion-channel protein. Obviously, the multitude of steps between binding of the transmitter molecule and opening or closing of the channel gate causes a significant delay in the response, so that the entire process may take as long as minutes. Most commonly, the mode of action mediated by metabotropic receptors is referred to as **indirect gating**.

Examples of ionotropic receptors are the nicotinic acetylcholine receptor, the $GABA_A$ receptor, the glycine receptor, one subclass of serotonin receptor, and different types of glutamate receptors. The class of metabotropic receptors includes receptors for norepinephrine and epinephrine, dopamine and neuropeptides, as well as the metabotropic glutamate receptor, the muscarinic acetylcholine receptor, and most serotonin receptors.

Nicotinic acetylcholine receptor: Molecular structure and function

From both a historical and a scientific point of view, the **nicotinic acetylcholine receptor** is of special interest because it was the first neurotransmitter receptor whose molecular structure was elucidated, and it remains one of the best-characterized receptors. Much of the pioneering work toward this molecular analysis was done by Shosaku Numa and co-workers in the 1980s, after Jean-Pierre Changeux (born 1936) and co-workers had isolated and purified this receptor from the electric organ of electric eels in the 1970s.

The nicotinic acetylcholine receptor forms a subclass of acetylcholine receptors, which can be activated not only by the natural ligand acetylcholine but also by the plant alkaloid **nicotine.** Thus, nicotine acts as an **agonist** of acetylcholine. The second subclass of acetylcholine receptors is characterized by its ability to bind **muscarine**, a toxin isolated from

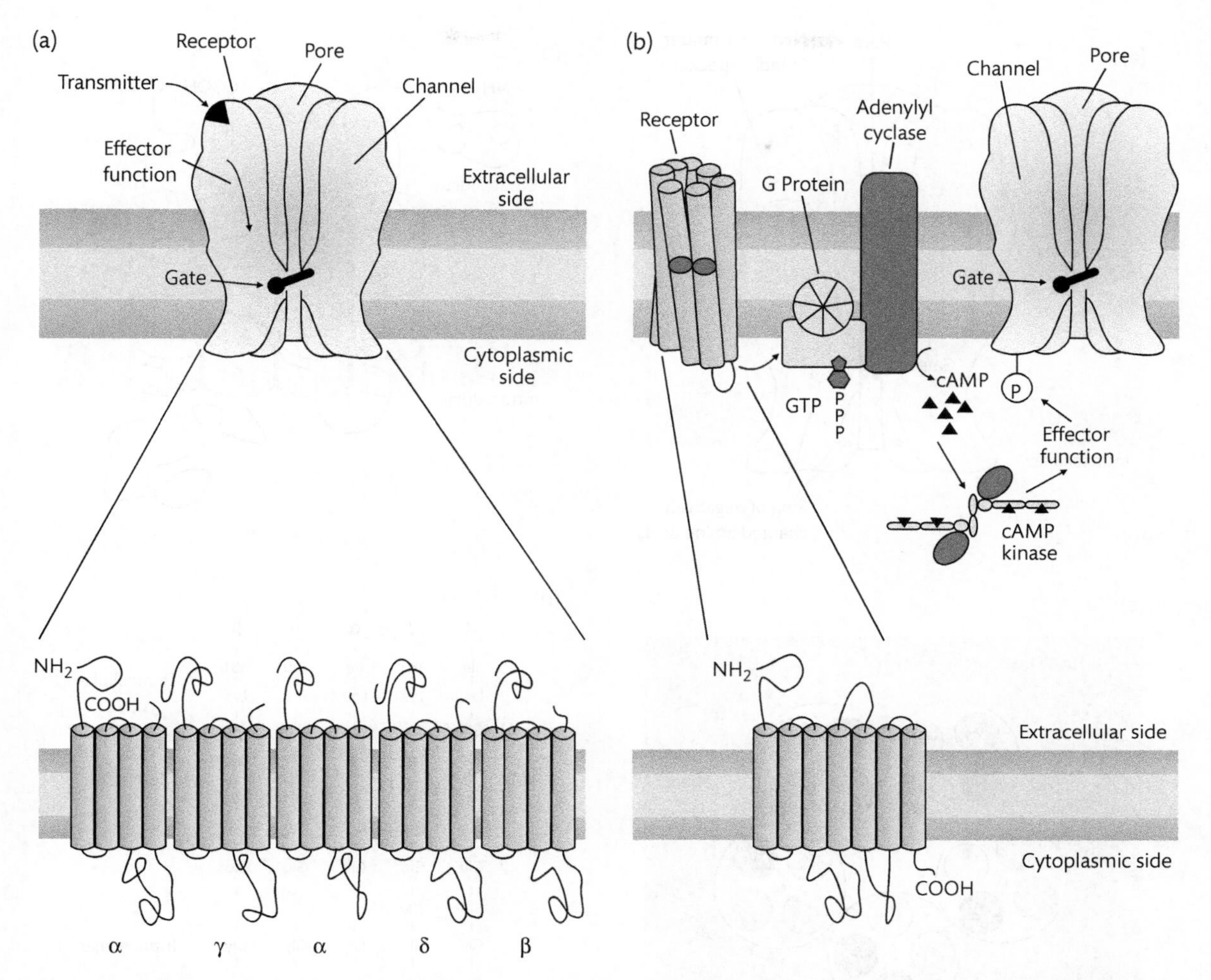

Figure 2.13 Ionotropic versus metabotropic receptors. (a) Ionotropic receptors integrate both the transmitter-binding function and the channel function in one macromolecule. In general, they are composed of five subunits, each of which consists of four membrane-spanning segments. (b) Metabotropic receptors are proteins distinct from the protein that mediates the channel function. Activation of the receptor triggers a cascade of intermediate steps (here, activation of G proteins and production of second messengers), which finally lead to conformational changes of the channel protein. Typically, metabotropic receptors are formed by a single protein consisting of seven membrane-spanning segments. (After: Kandel, E. R., Schwartz, J. H., and Jessell, T. M. (2000).)

certain mushrooms. Hence, this subclass is referred to as **muscarinic acetylcholine receptors.**

The nicotinic acetylcholine receptor has a molecular weight of approximately 290kDa and is composed of five subunits—two copies of an α-subunit and one copy each of β-, γ-, and δ-subunits (Fig. 2.14). These subunits assemble in the lipid bilayer of the cell membrane to form a ring that encloses a central pore.

Each subunit consists of four transmembrane-spanning segments. They are referred to as TM1–TM4. These segments are arranged such that the five TM2s, one contributed by each subunit, line the pore. Furthermore, the amino acids of the TM2 segments of the five subunits are aligned in such a way that three rings of negatively charged amino acids are formed. These rings are oriented toward the central pore of the channel. They are thought to function as a selectivity filter—while anions are repelled by the negatively charged amino acids of the rings, cations (mainly Na^+ and K^+) can pass through the central canal.

In addition, each nicotinic acetylcholine receptor protein has two binding sites residing in the extracellular domain. They are formed largely by six amino acids in the α-subunits and mediate

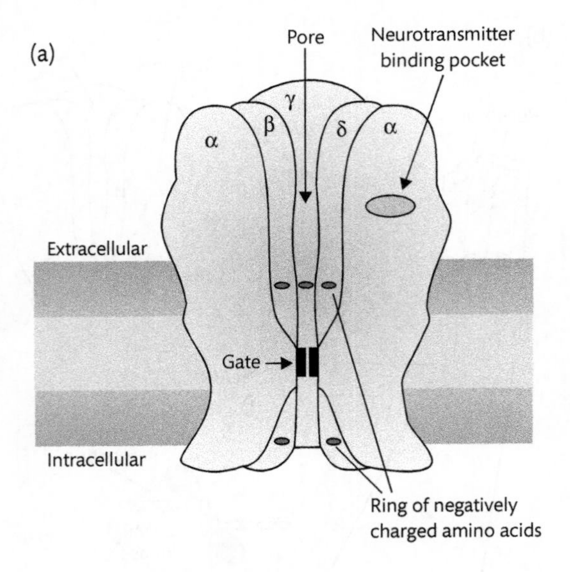

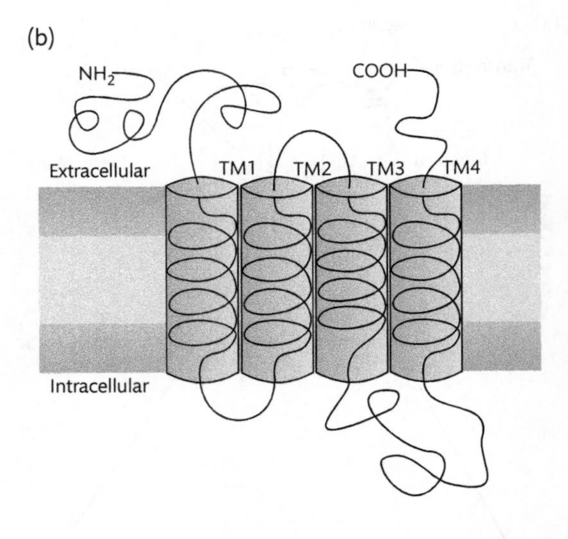

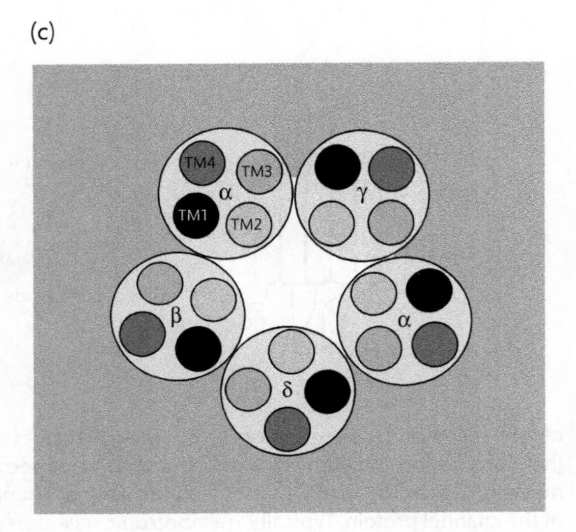

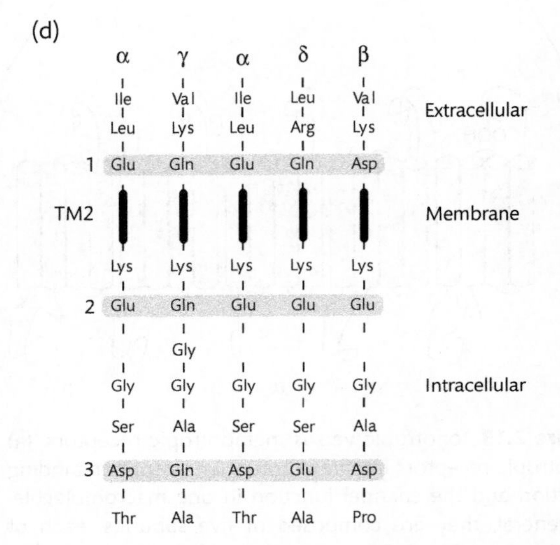

Figure 2.14 Molecular structure of the nicotinic acetylcholine receptor. (a) Vertical section showing the five subunits, designated α (two copies), β, γ, and δ, and their relative positions within the plasma membrane. In addition, the central pore, two of the three rings of negatively charged amino acids, and one of the neurotransmitter-binding pockets on one of the α-subunits are indicated. (b) The four transmembrane-spanning regions, termed TM1–TM4, of one subunit. (c) Top view of the five subunits, each with its four transmembrane-spanning segments. The five TM2 segments line the channel. (d) Amino acid sequences of the TM2 segments of the five subunits. In the native receptor protein, these amino acids are arranged such that they form three rings (labeled 1, 2, and 3) of negatively charged amino acids facing the channel. (After: Byrne, J. H. and Roberts, J. L. (2004).)

the binding of acetylcholine or agonists, such as nicotine. As soon as two molecules of such ligands bind to the nicotinic acetylcholine receptor, the channel opens with a very short (approximately 20μs) latency. This transition from a closed to an open position is associated with rotation of the TM2 segments.

Synaptic potentials

The main effect of the binding of the transmitter to the postsynaptic receptor is the generation of a **postsynaptic potential** (Fig. 2.15). If this potential is depolarizing, and thus brings the membrane potential closer to the threshold for generation of action potentials, it is called an **excitatory postsynaptic**

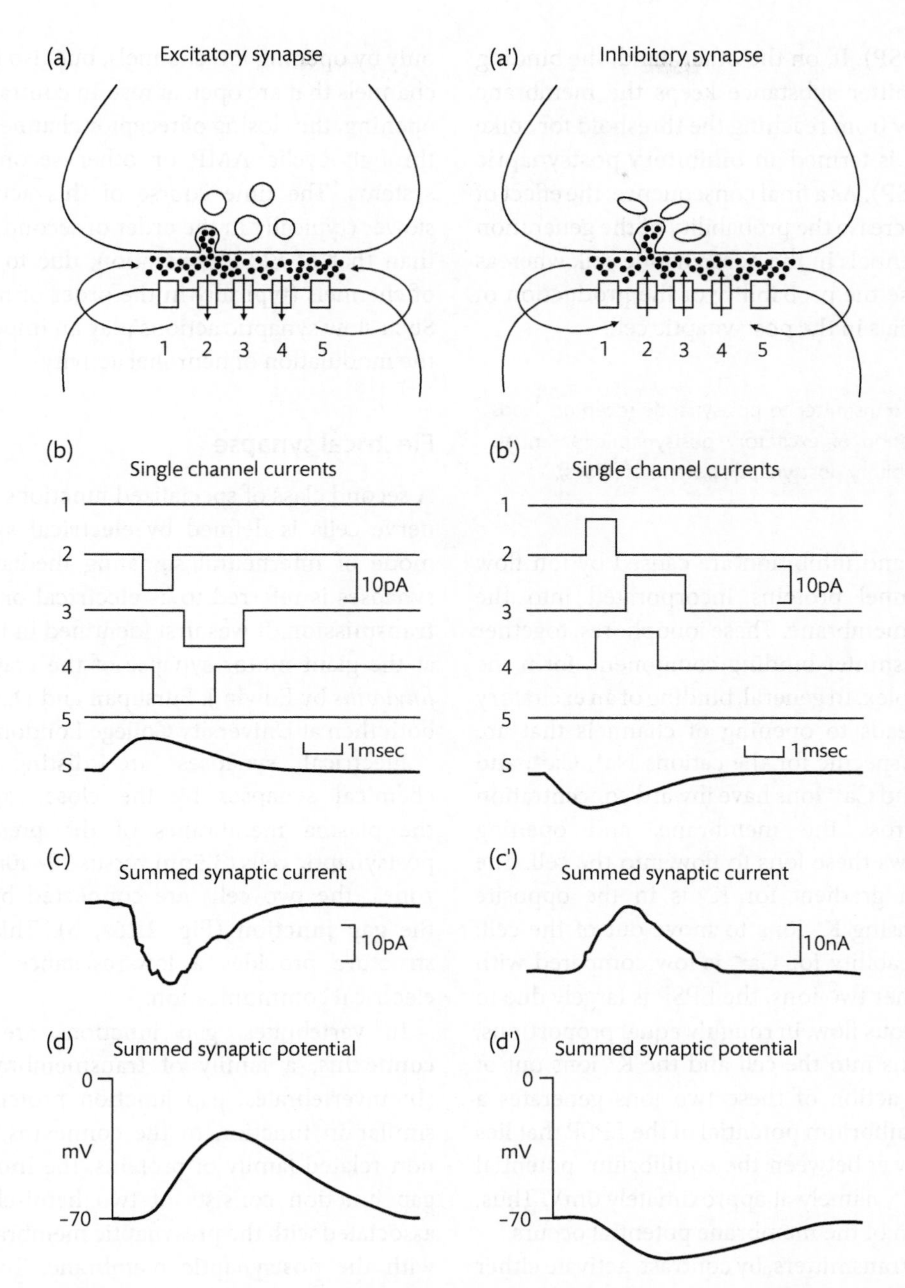

Figure 2.15 Generation of excitatory (a–d) and inhibitory (a′–d′) postsynaptic potentials (EPSPs and IPSPs, respectively). In each of the two cases, the transmitter is released into the synaptic cleft by fusion of the presynaptic vesicles with the presynaptic membrane (a, a′). The transmitter diffuses across the synaptic gap and binds to some postsynaptic receptors (here, the receptors labeled 2–4). This opens the conductance channels, which leads, at the excitatory synapse, to inward flow of positive ions, and to an outward flow across neighboring parts of the postsynaptic membrane. Recordings of the activity of the individual channels by means of the so-called patch-clamp technique reveals opening of channels 2–4, as indicated by the step-wise increase in negative (b) and positive (b′) current. These currents of the individual channels add up to produce a summed synaptic current (c, c′). Finally, the summed synaptic currents generate the summed EPSP and IPSP, respectively (d, d′). (After: Shepherd, G. M. (1988).)

potential (EPSP). If, on the other hand, the binding of the transmitter substance keeps the membrane potential away from reaching the threshold for spike generation, it is termed an **inhibitory postsynaptic potential (IPSP)**. As a final consequence, the effect of EPSPs is to increase the probability of the generation of action potentials in the postsynaptic cell, whereas IPSPs decrease the probability of the production of action potentials in the postsynaptic cell.

> Binding of transmitter to postsynaptic receptors leads to the generation of excitatory postsynaptic potentials (EPSPs) or inhibitory postsynaptic potentials (IPSPs).

Excitation and inhibition are caused by ion flow through channel proteins incorporated into the postsynaptic membrane. These ionophores, together with the transmitter-binding component, form the receptor complex. In general, binding of an excitatory transmitter leads to opening of channels that are relatively nonspecific for the cations Na^+, Ca^{2+}, and K^+. The Na^+ and Ca^{2+} ions have inward concentration gradients across the membrane, and opening channels allows these ions to flow into the cell. The concentration gradient for K^+ is in the opposite direction, causing K^+ ions to move out of the cell. As the permeability for Ca^{2+} is low compared with that of the other two ions, the EPSP is largely due to the simultaneous flow, in roughly equal proportions, of the Na^+ ions into the cell and the K^+ ions out of the cell. The action of these two ions generates a combined equilibrium potential of the EPSP that lies roughly midway between the equilibrium potential for Na^+ and K^+, namely at approximately 0mV. Thus, depolarization of the membrane potential occurs.

Inhibitory transmitters, by contrast, activate either Cl^- or K^+ channels. As Cl^- ions are at much higher concentration on the outside of the cell than on the inside, opening of Cl^- channels leads to movement of Cl^- into the cell and thus to a net increase in the negative charge on the inside of the membrane. As a result, the membrane becomes hyperpolarized, producing an IPSP. A similar effect is achieved by an outward flow of K^+ ions.

Some transmitters can cause either EPSPs or IPSPs, depending on the type of ionophore opened in response to binding of the transmitter molecule. Furthermore, generation of EPSPs is possible not only by opening ion channels, but also by closing K^+ channels that are open at rest. In contrast to channel opening, the closing of receptor channels is achieved through cyclic AMP or other second-messenger systems. The time course of this action is much slower (typically in the order of seconds or minutes) than that of EPSP generation, due to the opening of channels (typically in the order of milliseconds). Such slow synaptic actions play an important role in the modulation of neuronal activity.

Electrical synapse

A second class of specialized junctions between two nerve cells is defined by **electrical synapses**. The mode of interneural signaling mediated by these synapses is referred to as **electrical** or **electrotonic transmission**. It was first identified in the late 1950s at the giant motor synapse of the crayfish *Astacus fluviatilis* by Edwin J. Furshpan and David D. Potter, both then at University College London, U.K.

Electrical synapses are distinguished from chemical synapses by the closer apposition of the plasma membranes of the presynaptic and postsynaptic cells (3.5nm versus 20–40nm). At these zones, the two cells are connected by a channel, the **gap junction** (Fig. 2.16a, b). This specialized structure provides a low-resistance pathway for electrical communication.

In vertebrates, gap junctions are formed by **connexins**, a family of transmembrane proteins. (In invertebrates, gap junction proteins, although similar in function to the connexins, belong to a non-related family of proteins, the **innexins**.) Each gap junction consists of two hemi-channels, one associated with the presynaptic membrane, the other with the postsynaptic membrane. Together these two connected hemi-channels (commonly referred to as **connexons**) bridge the two cells, thus resulting in a continuity of the cytoplasm. Each connexon is composed of six identical connexin molecules.

The six connexin molecules of each connexon are arranged such that a central pore is formed (Fig. 2.16c). Rotation of the connexin molecules leads to opening and closing of this pore. The diameter of the pore is approximately 1.5nm and is thus much larger than the pores of voltage-gated channels. This permits not only ions to diffuse between the two cells, but also larger molecules, such

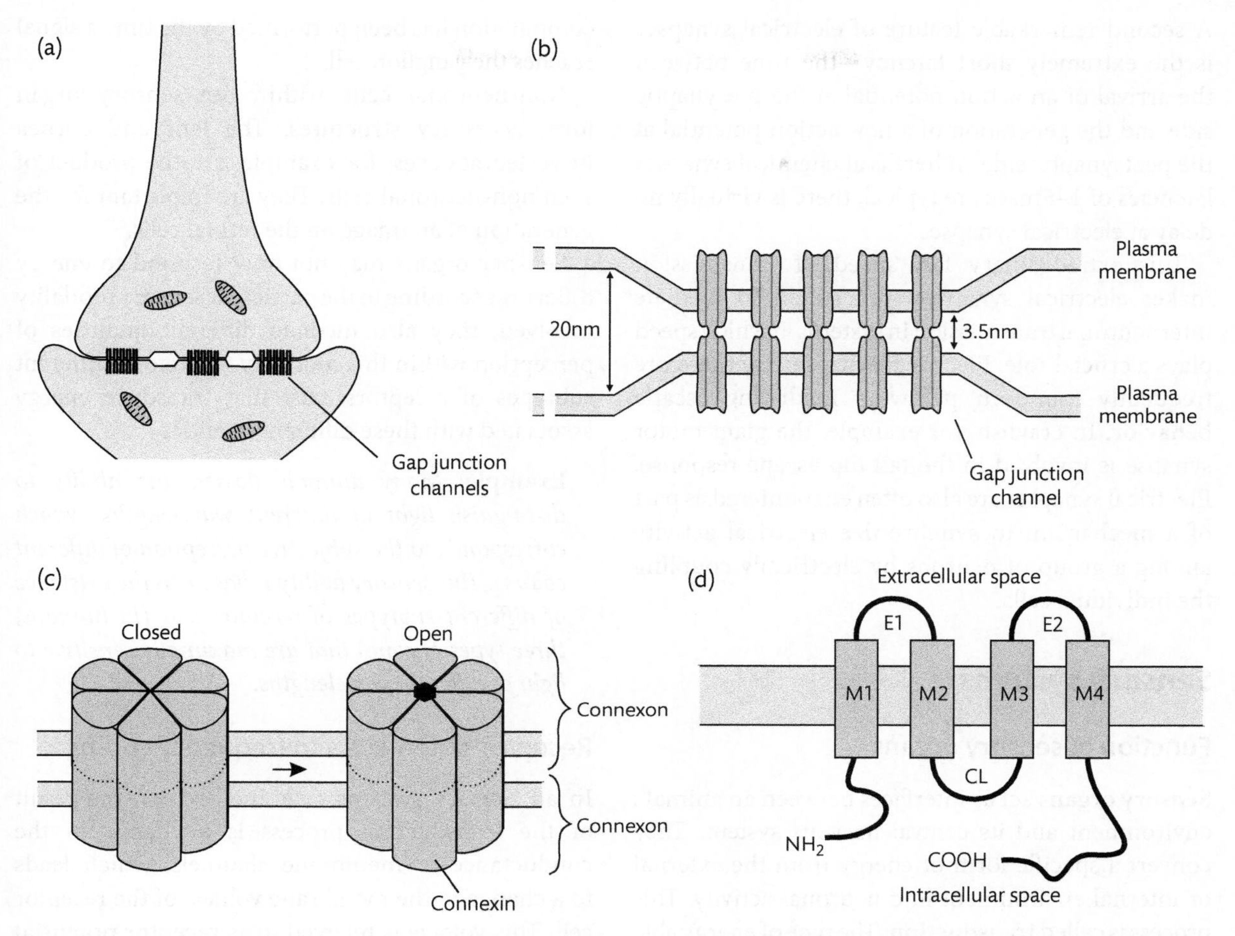

Figure 2.16 Structure of the electrical synapse. (a) Two cells connected by gap junction channels. (b) In the region of the gap junctions, the plasma membranes of the two cells are closely apposed, thereby reducing the width of the extracellular space from about 20nm to approximately 3.5nm. (c) Each gap junction channel consists of two hemi-channels. Each of these so-called connexons is formed by six connexin molecules. Rotation of the connexin subunits leads to opening and closing of the pore. (d) A single connexin molecule is composed of four transmembrane segments (M1–M4), two extracellular loops (E1 and E2), and cytoplasmic loops (CL), including the amino (NH_2) and carboxyl (COOH) termini. (Courtesy: G. K. H. Zupanc.)

as endogenous second messengers or experimentally applied fluorescent dyes.

Within each connexin molecule, the following three major domains can be distinguished (Fig. 2.16d):

- Four hydrophobic regions that span the plasma membrane.
- Two extracellular loops that appear to mediate the homophilic recognition of the two hemi-channels forming the gap junction.
- Several cytoplasmic regions, including the amino and carboxyl termini, which are presumably involved in modulation of the closing and opening of the channel pore.

Depolarization of the presynaptic cell by an action potential generates a positive charge on the inside of the presynaptic plasma membrane. Some of this current flows through the gap junction channel into the postsynaptic cell, resulting in local depolarization of the postsynaptic cell. If this depolarization exceeds the threshold, voltage-gated ion channels will open in the postsynaptic cell and an action potential will be produced.

As a consequence of this mechanism, transmission at electrical synapses is typically bidirectional, although some specialized gap junctions enable cells to conduct depolarizing current in only one direction. The latter are called **rectifying synapses.**

A second remarkable feature of electrical synapses is the extremely short **latency**—the time between the arrival of an action potential at the presynaptic side and the generation of a new action potential at the postsynaptic side. Whereas at chemical synapses latencies of 1–5msec are typical, there is virtually no delay at electrical synapses.

This extraordinary fast speed of transmission makes electrical synapses well suited to mediate interneuronal transmission in systems in which speed plays a crucial role. Electrical synapses are therefore frequently found in pathways mediating **escape behavior.** In crayfish, for example, the giant motor synapse is involved in the tail flip escape response. Electrical synapses are also often encountered as part of a mechanism to **synchronize** electrical activity among a group of neurons by electrically coupling the individual cells.

Sensory systems

Function of sensory organs

Sensory organs act as interfaces between an animal's environment and its central nervous system. They convert a specific form of energy from the external or internal environment into neuronal activity. This process is called **transduction.** The type of energy able to evoke such a neuronal response differs according to the **sensory modality**. In the auditory system of mammals, for example, hair cells of the cochlea transduce mechanical energy. In the visual system, rods and cones of the retina transduce light energy.

The cells within sensory systems that perform this task are specialized neurons called **receptor cells.** These cells transmit information to ganglion cells that send axons to the central nervous system. In some sensory systems, the axons of the ganglion cells are specialized and form the receptive structure. In other systems, such as the auditory system, the signals produced by the receptor cell are transmitted to a process of the ganglion cell. In the visual system, an interneuron is interposed between the retinal receptor cells and the ganglion cells. As one receptor cell typically sends information to more than one ganglion cell (**principle of divergence**), and one ganglion cell receives information from more than one receptor cell (**principle of convergence**), a considerable amount of computation has been performed by the time a signal reaches the ganglion cell.

Non-neuronal cells within the sensory organ form accessory structures. The lens and cornea in vertebrate eyes, for example, are the product of such non-neuronal cells. They are important for the generation of an image on the retinal cells.

Sensory organs may not only respond to energy differing according to the particular sensory modality involved; they also mediate different **qualities** of perception within this modality. Therefore, different subtypes of receptors exist that transduce energy associated with these different qualities.

> **Example:** *Many animals possess the ability to distinguish light of different wavelengths (which corresponds to the subjective perception of different colors). This sensory ability is based on the existence of different subtypes of receptor cells (in humans, three types of cone) that are maximally sensitive to light of different wavelengths.*

Receptor potentials and frequency coding

In all sensory systems examined so far, the result of the transduction process is a change in the conductance of membrane channels, which leads to a change in the membrane voltage of the receptor cell. This voltage is referred to as **receptor potential** or **generator potential**. Like the EPSPs and IPSPs at chemical synapses, it is graded.

The individual steps of the transduction process are especially well studied in the visual system. Absorption of photons (the elementary physical units of light) by receptor cells triggers a cascade of biochemical processes. At the end of this cascade, a massive closure of Na^+ channels occurs, causing a decrease in the resting potential (typically, –20 to –40mV, as measured in complete darkness) and thus hyperpolarization.

In other sensory systems, the transduction of energy may lead to an opening of Na^+ channels, and therefore to depolarization. And in still other systems, the response can even be biphasic, that is, one type of stimulation causes a hyperpolarizing response, whereas a different type of stimulation elicits a depolarizing response from a receptor cell.

In any sensory system, the process of transduction involves considerable amplification of the energy associated with the stimulus.

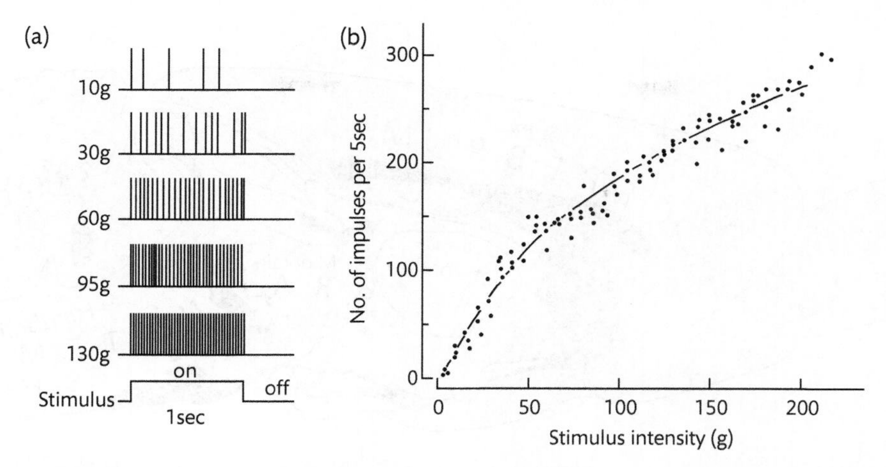

Figure 2.17 Pressure receptor in the foot of a cat. Pressure of various strengths is applied for 1sec. Impulse activity is recorded from afferent fibers. (a) Examples of recordings. The stimulus intensity is indicated on the left. (b) The number of impulses produced in response to a pressure stimulus increases with increasing stimulus intensity. (After: Penzlin, H. (1980).)

Example: *The energy of one photon of 500nm wavelength, which is equivalent to* 4×10^{-19}*J, is sufficient to excite an isolated rod cell. The receptor response consists of changes in current of roughly 1pA and in potential of roughly 1mV, and lasts for several seconds. As the energy is defined as the product of current, voltage, and time (1 joule = 1 ampere · volt · second), this change in energy is equivalent to roughly* 1×10^{-14}*J. Thus, amplification by at least a factor of 10 000 has occurred!*

➤ J = joule; A = ampere; V = volt; 1pA (picoampere) = 10^{-12}A.

Upon transmission to second-order neurons, the graded receptor potential is translated into a series of impulses (Fig. 2.17), a mechanism commonly referred to as **frequency coding.** Thus, the amplitude of the receptor potential is closely related to the rate of action potentials in the second-order neurons. The information contained in these impulses is carried to the central nervous system for further processing.

➤ Receptor cells within sensory organs transduce the energy associated with a stimulus into a receptor potential. Upon transmission to second-order neurons, the graded receptor potential is translated into a series of impulses.

Neurobiological approaches in neuroethology

Many of the methods used in neuroethology are identical to those employed in other neurobiological disciplines. What is different compared with other disciplines is the type of aspects addressed by applying these approaches. Typically, these aspects center around the question of how certain structural or physiological properties of neural networks or individual neurons relate to the production of a specific natural behavior. In this and the following sections, some of the major techniques that are routinely used in neuroethological investigations will be described.

Mapping of brain structure

In 1894, the German psychiatrist and neuropathologist Franz Nissl (1860–1919) introduced basic aniline dyes as a new stain to identify neuronal cell bodies (more precisely, DNA and RNA) in histological material. The method has become known as **Nissl staining** and has enabled investigators to identify distinct cell groups in histological sections (Fig. 2.18). The construction of maps based on Nissl-stained sections is an essential prerequisite for exploring those structures of the central nervous system that are involved in the generation of a specific behavior. This **cytoarchitectonic mapping**

(a)

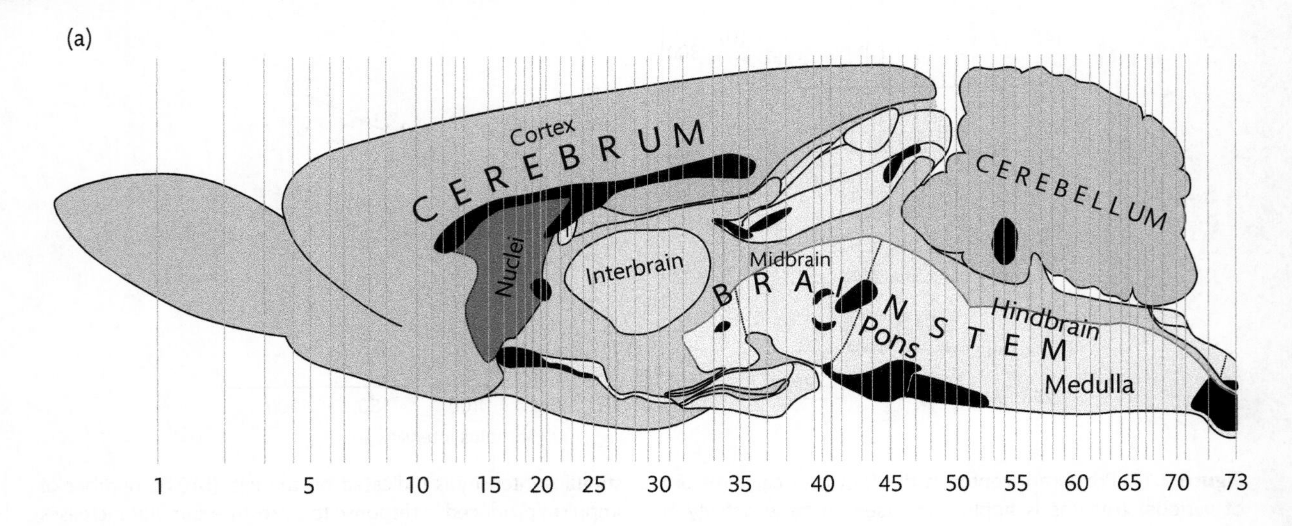

(b)

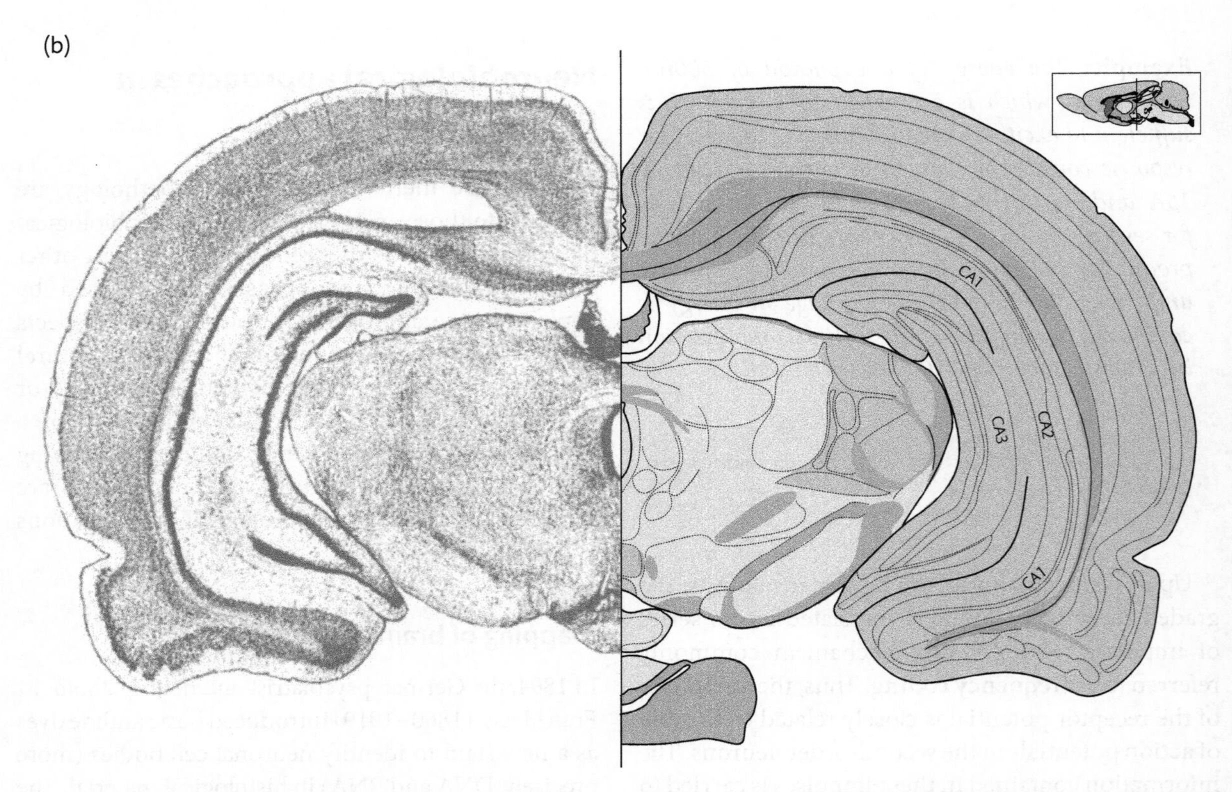

Figure 2.18 Example of a brain map informed by cytoarchitectonic features. To construct this atlas, the brain of an adult male rat was sectioned in the transverse plane, resulting in a total of 556 sections, 30μm or 40μm thick. The sections were stained with the Nissl stain thionin. (a) Out of the 556 transverse sections, 73 representative levels along the longitudinal axis were chosen for constructing the atlas, as indicated in this sagittal view of the brain. Only major subdivisions are labeled. (b) Transverse section at level 37. Photomicrograph (left) and interpretation (right). Cell groups are outlined with dashed lines. For sake of simplicity, only three of such cell groups are indicated – CA1, CA2, and CA3. These three fields form the Cornu Ammons (CA; also called Ammon's horn), a substructure of the hippocampal formation. (After: Swanson, L. W. (2004).)

has, in recent years, increasingly been combined with **expression data** collected at cellular resolution, and the result been used for the three-dimensional reconstruction of major parts of the central nervous system, including whole brains, of a variety of species. Industrial-scale efforts have yielded a number of high-quality **brain and spinal cord atlases.** A prominent example is the web-based, three-dimensional Allen Brain Atlas that combines high-resolution cytoarchitectonic data with information about the expression pattern of approximately 10 000 genes throughout the adult mouse brain. These global genomic data sets are generated by automated high-throughput procedures for *in situ* hybridization and data acquisition. These atlases, as well as their search and visualization tools, can be accessed via the Allen Brain Atlas portal (http://brain-map.org/).

Cytoarchitectonics Sub-discipline of neuroanatomy that uses Nissl staining for parcellation of the brain into distinct areas.

Elucidation of neuronal connections

A term often encountered in the neurobiological literature is that of the **projection** of a neuron. As noted in the section 'Neurons and nervous systems' (earlier), this term describes the route taken by an axon from its associated cell body to the target region. Information about this route allows neuroethologists to understand how different brain areas involved in the control of a particular behavior are connected to one another. Identification of this structural correlate in turn provides the basis to appreciate how these brain areas interact with one another at a physiological level. Large-scale efforts are currently made to elucidate the complete, brain-wide neural connectivity of select organisms. Such information defines the **connectome** of a given organism.

Connectome Complete description of the structural connectivity of the nervous system of an organism.

Most commonly, elucidation of such neural connections is achieved by neuroanatomical means, especially by the application of **neuronal tract-tracing techniques.** A variety of such techniques are available, many of which involve the application of tracer substances. Broadly defined, these substances fall into two categories, anterograde and retrograde tracer substances.

Anterograde tracer substances are taken up by neuronal somata and transported in the anterograde direction (i.e. in the direction of the normal information flow) within the axon to the terminal region. Today, a large variety of such tracer substances is available. Some of them, such as fluorescently labeled dextrans, can be visualized directly using fluorescence microscopy after cutting thin sections through the tissue of interest. Other substances, such as horseradish peroxidase, require a histochemical or enzymatic reaction to visualize the labeled neurons.

Retrograde tracer substances are taken up by axons and preferentially transported in the retrograde direction from the axonal terminal region toward the cell body. Sometimes, the same substance employed for anterograde tracing can be used, if it does not exhibit a significant preference for transport in the anterograde direction. Also, after having reached the soma, some retrograde tracer substances enter the dendritic processes and may thus provide valuable information, not only about the structure of the soma, but also about the dendritic morphology of the labeled neuron.

➤ Connections between different brain areas are commonly revealed using anterograde and retrograde tracer substances.

To unambiguously reveal the connections of a brain nucleus with other brain regions, both anterograde- and retrograde-tracing experiments are required. For example, the results of an anterograde-tracing experiment in which the tracer substance was applied to a brain nucleus A suggested a projection of this nucleus to a brain nucleus B. To verify this projection, in a second step a retrograde tracer will be applied to area B. If A, indeed, projects to B, and no other brain nucleus projects to B, this experiment would show labeled somata in A, but nowhere else. This second part of the experimental study is crucial because axons running through a nucleus (so-called **fibers-of-passage**) may lead to erroneous interpretation of a single tracing experiment. In our example, fibers-of-passage running through nucleus A might have labeled an axonal terminal region in B after tracer application to A, although the somata of these fibers do not originate from cell bodies in A. The retrograde-tracing experiment would, in this case, reveal the location(s) of the somata whose axons travel through A to B.

Figure 2.19 shows a labeled pyramidal neuron in layer II of the cortex of a rat, three days after application of the tracer substance biotinylated dextran amine to the visual cortex. As the photomicrograph demonstrates, neuronal tract tracing experiments can not only identify the location of traced neurons, but also reveal many of the fine morphological details, such as axonal collaterals or dendritic spines.

A powerful non-invasive approach for neural-circuit tracing developed in recent years is based on the expression of markers, such as fluorescent proteins, under the control of cell-type-specific regulatory elements in transgenic animals. In contrast to traditional tracers, which typically label only a fraction of the cell population, and produce significant variation among experiments, the **genetically encoded tracers** yield fibers of the desired cell type specifically and completely.

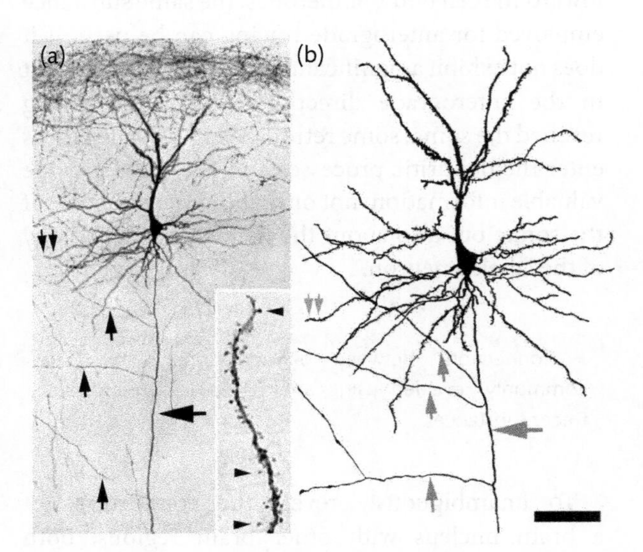

Figure 2.19 Neuron revealed through neuronal tract tracing. The biotinylated dextran amine, commonly known as BDA, was injected extracellularly into the visual cortex of a rat. This tracer substance is transported in both anterograde and retrograde direction within neurons. Three days after the injection, the pyramidal neuron shown in the photomicrograph in (a) was labeled in layer II of the cortex. Its axon (large horizontal arrow) gives rise to several collaterals (vertical arrows). Part of a dendrite, indicated by the double arrows, is rotated 90° clockwise and shown at higher magnification in the inset. This photograph demonstrates the presence of numerous dendritic spines. Three of them are depicted by arrowheads. A drawing of the pyramidal neuron is shown in (b). Scale bar: 75µm in a, 50µm in b, and 10µm in the inset. (© Ling, C., Hendrickson, M. L., and Kalil, R. E. (2012).)

Another class among the 'second-generation' tracers are **neurotropic viruses**. Their usage as tracer agents takes advantage of their intrinsic capability to infect neurons and replicate after entering the nucleus of their host cells. Most importantly, these viruses can spread trans-synaptically, a property exploited in tracing experiments to reveal connections across multiple neurons.

Neurotropic virus Virus with an affinity for nervous tissue.

One problem inherent to all of the above techniques is that, typically, many neurons are labeled. The use of monochromatic labels, as commonly employed, makes it then difficult, if not impossible, to trace individual fibers. A solution developed in recent years is to introduce three fluorescent proteins as intrinsic neuronal labels in transgenic animals, and to stably express them in a mosaic manner, yielding approximately 100 colors in these neurons. This multicolor ('**Brainbow**') approach enables investigators to distinguish individual neurons, instead of having to rely on information derived from analysis of frequently rather ill-defined subsets of neuronal populations carrying a monochromatic label.

Correlation of morphological properties and physiological activity of single cells

One of the central goals of behavioral neurobiology is to correlate morphological and physiological data at the cellular level, and relate this information to sensory perception and motor output at the organismal level. An important step towards this goal, at the single cell level, was done when Antony O.W. Stretton and Edward A. Kravitz of Harvard Medical School described in 1968 a new method for **intracellular staining of individual living neurons** in the lobster abdominal ganglia. This method is based on the injection of fluorescent dyes, such as

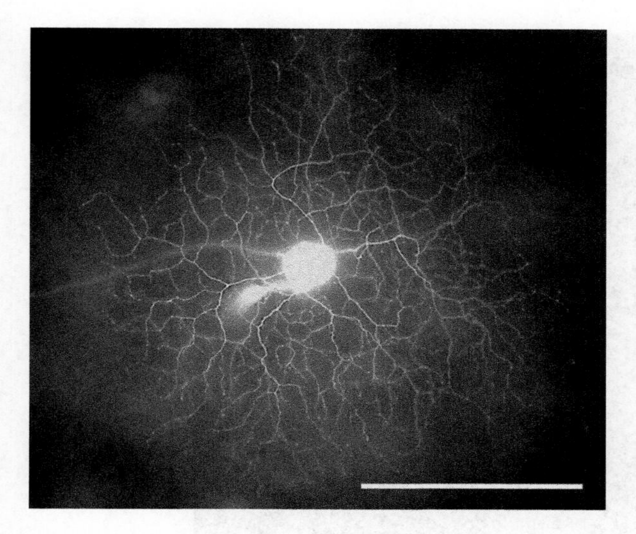

Figure 2.20 Ganglion cell in the rabbit retina injected with Lucifer Yellow. Scale bar, 200μm. (From: Stasheff and Masland (2002).)

Procion Yellow or Lucifer Yellow, from micropipettes into living cells. The dye diffuses within the cell, so that many of the cell's fine details can be revealed, including axonal and dendritic processes, and synaptic boutons (Fig. 2.20). Correlation of the morphology of an intracellularly stained neuron with its electrical activity is achieved by combining intracellular electrical recording with intracellular dye injection.

Mapping of brain activity

Like in the area of global mapping of brain structure, new physiological techniques have been developed in recent years with the aim to generate **global activity maps** with cellular resolution. These techniques capitalize on the advantages of optical reporters of neural activity. Although assessing neural activity through measurement of membrane potential is an obvious choice, the use of optical reporters to achieve this goal has been problematic, due to a number of constraints. Therefore, most researchers have turned to **fluorescent calcium imaging** as a proxy for measuring spiking activity of neurons.

Since calcium ions enter neurons during the firing of action potentials and during synaptic input, neural activity can be assessed through changes in intracellular free calcium. During activation, the cytosolic calcium concentration can rise from 100nM to approximately 1000nM. At the same time, calcium transiently accumulates in the soma, thereby amplifying the signal. Together with the brightness of the calcium indicators, this amplified response makes it possible to monitor the electrical activity simultaneously in thousands of neurons.

Both synthetic indicators and genetically encoded indicators are available as **fluorescent calcium sensors**. A major advantage of genetically encoded calcium sensors is that they can be targeted to specific cell types, and be stably expressed by **viral infection** or **transgenesis** for chronic, non-invasive imaging.

Figure 2.21 shows an example of whole-brain images of neural activity with cellular resolution. The images were recoded from larval zebrafish, and the activity was reported by a genetically encoded calcium indicator.

Transgenesis Process of introducing a foreign gene into the genome of an organism. This organism will exhibit some new properties and transmit these properties to its offspring.

In contrast to the microelectrodes used for extracellular recording, optical probes can be several millimeters away from the imaged area to still achieve single-cell resolution. This is of critical advantage for whole-brain imaging, which ultimately will provide an accurate global picture of the activity of the neural ensembles involved in performing certain behavioral functions. Although this approach is still in its infancy, activity map data collected while animals performed rather simple, reflex-like behaviors have revealed elaborate networks of active cell bodies and fiber tracks—far more widely distributed than previously thought. Such findings underscore the importance of global physiological activity maps for gaining a better understanding of how the brain, as a whole, processes sensory stimuli, controls the generation of motor patterns, and integrates these two processes.

Localization of tissue constituents

At the beginning of neuroethology, exploration of the structural correlates of behavior was often limited to the identification of individual neurons or neuronal assemblies involved in the control of a specific behavior. As we have seen above, this is commonly achieved through neuronal tract tracing, or a combination of physiological approaches

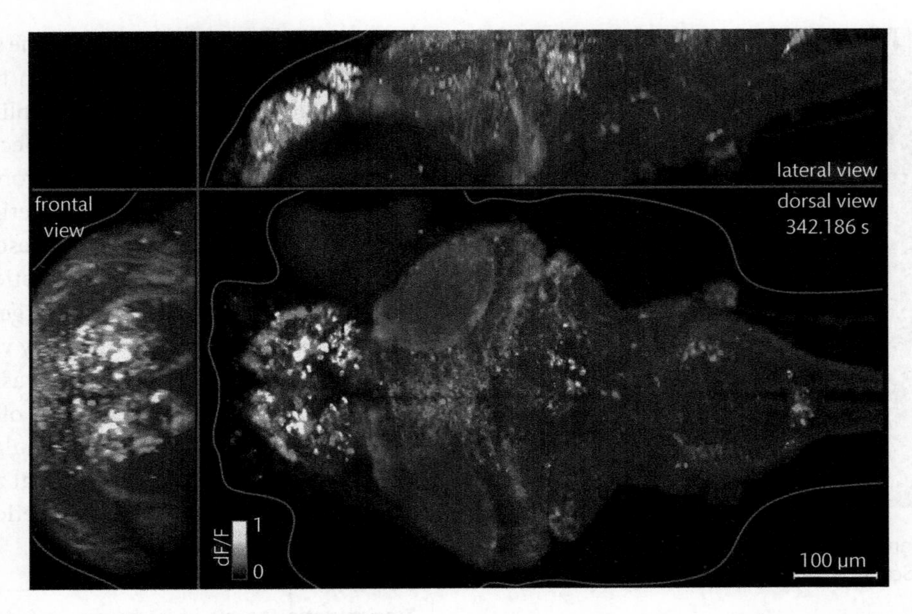

Figure 2.21 Imaging with cellular resolution of neural activity in the whole brain of larval zebrafish. The brain activity is reported through a genetically encoded calcium indicator, and recorded from the fish without external stimulation. Changes in neural activity are indicated by changes in fluorescence activity (dF/F)—cells emitting brighter signals exhibit larger neural activity. The three views show frontal, lateral, and dorsal projections of whole-brain functional activity. (After: Ahrens, M.B., Orger, M.B., Robson, D. N., Li, J. M., Keller, P. J. (2013). © 2013, Rights Managed by Nature Publishing Group.)

with intracellular labeling techniques. While identification of the network components is still a major goal of neuroethological research, studies have been greatly extended since the 1980s to further characterize these components, especially in terms of their biochemical constituents.

Let us, for example, assume that we have obtained physiological and anatomical evidence that a certain neuronal cell cluster is crucially involved in the processing of sensory information relevant for the execution of a behavior. In the next step, we would like to know which neurotransmitters, neuromodulators, receptors, and second-messenger systems are involved in this process. Two methods to address this issue are immunohistochemistry and *in situ* hybridization. The principles of these two techniques will be described in the following two sections.

Immunohistochemistry

Immunohistochemistry is based on the identification and localization of a cellular constituent (a so-called **antigen**) in a tissue section by labeling with a specific antibody (Fig. 2.22). The antibodies are generated by injecting the purified antigen (or a portion of the synthesized molecule) into a host animal (e.g. a rabbit). The immune system of this host animal recognizes the injected substance as a foreign compound and produces antibodies against it. When these antibodies are applied to a tissue section, they will bind to sites that contain the antigen.

The bound **primary antibodies** are characterized by two domains: one specific to the antigen and another specific to the host animal. As a consequence of the latter feature, antibodies generated in a particular host animal, even if directed against different antigens, share a so-called constant-fragment domain. To visualize the bound antibodies, **secondary antibodies** are used that bind to the host-specific constant-fragment part of the primary antibody. If, for example, the primary antibody was raised in a rabbit, anti-rabbit secondary antibodies are required for binding to the primary antibody. Secondary antibodies are commonly conjugated to a **fluorescent label** (a fluorophore) or an **enzyme** that catalyzes an enzymatic histochemical reaction to generate a visible reaction product. When excited

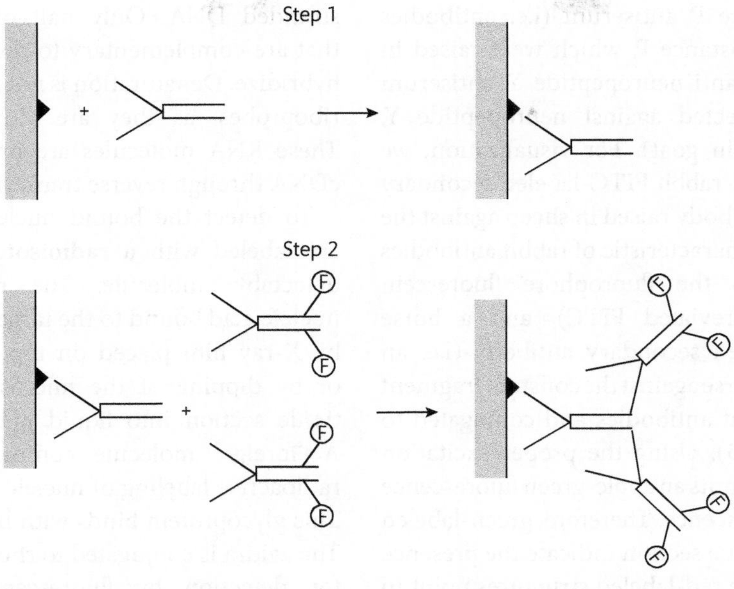

Figure 2.22 Immunohistochemical staining using secondary antibodies conjugated to a fluorophore. In the first step, the primary antibody binds specifically to a tissue site containing the antigen (symbolized by the filled triangle). In the second step, fluorescently labeled secondary antibodies bind to the constant fragment of the primary antibody. The fluorophore is indicated by the encircled 'F.' As more than one secondary antibody can bind to a primary antibody, enhancement of the signal results. (Courtesy: G. K. H. Zupanc.)

with a certain wavelength, the fluorophore emits a specific fluorescent color, which can be viewed using a fluorescence microscope (Fig. 2.23). Reaction products of enzymatic reactions are directly visible under bright-field microscopes.

> ➤ Immunohistochemical labeling involves binding of a primary antibody to an antigenic site in the tissue section, followed by secondary (and sometimes tertiary) reactions to visualize the binding site. It allows the researcher to study cellular constituents.

Demonstration of more than one antigen in the same tissue preparation can be achieved through **multiple immunostaining techniques**. Most commonly, these methods involve the use of antibodies raised against different antigens in different host animals. Thus, for visualization, secondary antibodies conjugated to different chromophores (molecules that are responsible for the antibody's color) and directed against the host-specific constant portion of the respective antibodies can be employed. As a result, each of the different antigens is indicated by a different color.

To illustrate the multiple immunostaining technique, let us assume we are interested in demonstrating the existence of two neuropeptides—substance P and neuropeptide Y—in the same tissue section. We could then apply to the tissue section a

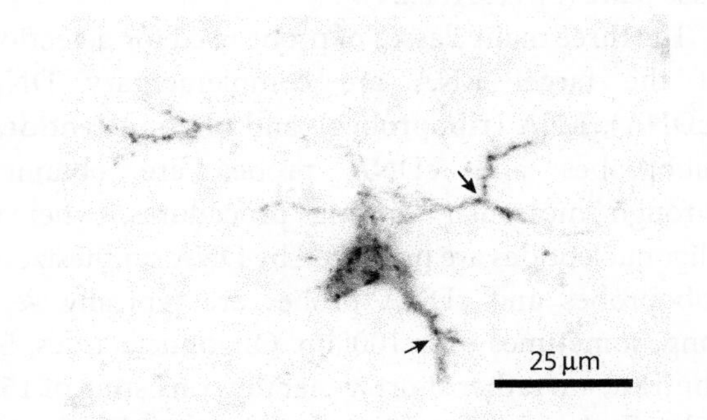

Figure 2.23 Image of a cell (presumably an astrocyte) in the cerebellum of a teleost fish labeled with primary antibodies raised against the neuropeptide somatostatin, followed by binding of secondary antibodies conjugated to the fluorescent dye Cy3. The morphology of the labeled cell body and its processes, including the bifurcation pattern (arrows), is clearly visible. The cytoplasm of the cell contains packages of immunoreactive material. (Courtesy: G. K. H. Zupanc.)

rabbit anti-substance P antiserum (i.e. antibodies directed against substance P, which were raised in rabbit) and a goat anti-neuropeptide Y antiserum (i.e. antibodies directed against neuropeptide Y, which were raised in goat). For visualization, we may use a sheep anti-rabbit FITC-labeled secondary antibody (i.e. an antibody raised in sheep against the constant fragment characteristic of rabbit antibodies and conjugated to the fluorophore fluorescein isothiocyanate, abbreviated FITC), and a horse anti-goat Cy3-labeled secondary antibody (i.e. an antibody raised in horse against the constant fragment characteristic of goat antibodies and conjugated to the fluorophore Cy3). Using the proper excitation wavelengths, FITC emits an apple-green fluorescence and Cy3 a red fluorescence. Therefore, green-labeled structures in the tissue section indicate the presence of substance P, while red-labeled structures point to the existence of neuropeptide Y.

In situ hybridization

***In situ* hybridization** is the method of choice to study the distribution and density of particular sequences of nucleic acids, most commonly of messenger RNA (mRNA), in tissue. This technique was first employed in the late 1960s and early 1970s. It is based on the ability of single-stranded nucleic acid molecules to form a double-stranded duplex with any nucleic acid that contains a sufficient number of complementary base pairs (bp) (Fig. 2.24).

The three main classes of probe used for detection of the target RNA are complementary DNA (**cDNA**), RNA (**riboprobes**), and **oligonucleotides**. Riboprobes and cDNA probes are obtained through molecular cloning procedures, whereas oligonucleotides are produced by DNA synthesizers. Riboprobes and cDNA probes are typically very long, sometimes over 1000bp. Oligonucleotides, by contrast, are rather short, generally consisting of 15–20bp. Both are targeted at only a portion of the entire RNA sequence.

As the cDNA with its 'insert' (i.e. the region complementary to the target mRNA) is usually contained within plasmid DNA and thus is double-stranded, the insert DNA must be cut out of the plasmid DNA by means of restriction enzymes and denatured by boiling to obtain single-stranded DNA. Only half of the strands—those that are complementary to the mRNA—are able to hybridize. Denaturation is not necessary when using riboprobes, as they are already single-stranded. These RNA molecules are produced from cloned cDNA through reverse transcription procedures.

To detect the bound nucleic acid probes, they are labeled with a **radioisotope** or an otherwise detectable molecule. The radioactively labeled nucleic acid bound to the target mRNA is visualized by X-ray film placed on top of the tissue section or by dipping of the microscopic slide with the tissue section into liquid photographic emulsion. A 'foreign' molecule commonly used for non-radioactive labeling of nucleic acid probes is **biotin**. This glycoprotein binds with high affinity to avidin. The avidin is conjugated to rhodamine, for example, for detection by fluorescence microscopy. For detection by bright-field microscopy, the biotin can be conjugated to peroxidase, which is then employed in an enzymatic reaction.

The *in situ* hybridization technique allows the researcher to study the **expression of specific genes** by single cells. It provides information about the distribution of these cells and about the level of expression. As mRNA is restricted to the somatic regions of cells, only cell bodies are labeled by the *in situ* hybridization technique. The level of expression can be assessed, at least in semi-quantitative terms, by the density of label associated with the cell expressing the respective gene.

> ➤ *In situ* hybridization involves the hybridization of a labeled probe complementary to an mRNA sequence of interest. This technique allows the researcher to study the expression of specific genes in cells at the mRNA level.

The information provided by *in situ* hybridization complements the results of immunohistochemical procedures. While *in situ* hybridization aims to detect mRNA, immunohistochemistry is used to detect the resulting cellular protein. There are many instances in which the mRNA level does not correlate well with the protein level. In some extreme cases, high levels of mRNA but no corresponding protein have been detected. This can be interpreted as an indicator of a high level of gene expression, paralleled

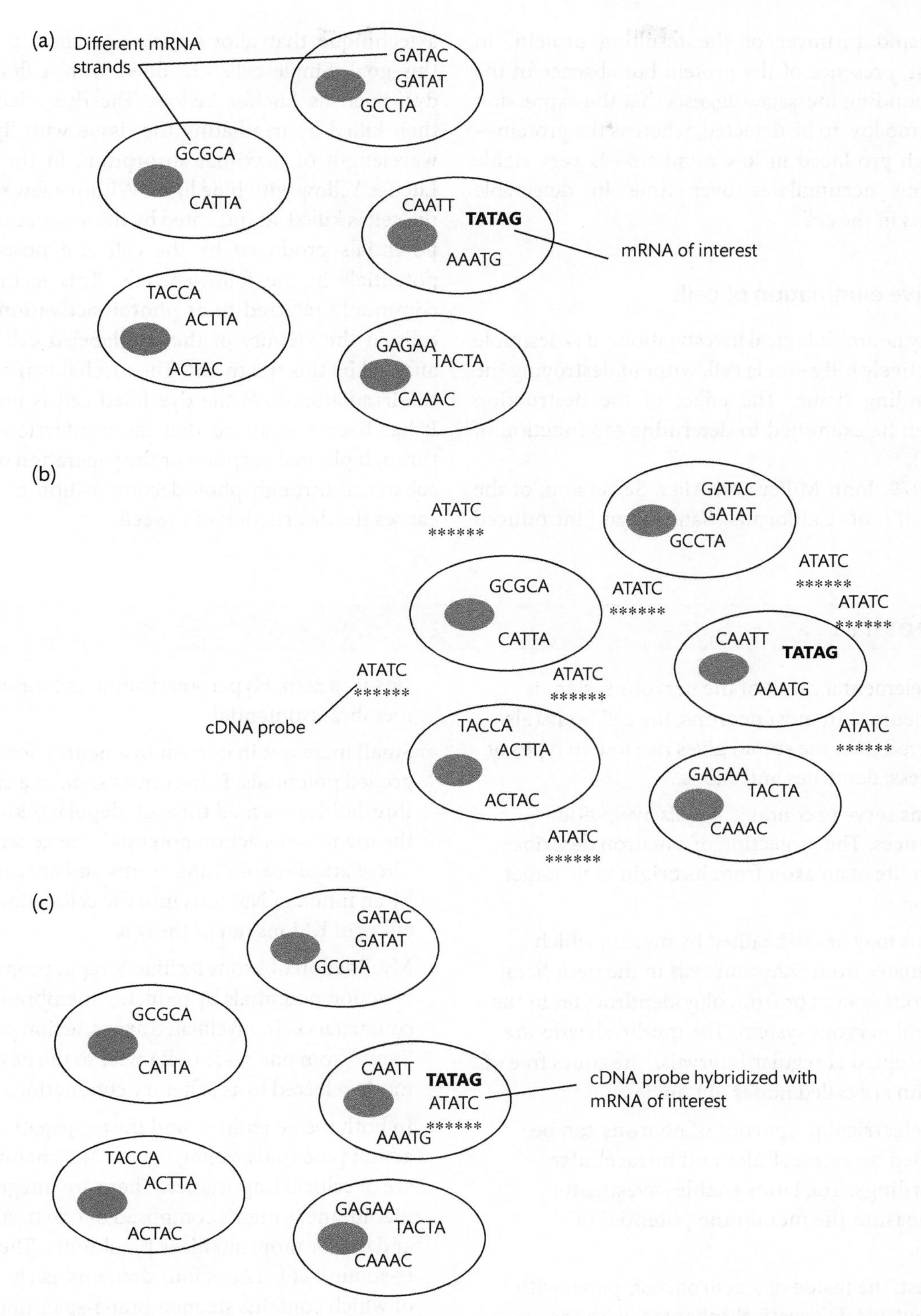

Figure 2.24 Principle of *in situ* hybridization using cDNA probes. (a) Although different cells may express different genes, and thus transcribe different mRNA molecules, the *in situ* hybridization technique enables the researcher to label specifically the mRNA of interest (indicated by bold typeface). (b) Fixed cells are permeabilized and incubated with a cDNA probe, whose sequence is complementary to the mRNA of interest. The probe is labeled (indicated by *) for later identification. (c) The cells are washed to remove unbound probe. The remaining cDNA probe has hybridized to the mRNA of interest. This allows the researcher to identify and localize cells expressing that mRNA. (Courtesy: J. Crowe.)

by a rapid turnover of the resulting protein. In contrast, presence of the protein but absence of the corresponding message suggests that the expression level is too low to be detected, whereas the protein—although produced in low numbers—is very stable and thus accumulates over time in detectable amounts in the cell.

Selective elimination of cells

In many neurobiological investigations, it is desirable to selectively kill a single cell, without destroying the surrounding tissue. The effect of the destruction can then be examined to determine the function of this cell.

In 1979, John Miller and Allen Selverston, of the University of California, San Diego, introduced a technique that allows the researcher to achieve this goal. Single cells are filled with a fluorescent dye, such as Lucifer Yellow. The dye-filled cell is then killed by irradiating the tissue with light of a wavelength of maximal absorption, in the case of Lucifer Yellow with blue light. Within a few minutes, the cell is killed, as indicated by the absence of action potentials produced by the cell and postsynaptic potentials in the follower cells. This technique is commonly referred to as **photoinactivation**. Other cells in the vicinity of the dye-labeled cell are not affected by this treatment. The mechanism by which the irradiation kills the dye-filled cell is unknown. It has been speculated that the production of heat through photoabsorption or the generation of a toxic substance through photodecomposition of the dye causes the destruction of the cell.

Summary

- The elementary unit of the nervous system is the neuron. In most neurons, the cell body (also referred to as the soma) gives rise to two types of process, dendrites and axons.
- Axons serve to conduct signals over wide distances. The projection of a neuron describes the route of an axon from its origin to its target region.
- Axons may be ensheathed by myelin, which originates from Schwann cells in the peripheral nervous system or from oligodendrocytes in the central nervous system. The myelin sheaths are interrupted at regular intervals. The zones free of myelin are called nodes of Ranvier.
- The electrical properties of neurons can be studied by extracellular and intracellular recordings. The latter enable investigators to measure the membrane potential of a cell.
- At rest, the inside of a neuron, compared with the outside, is negatively charged, with the difference in the range of −60 to −80mV. This so-called resting potential is caused by a differential distribution of ions across the plasma membrane. Depolarization brings this membrane potential closer to zero. Hyperpolarization increases the membrane potential.
- Small increases in current in a neuron lead to graded potentials. However, as soon as a critical threshold is reached through depolarization of the membrane, action potentials are generated. These are all-or-nothing events and are caused by an influx of Na^+ ions into the cell, followed by efflux of K^+ ions out of the cell.
- Myelination of axons facilitates rapid propagation of action potentials by reducing membrane conductance. In myelinated axons, action potentials 'jump' from one node of Ranvier to the next one, a mode referred to as saltatory conduction.
- In both the generation and the propagation of action potentials, voltage-gated Na^+ channels are of critical importance. These are integral membrane proteins composed of one α-subunit and one or more auxiliary β-subunits. The α-subunit consists of four domains, each of which contains six membrane-spanning segments. The transmembrane-spanning S5 and S6 segments play a crucial role in the formation of the channel pore, whereas the S4 segment is likely to act as the voltage sensor.

- During generation of an action potential, three states of the voltage-gated Na^+ channel can be distinguished: activated—the channel is open; inactivated—the channel is closed; and deactivated—the channel is closed but ready for opening.
- Synapses are contact points between two neurons. Chemical synapses are polarized in that a transmitter substance is released from the presynaptic cell into the synaptic gap. Upon diffusion to the postsynaptic cell, the transmitter molecules bind to specific receptors that are linked to ion channels. Opening or closing of these channels leads to the generation of postsynaptic potentials, which can be excitatory or inhibitory.
- Important types of transmitter substances are acetylcholine; amino acid transmitters, such as glutamate, glycine, and GABA; biogenic amines, including catecholamines, serotonin, and histamine; and neuropeptides.
- Based on the link between the transmitter-binding site of the postsynaptic receptor and the channel protein, two classes of neurotransmitter receptors are defined. In ionotropic receptors, the receptor domain forms an integral part of the ion channel. The response of these receptors is fast. In metabotropic receptors, the receptor function is mediated by a protein separate from the channel protein. Binding of transmitter molecules to the receptor protein is, typically, followed by the production of second messengers, which, in turn, initiate conformational changes in the channel protein. The response of these receptors is rather slow.
- One of the best-examined ionotropic receptors is the nicotinic acetylcholine receptor. It forms one of the two subclasses of acetylcholine receptors and is distinguished by its ability to bind nicotine, which acts as an agonist of acetylcholine. This receptor is composed of five protein subunits, each of which consists of four transmembrane-spanning segments. The pore is formed by the second transmembrane-spanning segment, called TM2, of each subunit. Three rings of negatively charged amino acids function as a selectivity filter; they repel negatively charged ions, but allow cations to pass through the central canal. Binding of two molecules of a ligand to the acetylcholine receptor induces rotation of the TM2s, and thus opening of the channel.
- Electrical synapses are formed by close apposition of the membranes of two cells and by a channel, the gap junction, connecting these cells. In vertebrates, the channel protein belongs to the family of connexins. Rotation of the connexin molecules leads to opening and closing of the channel pore. Typically, transmission at electrical synapses is bidirectional, although rectifying electrical synapses exist as well. A distinct feature of electrical synapses is their extremely short latency in transmitting electrical signals from the presynaptic cell to the postsynaptic cell.
- Sensory organs act as interfaces between an animal's environment and its central nervous system. The result of the transduction process is the generation of a receptor potential, which is translated into a series of impulses ('frequency coding').
- Parcellation of the brain into distinct areas is commonly achieved through Nissl staining. This technique also forms the basis for the construction of global anatomical brain maps.
- Neuronal projections can be elucidated through neuronal tract-tracing methods. Many of these techniques employ anterograde and retrograde tracer substances. More recently, non-invasive approaches for neural-circuit tracing have been developed based on genetically encoded tracers and neurotropic viruses.
- Correlation of morphological and physiological data can be achieved by combining intracellular dye injection with intracellular recording techniques.
- Global brain activity can be mapped using fluorescent calcium sensors.
- The two major techniques available to reveal tissue constituents are immunohistochemistry and *in situ* hybridization. They provide information about tissue-specific antigens and expression of specific genes, respectively.

The bigger picture

The establishment of new concepts and the development of new experimental approaches and techniques are inseparably connected in neurobiology. A telling example is the cell theory, as detailed in this chapter. Although postulated as early as in the first half of the nineteenth century, it took more than a century to firmly establish this concept. Milestones towards this achievement were the ability to discern individual cells in histological sections, the visualization of the structure of chemical synapses, and the demonstration of the unidirectionality of neuronal conduction at such synapses. Each of these critical steps was possible only through the development and application of new techniques—the Golgi silver impregnation of histological sections, the ultrastructural analysis by employing electron microscopy, and intracellular recordings using microelectrodes. Quite characteristically, neurobiology draws new approaches often from a multitude of biological and non-biological disciplines, including chemistry, physics, engineering, and mathematics. This interdisciplinarity—as challenging as it might be to learners and investigators alike—will, no doubt, continue to fuel progress in neurobiology.

Recommended reading

Byrne, J. H. and Roberts, J. L. (2009). *From Molecules to Networks: An Introduction to Cellular and Molecular Neuroscience*, 2nd edn. Elsevier/Academic Press, Amsterdam.

A very good introduction to the biochemical and biophysical properties of neurons and neural networks. Many parts of the book are targeted at a more advanced readership.

Kandel, E. R., Schwartz, J. H., and Jessell, T. M. (2000). *Principles of Neural Science*, 4th edn. McGraw-Hill, New York.

Classic textbook providing comprehensive information at a more advanced level.

Polak, J. M. and Van Noorden, S. (1997). *Introduction to Immunocytochemistry*, 2nd edn. BIOS Scientific Publishers, Oxford.

Although brief, this excellent introductory text covers all major theoretical and practical aspects of immunocytochemistry.

Shepherd, G. M. (1994). *Neurobiology*, 3rd edn. Oxford University Press, New York/Oxford.

A classic introductory textbook covering all major aspects of neurobiology. Especially recommended to readers with little prior knowledge in the neurosciences and to those who would like to get a broad overview of neuroscience.

Zigmond, M. J., Bloom, F. E., Landis, S. C., Roberts, J. L., and Squire, L. R. (eds) (1999). *Fundamental Neuroscience*. Academic Press, San Diego/London.

With more than 1500 pages, this is more a reference work than an introductory textbook. The major focus is on mammalian systems, including human.

Zupanc, G. K. H. (2017) Mapping brain structure and function: Cellular resolution, global perspective. *Journal of Comparative Physiology A* **203**:245–264.

This article reviews the 2000-year-old effort to explain brain function from a global perspective, and describes how recent technological advances in anatomical brain mapping and physiological activity mapping enable researchers to relate the constructed high-resolution maps to the behavior of the organism.

Short-answer questions

2.1 Describe the structural organization of the nervous system, as defined by the cell theory, and contrast this notion with the reticular theory.

2.2 What is the projection of a neuron? Define this term in one sentence.

2.3 Complete the following sentence: The major approach used by neuroanatomists to reveal neuronal connections in the nervous system is called .... It involves the use of substances transported in.................................. direction or in.................................. direction within the neuron.

2.4 Complete the following sentence: Myelin is produced by.................................. [specify: neurons OR glia]. Specifically, these myelin-producing cells are the.................................. [specify cellular subtype] in the central nervous system and the.................................. [specify cellular subtype] in the peripheral nervous system.

2.5 Complete the following sentence: The node of Ranvier is characterized by a high density of .... [specify ion type] channels. These channels play a critical role in the generation of .... [specify type] potentials. Since these potentials appear to 'jump' from one node of Ranvier to the next one, this phenomenon is referred to as ....

2.6 What is the major pathological change that occurs in the cellular structure of neurons in patients suffering from multiple sclerosis?

2.7 Choose the correct answer: Neurons have a resting potential of approximately

a) 0mV

b) +60mV

c) −60mV

2.8 The resting potential of neuronal cells is largely caused by differential distributions of ion species across the plasma membrane. What are the three major ion species contributing to this effect? Which equation provides a theoretical quantitative estimate of the resting potential, when considering only ONE of these three ion species?

2.9 Define in one sentence each the following two terms:

a) Depolarization

b) Hyperpolarization

2.10 Briefly describe how transient changes in conductivity of specific ion channels cause the two major phases of action potentials.

2.11 Complete the following sentence: Chemical synapses are characterized by a synaptic gap approximately...................................... (value PLUS unit) wide, the presence of.................................... [specify subcellular structure] within the presynaptic bouton which contain the transmitter substance, and the presence of.. [name of group of proteins] on the postsynaptic membrane to which the transmitter molecules bind.

2.12 What is the name of the transmitter that commonly mediates communication at the neuromuscular junction?

2.13 Indicate whether the following transmitters exert, most commonly, an excitatory or an inhibitory function:

a) L glutamate: excitatory OR inhibitory.

b) γ amino butyric acid: excitatory OR inhibitory.

2.14 The response of neurons or muscle cells to neurotransmitters falls into two broad categories, direct gating and indirect gating. Direct gating is mediated by .... [specify type] receptors. Indirect gating is mediated by .... [specify type] receptors.

2.15 What is the distinct property of ionotropic receptors, compared to metabotropic receptors?

2.16 Specify the two types of acetylcholine receptors.

2.17 Which are the three states that voltage-gated sodium channels can assume?

2.18 What does the term 'transduction' in sensory systems refer to?

2.19 Why does Nissl staining of histological sections not allow investigators to decide whether the nervous system is organized in agreement with the cell theory or the reticular theory?

2.20 What combination of two methods allows investigators to correlate the morphological properties and the physiological activity of single neurons?

2.21 What approach do investigators frequently use as a proxy for measuring spiking activity of neurons to generate global brain activity maps?

2.22 Summarize the design of an immunohistochemical labeling experiment through which the peptides cholecystokinin and neuropeptide Y can be simultaneously detected in tissue sections.

2.23 In an experiment, you would like to localize the neuropeptide somatostatin in axon terminals. What method will you use–immunohistochemistry or *in situ* hybridization? Justify your choice.

2.24 **Challenge question** Interdisciplinary research in neurobiology is often conducted not by single investigators but by teams of specialists, each of them applying his/her distinct expertise towards the shared goal. What do you anticipate as a major challenge in such a collaborative effort? How could this problem be met?

2.25 **Challenge question** The end-plate potential of approximately 70–80mV at the neuromuscular junction is the result of a build-up of hundreds of very small end-plate potentials associated with the quantal release of acetylcholine. Such release can also occur in the absence of stimulation of the presynaptic neuron. Sketch an idealized frequency-distribution histogram of these so-called miniature end-plate potentials. How could the amplitude of a single miniature end-plate potential be determined?

Essay questions

2.1 'Modern cellular analysis of the organization of the nervous system has provided evidence not only in favor of the cell theory, but also in support of the reticular theory.' Respond to this seemingly provocative statement.

2.2 Describe the cellular events leading to the generation of an action potential at a node of Ranvier. What is the role of voltage-gated Na^+ channels in these processes?

2.3 Describe the basic structure and function of chemical and electrical synapses. Under what behavioral conditions is one type better suited than the other?

2.4 Suppose a certain brain region called the 'nucleus controllaris' in an animal is suspected of being involved in the motor control of a specific behavior. Design physiological experiments through which you could verify this hypothesis.

2.5 The experiments you suggested in your answer to Question 2.4 have indeed supported the hypothesis that the nucleus controllaris is crucially involved in the control of the specific behavioral pattern. How could you elucidate possible interactions of this brain region with other areas of the central nervous system? How could you identify transmitters and neuromodulators mediating input to the nucleus controllaris?

Advanced topic Multiple sclerosis: From cause to cure

Background information

Myelin performs a number of important functions in the nervous system, including facilitation of the rapid conduction of action potentials, and stabilization and organization of the cytoskeleton of axons. Considering these functions, it is not surprising that diseases leading to demyelination (i.e. loss of myelin sheaths from around axons) cause severe neurological disability. The most common of such diseases is multiple sclerosis. Elucidation of the pathogenesis of this disease provides important insights not only into the development of effective therapeutic strategies to cure this disease, but also in gaining a better understanding of the mechanisms of myelination and the function of myelin in the normal nervous system.

Essay topic

In an extended essay, describe the epidemiology and clinical course of multiple sclerosis, and summarize the genetic and environmental factors that have been proposed as causes of this disease. Furthermore, discuss possible approaches to induce remyelination of demyelinated axons.

Starter references

Franklin, R. J. M. and ffrench-Constant, C. (2008). Remyelination in the CNS: From biology to therapy. *Nature Reviews Neuroscience* **9**:839–855.

Trapp, B. D. and Nave, K.-A. (2008) Multiple sclerosis: An immune or neurodegenerative disorder? *Annual Review of Neuroscience* **31**:247–269.

To find answers to the short-answer questions and the essay questions, as well as interactive multiple choice questions and an accompanying Journal Club for this chapter, visit **www.oup.com/uk/zupanc3e**.

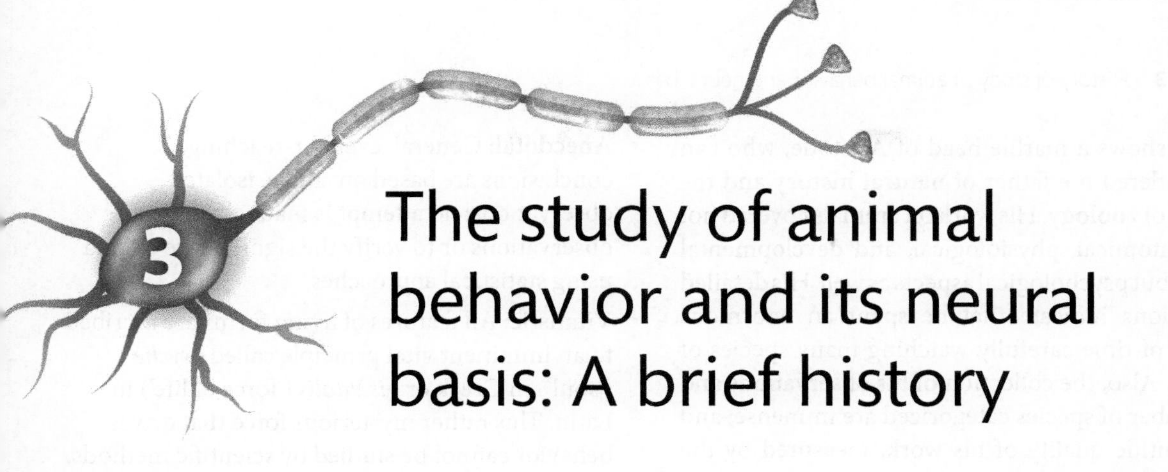

The study of animal behavior and its neural basis: A brief history

Introduction

The interest of humans in animal behavior is certainly as old as humankind itself. Especially before the rise of civilization, knowledge about the behavior of animals was often a matter of survival. However, in spite of this natural curiosity, it took surprisingly long to establish a scientific discipline that applied the rigor of empirical research to the study of animal behavior. One of the reasons for this rather slow and cumbersome development was that, until modern times, such studies tended to be compromised by **anthropocentric** approaches—animal behavior was interpreted from a human point of view, and, all too often, research results were used to 'prove' the superiority of human behavior. Neuroethology, a conception that was possible only after the advent of both well-defined ethological concepts and sophisticated neurobiological techniques, originated even later, and was firmly established only toward the end of the twentieth century.

> **Anthropocentrism** The view that places humans as the most important element in the center of the universe, and interprets the world exclusively in terms of human values and experience.

A true understanding of present developments is possible only by knowing the past. In this chapter, we will, therefore, take a closer look at the historical development of the study of animal behavior and neuroethology, including some of their major figures.

The roots of the study of animal behavior

Aristotle and the Middle Ages

First attempts to produce a systematic account of animal behavior can be dated back to the Ancient Greeks, especially to Aristotle (384–322 bc).

Fig. 3.1 shows a marble head of Aristotle, who can be considered the father of **natural history** and the founder of **zoology**. His work on animals covered not only anatomical, physiological, and developmental aspects, but psychological aspects as well. His detailed descriptions indicate that he spent an enormous amount of time carefully watching many species of animals. Also, the collection of his observations and the number of species categorized are immense; and the scientific quality of his work, measured by the standards of his time, can be considered very high.

The approach Aristotle employed is characterized by the following adjectives:

- **Inductive**: General conclusions are drawn from observations of particular instances.
- **Comparative**: Particular phenomena are analyzed in various animal species and compared with each other.

Figure 3.1 Marble head of Aristotle in the Kunsthistorisches Museum in Vienna (Austria). (Courtesy: Kunsthistorisches Museum, Vienna, Austria/Bridgeman Art Library.)

- **Anecdotal**: General, even far-reaching, conclusions are based on single, isolated observations; no attempt is made to replicate observations or to verify the significance of data using statistical approaches.
- **Vitalistic**: All features of living forms are ascribed to an imminent vital principle called *psyche* ('soul') in Greek or *vis vitalis* ('force of life') in Latin. This rather mysterious force that drives behavior cannot be studied by scientific methods.
- **Teleological**: Behavior, like the morphological properties of the organism, is explained as the expression of an all-pervading design of a *telos* (Greek: purpose, goal). Obviously, such an approach cannot explain features of biological phenomena that are not optimized to serve a certain purpose. However, as we know today, behavior, similar to morphological structures, also reflects its evolutionary history and not just its purpose. As a striking example of imperfection in design, we will discuss, in Chapter 8, the neural implementation of the jamming avoidance response of some weakly electric fish.
- **Anthropomorphic**: Analysis of behavior is performed from an *anthropos* (Greek: human) point of view, rather than from an objective one. This becomes evident, for example, when certain character traits of human beings, such as jealousy, courage, and nobility, are attributed to animals.

> ➤ Aristotle's approach, although inductive and comparative on the one side, is also anecdotal, vitalistic, teleological, and anthropomorphic on the other.

While the former two approaches are also characteristic of modern ethology, the latter four are nowadays regarded as rather unscientific. However, as we will see, they have dominated the study of animal behavior for more than 2000 years.

The period following Aristotle, for almost 2000 years, witnessed not only a halt in the initial progress made in the study of animals and their behavior, but even a decline in this area, as well as in sciences in general. During the 'dark' Middle Ages, experimentation and scientific thinking were largely suppressed. Also, mainly influenced by the rising Christian theology, a dichotomy was drawn between the 'rational soul' of humans and the 'sensitive soul' of beasts. The authority of the writings of Aristotle on natural history was mainly used to find evidence of the superiority of humankind, rather than viewing his texts as a stimulus for further observations.

A remarkable exception to this development was Albertus Magnus (*c.* 1206–1280), a German Dominican bishop, scientist, philosopher, and theologian. Although devoted to the reconciliation of reason and faith through the fusion of Aristotelianism and Christianity, he was also—in contrast to most of his contemporaries—critical of the work of Aristotle, and emphasized the importance of independent observations and experimentation. As he put it, '*The aim of natural science is not simply to accept the statements of others, but to investigate the causes that are at work in nature.*' Based on this principle, he made—during numerous excursions—many biological observations of high originality, especially in the field of zoology.

Mind-body dualism

Although Albertus Magnus was considered an extraordinary genius by his contemporaries and by posterity, his work could not prevent the adherence to Aristotle as a source of authority by most scholars in the following centuries. This attitude was only gradually overcome in the sixteenth and seventeenth centuries. These, as well as the following centuries, were marked by an abundance of new discoveries, especially in the physical sciences. In the biological sciences, emphasis was on classification, morphology, physiology, and embryology. Despite this scientific awakening, the impact on the study of animal behavior as a scientific discipline was rather small. However, triggered by the enormous success in applying the laws of physics to explain not only physical phenomena but also biological processes—such as the circulation of blood, as done by the English physician William Harvey (1578–1657)—a revolt began against the traditional vitalistic interpretation of the organism. The chief exponent of this movement was the French philosopher, mathematician, and scientist René Descartes (1596–1650; Fig. 3.2). His philosophical work included the attempt to explain bodily processes purely on **mechanistic** grounds. Animals were viewed as natural machines. Mind,

Figure 3.2 René Descartes. (Courtesy: The Popular Science Monthly (1890). Artist unknown.)

which included conscious and psychic functions, was assumed to be completely separate from matter, a concept referred to as **mind-body dualism**.

> **Mind-body dualism** The philosophical concept that mind and body are fundamentally different kinds of entities. According to this theory, perception, emotion, thoughts, and other 'mental' phenomena are supernatural and not the result of brain action.

The new era in the study of animal behavior

Evolutionary theory and comparative approaches

Despite its mechanistic outlook, Descartes' doctrine did not cause conflict with the orthodox theology of his times, as it did not challenge man's primacy in terms of his mental and psychic capabilities. It took another 200 years to seriously question the distinct position of humans among living organisms. This was done by the **evolutionary theory** of Charles Darwin (1809–1882) (Fig. 3.3). His ideas were published in various books, including *On the Origin of Species by Means of Natural Selection* (1858), *Variation of Animals and Plants under Domestication* (1868), *Descent of Man* (1871), and *The Expression of the Emotions in Man and Animals* (1872). As a central dogma, the evolutionary theory proposes a continuity of both morphological and behavioral characteristics within the living world, including man. These publications triggered an enormous interest in behavioral observations from a **comparative** perspective, a theme that would become central to ethology. Moreover, the initially still anecdotal, and sometimes even anthropomorphic, approach was gradually replaced by objective and systematic methods. Important figures in this development were Douglas Alexander Spalding and Conway Lloyd Morgan.

Figure 3.3 Charles Darwin in 1881. (Courtesy: Bettmann CORBIS.)

➤ Darwin: Principles of evolution apply not only to morphological characteristics, but also to behavior.

Experimental and objective approaches

Douglas Alexander Spalding, who was born in London to working-class parents, in 1841, and died in 1877 from tuberculosis, had only a few years of scientific activity. His work received wide recognition for a short time after his death, but was then mostly forgotten for more than half a century, until it was rediscovered in the 1950s. Spalding combined his original behavioral observations with carefully designed experiments, thus proceeding far beyond the anecdotal approach that dominated the study of animal behavior in the nineteenth, and even part of the twentiethth, century. His major contributions were to the **development of behavior**. By using a self-constructed incubator, he hatched chicken eggs to study the influence of visual and acoustic experiences on the maturation of behavior. Spalding is also credited with the first experimental study of the **following response** and the **critical period** of birds, which laid the foundations of the **imprinting** concept of ethology. Furthermore, he worked on the nature of instinct and anticipated, to a certain degree, the **releaser** concept. The latter has been seminal to both ethology and neuroethology.

Among those who were greatly inspired by Spalding's work, especially by his experimental approach, was Conway Lloyd Morgan (1852–1936) (Fig. 3.4) of the University of Bristol in the U.K. His contributions had a major impact on the further development not only of ethology, but also of psychology. In particular, he stressed the need for operational definitions and for replication of experiments—requirements that are obvious today, but not at Morgan's time when anecdotal and subjective approaches were still widely used. This encompasses what has become known as **Morgan's Canon**, published in 1894. It is summarized in his own words as follows: '*In no case may we interpret an action as the outcome of the exercise of a higher psychical faculty, if it can be interpreted as the outcome of the exercise of one which stands lower in*

Figure 3.4 Conway Lloyd Morgan. (Courtesy: University of Bristol Special Collections.)

the psychological scale.' In other words, a behavioral pattern of an animal should not be interpreted in terms of 'higher' intellectual capabilities, if a 'simpler' explanation is possible as well. The major significance of this requirement has been to help avoid biased and anthropomorphic interpretation of animal behavior. On the other hand, strict adherence to this (unproved) 'law' could easily lead to misinterpretation of behavior. The goal of any study of animal behavior should be to explain the complex mechanisms of behavior correctly—even if this may not be the simplest interpretation.

Example: *In conditioning experiments, pigeons were trained to discriminate between two-dimensional patterns rotated at different angles in the plane of presentation and mirror images of these forms (Fig. 3.5). After a learning phase, they performed very well—actually similar to humans in such experiments. These results have been used as evidence for the existence of intelligence in pigeons. An alternative explanation, following Morgan's Canon, is that pigeons, as airborne animals, are frequently confronted with the problem of viewing ground-based objects at different angles. They have, therefore, developed sensory mechanisms enabling them to readily identify objects, even if these objects—for example, the projection of trees or houses—are viewed under different rotational angles.*

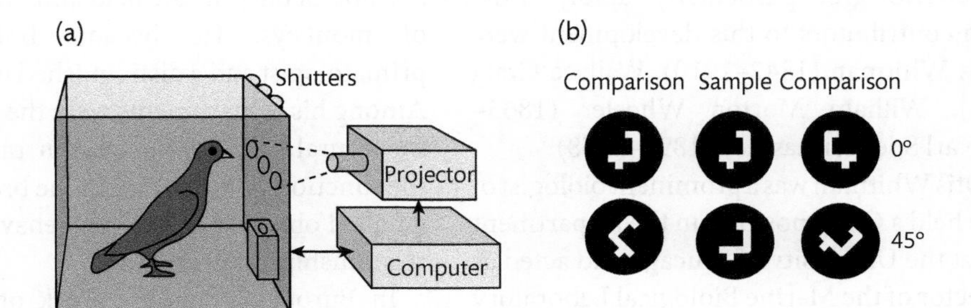

Figure 3.5 Experimental design used by Valerie D. Hollard and Juan D. Delius to test the ability of pigeons (*Columba livia*) to recognize visual forms that are identical to, or a mirror image of, a sample form. (a) Skinner box with three keys. Stimuli were displayed through a projector on the individual keys. A computer controlled the events within the experimental sessions and recorded the data. As part of the operant conditioning paradigm, the pigeons had to identify correctly either comparative stimuli that were identical to a sample stimulus (half of the pigeons used in the experiments were trained and tested according to this schema), or comparative stimuli that were mirror images of the sample stimulus (the other half of the pigeons used in the experiments were trained and tested according to this reverse schema). (b) Examples of stimulus sets to test rotational invariance. In the first row, the left comparison stimulus is identical to the sample, while the right comparison stimulus is a mirror image of the sample. In the second row, the left comparison stimulus is a mirror image of the sample, while the right comparison stimulus is identical to the sample (but rotated 45° clockwise). (Hollard, V. D., and Delius, J. D. Rotational invariance in visual pattern recognition by pigeons and humans. Science. 1982 Nov 19; 218(4574): 804-6. Reprinted with permission from AAAS.)

In the twentieth century, especially within the first 50 or 60 years, the study of animal behavior took two quite separate routes. One of these developments culminated in Europe in the establishment of **ethology** as a biological discipline. The other led to the foundation of several psychological schools, such as **comparative psychology** and **behaviorism,** in North America. All three disciplines are characterized by the attempt to use an objective approach. The main difference between them lies in the conditions under which they observe and analyze behavior: Ethology stresses the importance of observations of animals under **natural** (or **semi-natural) conditions,** whereas the more psychologically oriented schools prefer to analyze behavior under the stringent conditions of the laboratory.

Ethology The biological discipline that studies natural animal behavior in a broad range of organisms by using objective and comparative approaches.

➤ Douglas Spalding and Conway Lloyd Morgan introduced experimental and objective approaches to the study of animal behavior.

The rise of ethology

In the first decades of the twentieth century, the study of animal behavior grew particularly rapidly in the U.S.A. Main contributors to this development were Charles Otis Whitman (1842–1910), Wallace Craig (1876–1954), William Morton Wheeler (1865–1937), and Karl Spencer Lashley (1890–1958).

Charles Otis Whitman was a prominent biologist of his time. He held a Chair position in the Department of Zoology at the University of Chicago and acted as the first director of the Marine Biological Laboratory at Woods Hole, Massachusetts. Although in his lifetime he published just two papers in the field of animal behavior, he exerted a major influence on the development of ethology. Already in 1898, he wrote that '*... instincts and organs are to be studied from the common viewpoint of phyletic descent.*' The results of his detailed research on the behavior of pigeons were published posthumously.

Among Whitman's students were Craig and Wheeler. Wallace Craig continued the studies of his mentor on the behavior of pigeons. He was the first to distinguish between **appetitive behavior** and **consummatory action.** According to this classic ethological concept, changes in an animal's internal state (e.g. in water balance caused by dehydration) are sensed by the brain and result in a build-up of 'drive.' This build-up shows itself externally as a state of agitation called appetitive behavior, which involves a search for a suitable external stimulus (e.g. water). When this stimulus is encountered, it triggers a consummatory action (e.g. drinking). This leads to a reduction, and finally to a cessation, in the corresponding appetitive behavior.

William Morton Wheeler was the first to use the term 'ethology' in the English literature, in a paper published in *Science* in 1902. He became widely recognized for his detailed descriptions of the social life of insects. Wheeler also developed a classification scheme of insect societies. Among his achievements related to animal behavior was the discovery of **trophallaxis.** This term describes the donation of salivary secretions by larvae of social wasps to their adult, winged sisters. The secreted substances are important means of recognition and communication within colonies. As such, his discovery prepared the grounds for investigation of **pheromones.**

Karl Spencer Lashley (Fig. 3.6) worked in his early career with the later founder of behaviorism, John B. Watson (see 'Mechanistic schools,' below) on the homing of birds in the field and on the development of monkeys. He became both a renowned primatologist and a distinguished neurophysiologist. Among his achievements were the attempts to reveal the neural basis of learning in rats and to analyze the function of the cortex in the brain. He also made detailed observations of the behavior and the social relationships in primates.

In Europe, it was the work of Oskar Heinroth (1871–1945) that especially inspired many. Heinroth (Fig. 3.7), a distinguished ornithologist and director of the Berlin Zoo, proposed that **behavioral patterns can be used to analyze systematic relationships between species**—just as morphological patterns can be used for taxonomic classification. In an influential paper published in 1911, Heinroth furthermore demonstrated that young goslings follow the first moving object they see after hatching—a rediscovery of the phenomenon already described by Spalding

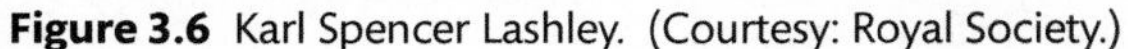

Figure 3.6 Karl Spencer Lashley. (Courtesy: Royal Society.)

Figure 3.7 Oskar Heinroth. (Courtesy: AKG, London.)

in 1873. The young geese subsequently follow this object with preference to any other object. Normally, this object is the parent goose. Goslings reared by a human foster parent thus prefer this person to any conspecific. Heinroth called this phenomenon **imprinting**. In the same paper, he was also the first to use the term 'ethology' in the modern sense, namely, as the 'study of natural behavior.'

Another influential scholar of this pre-ethological time was Jakob von Uexküll (1864–1944; Fig. 3.8). Based on his wealth as a 'Baltic Baron,' he largely funded his studies himself, which enabled him to maintain an exceptional degree of independence. His non-conformism was also expressed in many of his ideas, including the proposal that the subjective world in which an animal lives is quite different from the objective physical environment. He called this subjective world **Umwelt** (plural: **Umwelten**). Although the term, literally translated from German, means 'environment,' we will, in agreement with other authors, use 'Umwelt' to indicate an animal's subjective world, and the term 'environment' to denote the objective physical world in which the organism lives. This concept has had a significant impact not only on sensory physiology, but also on the rising ethology.

Example: *The visual spectrum of honey bees is, compared with that of humans, shifted into the UV range. Bees make use of this sensory ability when searching for food, as some flowers have markings on their petals, the so-called* ***honey guides*** *(Fig. 3.9), visible only in the UV range of light, which guide the bee to the flower's source of nectar. On the other hand, honey bees are blind in the red range of light. Their Umwelt is thus quite different from that of humans, although both humans and bees live in the same physical environment.*

Figure 3.8 Jakob von Uexküll. (Courtesy: Wikimedia Commons; original source appears to be Jakob von Uexküll-Archiv für Umweltforschung und Biosemiotik der Universität Hamburg.)

The final establishment of ethology as a new and independent discipline was achieved by Konrad Lorenz (see Box 3.1) and Niko Tinbergen (see Box 3.2) between the 1930s and the 1950s. They are, therefore, considered the founders of ethology. Their major accomplishment was to provide a greater conceptual framework for their own observations and experiments, as well as for those of others, including Heinroth, Whitman, Craig, and Morgan. Many concepts of this early theoretical framework centered on the nature of **innate behavior**. Milestones in this development were two publications by Tinbergen. In a seminal paper entitled *On aims and methods of ethology*, published in 1963, he defined what became known as Tinbergen's Four Questions—the need to study both proximate causes (neural/physiological mechanisms and development) and ultimate causes (adaptive value and phylogeny) of behavior. The second milestone publication was *The Study of Instinct*, which appeared in 1951 and comprised the first textbook of ethology.

➤ Ethology, as the study of the natural behavior of animals, was founded by Konrad Lorenz and Niko Tinbergen, who provided a conceptual framework for this new biological discipline.

(a)

(b)

Figure 3.9 Honey guides. (a) The drawing was made after a photograph taken on regular film. The flowers appear white, without any distinct pattern. (b) The same flowers taken through a photographic system sensitive to ultraviolet light. Now a striking color pattern, the so-called honey guide, is revealed. These two records approximate the visual pattern perceived by a human and a honey bee, respectively, and thus demonstrate the differences in the Umwelt between the two species. (After: McFarland, D. (1993).)

Box 3.1 Konrad Lorenz

Konrad Lorenz with graylag geese imprinted to follow him, rather than their mother. (Courtesy: Nina Leen/Timepix.)

Konrad Zacharias Lorenz is, together with Niko Tinbergen, regarded as the founder of ethology. He was born in Vienna (Austria) in 1903. Like his father, a famous orthopedic surgeon, Lorenz studied medicine, first at Columbia University in New York and then at the University of Vienna. It was during his time at the Institute of Anatomy in Vienna, working with the well-known anatomist Ferdinand Hochstetter, that he became exposed to the idea of revealing evolutionary descent by comparing homologous morphological structures of various animals. Using the underlying principle, Lorenz soon introduced a similar approach to the study of animal behavior to elucidate evolutionary traits by comparing homologous behavioral patterns.

Despite his training in medicine, recognized by the award of an M.D. degree in 1928, his greater passion was for biology, watching birds and fish in particular. During these years, he conducted most of his studies in his parents' home—actually more a kind of a castle built on a huge property in Altenberg, a small village near Vienna. His investigations resulted in the award of a Ph.D. in zoology in 1933. It was around that time that Lorenz met the German ornithologist Oskar Heinroth and the Dutch naturalist Niko Tinbergen, the latter being considered the co-founder of ethology. Influenced by these two men, Lorenz published in the 1930s a series of key papers, especially on avian behavior. These publications laid the foundations for the definition of several key concepts in ethology.

In 1940, Lorenz was appointed to the Chair of Comparative Psychology at the University of Königsberg, an eminent position that goes back to the German philosopher Immanuel Kant as its first holder. One year later, Lorenz was drafted into military service. Evidence discovered after Lorenz's death revealed that, while he was stationed in Posen (present-day Poznań, Poland), he participated as a military psychologist and a member of the Office of Race Policy in a psychological examination of German-Polish *Mischlinge* ('half-breeds'). During the time of the Third Reich, he also published several papers in which he claimed that civilizing processes, like domestication in animals, leads to a genetically conditioned degeneration of human societies. He was convinced that such decay could be prevented through 'deliberate, scientifically underpinned race policy.' These and other statements, combined with explicitly expressed sympathies for the Nazi racial ideology, made Lorenz a politically controversial figure during his lifetime and even more so after his death, when previously unknown details about his role as a supporter of the Nazi regime became public.

In 1944, Lorenz was taken prisoner by the Russian army and released only in 1948. During his imprisonment, he wrote the manuscript to a book that, when published 25 years later under the title *Behind the Mirror: A Search for a Natural History of Human Knowledge*, became a bestseller.

Lorenz's major contribution after the Second World War was the establishment, in Seewiesen near Munich, of the Max Planck Institute for Behavioral Physiology, of which he was director from 1961 to 1973. In addition to his enormous influence on the development of ethology, an achievement for which he received—together with Niko Tinbergen and Karl von Frisch—the Nobel Prize in 1973, he published a large number of bestsellers. They include popular books, such as *King Solomon's Ring* and *Man Meets Dog*, as well as *On Aggression*, a highly controversial attempt to explain human aggression as the result of a spontaneously active drive.

Konrad Lorenz died in Altenberg in 1989.

Mechanistic schools

At the same time that ethology gradually grew, several more mechanistically oriented schools emerged. Their members were often affiliated with psychology rather than biology departments. All of these different schools shared with ethology the objective approach toward the study of behavior. In contrast to ethology, their emphasis was, typically, on laboratory-based experiments, often conducted on only one or a few 'model systems' (frequently rats or pigeons).

One of the first of these mechanistic schools was established by Jacques Loeb (1859–1924; Fig. 3.10) with his **theory of tropism**. Loeb was born in Mayen (present-day Rhineland Palatinate). He first studied philosophy at the University of Berlin, but dissatisfied with metaphysical philosophy which could not provide answers to questions that were critical to the understanding of human nature—such as whether there is free will—he went to the University of Strasbourg to work with Friedrich Leopold Golz in neurology. In 1886, he became assistant to Adolf Fick, Professor of Physiology at the University of

Box 3.2 Niko Tinbergen

Niko Tinbergen. (Courtesy: The Nobel Foundation.)

Among the 'fathers' of ethology, Konrad Lorenz has often been characterized as the theorist who formulated major ethological concepts based on observations of animals kept under semi-natural conditions in captivity. In contrast, Niko Tinbergen was primarily a field biologist who carefully collected an impressive amount of data on the **natural behavior of animals in the wild**, and who demonstrated a remarkable ability to devise simple, yet highly informative, **field experiments**.

Nikolaas ('Niko') Tinbergen was born in 1907 in The Hague, in The Netherlands. While still in school, he developed a keen interest in bird watching and animal photography. However, when studying biology at the University of Leiden, he did not impress his professors—he often missed classes because he went bird watching or played hockey instead. At the latter he did so well that he even became a member of the Dutch national hockey team.

In his doctoral thesis, Tinbergen studied the behavior of bee-hunting digger wasps. This work formed the basis for subsequent important publications on wasps' homing and hunting behavior, as well as their ability to learn landmarks. Following the completion of his graduate studies, Tinbergen went, with his wife Elisabeth whom he had just married, on an unusual honeymoon—a Dutch expedition to the Scoresby Sound region of East Greenland. This one-year trip resulted in the publication of monographs on the snow bunting and the red-necked phalarope.

Tinbergen spent the period between 1933 and 1942 at the University of Leiden, where he carried out a number of influential studies on the behavior of various animals, including digger wasps, sticklebacks, and herring gulls. In 1938, Konrad Lorenz invited Tinbergen to work with him at his home in Altenberg, where they subsequently produced a classic piece of research on the **egg-retrieval response of nesting geese**.

During the years 1942–1945, Niko Tinbergen was imprisoned in a German internment camp, after he and other faculty members of the University of Leiden had protested against the dismissal of Jewish professors. Despite all the difficulties during those years, he wrote several children's books and prepared the draft of *Social Behavior in Animals*, which was published in 1953.

After the war, he returned to Leiden where he became a full professor at the university in 1947. In 1953, he accepted a lectureship at the Zoology Department of the University of Oxford. This move tremendously accelerated the establishment in Great Britain of the new field of ethology, until then dominated by continental scientists. Elected to a Chair in Animal Behavior, in 1966, Tinbergen stayed in Oxford until the end of his life, in 1988. During this period at Oxford, he broadened his research interests to include studies on human behavior. His ethological investigations on **childhood autism**, conducted in collaboration with his wife, have generated significant interest from psychologists.

Tinbergen's pioneering work, which included the publication of the first textbook of ethology (*The Study of Instinct*) in 1951, was recognized by the award of the Nobel Prize for Medicine and Physiology, together with Konrad Lorenz and Karl von Frisch, in 1973.

Würzburg. There, he met the botanist Julius von Sachs, whose experiments on **tropisms** in plants had a profound impact on Loeb's future work. In 1891, he immigrated to the U.S.A., where he taught briefly at Bryn Mawr College, but subsequently held major positions at the University of Chicago, the University of California at Berkeley, and the Rockefeller Institute for Medical Research (now Rockefeller University). He died in 1924 in Hamilton, Bermuda.

Tropism Involuntary orientation of an organism in response to an environmental stimulus. Such orientation involves a turning movement or differential growth.

Figure 3.10 Jacques Loeb (Courtesy: National Library of Israel, Schwadron collection.)

Loeb became one of the most eminent scientists of his time, particularly after he had performed experiments in which he demonstrated that echinoderm larvae could be chemically stimulated to develop in the absence of fertilization. Besides his seminal studies in development, Loeb's work had also a profound impact on research in animal behavior and nervous system function. Using a similar, rigorous experimental approach as von Sachs did in plants, Loeb attempted to explain motions of animals on the basis of **tropisms**, rather than as reflexes, as was the dominant view at that time. It was this objective, experimental approach based on physico-chemical theory that deeply impressed many scientists of his time, as people welcomed Loeb's theory as a countermovement to the previously dominating vitalistic, anthropomorphic, and anecdotal approaches. However, his attempt to explain animal and human motions in terms of tropisms turned out to be insufficient as a universal mechanism underlying behavior.

While the theory of tropism did not survive its major proponent, another mechanistic theory, the **reflex theory** of Ivan Petrovich Pavlov, did. Pavlov (1849–1936) (Box 3.3) was a Russian physiologist

Box 3.3 Ivan Pavlov

Ivan Pavlov. (Courtesy: U.S. National Library of Medicine.)

Ivan Pavlov's name is inseparably linked to the concept of classical conditioning. Through his monumental experimental study of this process, he laid the foundation for an objective exploration of behavior—arguably, more so than any other scholar of the twentieth century.

Ivan Petrovich Pavlov was born as the son of a priest in 1849, in Ryazan, a provincial city located 200 km southeast of Moscow. There, he attended first a church school, and then the theological seminary. However, influenced by progressive ideas, he abandoned religion and became an atheist, and left the seminary without graduating. In 1870, he enrolled in natural science at the University of St. Petersburg. After completing the program, he continued his studies at the Academy of Medical Surgery, where he was awarded the degree of doctor of medicine in 1879. Over the following years, as an assistant, he managed the Laboratory of Physiology of the famous Russian clinician Sergei P. Botkin, while, at the same time, working on his dissertation. In 1883, he defended his thesis entitled *The centrifugal nerves of the*

. . . continued

heart, in which he demonstrated that the vagus nerve with its fibers slows the heartbeat and weakens the contractions, whereas the fibers of the sympathetic nerve accelerate the heartbeat and strengthen the contractions. Supported by a scholarship from the Military Medical Academy, he studied, between 1884 and 1886, in Germany in the laboratories of two world's leading physiologists—Rudolf Heidenhain at the University of Breslau (now Wrocklaw, Poland) and Carl Ludwig at the University of Leipzig. Pavlov regarded them as the prototypes of 'ideal scientists', who devoted their entire lives to the quest for truth. It was the same mission that also defined his own life and scholarly work. In 1890, the Imperial Institute of Experimental Medicine opened in St. Petersburg, as the first Russian research institution in biology and medicine. Pavlov accepted the offer to head the Department of Physiology—a task that he performed until the end of his life in 1936—while holding professorial appointments at the Military Medical Academy during the first 35 years of that time period. Under Pavlov's leadership, the Institute of Experimental Medicine became a large-scale enterprise and one of the world's foremost centers of physiology.

During the first decade of his tenure at the Institute, the focus of Pavlov's research was on the **physiology of digestion**. By developing a new surgical method that enabled him to create **fistulas**—artificial openings in various organs—he was able to analyze the function of the organs involved in digestion, while keeping the remaining organism intact. This 'chronic' method overcame the limitations of the up-to-then-employed approach of 'acute' vivisection. In recognition of his achievements in this area, Pavlov was awarded the Nobel Prize in Physiology or Medicine in 1904.

While studying the physiology of digestion, Pavlov noticed that dogs started to drool as soon as they saw the white coats of the staff who fed them, without necessarily receiving food. By implanting fistulas into the ducts of the salivary glands, he was able to use the amount of saliva as a quantitative measure to analyze this 'psychic reaction', as he initially called it. Later, he replaced this term by *uslovnyi refleks*, the most appropriate translation of which is **conditional reflex** (although the less adequate translation as **conditioned reflex** is now commonly accepted). Pavlov used the term to distinguish the provisional nature of the association of these reflexes from other reflexes thought to be unvarying. The discovery that reflexes underlie 'psychic' actions made possible hitherto unknown avenues in science—foremost the mechanistic study of behavior. Over the last three decades of his life, he expanded these initial experiments to a systematic investigation of the phenomenon of **conditioning** and of, what he believed, related behavioral processes, such as generalization, discrimination, and extinction, which play critical roles in the etiologies and therapies of human psychoses.

To gather sufficient experimental data, Pavlov ran his research operation like a factory—by the end of his life he had three separate laboratories with several hundreds of scientists and technicians. Through his achievements, Pavlov had become a national hero, and particularly towards the end of his life he received massive public funding. His attitude towards communism and the Soviet government was ambivalent. Whereas on some occasions he publicly praised the 'great social experiment' of the Soviet state, on other occasions he denounced communism. In a letter to Joseph Stalin, he protested that 'on account of what you are doing to the Russian intelligentsia—demoralizing, annihilating, depraving them—I am ashamed to be called a Russian!' However, despite having saved numerous people from the gulag through his courageous actions, in his personal relationships he was feared by many for his authoritarian style, choleric temperament, and dogmatic attitude.

Ivan Pavlov died in Leningrad in 1936, at the age of 86. While his casket was displayed for public viewing, 100 000 mourners paid their respect, and over the following months mass meetings throughout the Soviet Union were organized to celebrate the country's greatest scientist.

who was awarded the Nobel Prize for Medicine in 1904 for his work on the physiology of digestion. His influence was especially pronounced in the U.S.A. after the publication of the English translation of his book *Conditioned Reflexes*. The discovery that made Pavlov famous was the phenomenon of **classical conditioning**, sometime also referred to as **Pavlovian conditioning**. It involves the formation of an association between a **neutral stimulus** and an **unconditioned stimulus**. The unconditioned stimulus naturally elicits a reflexive, **unconditioned response**. Repeated pairings of the two stimuli leads to an association between the neutral stimulus and the unconditioned stimulus, so that the initially neutral stimulus is able to elicit the unconditioned response, without any presentation of the unconditioned stimulus. The unconditioned response has now become the **conditioned response**.

Example: *In Fig. 3.11, a popular setup for classical conditioning of dogs is shown. Although often attributed to Ivan Pavlov, this apparatus is in fact a more sophisticated device developed by the German physician and physiologist Georg Friedrich Nicolai, based on simpler versions that were used by Pavlov. A hungry dog is restrained by a harness. A salivary fistula is implanted to collect saliva. The amount of saliva produced and the salivary flow are used as measures of the behavioral response of the dog. When the dog is presented with food (unconditioned stimulus), it salivates as a reflexive, unconditioned response. Then, upon repeated occasions, a normally neutral sensory stimulus, such as the ticking of a metronome, is presented just before stimulation with the food. This stimulation regime leads to a pairing of the neutral stimulus and the unconditioned stimulus. After a certain number of such paired presentations, the ticking of the metronome alone is sufficient to elicit salivation, even in the absence of food. Classical conditioning has taken place, and the neutral stimulus has become the conditioned stimulus. Upon presentation of the latter, salivation as a conditioned response is shown by the dog.*

For the acquisition of a conditioned response, it is important that the conditioning stimulus is presented prior to the unconditioned stimulus, and that the temporal pairing occurs within a short period of time. Typically, the length of this **inter-stimulus interval** must not exceed a few seconds.

In many of his experiments, Pavlov used a metronome, a device used by musicians that produces regular aural pulses at variable tempi. He preferred the ticking of the metronome over many other sensory stimuli because it allowed him to precisely control both the quality and the duration of the stimulus, and thus to study the effect of these stimulus properties on conditioning. Contrary to the frequent claims in the literature, Pavlov probably never employed the sound of a bell for the conditioning experiments—simply because of the difficulty of controlling its stimulus properties.

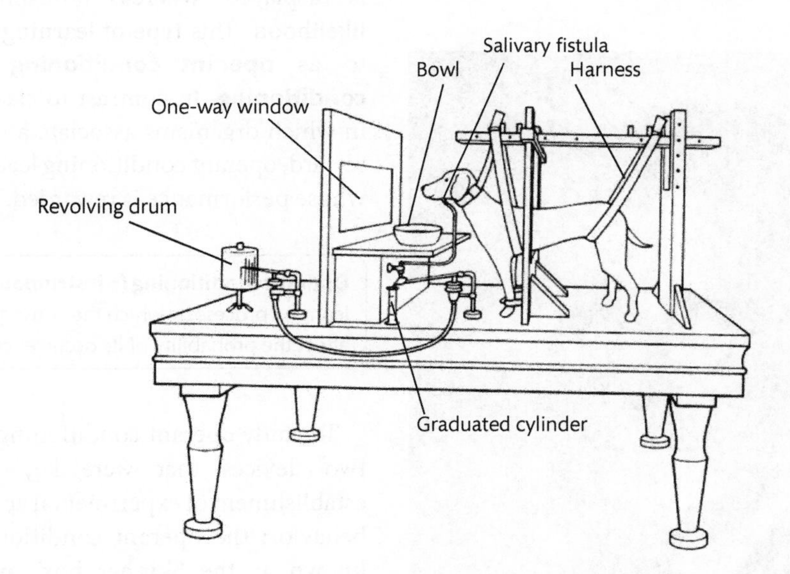

Figure 3.11 Apparatus for classical conditioning of dogs. This device, a modification of simpler setups used by Ivan Pavlov, was designed by the German physician and physiologist Georg Friedrich Nicolai. A hungry dog is restrained by a harness. During the training phase, presentation of food (unconditioned stimulus) in the bowl is paired with presentation of an initially neutral stimulus. After some time, the dog starts to salivate upon presentation of the latter stimulus, even if no food is given. At this point, association between the unconditioned stimulus and food has been formed, and the formerly neutral stimulus has become the conditioned stimulus. For quantification of the dog's response, its saliva is collected through a tube (fistula) implanted into the salivary duct. The amount of saliva is measured using a graduated cylinder, and the salivary flow is continuously recorded by a revolving drum. The dog can be observed by the experimenter through a one-way window. (After: Nicolai, G. F. (1907) and McFarland, D. (1993).)

Pavlov's objective approach to studying observable behavior was received with great enthusiasm by many of his contemporaries. In the United States, it led to the establishment of **behaviorism**, a term first used by John B. Watson (1878–1958; Fig. 3.12) in his paper *Psychology as the Behaviorist Views It*, published in 1913. In this manifesto, Watson outlined the methodology of this new branch of psychology: '*Psychology as the behaviorist views it is a purely objective experimental branch of natural science. . . . Introspection forms no essential part of its methods, nor is the scientific value of its data dependent upon the readiness with which they lend themselves to interpretation in terms of consciousness. The behaviorist . . . recognizes no dividing line between man and brute.*'

> **Behaviorism** The psychological discipline that studies observable behavior in a select number of model organisms by using objective approaches. Its main emphasis is on the exploration of how environmental stimuli produce and modulate behavior.

Figure 3.12 John B. Watson in 1932. (Courtesy: Nickolas Muray/Getty Images.)

Watson had a major influence on many psychologists, particularly in the first half of the twentieth century. Among them was Burrhus Frederic Skinner (1904–1990), who became the most important figure of psychology of the twentieth century (Box 3.4). He shared with Watson the rejection of inner mental mechanisms as explanations for behavior. However, he did not deny the existence of 'higher mental processes', such as feelings, cognitive thought, and consciousness. To him, they were states of the body that were amendable to analysis in their behavioral expression.

One of the major achievements of Skinner was the definition of an organizing framework of behavior. He distinguished between respondent behavior and operant behavior. **Respondent behavior** is elicited by a specific stimulus in a reflex-like fashion. **Operant behavior** is, when it first occurs, emitted without any identifiable stimulus present. Operant behavior is **controlled by its consequences**—reinforcement or punishment affects the probability that this behavior will be exhibited again in the future. Reinforcement results in an increased likelihood that the behavior is displayed, whereas punishment reduced this likelihood. This type of learning process is referred to as **operant conditioning** or **instrumental conditioning**. In contrast to classical conditioning, in which organisms associate a novel stimulus with reward, operant conditioning leads to certain actions whose performance is rewarded.

> **Operant conditioning (= instrumental conditioning)** A learning process by which the consequences of a behavior affect the probability of its occurrence in the future.

To study operant conditioning, Skinner designed two devices that were key to the successful establishment of experimental approaches to analyze behavior: the operant conditioning chamber, also known as the Skinner box, and the cumulative recorder. The **Skinner box** enables investigators to manipulate the behavior of an animal under strictly controlled experimental conditions, and to generate a quantifiable behavioral outcome. An example of such a box is shown in Fig. 3.13(a). It consists of a chamber in which an animal (here, a rat) can manipulate an object (here, depressing a lever). Upon depression, a door may open to release a food pellet. After the rat

Box 3.4 B.F. Skinner

B.F. Skinner in 1964. (Courtesy: Nina Leen/Timepix.)

In surveys about the most eminent psychologists in the twentieth century, B.F. Skinner has been consistently ranked at, or near, the top of the list. Among the achievements that define his legacy, arguably his rigorous, objective approach to the experimental analysis of behavior might be the most important one.

Burrhus Frederic Skinner—better known as B.F. Skinner—was born in 1904 in Susquehanna, a small railroad town in Pennsylvania. He attended Hamilton College in Clinton, New York, from which he graduated with a B.A. in English literature. The following year, he spent time in Greenwich Village of New York City and in Paris, trying to establish himself as a writer. However, influenced by the works of Bertrand Russell, Ivan Petrovich Pavlov, and John B. Watson, he entered Harvard University for graduate study in psychology. He earned his Ph.D. in 1931, with his thesis entitled *The Concept of the Reflex in the Description of Behavior*. As part of his graduate research, he invented the experimental chamber that became famous as the **Skinner box**; and the **cumulative recorder**, a device for recording an animal's discrete actions, like pressing of a lever, over time.

During the following five years, Skinner continued to do research at Harvard University, partially supported by a prestigious Junior Fellowship of Harvard's Society of Fellows. His mentor was William J. Crozier, then head of the Department of General Physiology. A major focus of Crozier's research was on **tropisms**, which he studied under strictly controlled laboratory conditions so that reliable quantitative data could be collected, enabling him and his co-workers to describe the stimulus-response relationship in mathematical terms. The exposure to this research remained of lifelong influence on Skinner's own approach to behavioral analysis. Based on his experimental work over these years, mainly in the area of conditioned reflexes, he wrote his first book, *The Behavior of Organisms*, published in 1938. It became a classic and laid the foundation for a new school of psychology concerned with the experimental analysis of behavior.

Following faculty appointments at the University of Minnesota and at Indiana University, Skinner returned to Harvard in 1948, where he remained until the end of his life. There, he set up a major research laboratory, using largely pigeons as a model system to study operant behavior. At the same time, he expanded his work, foremost by applying the behavioral principles established in rats and pigeons to other areas. Skinner's work on **operant conditioning** has shown that the consequences of behavior affect the probability of its recurrence in the future. In analogy to the mechanisms that lead to the evolution of new morphological traits, Skinner applied the principle of **selection by consequences**, in a unifying theory, to two scenarios: the modification of the behavior of the individual, and the development of cultural practices within populations of people.

In the 1950s and 1960s, B.F. Skinner played a significant role in the development of **behavioral pharmacology**, not so much through his own research but instead through the activities of his students and co-workers; they applied to pharmacology the reliable behavioral methods developed in his laboratory. In addition, Skinner himself undertook great efforts to apply the behavioral principles derived from operant conditioning experiments to **programmed instructions**, including the development of so-called **teaching machines**. By using such machines, students could study at their own pace. By systematically reinforcing correct responses, their learning outcome was improved when compared to the results achieved through traditional instruction. Based on Skinner's **behavioral engineering** methods, training programs have been designed for applications in industry and the armed forces; and therapies

. . . *continued*

have been developed for special education, for example to help children with autism.

Skinner was not only the most eminent psychologist of his time, but also a highly acclaimed–and at the same time vehemently attacked–figure of national and international prominence. In 1948, he published *Walden Two*, a novel describing a utopian community that applies scientific principles of behavioral analysis to improving the condition of human life. Although during the first decade only a few hundred copies per year were sold, sales rocketed in the 1960s. At the turn of the millennium, Skinner's novel was translated into all major languages, and by then over two million copies had been sold.

A similarly successful–and controversial–book was *Beyond Freedom and Dignity*, which appeared in 1971 and remained on the *New York Times* Best Seller list for eighteen weeks. Since, according to Skinner's selection-by-consequences theory, reinforcers in an individual's environment determine the behavior of this individual, the actions of a person are predictable as long as science fully understands the causal laws of this determinism. Consequently, and as clearly articulated in this book, Skinner dismissed free will as an illusion. On the other hand, he proclaimed that accepting this proposition would make people more open to the application of the science of behavior to improve life, and thus for the creation of a more humane society.

Skinner's philosophical position has become known as **radical behaviorism**. Its central dogma is based on the insistence that behavioral analysis be restricted to the behavioral level, without appealing to 'mental' events for explanation. This approach reflects, at least in part, his rebellion against a philosophy of mysticism and a method of self-centered introspection, both of which dominated large parts of psychology at the beginning of the twentieth century. In Skinner's view, there was also no need to study physiological processes to fully understand behavior. On the other hand, although often misinterpreted, he did not deny the existence of 'higher mental processes,' such as feelings or thoughts, but he rejected resorting to inner states in the absence of external–behavioral–representation of such states. Likewise, he did not dismiss, as widely assumed, the importance of genetic factors in determining behavior. When Skinner received the first award given by the American Psychological Association for Outstanding Lifetime Contribution, he delivered to the audience at the Annual Convention in 1990 what turned out to be his final address. In it, he reiterated his position on the avoidance of the study of inner mental states to understand behavior. His speech culminated in labeling the then increasingly popular cognitive science as the 'creationism of psychology'.

Eight days after delivering his address, B.F. Skinner died of leukemia, at the age of 86.

has pressed the lever the first time by accident, it will learn quickly to perform this behavior in order to get rewarded with food. Thus, food serves in this type of experiment as a **reinforcer**.

Reinforcer A stimulus that increases the probability of a behavioral response in operant conditioning. Reinforcers can be either positive (such as reward) or negative (such as removal of a negative consequence).

To measure the behavioral outcome of operant conditioning over time, Skinner used the rate of responses. In the above example, the response is quantified by the rate at which the rat depresses the lever. To obtain an automatic recording of the rate of occurrence, Skinner designed a second device, the **cumulative recorder**. From its inception in the 1930s until the time when laboratory computers became available, this device was the most widely used instrument to measure behavior, and as such was key to the success in the experimental analysis of behavior. The cumulative recorder produces a graph known as a **cumulative record** (Fig. 3.13(b)). A slowly rotating cylinder pulls a roll of recording paper at a constant speed. A pen rides horizontally on the paper and is moved upward one small step upon each response produced by the animal. Thus, the horizontal (x) axis reflects the time of the experimental session, whereas the vertical position of the pen at a given time indicates the total number of responses generated during this session. The slope of the resulting curve indicates the response rate—a slope of zero (i.e., a line parallel to the x axis) indicates a lack of responses; a steep slope a high rate of responses; and a shallow slope a low rate of responses. Overall, the graph shows the cumulative number of responses as a function of time.

The operational definition of behavior, as used throughout the work of Skinner, was a logical consequence of his view of the valid scientific approach to the study of behavior. It is only the

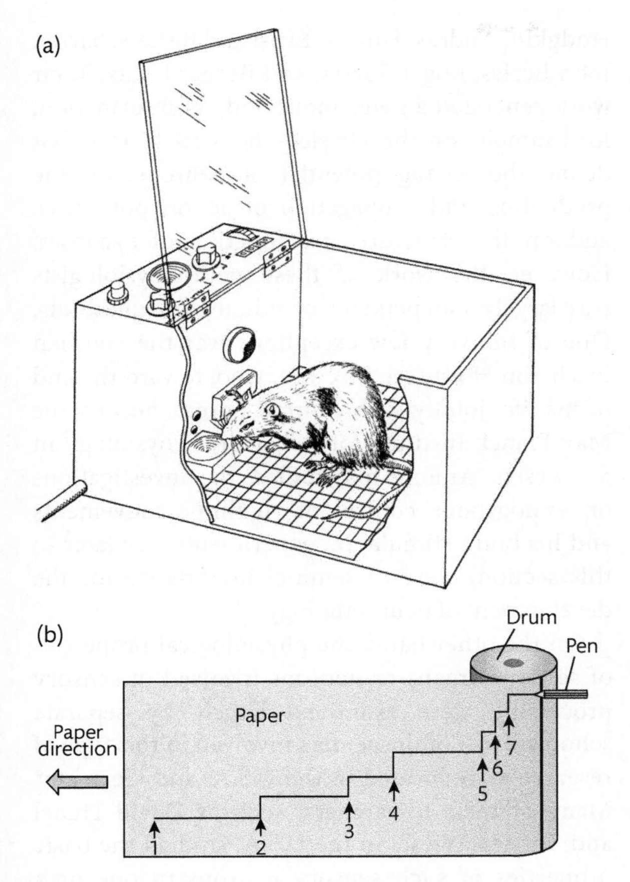

Figure 3.13 (a) Rat in a Skinner box. Upon pressing a lever, a pellet of food is delivered into the cup. (b) Cumulative record. A continuous roll of paper is fed across a metal plater at a constant speed, and a pen is riding on the paper. The pen draws a horizontal line if no response occurs, but steps vertically with each response (which, in (a), consists of pressing the lever). In the example shown, the experiment started when the pen was at 1, and the first response was made at time point 2. After some time had elapsed, two responses were emitted at time points 3 and 4. After another pause, a succession of three rapidly emitted responses occured at time points 5, 6, and 7. Courtesy: (a) After: Manning, A. (1972); (b) G.K.H. Zupanc.)

objectively measurable changes in behavior that can be analyzed, not subjective sensations, such as the feeling associated with the receipt of reward. One of the primary goals of behaviorism is, therefore, to identify reinforcement contingencies that result in changes in behavior. In contrast to Watson, Skinner did not aim at elucidating the neuronal or hormonal substrates underlying such behavioral processes. In his opinion, experimental analysis of all behavioral operations possible would be sufficient to finally lead to a complete understanding of behavior.

Although behaviorism and ethology share the objective approach to the study of behavior, it was behaviorism's emphasis on the importance of learning processes that caused considerable controversy, especially in the 1960s and 1970s, when the focus of ethology was primarily on innate behavior. At one extreme, some behaviorists claimed that any form of behavior is learned, thus denying the existence of genetic determinants. These behaviorists believed any animal or human to be born as a *tabula rasa* (a Latin term for 'blank writing tablet')—an idea that traces back to the empiricism of the British philosopher John Locke (1632–1704). According to this notion, it is experience (i.e. learning processes) that determines what is 'on the tablet.' This rather dogmatic view has led to an intense, sometimes even polemic, dispute over the relative importance of genes ('nature') and environment ('nurture'), known as the **nature-versus-nurture controversy**.

Although the impact of this dispute reached far beyond biology, research in recent years has increasingly indicated that the nature–nurture dichotomy reflects a rather inadequate understanding of the actual contributions of genotype and environment, and the interrelations of the two on the development of phenotypic traits, including behavior. Such studies have also shown that the environment affects subsequent development throughout life, even though this process may be particularly pronounced during critical periods of embryonic, neonatal, or juvenile ontogenesis. Importantly, 'environment' is not only the source of external cues—physical, biotic, social—but also of internal—chemical and electrical—signals. These environmental cues influence the expression of genes via **epigenetic** mechanisms. This process can provide an explanation for the long-standing observation that experience (in the broadest sense) may produce multiple morphological and behavioral phenotypes from a single genotype.

Epigenetics The study of the heritable alterations in a chromosome, without changes in the DNA sequence. Epigenetic alterations are stably transmitted through mitotic cell divisions within an organism, or through meiosis across generations. Several epigenetic processes have been identified, including DNA methylation and chromatin modification.

Whereas the failure of (radical) behaviorism to recognize the role of inheritance on behavior clearly impeded progress, its stringent empirical approach had, especially in North America, a positive impact on the further development of ethology. During its classic phase, ethological research was often restricted to descriptive investigations, and sometimes even major conclusions were drawn based upon judgmental and not very quantitative observations. However, later studies made extensive use of experimental approaches and statistical methods. Also, the initial emphasis on innate behavior gradually lost its importance, and many ethologists included learned behavior in their studies. Nevertheless, what continues to distinguish ethology from psychologically oriented disciplines is its attempt to study biologically relevant forms of animal behavior under conditions that are as natural as possible, and to include comparative approaches into its investigations.

➤ Mechanistic schools share with ethology the objective approach; however, in contrast to ethology, their emphasis is on laboratory-based experiments.

Relating neuronal activity to behavior: The establishment of neuroethology

The beginnings

As early as 1951, Niko Tinbergen wrote in *The Study of Instinct* that it is the job of the ethologist to carry the analysis of behavior down to the level studied by the physiologist. Despite his call, initial progress in this area of research, which is now known as **neuroethology**, was rather slow. One reason for this is that the problems encountered in relating the activity of individual neurons, or assemblies of neurons, to a biologically relevant behavior of the whole animal are tremendously complex.

Starting shortly after the beginning of the twentieth century, rapid progress was made in the elucidation of the physiological properties of neurons. This led to the establishment of strong neurophysiological schools, particularly in Great Britain and the U.S.A. Prominent leaders in this development were Charles Sherrington, Alan Hodgkin, Andrew Huxley, Ernst and Berta Scharrer, John Eccles, Roger Sperry, and Bernard Katz. Their work generated an enormous body of information, for example, on the physico-chemical factors that define the resting potential of neurons, on the production and propagation of action potentials, and on the structure and function of synapses. However, the work of these neurophysiologists was largely independent of ethological questions. One of the very few exceptions was the German Erich von Holst (see Box 3.5), who, toward the end of his life, jointly with Konrad Lorenz headed the Max Planck Institute for Behavioral Physiology in Seewiesen. Among his studies, the investigations on endogenous control of rhythmic movements and his brain stimulation experiments (see later in this section) were of seminal importance for the development of neuroethology.

On the other hand, the physiological properties of sensory organs or neurons involved in sensory processing were examined largely by separate schools. Most of the centers involved in this type of research were located in the U.S.A. and Germany. Many of their researchers, such as David Hubel and Torsten Wiesel in the U.S.A, studied the basic properties of such sensory neurons in one or a very few particularly favorable model systems, such as the responses of neurons to arbitrary, simple stimuli, and did not pursue the relationship of these properties to the natural stimuli. Yet, at the same time, schools with a strong zoological foundation did emerge. Their research was driven by the desire to link the physiological properties of sensory systems, in a variety of taxonomic groups, to behavioral function. Important figures in this movement were Karl von Frisch (see Box 3.6) in Germany, who established conditioning paradigms as powerful tools to explore the sensory capabilities of the whole animal; and Theodore H. Bullock (see Box 3.7) in the U.S.A, who, more than anyone else, championed the comparative approach and whose research on a large number of different taxonomic groups led, among others, to the discovery of pit organs in pit vipers and of electroreceptors in weakly electric fish and other electrosensory animals. Both von Frisch and Bullock established large schools from which many distinguished neuroethologists originated.

Box 3.5 Erich von Holst

Konrad Lorenz (left) and Erich von Holst (right) at the Max Planck Institute for Behavioral Physiology in Seewiesen. (Courtesy: Gerhard Gronefeld.)

Among the great scientists who have made outstanding contributions to the development of neuroethology, Erich von Holst is one of the most influential. His life was characterized by fundamental and highly original discoveries, enormous experimental skills, and absolute dedication to whatever he did. It was also he who pioneered physiological experiments conducted on whole animals, rather than on isolated organs.

Born in Riga in 1908, he studied in Kiel, Vienna, and Berlin. In his Ph.D. thesis, published in 1932, he analyzed physiological mechanisms controlling the movements of earthworms. According to the predominant view of his times, the peristaltic waves traveling in a rostro-caudal direction across the earthworm's body are caused by **reflex chains**: Mechanical deformation of one segment induces mechanical deformation of the following segment, and so on. If this theory were correct, the elimination of one or several segments, while leaving the ventral nerve cord intact, should lead to cessation of the contraction waves generated by the longitudinal muscles. However, von Holst's experiments demonstrated the opposite: The waves of contraction are transmitted through the ventral nerve cord even across gaps created by the experimenter in the body's segments. This suggests that the **pattern of movement is controlled by endogenous rhythms in the central nervous system**, rather than by reflex chains. He thus disproved the then very popular hypothesis that all behavior can be explained as a result of the action of reflex chains.

In 1946, von Holst was appointed to a Zoology Chair at the University of Heidelberg. In 1948, he became one of the directors of the Max Planck Institute for Marine Biology in Wilhelmshaven, and from 1957 on he headed the newly established Max Planck Institute for Behavioral Physiology in Seewiesen near Munich. The idea for the foundation of this center for behavioral physiological and ethological research dates back to 1936, when von Holst met Lorenz for the first time. This meeting was of historical importance. It was in particular the proposal of a **central pattern generator** (although von Holst did not use this rather modern term) that intrigued Lorenz, as he saw in this finding the physiological correlate of spontaneously occurring behavior—an idea vehemently rejected at that time by the school of behaviorism.

Among other research projects that made von Holst famous were his studies on **relative coordination**, that is, the influence of different groups of neurons on the temporal pattern of rhythmic activity; the **biophysics of avian flight**; and the **physiology of the vertebrate labyrinth**. Later in his life, he added more and more theoretical studies to these experimental investigations. One of the results of this work was the formulation of the **reafference principle**, a model to explain how animals and humans can distinguish moving retinal images caused by movements of objects from those generated by movements of the body. Together with his associate Horst Mittelstaedt, he proposed that the brain produces a 'copy' of the efferent command information controlling the body movement, and this efference copy is compared with the incoming (afferent) sensory information mediated by retinal stimulation. In the last years of his life, von Holst turned, once again, to a new research theme. Using **electrical brain stimulation** techniques, he studied mechanisms of neural control of behavior patterns in the chicken.

It is also characteristic of von Holst that, in addition to his scientific work, he spent many years of his life on a completely different subject. He was not only an excellent viola player, but he also designed and built himself revolutionary new violas. They were asymmetric instead of symmetric, a feature that combines excellent sound characteristics with the comfort of the rather small size of the instrument's body.

Suffering from heart disease since childhood, von Holst tried to achieve as much as he could in the time available. Yet many projects remained unfinished when he died in 1962, at the age of only 54.

Box 3.6 Karl von Frisch

Karl von Frisch. (Courtesy: Bettmann/CORBIS.)

Only a very few scientists have achieved what Karl von Frisch did—to become accepted as a leading figure by the scientific community, while reaching a degree of popularity usually reserved for artists and writers. The work upon which both von Frisch's scientific success and his popularity is based centers around the dance language of honey bees, although he also made pioneering discoveries in many other areas of sensory and behavioral physiology.

Karl von Frisch was born in Vienna, Austria, in 1886. His family, which included renowned professors and medical doctors, provided an intellectually stimulating environment for him and encouraged his leaning toward research and scholarship. This interest was further reinforced by the extended periods he spent in the family's summer home in Brunnwinkl on Lake Wolfgang, where he collected specimens for his 'little zoo' and made his first scientific observations. Later in his life, he frequently retreated there to perform his experiments in the peaceful and harmonious surrounding of the lake. And it was also Brunnwinkl that provided him refuge during the two world wars.

After completion of secondary school and yielding to his father's request, he enrolled in 1905 at the medical school of the University of Vienna. However, disenchanted with medicine, he transferred after two years to the Zoological Institute of the University of Munich, Germany, to study under the famous Richard von Hertwig. In one of the frequent excursions to the Dolomites, von Frisch was assigned to study the behavior of solitary bees. This marked the beginning of a lifetime interest.

In 1910, von Frisch obtained his Ph.D. degree with a thesis on color adaptation and light perception in minnows. After various academic positions, he was appointed Professor of Zoology and Director of the Institute of Zoology at the University of Munich in 1925. He remained affiliated with this internationally renowned institution until his retirement in 1958, except for a short interruption after the Second World War.

Among his epoch-making contributions to biology were the demonstration of **color vision in bees** and **hearing in fish**, the discovery of ***Schreckstoff*** (alarm substance) **in minnows**, and a detailed analysis of the **bee language**, including the round and waggle dances. Many of his discoveries disproved up-to-then widely accepted dogmas—such as the belief, at the beginning of the twentieth century, that fish and all invertebrates are color-blind. It was also von Frisch who introduced **conditioning paradigms** to sensory physiology. These paradigms have proven to be extremely powerful tools that use the natural behavior of animals to examine their sensory capabilities.

Von Frisch was not only an ingenious scholar, but also an excellent science communicator. His books are among the best ever written in biology and include *The Dancing Bees: An Account of the Life and Senses of the Honey Bee* and *Man and the Living World*.

In 1973, at the beginning of his 88th year, von Frisch was, together with Konrad Lorenz and Niko Tinbergen, awarded the Nobel Prize for Physiology and Medicine. He died in Munich in 1982 at the age of 95.

However, due to technical limitations, the work of both the neurophysiologically oriented schools and the sensory physiology groups was restricted to the analysis of a few elements within the entire neural chain involved in the sensory processing and the generation of the motor output of a given behavior. Thus, it did not lead to an integrative understanding of the neural mechanisms underlying a specific behavior.

Another reason for the initially slow development of neuroethology was that several—in themselves

Box 3.7 Theodore Holmes Bullock

Theodore H. Bullock. (Courtesy: Scripps Institute of Oceanography.)

One of the fathers of neuroethology, whose efforts were instrumental in the establishment of this scientific discipline in North America, was Theodore ('Ted') Holmes Bullock. Bullock was born to American Presbyterian missionaries in 1915 in Nanking, China, where he grew up until the age of 13. For his undergraduate education, he attended Pasadena Junior College and the University of California at Berkeley, majoring in zoology. His Ph.D. thesis, also conducted at Berkeley, was on the anatomy and physiology of acorn worms. After four years at Yale University, with summers at the Marine Biology Laboratory at Woods Hole on Cape Cod, he assumed an Assistant Professorship at the University of Missouri School of Medicine at Columbia, followed by faculty positions at the University of California at Los Angeles and the University of California at San Diego.

Among the features that distinguished Ted Bullock were his exceptionally diverse research interests, covering not only numerous research themes, but also a huge number of taxonomic groups. This provided the basis for a comprehensive analysis of **brain evolution**. One of the themes to which Bullock returned, time and again, was the **importance of direct-current and low-frequency electric fields for intercellular communication** in the brain. This phenomenon is still largely unexplored in non-human species, partially because many of the corresponding concepts and techniques, such as electroencephalography, have primarily been developed for application to the human brain. However, Bullock's research suggested that the effects of such fields arising from individual cells, or synchronized populations of cells, may be significant in modulating those mediated by conventional forms of intercellular communication, which employ all-or-none electrical impulses or chemical transmitter substances and neuromodulators. Other pioneering scientific contributions of Bullock include the discovery of two new sensory modalities: the **facial pit of pit vipers as an infrared receptor**, which enables the snakes to detect temperature changes caused by prey animals whose temperature differs from that of the background surfaces; and **electroreceptors in weakly electric fish** and others, which respond even to extremely weak electric signals with high specificity (see Chapter 8).

It is also characteristic of Bullock that, throughout his life, he was a true cosmopolitan. He never ceased to encourage international collaborations and the exchange of ideas across nations. Furthermore and rather unusually, he stimulated others to enter new, promising research areas, without necessarily pursuing further work in this field himself. This is exemplified by the pioneering work of Ulla Grüsser-Cornehls and Otto-Joachim Grüsser on **complex recognition units** in the frog's optic tectum (see Chapter 7), which was initiated during a visit by the two German researchers to the laboratory of Bullock in California.

In addition to his academic work, Ted Bullock served as president of several societies, including the American Society of Zoologists, the Society for Neuroscience, and the International Society for Neuroethology. His book *Structure and Function in the Nervous Systems of Invertebrates*, co-authored by Adrian Horridge, is widely regarded as the most comprehensive and authoritative review of this topic ever written.

Ted Bullock died in San Diego, in 2005, at the age of 90.

quite promising—attempts to relate the action of neuronal assemblies to the behavior of the whole animal remained, for a considerable amount of time, rather isolated and were only reluctantly used by others. This was, for example, the case with the **focal brain stimulation technique**. This approach was developed and championed by Walter Rudolf Hess (1881–1973; Fig. 3.14) of the University of Zurich, Switzerland, who was awarded the Nobel Prize in Physiology and Medicine in 1949. Using the focal brain stimulation technique, he examined in great detail how regions within the diencephalon

Figure 3.14 Walter Rudolf Hess. (Courtesy: Nobel Foundation.)

control vegetative functions and various behaviors of cats.

The power of this approach for neuroethological research was demonstrated in the late 1950s and the beginning of the 1960s, by Erich von Holst, through an extensive series of experiments. Von Holst showed that stimulation of certain brain areas in chickens can evoke specific behavioral patterns. The exact type of behavior elicited, and the degree of expression, depend upon both the intensity of the current applied and the presence or absence of an external stimulus. Stimulation of the hypothalamus in an alert chicken, for example, triggers an attack on a stuffed weasel presented to it. The intensity of this attack behavior depends upon the current applied—without electrical stimulation, the stuffed weasel evokes no special response.

> ➤ The 'fathers' of neuroethology: Karl von Frisch, Erich von Holst, and Theodore H. Bullock.

The breakthroughs

Significant advances towards the establishment of neuroethology as a scientific discipline were made only in the 1970s and 1980s. This success was due to both the advent of new neurobiological methods and the focus on simple and robust forms of behavior. The new methods allowed researchers to routinely trace neural pathways by employing a variety of *in vivo* and *in vitro* techniques; to characterize individual neurons through immunolabeling and *in situ* hybridization; to perform intracellular recordings and combine them with intracellular labeling techniques; and to apply selective agonists and antagonists of transmitters to characterize the physiological and pharmacological properties of neurons. The list of behaviors investigated by neuroethologists was rather short and included, in particular, rhythmic motor patterns. Even movements of internal organs, such as those of the intestines, were intensively studied. Although the latter types of behavior are hardly considered worth studying by the majority of ethologists, their analysis led to the formulation of many useful concepts, such as those of identified neurons, central pattern generators, and modulators.

The major breakthroughs achieved making use of these new approaches included the work of Jörg-Peter Ewert on neural correlates of prey recognition in toads (see Chapter 7), of Walter Heiligenberg on the jamming avoidance response in weakly electric fish (see Chapter 8), of Mark Konishi on sound localization in barn owls (see Chapter 7), of Eve Marder on the modulation of the motor pattern of the stomatogastric ganglion in decapod crustaceans (see Chapter 9), and of Franz Huber on auditory communication in crickets (see Chapter 12).

> ➤ A breakthrough in the establishment of neuroethology was possible by focusing on simple and robust forms of behavior, and by applying modern neurobiological methods to explore the entire chain of sensory and neural mechanisms underlying these behaviors.

The formal establishment of neuroethology as a new discipline was completed by the appearance of its first textbook, entitled *Neuroethology*, by

Jörg-Peter Ewert (first published in German, in 1976, and in English, in 1980) and by the foundation of the **International Society for Neuroethology**, which held its first congress in Tokyo in 1987.

The future

In 1975, Edward O. Wilson set forth, in his epoch-making book *Sociobiology: The New Synthesis*, his vision of the future of behavioral biology. His prediction was that '*both [ethology and comparative psychology] are destined to be cannibalized by neurophysiology and sensory physiology from one end and sociobiology and behavioral ecology from the other.*'

What is the status of this development nearly half a century later? And where will neuroethology head in the next 50 years?

As predicted by Wilson, neuroethology has, indeed, increasingly incorporated **neurophysiological** and **sensory physiological** techniques and concepts, to explore the neural basis of behavior, frequently down to the level of single neurons or even single types of channels. As the data generated through such experiments become more and more complex, combining the information to produce testable models will be essential. In this development, **computational neuroscience**, a flourishing theoretical discipline, will play an important role.

In addition, **neuroendocrinology** has also gained significantly, in terms of its impact upon neuroethology. Its influence is likely to increase even further due to the enormous advances made in the molecular characterization of hormones, peptidergic releasing factors, and their receptors. This has triggered the generation of novel pharmacological receptor agonists and antagonists that, as yet largely unexplored, provide exciting tools to investigate the involvement of specific hormonal and peptidergic factors, as well as their receptors and receptor subtypes, in behavioral processes. The numerous investigations in this area have also made the seemingly clear-cut boundary between endocrine factors and transmitters become more graded and continuous (see Chapter 2).

The largest gain over the next decades is to be expected from **molecular biology**. In some favorable cases, it has now become possible to pinpoint the execution of certain behaviors of an animal down to the expression of single genes, or even single point mutations of a gene.

Example: *Naturally occurring strains of the nematode* Caenorhabditis elegans *are distinguished by their difference in feeding behavior. While worms of one group are solitary feeders, move slowly on a bacterial lawn, and disperse across it, individuals of the other group feed socially, move rapidly on bacteria, and aggregate together (Fig. 3.15). As has been shown by Mario de Bono and Cori Bargmann, then at the University of California, San Francisco, this behavioral difference is due to two naturally occurring alleles of a single gene that differ by a single nucleotide. This gene, called* npr-1, *encodes a G protein-coupled receptor resembling the receptors of the neuropeptide Y family; it is referred to as NPR-1. One of the resulting two isoforms of this receptor, NPR-1 215F, contains the amino acid phenylalanine (designated 'F' in the one-letter code of amino acids) at position 215; this isoform is found exclusively in social strains. In the second isoform, called NPR-1 215V, the phenylalanine at position 215 is replaced by a valine (designated 'V'); this isoform is found exclusively in solitary strains. Position 215 is in the third intracellular loop of the putative receptor. This region is important for G protein coupling in many seven-transmembrane receptors. Mutations in this loop are likely to result in changes in the presumptive neuropeptide pathway, which appear to generate the natural variation in behavior. The hypothesis that the difference in behavior is caused by this mutation in the NPR-1 protein has received strong support by transgenic experiments. They demonstrated that an NPR-1 215V transgene can induce solitary feeding behavior in a wild-type social strain.*

Does the increasing importance of molecular approaches and the focus of neuroethology on a few selected model systems and (seemingly) simple behaviors make behavioral research redundant? Certainly not! Neuroethological studies continuously generate an abundance of specific behavioral questions that wait to be answered through ethological approaches. On the other hand, the results of such behavioral investigations, based on observations of and experiments on natural behavior often also stimulate further research at the neurobiological level. For these studies, molecular

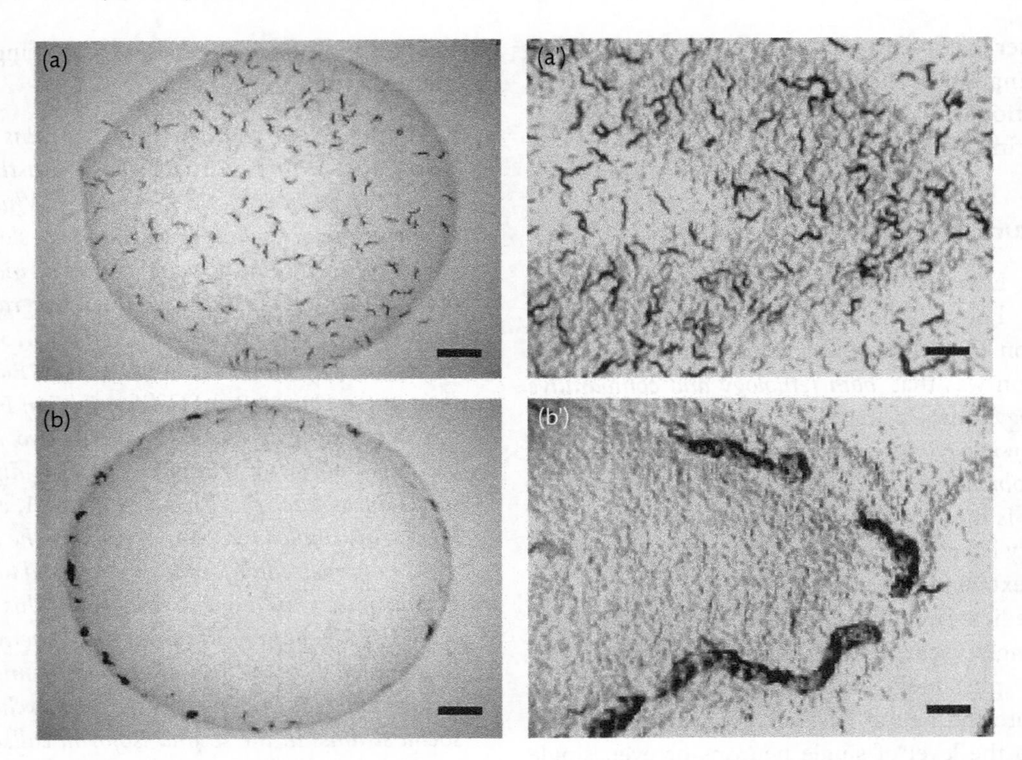

Figure 3.15 Solitary and social behavior in the nematode *C. elegans*. (a) Animals of the N2 wild strain, which exhibit solitary behavior by dispersing evenly across a bacterial lawn and browse alone. (a′) Solitary nematodes of this strain shown at higher magnification. (b) Animals of the *npr-1* genotype, which exhibit social behavior by aggregating together into clumps and accumulating at the edge of the lawn where the bacteria are more abundant. (b′) Social nematodes of this genotype shown at higher magnification. Calibration bars: 2.5mm in (a) and (b) and 1mm in (a′) and (b′). (Reprinted from: Mario de Bono and Cornelia I. Bargmann (1998), Natural variation in a neuropeptide Y receptor homolog modifies social behavior and food response in *C. elegans*, *Cell* **94**:679–689. © 1998, with permission from Elsevier.)

biology and neurobiology provide powerful tools through which behavior can be studied.

> ➤ A distinctive feature of neuroethology is the attempt to gain an integrative understanding of how the brain controls behavior in the whole animal.

Like many areas of biology, neuroethology is on the verge of a new era. The current rapid development in computational neuroscience, neuroendocrinology, and molecular genetics will provide exciting opportunities for neuroethology. However, the success of neuroethology will not be determined simply by the application of the corresponding techniques. Rather, its strength lies in its ability to integrate across the different levels of organization—molecular, subcellular, cellular, tissue, and organismal—to advance our understanding of neural mechanisms of natural behavior. It is this integrative, question-driven approach that distinguishes neuroethology from other disciplines.

Summary

- The study of animal behavior has its roots in natural history, which can be traced back to Aristotle. However, until the end of the nineteenth century, behavioral research was dominated by anecdotal approaches, as well as vitalistic, teleological, and anthropomorphic interpretation. Investigations were concerned mainly with finding evidence for the superiority of humankind, thus drawing a sharp dichotomy between the 'rational soul' of humans and the 'sensitive soul' of beasts.

- A more objective approach toward the study of behavior emerged only at the time of the announcement of Charles Darwin's evolutionary theory. His publications proposed a continuity of both morphological and behavioral characteristics within the living world, including man. This triggered an enormous interest in behavioral observations from a comparative perspective, thus marking the beginning of ethological research.
- In ethology, comparative observations were typically conducted under natural or semi-natural conditions. This was supplemented by the application of experimental approaches, as well as by the use of operational definitions and objective interpretations. Final establishment of ethology as an independent discipline within biology was achieved by Konrad Lorenz and Niko Tinbergen in the period between the 1930s and the 1950s. Their merit lies in having placed the observational and experimental data accumulated at that time within a conceptual framework.
- Parallel to the emergence of ethology, several mechanistically oriented schools developed. The theory of tropism of Jacques Loeb attempted to explain behavior in terms of tropisms—involuntary orienting movements. The reflex theory of Ivan Petrovich Pavlov had a pronounced effect particularly on the development of behaviorism. One direction within behaviorism led by John B. Watson proposed to explain behavior in physiological terms based on reflexes as the elementary behavioral unit. A second direction within behaviorism established by Burrhus Frederick Skinner stressed the importance of experimental manipulations and operational definitions to understand behavior. The main emphasis of this research direction was on learned behavior. The denial of innate components of behavior by some of the proponents of behaviorism resulted, for a considerable length of time, in a sharp dispute with ethology, known as the 'nature-versus-nurture controversy.'
- Neuroethology emerged only at the end of the twentieth century when scientists started to apply modern neurobiological concepts and techniques to the elucidation of neural mechanisms underlying simple forms of behavior in animals. Important contributions to this development originated from both neurophysiology and sensory physiology. Eminent figures in the early stages of this development were Karl von Frisch, Erich von Holst, and Theodore H. Bullock.

The bigger picture

It can be argued that no other natural phenomenon has fascinated humans as much as behavior, foremost their own. Given this genuine interest, it has taken surprisingly long to develop objective approaches to study human and animal behavior—much longer than it took to deduce the laws of physics or to analyze morphological structures of humans and animals. To understand this delay in the application of objective scientific principles to behavioral research, it is vital to review the history of the study of behavior. The important lesson to be learned from this exercise is that, throughout history, scientific goals and research methodologies have been intimately linked to societal and cultural trends; current and future developments are no exception to this rule. Knowing more about the history of science will, therefore, help us to better understand ourselves and societies. Perhaps equally important, the history of science reminds us that we benefit from the achievements of those who came before—in the behavioral sciences, we stand on the shoulders of such giants like Aristotle, Charles Darwin, Ivan Pavlov, B.F. Skinner, Konrad Lorenz, and Karl von Frisch. Knowing history is, thus, also a means to give proper credit to them.

Recommended reading

Klopfer, P. H. (1974). *An Introduction to Animal Behavior: Ethology's First Century*, 2nd edn. Prentice-Hall, Englewood Cliffs/New Jersey.

An introduction to ethology written from a historical perspective, containing a wealth of information on the development of scientific disciplines studying animal behavior.

Thorpe, W. H. (1979). *The Origins and Rise of Ethology: The Science of the Natural Behaviour of Animals*. Heinemann Educational Books, London.

A vivid historical and personal account of the study of animal behavior, written by a British ethologist who contributed to this history himself.

Todes, D. P. (2014). *Ivan Pavlov: A Russian Life in Science*. Oxford University Press, New York.

This book is a marvelously written scholarly biography of a great scientist. In addition, its author has succeeded in presenting the work and life of Pavlov in a captivating way within the context of Russian history, spanning the times from late imperial Russia through the civil war to Stalin's dictatorship.

Short-answer questions

3.1 What are anthropocentric and anthropomorphic approaches, and why have they impeded progress in the study of animal behavior?

3.2 Define the inductive approach to research. Provide an example of this approach from the study of animal behavior.

3.3 What is mind–body dualism?

3.4 In what way has the evolutionary theory of Charles Darwin advanced animal behavior research?

3.5 Define Jakob von Uexküll's concept of Umwelt. To illustrate this concept, compare color vision in humans and honey bees.

3.6 Who are the two founders of ethology (first and last names)?

3.7 Complete the following sentence: By studying the physiology of .... in dogs, the Russian physiologist.. (first and last name) discovered, at the beginning of the twentieth century, ... conditioning, also known as .... conditioning. This discovery laid the foundation for the development of the American psychological school of..

3.8 Name (first and last names) the two most important scholars of behaviorism in the twentieth century.

3.9 What is operant conditioning?

3.10 What is a Skinner box? What is it used for?

3.11 How is an animal's response indicated on a cumulative record?

3.12 What is the nature-versus-nurture controversy?

3.13 List three characteristics that distinguished ethology and behaviorism in the earlier stages of their development.

3.14 'Innovation in science frequently comes from outside a given discipline.' To support this statement, describe how a new physiological technique, developed originally outside of neuroethology, advanced research in this area.

3.15 **Challenge question** In what sense did the theory of evolution legitimize John Watson's postulate that research on animals, as conducted by behaviorists, can be used to understand human behavior?

Essay questions

3.1 *'It is one of the ironies of the history of science that Aristotle's writings, which in many cases were based on first-hand observations, were used to impede observational science.'* Making reference to this quote from the University of California Museum of Paleontology's website, describe Aristotle's approach to exploring nature, and discuss the influence of the underlying principles on the historical development of the study of animal behavior.

3.2 Discuss the major historical reasons for the very slow progress made in applying an objective approach to the study of animal behavior. At what point was this difficulty finally overcome? Name at least three scientists who have made major contributions toward the establishment of the study of animal behavior as an empirical scientific discipline, and briefly describe their achievements.

3.3 Two major scientific disciplines that had an enormous impact on the study of animal and human behavior in the twentieth century are behaviorism and ethology. What do these disciplines have in common, in what respect do they differ, and what approaches do they use? Name at least two prominent figures in each of the two disciplines, and briefly describe their major contributions.

3.4 What is the 'nature-versus-nurture controversy'? Discuss the significance of this academic dispute in the context of the development of the various scientific disciplines that study animal and human behavior.

3.5 Sketch the historical development of neuroethology. In your essay, include a brief summary of the work of two prominent scientists who have greatly influenced the establishment of neuroethology as an independent discipline.

Advanced topic The role of classical learning for food consumption

Background information

The concept of classical conditioning, discovered by Ivan Pavlov and his school, has proved to be one of the most powerful theories in the history of the behavioral sciences. An association between arbitrary external cues and certain behaviors plays an important role in all animals, including humans. Studies have shown that such a process may be instrumental in promoting eating, even if the animal is in a sated state. Such acquired motivational properties can dominate regulation by energy depletion signals, and thus exert maladaptive effects that contribute to overeating and obesity.

Essay topic

In an extended essay, describe how conditioned cues can elicit feeding in sated animals. Discuss the behavioral paradigm used in these studies within the context of classical conditioning. Summarize the evidence that suggests that this conditioned response is mediated by a brain network formed by the amygdala, lateral hypothalamus, and medial prefrontal cortex. Discuss the possibility of learned motivational control of eating in humans.

Starter references

Petrovich, G. D. and Gallagher, M. (2007). Control of food consumption by learned cues: A forebrain-hypothalamic network. *Physiology & Behavior* **91**:397–403.

Weingarten, H. P. (1983). Conditioned cues elicit feeding in sated rats: A role for learning in meal initiation. *Science* **220**:431–433.

To find answers to the short-answer questions and the essay questions, as well as interactive multiple choice questions and an accompanying Journal Club for this chapter, visit **www.oup.com/uk/zupanc3e**.

Orienting movements

Introduction

Animals orient their body in a specific way relative to the environment. To maintain orientation after disturbance, or to change position in certain behavioral situations, the animal needs to continuously collect and process relevant stimuli from the environment and translate them into proper behavioral action.

In this chapter, we will examine how animals manage to perform this process of translation. We will first look at the diversity of orienting movement in animals and summarize the rules defining how to classify and name these movements. Then we will describe in more detail the cellular and neural mechanisms governing spatial orientation. We will first focus on a particularly remarkable organism—an animal that, although completely lacking a nervous system, is nevertheless able to produce orienting movements, with high levels of sophistication, in response to environmental stimuli. In the second part, we will have a closer look at how vertebrates detect and maintain an equilibrium position using the force of gravity as a frame of reference.

Key concepts

The orienting responses of animals are extremely diverse. A first major attempt to classify the different types of these behaviors was undertaken by the German zoologist Alfred Kühn (1885–1968) and is described in his book *Die Orientierung der Tiere im Raum* ('Spatial Orientation of Animals'), published in 1919. His system was extended by Gottfried S. Fraenkel (1901–1984) and Donald L. Gunn in their book *The Orientation of Animals*, published in 1940. A number of aspects of the classification system of these three authors are still widely applied, while others have since been modified. In particular, it has become increasingly evident over the last decades that orientation movements involve not just one sense but multisensory guidance.

In general, scientists distinguish between primary and sensory orientation. **Primary** (or **positional**) **orientation** involves control of body posture, whereas **secondary orientation** controls the response of an animal with respect to a particular stimulus from the environment. Secondary orientation is subdivided into the following two major types of orienting response:

1 **Taxis** defines an orienting movement in freely moving animals (in sessile animals, such a response is referred to as **tropism**). The body of the animal is oriented, and possibly also moved, in a particular direction with respect to the source of stimulation.

Taxis Orienting reaction or movement in freely moving organisms directed in relation to a stimulus.

2 **Kinesis** involves a behavioral response in which the animal's body is not oriented with respect to the source of stimulation. The animal's main response involves a change in speed of movement and/or in rate of turning. This undirected movement alters the position of the animal in relation to the source of the stimulus.

Example: *Woodlice (Porcellio scaber) are small isopods that cannot tolerate prolonged periods of dryness. They therefore tend to aggregate in moist places, where they are rather motionless. In contrast, in dry areas they exhibit increased locomotor activity, which enhances the probability of their finally reaching damper sites. Thus, and as first suggested by Donald Gunn and associates of the University of Birmingham, UK, in the 1930s, the tendency of woodlice to aggregate in moist places can solely be explained by proposing that locomotor activity is controlled by relative humidity. The assumption of a directed movement, that is, a taxis behavior of the animal toward areas of higher humidity, is not necessary. Rather, we can classify the response of woodlice as a kinesis behavior.*

When applying the term 'taxis' to describe the directed movement of an animal, a prefix is added to define the nature of the stimulus. For example, 'phototaxis' indicates a reaction to light (Greek *photos* = light). The possible addition of the modifiers 'positive' and 'negative' indicates the animal's mode of reaction to the stimulus. 'Positive' means an orienting movement toward the source of stimulation, while 'negative' points to an orienting movement in the opposite direction.

Example: *Positive phototaxis indicates an orienting movement of an animal towards light.*

Other taxis movements include the following:

- **Anemotaxis:** Orienting movement in relation to the wind direction.
- **Chemotaxis:** Orienting movement triggered by a chemical stimulus.
- **Galvanotaxis:** Orienting movement with an electric current acting as a directing stimulus.
- **Geotaxis:** Orienting movement in reference to gravity.
- **Phonotaxis:** Orienting movement elicited by an acoustic stimulus.
- **Rheotaxis:** Orienting movement directed by the current of water.
- **Thermotaxis:** Orienting movement induced by heat.
- **Thigmotaxis:** Orienting movement with a rigid surface acting as a directing force.

A certain degree of deviation from this terminology occurs in some other terms used to describe taxis movements, such as phobotaxis, which indicates an orienting response of an animal induced by an aversive stimulus.

Model system: Taxis behavior in *Paramecium*

The response of an organism to certain environmental stimuli does not necessarily require highly developed sensory organs, such as eyes or ears. Even the presence of a nervous system is not necessary. A thorough analysis of the behavior of such 'lower' organisms started at the end of the nineteenth century, mainly through the research of the German physiologist **Max Richard Constantin Verworn** (1863–1921) (Fig. 4.1) and the American zoologist and geneticist **Herbert Spencer Jennings** (see Box 4.1). The widely-studied animal-like unicellular protist *Paramecium caudatum*, for example, exhibits a wide range of orienting behaviors—despite the fact that this ciliate protozoan lacks sensory cells and neurons. Chemotaxis is one of these orienting behaviors. This reaction can be observed when paramecia (singular:

Figure 4.1 Max Richard Constantin Verworn. (Courtesy: Peter Matzen/Voit Collection.)

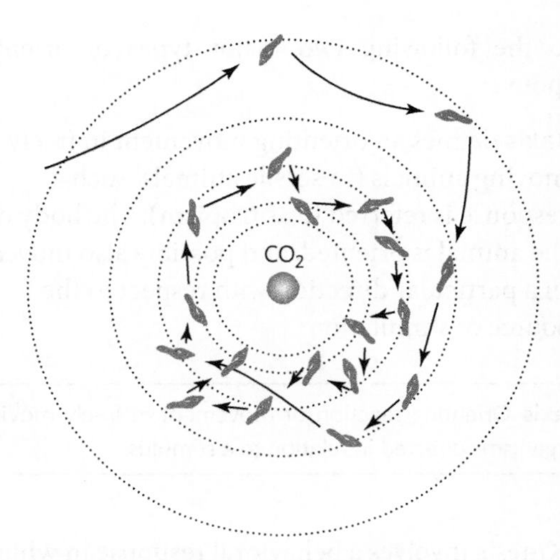

Figure 4.2 Chemotaxis response of a paramecium to a bubble of CO_2. The bubble is surrounded by a pH gradient gradually increasing with increasing distance from the center. Zones of equal acidity are indicated by dotted concentric rings. Due to avoidance of zones of both high and low acidity, the paramecium tends to remain in a zone of optimal pH. (After: Keeton, W. T. (1980).)

paramecium) encounter a bubble of carbon dioxide (CO_2) or a drop of diluted acetic acid (Fig. 4.2). Due to diffusion, there are circular zones surrounding the CO_2 bubble or the drop of acetic acid that differ in acidity. Near the center of the bubble or the drop, the acidity is rather high, that is, the **pH** is low. With increasing distance from the center, the pH gradually increases. If the paramecium gets too close to the CO_2 bubble or the drop of acetic acid, it reacts by stopping, backing away, and swimming forward again. These movements are accompanied by **gyrations** or revolutions around the longitudinal axis of the cell. Paramecia also show a negative taxis response to zones of low acidity. Thus, through the avoidance of zones of high and low acidity, the paramecium is finally 'trapped' in a zone of optimal pH.

pH Method of quantitatively expressing the concentration of H_3O^+ ions in a solution. The pH is defined as the negative logarithm to the base 10 of the hydrogen ion concentration. Pure water is neutral and has a pH of 7. A pH of less than 7 indicates an acidic solution and a pH of greater than 7 a basic (alkaline) solution.

Gyration Revolution around the longitudinal axis.

➤ Orienting behavior does not depend on the existence of a nervous system. The unicellular protist *Paramecium caudatum*, for example, exhibits several taxis reactions, including chemotaxis and galvanotaxis.

It is thought that this chemotactic response is an adaptation to the feeding behavior of paramecia. One of their major food sources is decaying bacteria, which generally lower the pH of the surrounding water. Therefore, a positive chemotactic reaction toward zones of mildly acidic pH does help the paramecia to localize these bacteria. On the other hand, by exhibiting a negative chemotactic response to zones of high acidity, they protect themselves from being damaged.

Cellular mechanisms of taxis behavior in paramecia

The cellular mechanisms mediating taxis behavior in ciliated protozoans have been relatively well

Box 4.1 Herbert Spencer Jennings

Herbert Spencer Jennings. (Courtesy: Bettman/CORBIS.)

While his name has become largely forgotten at the beginning of the twenty-first century, the central principle that guided his work—to perform an objective, experimental analysis of behavior—laid the foundation for modern behavioral research. It is the merit of the American **Herbert Spencer Jennings** to have established this principle at a time when many behavioral studies were still dominated by subjective and rather anecdotal approaches.

Jennings lived from 1868 to 1947. After graduating from high school, he taught for some time in country schools and at a small college. Jennings received his undergraduate training at the University of Michigan, where he also started to study rotifers and protozoans—the organisms that remained the subject of choice in most of his investigations. Following a year of graduate study at Michigan, he attended Harvard, where he received his Ph.D. in 1896 with a descriptive study of the development of rotifers. In 1897, he went to Europe to work with the then well-known German physiologist Max Verworn (Fig. 4.1) in Jena, and to spend some time in Italy at the Naples Biological Station. After a decade of temporary positions, in 1906, Jennings joined the faculty of Johns Hopkins University, where he remained until his retirement in 1938. Soon after arrival at Johns Hopkins, he stopped his behavioral work and devoted the rest of his life to genetics, particularly the study of inheritance in protozoans.

Jennings' behavioral studies and theoretical concepts, summarized—among others—in his classic book entitled *Behavior of the Lower Organisms* (published in 1906), had a profound influence not only in biology, but also on the development of behavioristic psychology. Among his students was John B. Watson (see Chapter 3), the founder of the school of behaviorism, who extended many of the principles developed by Jennings, through the study of lower organisms, to the behavioral analysis of man. However, in contrast to many prominent behaviorists, Jennings denied the general applicability of the hypothesis that external stimuli are the sole determinants of behavior. Instead, he emphasized the complexity and variability of behavior and the importance of internal factors. This zoological way of thinking, combined with the objective approach he applied toward the study of behavior, put Herbert Spencer Jennings many years ahead of most of his contemporaries.

studied for two orienting movements, phobotaxis and galvanotaxis. The motor responses observed in both behaviors are mediated by **cilia**, subcellular organelles formed by microtubules and of similar structure to those of other eukaryotes. Their light microscopic and electron microscopic appearances are shown in Fig. 4.3. We owe much of the information on how proper stimulation of a paramecium leads to a corresponding taxis behavior to **Roger Otto Eckert** (1934–1986) of the University of California, Los Angeles. Together with his collaborators, he discovered several of the key mechanisms that control movement in *Paramecium*.

Cilia (singular: cilium) Hair-like appendages of many kinds of cells with a bundle of microtubules at their core. The microtubules are arranged such that nine doublet microtubules are located in a ring around a pair of single microtubules. This arrangement gives them the distinctive appearance of a '9 + 2' array in electron micrographs of cross-sections.

In the following, we will first look at the subcellular mechanisms that mediate changes in movement of the cilia as part of the phobotactic

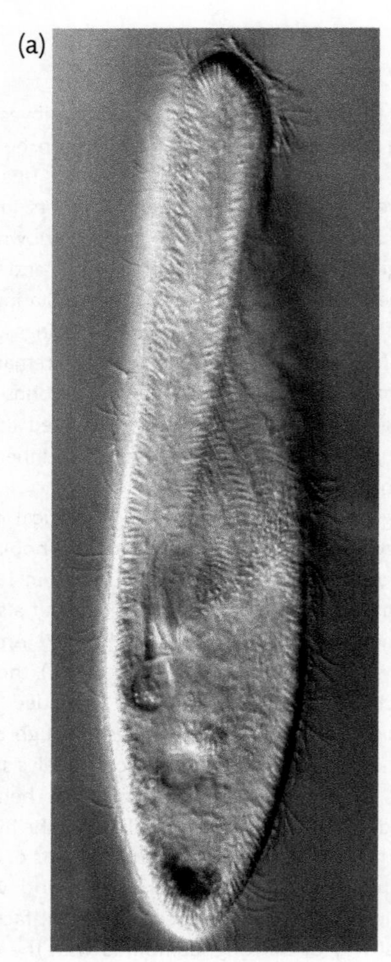

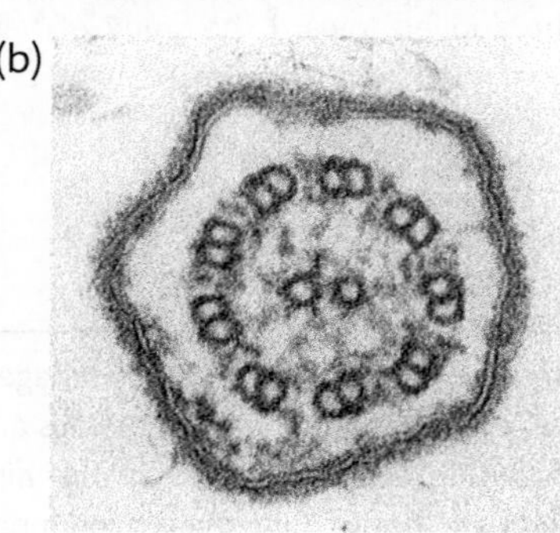

Figure 4.3 Light microscopic and electron microscopic appearance of cilia of a paramecium. The cilia are formed by numerous outward projections of the cell membrane. (a) At the light microscopic level, the cilia are visible as hair-like structures covering the whole surface of the paramecium. (b) An electron micrograph of a cross-section through a cilium reveals that the core structure is formed by nine doublet microtubules arranged in a ring, around a pair of single microtubules. This array of microtubules is surrounded by the plasma membrane. (After: Machemer, H. (1988a).)

response of *Paramecium*. Then, in the next section, we will extend these insights using a second taxis behavior, galvanotaxis, and model the cell's response to electric field stimulation by applying information collected through experiments.

Phobotaxis

Phobotaxis can be elicited as a natural behavior in paramecia, as well as in many other ciliates, by mechanical stimulation, for example, through touching of the cell membrane with an object. Without stimulation, the cilia move effectively in a posterior and right direction, thus resulting in a left, forward gyration of the paramecium. Mechanical deformation of the posterior cell end augments the normal ciliary rate and reorients the beat direction clockwise toward the posterior pole; this leads to movement straight forward. In contrast, mechanical deformation of the anterior cell end depresses the normal beat rate and reorients the beat direction counterclockwise close to the anterior pole. Thus, the paramecium shows a backward locomotor response, often accompanied by a right gyration. These movements lead to coordinated phobotactic responses away from the source of potentially harmful mechanical impact.

These behavioral responses are initiated by local activation of mechanosensitive channels permeable only to certain ions; they are concentrated at the anterior and posterior poles of paramecia. The following cellular events, schematically shown in Fig. 4.4, can be distinguished, depending on the site of stimulation:

- Mechanical deformation of the posterior end of the cell activates mechanosensitive K^+ channels, leading to an efflux of K^+ ions, and thus to a K^+-dependent hyperpolarizing receptor potential.
- Mechanical deformation of the anterior end of the cell activates mechanosensitive Ca^{2+} channels, leading to an influx of Ca^{2+} ions, and thus, to a Ca^{2+}-dependent depolarizing receptor potential.

Electrophysiological experiments in a number of ciliates have shown that the local receptor potentials spread rapidly and almost without any decay across the somatic membrane, so that the membranes

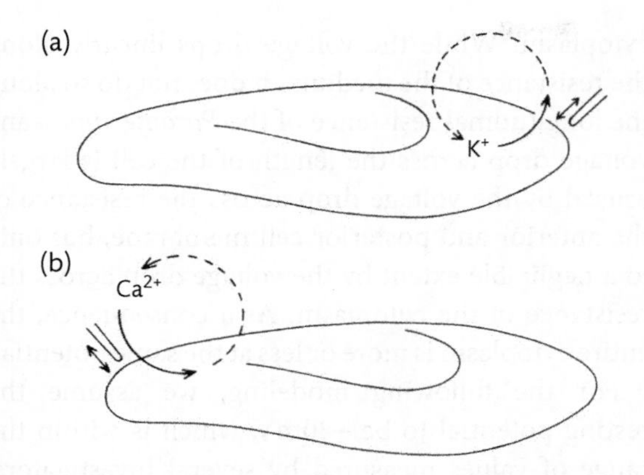

Figure 4.4 Changes in membrane potential induced by mechanical stimulation of *Paramecium species.* (a) Mechanical stimulation of the posterior end activates mechanosensitive K^+ channels. This leads to an efflux of K^+ ions, and thus to a hyperpolarization of the membrane. (b) Mechanical stimulation of the anterior end activates mechanosensitive Ca^{2+} channels, which are concentrated in this region. The resulting influx of Ca^{2+} ions causes a depolarization of the membrane. (After: Machemer, H. (1988a).)

of almost all cilia are instantly hyperpolarized or depolarized, respectively. In the ciliary membrane, the depolarizing receptor potential activates voltage-sensitive Ca^{2+} channels, leading to an influx of Ca^{2+} ions into the cilium. This results in increased depolarization, which causes an opening of further Ca^{2+} channels and an influx of Ca^{2+} ions through positive feedback. Due to this mechanism, the intraciliary Ca^{2+} concentration transiently increases from resting levels near 10^{-7}M to approximately 10^{-4}M. This elevation of intraciliary Ca^{2+} concentration triggers, at a subcellular level, a reversal of the ciliary activity. Theoretical considerations have shown that an increment of just 100 Ca^{2+} ions per 10μm length of a cilium is sufficient to cause a rise in the Ca^{2+} concentration from 10^{-7} to 10^{-6}M.

In contrast to the depolarization-induced ciliary activation, the coupling between mechanosensitive ion channels in the membrane and the hyperpolarization-induced ciliary activation is less well understood. At present, experimental data favor the hypothesis that the hyperpolarizing receptor potential causes, through an as yet unknown mechanism, a downregulation of the intraciliary Ca^{2+} concentration, which then leads to a reorientation of the ciliary power stroke. Thus, according to this hypothesis, Ca^{2+} acts as a universal messenger of ciliary electromotor coupling, mediating the fast phobotactic responses of paramecia by up- and downregulation of intraciliary Ca^{2+}.

Galvanotaxis: Modeling of a behavioral response

If direct current is flowing in a solution, paramecia will show a characteristic orienting behavior in relation to the resulting gradient—they will swim, with their anterior pole first, toward the **cathode**. Reversal of the polarity of the electric field causes a reversal of the swimming direction of the paramecium. This orienting movement toward the cathode is referred to as **galvanotaxis.**

Cathode Negative pole of a voltage source.

Although it is doubtful that galvanotaxis ever occurs under natural conditions, the detailed analysis of this behavioral response has provided an excellent insight into how the transduction of an external stimulus is coupled to a motor response. For the sake of simplicity, we will assume that a paramecium swims between the two metal plates of a **parallel plate capacitor**, as shown in Fig. 4.5. The polarity of the capacitor is indicated by the '+' (**anode**) and the '–' (cathode) symbols. Between the two electrodes, separated by a distance of 1cm, a voltage of 2V is applied. This voltage of 2V then drops over 1cm or, assuming homogeneous conductivity within the solution, 0.04V (or 40mV) drops over the 200μm length of a paramecium. In other words, we have applied an electric field with an intensity of 2V/cm; the orientation of the electric field lines is perpendicular to the plates.

Parallel plate capacitor Composed of two parallel metal plates isolated against each other. They have equal but opposite charges. By definition, the electric field lines arise from the positively charged plate (anode) and run perpendicular to the plates to the negatively charged plate (cathode).

Anode Positive pole of a voltage source.

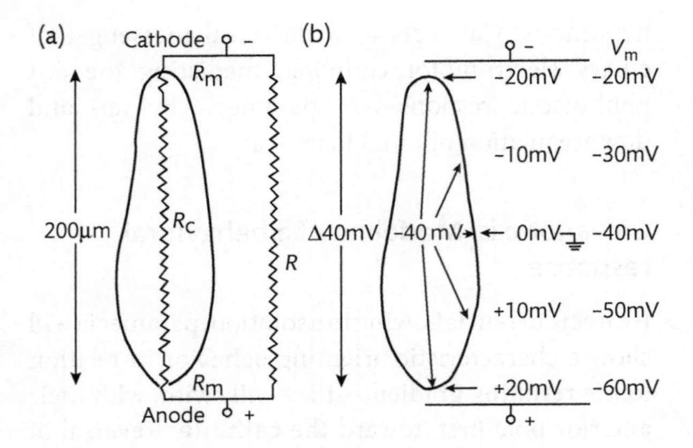

Figure 4.5 (a) Transcellular resistance of a paramecium. The resistance across the cell is equivalent to the sum of the resistances of the anterior half of the cell membrane, $R_{m\text{-anterior}}$, of the cytoplasm, R_c, and of the posterior half of the cell membrane, $R_{m\text{-posterior}}$. If a paramecium is brought into the electric field of a parallel plate capacitor such that its longitudinal axis is oriented parallel to the field lines, the transmembrane resistance and the resistance of the medium will be in parallel. (b) Polarization of a paramecium in an electric field. The transmembrane potential is equivalent to the difference of the potential within the cell and the potential of the external voltage gradient. As a result, the membrane potential is reduced (depolarized) toward the end closest to the cathode, and increased (hyperpolarized) toward the end closest to the anode. (After: Machemer, H. (1988a).)

In such an electric field, a paramecium is subject to electrical stimulation, because it represents a spatially extended resistor arranged in parallel to the resistance of the medium. If the longitudinal axis of the paramecium is oriented parallel to the field lines, then the transmembrane resistance and the resistance of the medium are in parallel over the entire length of the cell. The resistance across the cell is the sum of the following three partial resistances, arranged in series:

1 The resistance of the anterior half of the cell membrane, $R_{m\text{-anterior}}$ (approximately 8×10^7ohm).
2 The resistance of the cytoplasm, R_c (approximately 1.5×10^5ohm).
3 The resistance of the posterior half of the cell membrane, $R_{m\text{-posterior}}$ (8×10^7ohm).

These figures demonstrate that the resistance of the membrane of each of the two halves is roughly 500 times higher than the resistance of the cytoplasm. While the voltage drops linearly along the resistance of the medium, it does not do so along the longitudinal resistance of the *Paramecium*—any voltage drop across the length of the cell is largely caused by the voltage drop across the resistance of the anterior and posterior cell membrane, but only to a negligible extent by the voltage drop across the resistance of the cytoplasm. As a consequence, the entire cytoplasm is more or less at the same potential.

For the following modeling, we assume the resting potential to be –40mV, which is within the range of values measured by several investigators. We will, furthermore, assume that, in the case of a parallel orientation of the longitudinal axis of the paramecium and the electric field lines, roughly half of the assumed total voltage drop of 40mV across the 200µm length takes place at the resistances of each of the two membrane portions of the anterior and posterior halves of the paramecium.

To examine the effect of the voltage gradient on the polarization pattern of the membrane of the paramecium, we arbitrarily set the potential of the medium at the level halfway between the anterior and posterior pole at a reference value of 0mV. Then, relative to this reference point, the potential of the medium at the level of the cell pole closest to the cathode is –20mV, and at the level of the cell pole closest to the anode is +20mV. Thus, the voltage difference between these two points is, as expected, 40mV. As mentioned above, the potential within the cell is assumed to be uniformly –40mV.

The difference in voltage drop along the resistance of the medium and the longitudinal resistance of the cell has an important consequence for the potential drop across different parts of the membrane—while at the cathodal end the membrane potential is only –20V [–40mV–(–20mV)], at the anodal end the membrane potential is –60mV [–40mV–(+20mV)]. Thus, the part of the cell membrane closest to the cathode is depolarized, whereas the membrane portion closest to the anode is hyperpolarized.

The polarization of the cell membrane of a paramecium bears consequences for the beat direction of the cilia. As we have seen above, depolarization of the membrane results in the cilia beating toward the anterior pole of the cell, while hyperpolarization causes the cilia to beat toward the posterior pole. Thus, the cilia in the segments of the cell that are depolarized

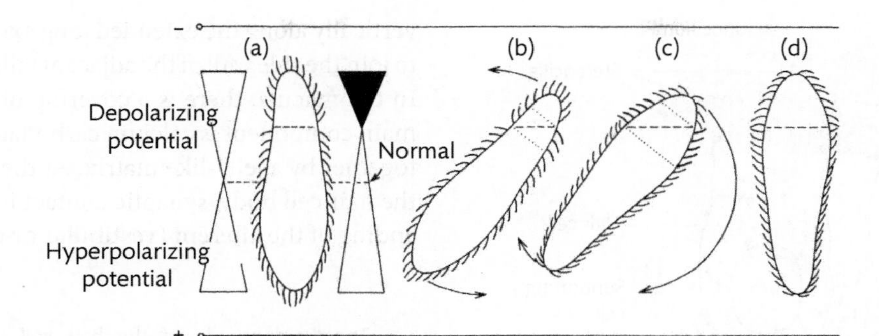

Figure 4.6 Orientation of a paramecium in a homogeneous electric field of a parallel plate capacitor. (a) Homodromic orientation. The longitudinal axis is oriented parallel to the field lines traveling from the anodal plate in a direction perpendicular to the cathode. The anterior end of the paramecium is oriented toward the cathode. Those cilia that are subject to a depolarizing potential greater than approximately 5mV beat toward the anterior pole. The remaining cilia beat in the opposite direction. This orientation is stable because an equal number of cilia on either side of the longitudinal axis are activated; it results in movement of the paramecium toward the cathode. (b, c) Deviation from this homodromic orientation causes an asymmetry in the beat orientation of the cilia within the zone marked by the dotted lines, and thus a turning movement as indicated by the arrows, until the stable homodromic orientation of (a) is adopted. (d) Antidromic orientation, with the posterior end of the paramecium closest to the cathode. This orientation is unstable because, through an intermediary position similar to the one shown in (c), any deviation leads automatically to the homodromic orientation of (a). (After: Machemer, H. (1988a).)

or hyperpolarized, respectively, beat in the opposite direction. As the beating toward the anterior end takes place only at depolarization values above approximately 5mV, the portion of the cell that beats toward the posterior end is larger than the portion beating toward the anterior end. Overall, this causes the paramecium to adopt a position with its anterior end oriented toward the cathode (**homodromic orientation**), as any deviation from this orientation will automatically be corrected by the asymmetric number of cilia on either side of the depolarized segment beating in the anterior direction, or of the hyperpolarized segment beating in the posterior direction, respectively (Fig. 4.6). Simultaneously, due to the larger number of cilia beating in the posterior direction, the paramecium will swim, with its anterior end first, toward the cathode.

A possible **antidromic orientation**, that is, an orientation of the paramecium with its longitudinal axis parallel to the field lines, but with the posterior end positioned closest to the cathode, and the anterior end closest to the anode, would seemingly result in the opposite movement. However, this orientation is—in contrast to the homodromic orientation—unstable, so that any deviation automatically leads to a homodromic orientation and a 'correct' galvanotactic response of the cell.

Model system: Geotaxis in vertebrates

The vast majority of animals assume a preferential position relative to the earth's force of gravity. An obvious prerequisite of this **geotactic response** is that the animal is capable of **gravireception.** In vertebrates, the sensory organ mediating this sensory ability is the **otolith organ** of the inner ear. Together with the **semicircular canals** (which detect rotational movements of the head), it forms the **vestibular organ.** In mammals, the otolith organ gives rise to the **cochlea,** which is the sensory organ involved in hearing. All three organs—the otolith organ, the semicircular canals, and the cochlea—together form the **labyrinth.**

> **Labyrinth** Otolith organ plus semicircular canals plus cochlea.

The otolith organ, whose structure and function will be examined in more detail below, consists of a patch (called the **macula**) of sensory cells and a covering matrix. Since the sensory cells have fine 'hairs' at their tips, they are commonly referred to as **hair cells.** Hair cells are also a central component of the semicircular canals and the cochlea, as well as of the lateral line

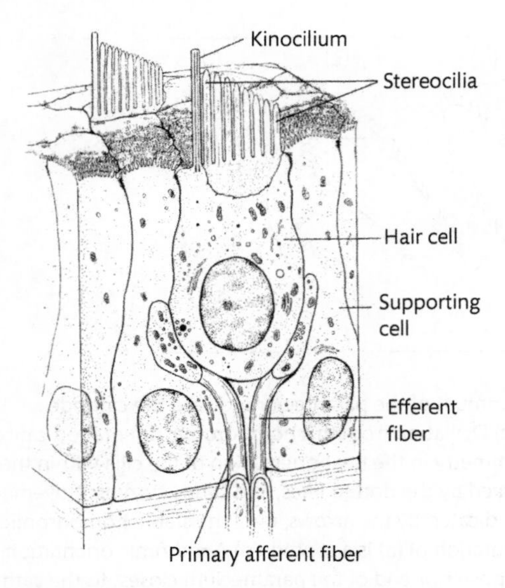

Figure 4.7 Hair cells of the vestibular system. The characteristic feature of these cells is the presence of hair-like projections at their apical surface, which are formed by several rows of stereocilia and a single kinocilium. The hair cells are embedded into a matrix of supporting cells covered with microvilli at their apical surface. At its base, each hair cell is innervated by an afferent process of a vestibular ganglion cell and an efferent process originating from cells in the brain stem. The efferent process mediates input to regulate the activity of the hair cell, whereas the primary afferent fiber provides an output channel to convey sensory information to the brain. (After: Kandel, E. R. and Schwartz, J. H. (eds) (1985).)

system of fishes and aquatic amphibians. As shown in Fig. 4.7, the 'hairs' of each hair cell are composed of two types of cellular projection: one **kinocilium,** which in mammals disappears soon after birth, and a few dozen **stereocilia.** Whereas the kinocilium is a true cilium exhibiting the typical 9 + 2 microtubuli structure in cross-sections, the stereocilia are—contrary to what their name implies—not cilia, but microvilli composed of an actin cytoskeleton ensheathed by a tube of plasma membrane.

The stereocilia are hexagonally packed and increase in length with increasing proximity to the kinocilium, which, if present, is always found at the tallest extreme of the bundle of stereocilia. This arrangement allows the researcher to arbitrarily define an axis of polarity, running from the row of the shortest stereocilia to the kinocilium (or the row of the tallest stereocilia).

The stereocilia are extensively cross-linked. One type of connection known as the **tip link** emerges from the tip of the stereocilium and runs almost vertically along the extended long axis of the hair cell to join the side wall of the adjacent taller stereocilium. In the macula, there is a covering of **otoliths.** Their main component is calcium carbonate crystals glued together by a jelly-like matrix. At the basal region of the hair cell body, synaptic contact is made with the ending of the **afferent** (**vestibular** or **eighth**) **nerve.**

➤ The 'hair' bundle of the hair cell is composed of stereocilia (which are actually microvilli) and a kinocilium (which is a true cilium).

Adequate stimulus

For a long time, it had remained unclear whether the mode of operation of the otolith organ depends on **pressure force** exerted by the 'weight' of the otolith on the hair cells of the macula, or whether **shear force** (resulting in a deflection of the hair cells) provides the **adequate stimulus.** This question was addressed, and finally answered, by a series of experiments published by Erich von Holst (see Box 3.5) in 1950.

Pressure and shear forces Contact interactions between different material bodies take place across contact surfaces. Any contact force acting on a surface can, regardless of its direction, be resolved into its normal (= perpendicular to the surface) component and its tangential (= parallel to the surface) component. The normal component is referred to as the pressure force, and the tangential as the shear (or shearing) force.

Adequate stimulus Form of stimulation that elicits an optimal response from a sensory receptor or a sense organ. Most commonly, this stimulus requires minimal energy to cause excitation. For example, the adequate stimulus for visual organs is light of specific wavelengths. Although other sensory modalities may also stimulate eyes or photoreceptors, only light can elicit responses at minimal energies.

The experiments of von Holst were based on the idea of increasing the weight of the otoliths and evaluating the effect of this experimental manipulation on the body's orientation. An increase in otolith weight raises both the pressure and the shear force, but, as we will see below, not necessarily to equal degrees. If, upon exposure to such changes, an animal in its normal position tries to hold the shear force constant, then the experiment would indicate that this parameter serves as the adequate stimulus. If, on the other hand, an animal tries to compensate for the pressure force, then this response would, as von Holst argued, demonstrate that the animal uses the latter parameter to define its position relative to gravity.

To provide the experimenter with precise measurements, von Holst used two popular aquarium fishes, the tetra (*Gymnocorymbus* sp.) and the angelfish (*Pterophyllum* sp.). The body of these two species is laterally compressed, which facilitates measurements of the angle the fish assumes in relation to gravity. Moreover, both fishes are no more than a couple of centimeters long, which makes it possible to place the aquarium with the fish into a centrifuge and, thus, experimentally increase the weight of the otolith.

> ➤ The gravitational field to which the otolith organ is exposed can be increased by placing the animal into a centrifuge and modifying the speed of rotation.

In addition to gravity, the incident angle of light also determines the position of the fish. Normally, light comes straight from above, so that both the system mediating the **dorsal light reaction** and the mechanism underlying the response to the gravitational field tell the fish to adopt a vertical position. When light is presented from the side, the two positional mechanisms provide conflicting information. The result is a compromise, as shown in Fig. 4.8. Now, the fish assumes a tilted posture, with the dorsal part of its body oriented somewhat towards the source of light. The exact angle at which the fish deviates from the vertical position—which is sensed by the otolith organ—depends, among other factors, on the intensity of the light. When both otoliths are removed, the fish relies solely on the light information and completely ignores the gravitational field.

To exclude the effect of light, von Holst kept the incident angle of light constant throughout the experiments. When he doubled the weight of the otoliths, by increasing gravity from 1*g* to 2*g* in the centrifuge, the fish adopted a position that showed its attempt to hold the shear force constant (Fig. 4.9). Thus, as suggested by these whole-animal

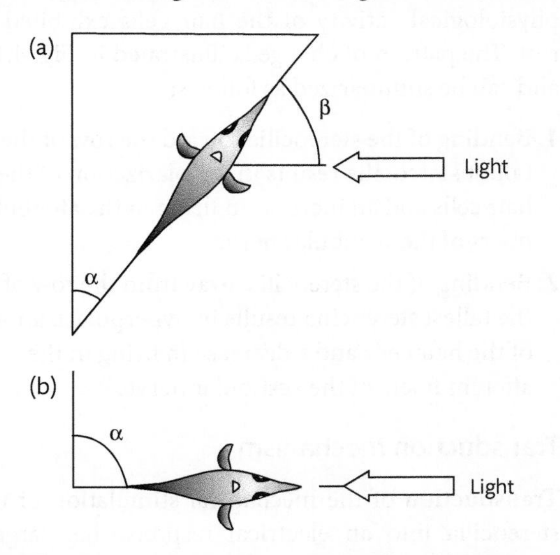

Figure 4.8 The orientation of the body of an angelfish is determined largely by gravity and the incident angle of light. (a) When light is presented to an intact fish from the side, the fish assumes a tilted position, with the dorsal part of its body oriented at a certain angle toward the source of light. (b) When the otoliths on both sides of the head are removed and an identical stimulus is presented, the fish orients, exclusively, towards the light source. (After: von Holst, E. (1950).)

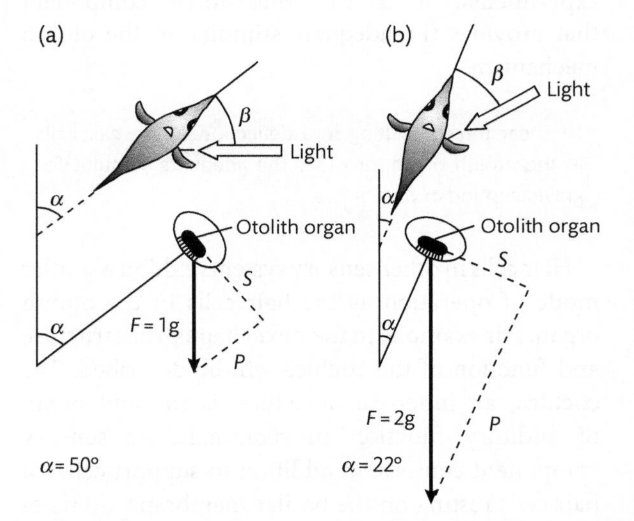

Figure 4.9 Experimental design employed by Erich von Holst to decide whether shear or pressure force is used by an angelfish to detect gravity. (a) When light is presented from the side, the fish assumes a tilted position, with its back oriented somewhat toward the light source. Below the fish, the corresponding orientation of the otolith is drawn. The gravitational force, *F*, acting upon the otolith has two components—a shear force, *S*, and a pressure force, *P*. (b) To increase the gravitational force from 1*g* to 2*g*, the fish was placed into a centrifuge. Now the angle α that the fish maintains relative to the force of gravity, *F*, reveals that the shear force, *S*, is held constant, while the pressure force, *P*, increases. Note that the incident angle of light is adjusted in the second experiment, such that it is identical to the angle in the first experiment. (After: von Holst, E. (1950).)

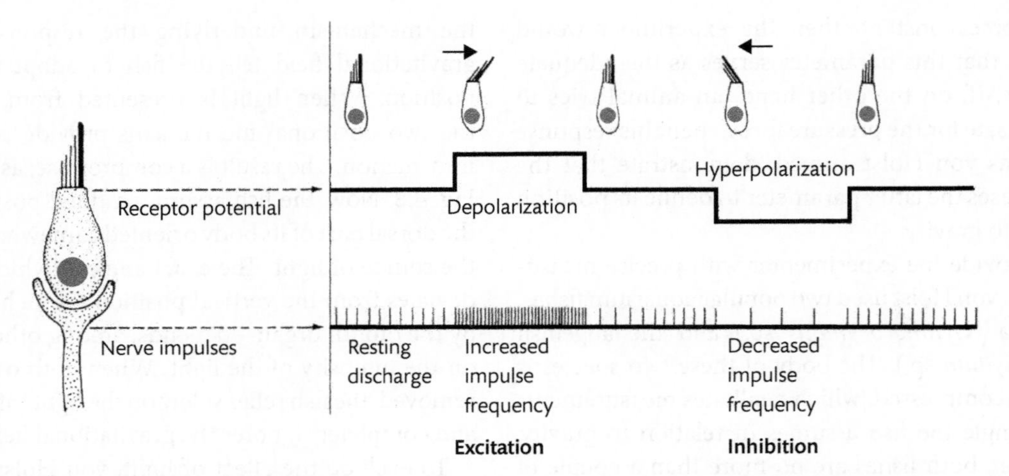

Figure 4.10 Physiological response of a hair cell to stimulation. The receptor potential of the cell is measured intracellularly and the rate of firing in the afferent fibers of the eighth nerve by extracellular techniques. Bending of the stereocilia toward the row of the tallest stereocilia causes depolarization in the hair cell and an increase in spike frequency in the vestibular nerve, compared with the resting discharge. Conversely, bending of the stereocilia away from the row of the tallest stereocilia results in hyperpolarization and a decreased impulse frequency. (After: Flock, Å. (1965).)

experiments, it is the shear-force component that provides the adequate stimulus to the otolith mechanism.

➤ Shear force, resulting in a deflection of the hair cells in the otolith organ, provides the adequate stimulus for gravireception in vertebrates.

Hair cells in other sensory systems exhibit a similar mode of operation as the hair cells in the otolith organ. For example, in the next chapter the structure and function of the cochlea will be described. The cochlea, an inner-ear structure, is the end organ of auditory function in mammals. Its sensory component consists, in addition to support cells, of hair cells resting on the basilar membrane. Some of the stereocilia of these hair cells are embedded in the tectorial membrane (for details, see Chapter 5). Pressure waves reaching the inner ear cause up-and-down vibrations of the basilar membrane and the tectorial membrane. However, the two membranes pivot about different hinging points, which results in tilting movements in opposite direction of the hair cell stereocilia embedded in the tectorial membrane and the hair cell bodies resting on the basilar membrane. Again, it is the shear force that acts as an adequate stimulus to transduce mechanical energy into a receptor potential generated by the hair cells.

Physiological properties of hair cells

The observation that shear force provides the adequate stimulus for the hair cells is reflected by the physiological properties of the hair cells. Bending of the stereocilia leads to a characteristic change in the physiological activity of the hair cells exhibited at rest. The pattern of change is illustrated in Fig. 4.10, and can be summarized as follows:

1. Bending of the stereocilia toward the row of the tallest stereocilia results in depolarization of the hair cells and an increase in firing in the afferent fibers of the vestibular nerve.
2. Bending of the stereocilia away from the row of the tallest stereocilia results in hyperpolarization of the hair cells and a decrease in firing in the afferent fibers of the vestibular nerve.

Transduction mechanism

Transduction of the mechanical stimulation of the stereocilia into an electrical response has largely been elucidated through the work of James Hudspeth (now at Rockefeller University in New York) and David Corey (now at Harvard Medical School in Boston, Massachusetts). As their micromanipulation experiments performed on isolated sensory epithelium have demonstrated, transduction is mediated directly by **mechanically sensitive channels.** This mode is in contrast to many other sensory receptor cells, such as photoreceptors and olfactory neurons, which employ cyclic nucleotides or other second messengers. The latter strategy has the advantage that the stimulus signal can be amplified. Furthermore, feedback within the metabolic pathway provides the opportunity for gain control, resulting, for example, in adaptation and desensitization. On the other hand, transduction without the intervention of a second messenger results in a much higher speed of response. Hair cells operate not only more quickly than other sensory receptor cells, but even faster than neurons themselves. In the fastest hair cells found, the delay between the deflection of the hair bundle and the onset of a receptor current has been estimated to be only 10μsec! Another remarkable feature of hair cells is the high sensitivity of their response. Studies have shown that a deflection of the tip of the stereocilia by as little as ±0.3nm, corresponding to ±0.003°, may be sufficient to trigger a response.

➤ Transduction of mechanical stimulation of stereocilia of hair cells into an electrical response is mediated directly by mechanically sensitive channels.

➤ **Nanometer (nm)** 1nm = 10^{-9}m.

Most transduction channels are concentrated near the tip of the stereocilia, rather than being located at the bases. The number of channels is probably quite low. In the bullfrog (*Rana catesbeiana*), the hair bundles of the sacculus have 100–250 transduction channels each. As each of these hair cells has about 60 stereocilia, assuming even distribution through the bundle, each stereocilium would have only two to five channels.

The transduction channels are relatively non-selective for cations, but, due to the high concentration of K^+ in the ionic milieu around the hair bundle, the major carrier for the transduction current is K^+. The channel gate is connected to the adjacent taller stereocilium by tip links, elastic filaments that exhibit spring-like properties (Fig. 4.11). During displacement toward the tallest stereocilium, the tip links provide the tension to open the transduction channels. When deflection occurs in the opposite direction, the tip-link tension slackens, thus increasing the probability that the channels close.

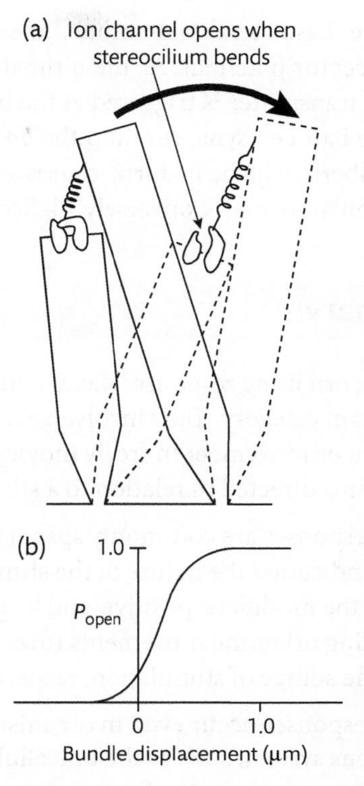

Figure 4.11 (a) Tip-link model for transduction. A mechanical stimulus originating from the bending of the 'hairs' of the hair cells acts directly on a mechanically sensitive ion channel in the stereocilium, by modifying the tension in the 'gating spring' attached to the channel gate. Displacement of the bundle toward the tallest stereocilia stretches these tip links and thus increases tension. (b) Sensitivity curve relating displacement of the hair bundle to probability of opening (P_{open}) of the transduction channels. The diagrams also demonstrate the enormous sensitivity of hair cells. Displacement of only one-third of a micrometer, corresponding roughly to the diameter of one stereocilium, already provides a saturating stimulus. Note that a certain fraction of channels is open at rest. (After: Pickles, J. O. and Corey, D. P. (1992).)

In an unstimulated hair cell, a small portion (approximately 10–25%) of the transduction channels are already open. The resulting inward, positively charged flux partially contributes to the depolarizing resting potential of about 260mV. Deflection of the hair bundle toward its tall edge opens additional channels, and the influx of cations produces a depolarizing voltage charge. This initial event activates a variety of voltage-dependent channels across the basolateral membrane, thereby setting up the receptor potential. As the ultimate result, the release of transmitter is triggered at the basal region, where the hair cell synapses onto the endings of the afferent fibers, which, in turn, causes excitation in the vestibular nerve. Conversely, deflection of the stereocilia toward the short edge of the hair bundle closes transduction channels that are open at rest, thus leading to a reduced influx of cations into the cell. As a consequence, hyperpolarization occurs, resulting in a decrease in transmitter release and inhibition of firing in the fibers of the vestibular nerve.

Summary

- Among orienting responses, taxes constitute an important category. They involve an orienting reaction or movement in freely moving organisms directed in relation to a stimulus.
- Taxis responses are commonly specified by a prefix indicating the nature of the stimulus, and by the modifiers 'positive' and 'negative' indicating orienting movements toward or away from the source of stimulation, respectively.
- Taxis responses occur even in organisms without a nervous system, such as the unicellular protist *Paramecium* species, the first model system described in detail in this chapter. This ciliate protozoan exhibits, among others, chemotaxis, phobotaxis, and galvanotaxis.
- The taxis behaviors in paramecia are mediated by cilia, whose stroke pattern is largely determined by the intraciliary Ca^{2+} concentration.
- Mechanical stimulation of the posterior end of a paramecium leads to a hyperpolarizing receptor potential, whereas such stimulation of the anterior end results in a depolarizing receptor potential. These differential receptor potentials cause transient up- and downregulation of the intraciliary Ca^{2+} concentration, respectively, which, in turn, lead to opposite ciliary activities.
- In an electric gradient, paramecia exhibit galvanotactic behavior by swimming toward the cathode, with their anterior pole first. This behavior is triggered by a differential polarization of those parts of the cell that are closest to the cathode and the anode, respectively, and thus, by the generation of a differential beat pattern of the cilia in different parts of the cell.
- The focus of the second model system is on geotaxis in vertebrates. This taxis behavior involves orientation of the body relative to gravity. Geotaxis requires that the animal be capable of perceiving the earth's gravitation field, a sensory capability referred to as gravireception.
- In vertebrates, gravireception is mediated by the otolith organ of the inner ear. This organ consists of a macula of sensory cells called hair cells. The 'hairs' at the apical region of each hair cell are composed of two types of projection, one kinocilium, and a few dozen stereocilia. The stereocilia are hexagonally packed and increase in length with increasing proximity to the kinocilium, which, if present, is found at the tall extreme of the bundle of stereocilia. A mechanically sensitive channel at the tip of each stereocilium is connected to the side wall of its adjacent taller neighbor, via a tip link. The hairs of the hair cells are covered by otoliths.
- Behavioral physiological experiments have shown that it is the shear force, and not the pressure force, exerted by the otoliths on the hair cells of the macula, that provides an effective physiological stimulus to signal positional changes of the head.
- Any tilting of the head, and thus of the otolith organ, causes a bending of the hairs, which leads to a characteristic physiological response such that (i) bending of the stereocilia toward the row of the tallest stereocilia results in depolarization of the hair cells and an increase in firing in the afferent fiber making synaptic contact with these cells; (ii) bending of the stereocilia away from the row of the tallest stereocilia results in hyperpolarization of the hair cells and a decrease in firing in the afferent fibers.
- Transduction of the mechanical stimulation of the stereocilia into an electrical response is mediated

by mechanically sensitive transduction channels concentrated near the tip of the stereocilia. During displacement of the stereocilium toward the adjacent taller stereocilium, the tip link provides the tension to open the transduction channel, leading to an influx of cations and, thus, to depolarization of the membrane of the hair cells, which is followed by an increase in firing in the afferent fiber. Conversely, deflection of the stereocilia toward the shorter edge of the hair bundle closes transduction channels that are open at rest, leading to a reduced influx of cations into the cell and, thus, to hyperpolarization, which is then followed by a decrease in firing in the afferent fiber.

The bigger picture

As we have seen in Chapter 3, the beginning of the twentieth century was a critical time in the development of the behavioral sciences. Scientists in different countries and from different disciplines sought to overcome the rather subjective, intuitive, and often anecdotal approaches that had dominated the study of behavior for over two thousand years. It, therefore, was a logic consequence that these researchers focused on simple behavioral patterns that exhibited robust response characteristics upon experimental stimulation of the organism. An obvious choice for researchers was the study of taxis behavior in unicellular protists—simple organisms without a nervous system. Not only did these studies fuel enormous progress in the development of stringent approaches applied to the analysis of behavior, but also they were subsequently seminal in exploring ionic mechanisms of excitation. In recent years, interest in taxis behaviors of protists has been revitalized, particularly by engineers who use them as bio-inspired model systems to derive design principles for miniature robots that deliver, for example, certain molecules to specific targets using autonomous control.

Recommended reading

Adler, J. (1987). How motile bacteria are attracted and repelled by chemicals: An approach to neurobiology. *Biological Chemistry Hoppe-Seyler* **368**:163–173.

An excellent starting point to learn more about the genetic and biochemical basis of chemotaxis behavior of bacteria. This was written by a biologist who originally wanted to study the neural mechanisms of behavior in animals. However, due to the lack of proper approaches available for the study of higher organisms at the time when he started his career, Adler turned to bacteriology, where he laid the foundation for today's research in this fascinating field.

Pickles, J. O. and Corey, D. P. (1992). Mechanical transduction by hair cells. *Trends in Neurosciences* **15**:254–259.

This well-written review article focuses on transduction of the mechanical stimulus into an electrical signal by hair cells.

Short-answer questions

4.1 Define, in one sentence, the term 'positive phototaxis'.

4.2 Attracted by the calling song of an invisible conspecific male, a female cricket walks to the source of sound. What type of taxis behavior does the female display?

4.3 What distinct ultrastructural feature characterizes a cilium?

4.4 Briefly describe the chemotactic reaction of paramecia when they encounter a bubble of carbon dioxide. What physical property of the water directs this behavior?

4.5 What ionic and physiological responses are triggered by mechanical stimulation of the anterior part of *Paramecium*? What ionic and physiological responses are evoked by such stimulation of the posterior end?

4.6 Define, in one sentence, the galvanotactic response of *Paramecium*.

4.7 What sensory organ mediates gravireception in vertebrates?

4.8 Complete the following sentence: Hair cells are the sensory receptors of the .... and the ...., which together form the vestibular organ in vertebrates. The 'hairs' of the hair cells are cellular projections of two types, namely .... and ....

4.9 Indicate the correct answer (by circling the correct option): Bending of the stereocilia toward the row of the tallest stereocilia results in <depolarization OR hyperpolarization> of the hair cells of the otolith organ and an <increase OR decrease> in firing in the afferent fibers of the vestibular nerve.

4.10 Transduction of the mechanical stimulation of the stereocilia into an electrical response is mediated by mechanically sensitive channels. What advantage do such channels have compared to transduction mechanisms that involve cyclic nucleotides of other second messengers? What are two disadvantages of employing mechanically sensitive channels, instead of second-messenger mechanisms, for transduction?

4.11 **Challenge question** *Caenorhabditis elegans* displays, among other taxis behaviors, chemotaxis. Outline the design principles of an experiment aimed at identifying mutants with altered chemotaxis behavior, compared to wild-type worms.

Essay questions

4.1 Describe how mechanical stimulation of paramecia leads, through proper activation of the cilia, to a phobotactic response.

4.2 What is galvanotaxis? How is this behavioral response mediated at the cellular level in paramecia?

4.3 The postural reaction of vertebrates is determined largely by the action of the vestibular system. What is the effective physiological stimulus? How is this stimulus transduced into a physiological response of the vestibular organ?

Advanced topic Physical modeling of galvanotaxis in *Paramecium*

Background information

Galvanotaxis involves orientation in a DC electric field of *Paramecium* toward the cathode. This orienting movement is mediated by specific changes in the beat pattern of the individual cilia, including a reversal of the beat direction of some cilia. This asymmetry in beat direction generates both a forward force and a backward force. If the *Paramecium* is tilted relative to the electric field lines, these two forces will result in a torque, which directs the *Paramecium* toward the cathode.

Essay topic

In an extended essay, describe the torque model that Naoko Ogawa and co-authors proposed in their 2006 paper (see 'Starter references' below) to link the microscopic ciliary beat pattern and the macroscopic behavior of *Paramecium* during galvanotaxis. Depending on your knowledge and training in physics or engineering, either restrict your essay to a qualitative description, or give a more detailed account of the bottom-up approach, based on systems theory, that the authors employed. Include in your essay how numerical simulations were used to test whether the modeled cell exhibits realistic behavior when stimulated by changes in the magnitude and direction of the electric field.

To compare the data of simulations with data obtained from experiments using real paramecia, the authors recorded the responses of *Paramecium* to electric fields using a high-speed tracking system. Describe the elements and the overall configuration of this system, as published in the paper by Ogawa and co-workers in 2005 (see 'Starter references' below). Why was the development of this tracking system critical for the comparison of the data? Were the data of the simulation experiments in agreement with the data of the biological experiments?

In a final section in your essay, discuss possible applications in which paramecia, controlled through engineering methods, could be used as smart micro-robots for microdelivery of substances, or as microsensors.

Starter references

Ogawa, N., Oku, H., Hashimoto, K., and Ishikawa, M. (2005). Microrobotic visual control of motile cells using high-speed tracking system. *IEEE Transactions on Robotics* **21**:704–712.

Ogawa, N., Oku, H., Hashimoto, K., and Ishikawa, M. (2006). A physical model for galvanotaxis of *Paramecium* cell. *Journal of Theoretical Biology* **242**:314–328.

To find answers to the short-answer questions and the essay questions, as well as interactive multiple choice questions and an accompanying Journal Club for this chapter, visit **www.oup.com/uk/zupanc3e**.

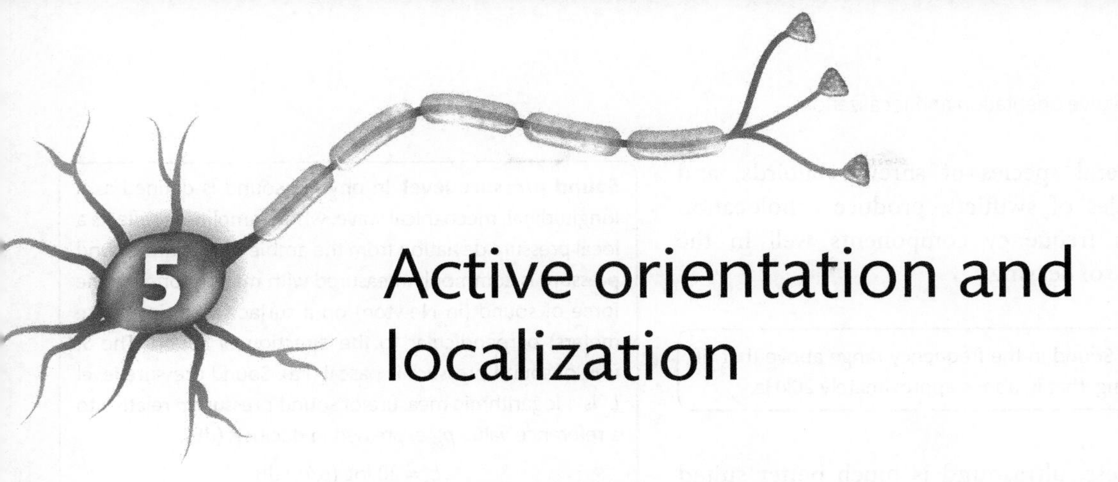

5 Active orientation and localization

Introduction

At least to some degree, all animals rely for orientation on sensory cues arising from the environment. We discussed such sensory stimuli and the resulting orienting behaviors in the last chapter. Some animals have, in addition, developed **active orientation mechanisms**. The distinctive feature of these mechanisms is that the animal itself produces a carrier signal for extracting information about the environment. Two such mechanisms have been discovered to date, **electrolocation** and **echolocation**. Whereas electrolocation will be covered in Chapter 8, 'Sensorimotor integration,' in this chapter we focus on echolocation.

Echolocation, as a mechanism to negotiate the animal's surroundings, is based on the emission of sound, its reflection by objects located within the emission beam, and the analysis of the returning echo. Echolocation was first demonstrated in **bats**, based on a number of critical findings described in detail later in this chapter. Besides its existence in certain bat species, echolocation has also been found among mammals in **toothed whales** (Odontoceti), certain **tenrecs** (Tenrecidae), and several species of **shrews** (Soricidae); as well as in two groups of birds—**oilbirds** (Steatornithidae) and several species of **swiftlets** (Apodiformes).

In the following, we will review some of the key findings of echolocation research in bats, by far the most intensively studied echolocating animal group. Readers interested in echolocation in toothed whales or in oilbirds and swiftlets will find two excellent starter references in the section 'Recommended reading'.

Key concepts

Echolocation is based on the production of high-intensity sound and the perception of the much less intense echo. Although sometimes misunderstood, echolocation does not necessarily involve the production of **ultrasound**. Some animals, such as the Old World fruit bats *Rousettus aegyptiacus*, *Rousettus leschenaultii*, and *Rousettus amplexicaudatus*, certain tenrecs, several species of shrews, oilbirds, and several species of swiftlets, produce echolocation signals with frequency components well in the human range of hearing.

Ultrasound Sound in the frequency range above that of human hearing, that is, above approximately 20kHz.

Nevertheless, ultrasound is much better suited for **detection of small objects** by analysis of the returning echo, than sound in the human frequency range. The reason for this frequency-dependence of object resolution is that, as a rule of thumb, an object reflects echoes only if its cross-section is at least approximately one-third the wavelength impinging on it.

Exercise: *The fundamental frequency of a man's voice is approximately 100Hz and that of a woman's approximately 200Hz. The constant-frequency component of a greater horseshoe bat is close to 80kHz. Calculate the wavelengths of sound of 100Hz and 80kHz, and compare the resolving power of these two signals when used for echolocation.*

Solution: *The wavelength is calculated according to the following formula:*

$$s = f \times \lambda$$

where s is the speed of sound (343m/sec), f is the frequency of sound (in Hz), and λ is wavelength of sound (in m). Thus, the wavelength of sound of 80kHz is approximately 4.3mm, and that of sound of 100Hz is approximately 3.4m. As a consequence, sound of 80kHz is reflected as an echo by objects as small as 1-2mm, whereas sound of 100Hz is reflected as an echo only by objects larger than approximately 1m.

On the other hand, ultrasound exhibits a much stronger **atmospheric attenuation** than sound in the frequency range of human hearing. For instance, a sound of 20kHz is attenuated by the atmosphere by 0.5 decibels per meter (dB/m). At 40kHz, the attenuation of **sound pressure level** is 1.2dB/m, and at 100kHz it is 3dB/m. This attenuation effect limits echolocation to frequencies below 150kHz and to distances between the bat and the object smaller than approximately 20m.

Sound pressure level In physics, sound is defined as a longitudinal, mechanical wave, whose amplitude reflects a local pressure deviation from the ambient pressure. Sound pressure is commonly measured with microphones as the force of sound (in Newton) on a surface area (in square meters) perpendicular to the direction of sound. The SI unit of sound pressure is pascal (Pa). Sound pressure level L_p is a logarithmic measure of sound pressure p relative to a reference value p_0, expressed in decibels (dB):

$$L_p = 20 \log (p/p_0) \text{ dB}$$

A logarithmic scale is used because the intensity perceived by humans is proportional to the logarithm of the actual intensity measured. In air, the commonly used reference value is 20µPa, which corresponds roughly to the threshold of human hearing. If sound pressure is doubled, the sound pressure level will increase by 6dB. A sound pressure level of approximately 120dB, relative to 20µPa, is generated by a jet engine at a distance of 100m.

Orientation over such a distance can be accomplished only by emitting extremely loud and highly directional calls. Although the narrowing of the beam leads to an increase in the intensity of the sound, it also limits the receptive acoustic field. To overcome the latter problem, bats have to generate multiple, successive calls in different directions. Similarly, since echolocation calls produce only 'stroboscopic' images of the bat's environment (each call yields one acoustic image), and because in most species the **duty cycle** is quite low (typically 4–20%), the bat remains 'in the dark' for most of the time. To compensate for at least part of this problem, the bat has to emit an enormous number of echolocation calls. Gerhard Neuweiler (1935–2008), of the Ludwig Maximilian University of Munich, Germany, estimated that, during an 11h-foraging flight, the horseshoe bat *Rhinolophus rouxii* produces approximately 400 000 echolocation calls. Although its duty cycle is, at 40–50%, very high compared to other bat species, it still fails to perceive acoustic information about its surrounding for more than half of the time!

Duty cycle Proportion of time during which a device is operated. Refrigerators, for example, do not run continuously but are switched on and off for a certain amount of time. If, for example, a refrigerator pump is on for 45min over a 60min period, then its duty cycle is 75%. In bat echolocation research, the duty cycle defines the relative 'on' time of the emitted echolocation calls.

Model system: Echolocation in bats

Bats form a large order, Chiroptera, among mammals, with approximately 950 different species. The two suborders are Megachiroptera and Microchiroptera. The Megachiroptera comprise about 150 species; they are, as the Greek prefix (*mega* = big) indicates, large bats. By contrast, the second suborder, Microchiroptera, are, as again the Greek prefix suggests (*micro* = small), small when compared with the Megachiroptera. The Microchiroptera consist of approximately 800 species.

All but one genus (*Rousettus*) of the Megachiroptera do not echolocate at all, and are frugivorous. Instead, they have developed large eyes and rather small ears, and possess excellent night vision.

Figure 5.1 Lazzaro Spallanzani. (From: Wikimedia Commons.)

Figure 5.2 Hamilton Hartridge. (© Godfrey Argent Studio.)

The beginnings of echolocation research

The beginnings of the study of bat echolocation date back to the end of the eighteenth century when several scientists conducted a series of experiments on how bats orient in the dark. Among them were the Italian Lazzaro Spallanzani (1729–1799) (Fig. 5.1), a professor at the Universities of Reggio, Modena, and Pavia, and the Swiss Charles Jurine (1751–1819), a member of the Geneva Natural History Society. Jurine had found that if he tightly sealed the ears of a bat with candle wax the animal helplessly collided with obstacles in its flight path. Spallanzani confirmed these results and extended them by demonstrating that blinded bats could avoid obstacles perfectly well in closed rooms, both during the day and at night. Based on these and many other experiments, Spallanzani concluded that the ears, rather than the eyes, of bats serve to orient the bat during flight.

The proposal of Spallanzani contrasted with the hypothesis of an influential contemporary, the French naturalist Georges Cuvier (1769–1832), who explained the bat's capability to orient in the dark through the presence of a nerve net on its wings that, he thought, acted as a highly sensitive sense of touch. Although Cuvier's explanation lacked experimental support, it was his hypothesis that was generally accepted for almost 150 years.

The situation began to change only in the first half of the twentieth century. In 1920, Hamilton Hartridge (1886–1976) (Fig. 5.2), then a Fellow of King's College at the University of Cambridge (United Kingdom), performed experiments in which threads were hung across a completely darkened room in which bats were flying. Despite carefully monitoring, Hartridge failed to find evidence that the bats ran into any of these threads. This finding excluded the possibility that they used the sense of vision for orientation under such conditions. Hamilton Hartridge proposed that the bats emit sound with wavelengths short enough (i.e., with frequencies above the upper frequency limit of human hearing) that these signals can be reflected

Box 5.1 Donald R. Griffin

Donald R. Griffin. (From: Gross, C. G. (2005).)

Donald R. Griffin's name is intimately linked to the term he coined in 1944—echolocation. Born in Southampton, New York, in 1915, he was admitted to Harvard University in 1934. While still an undergraduate student, he learned about the experiments that Lazzaro Spallanzani had conducted in 1793. These experiments had shown that blinded bats, although completely deprived of visual cues, can avoid obstacles and fly normally. Spallanzani had explained the bats' ability by the use of auditory cues to orient in the dark. However, his hypothesis received little acceptance by scientists until Griffin made a pioneering discovery.

When Griffin heard that a Harvard physics professor, George Washington Pierce, had developed a new apparatus capable of detecting sound above the frequency range of human hearing, he asked him whether he could use this device for studying bats. Together with a fellow student, Robert Galambos, he demonstrated that bats emit ultrasound and that they can detect—and avoid—obstacles by hearing the echoes. Later, Griffin did the first ultrasound recordings of free-flying bats in the field and discovered that they can catch small flying insects by echolocation, indicating that echolocation is more than just a collision warning system, as initially assumed. Griffin's popular presentation of these and related studies in his book *Listening in the Dark: The Acoustic Orientation of Bats and Men*, published in 1958, became a 'classic' and stimulated numerous investigators to work in the field of ultrasound production by animals and echolocation in bats.

Besides his research on echolocation, Griffin worked intensively on homing of birds. While still a graduate student at Harvard, and years before radiotracking became feasible, he displaced birds into unfamiliar territories and traced their routes back to their nests by following them, sometimes over many hours, with small airplanes. Later in his career, he became increasingly convinced that animals may experience conscious thoughts and subjective feelings. However, his publications in the area of 'cognitive ethology,' as he called this subdiscipline, were controversial and received little acceptance by the scientific community.

Between 1946 and 1986, Griffin held faculty positions at Cornell, Harvard, and Rockefeller. After his retirement, he continued to work at the Concord Field Station of Harvard. He died in Lexington, Massachusetts, in 2003.

even by small objects in the flight path, thereby forming 'sound pictures' (as he referred to the acoustic information produced) used by the bats to negotiate their surroundings.

The missing piece in the puzzle was evidently the demonstration that bats can produce and perceive ultrasound, and that they use the echoes of such signals for orientation. This demonstration was achieved in the late 1930s and early 1940s. In 1938, the physicist George Pierce and the undergraduate student Donald R. Griffin (see Box 5.1), both of Harvard University, published the results of a study in which they reported that bats can emit ultrasound. For this investigation, Griffin used a novel device, a so-called sonic detector, which had been invented by Pierce to detect ultrasound. Although in this initial study the ultrasound was interpreted as a call note or alarm signal, rather than a means of avoiding obstacles in the flight path, subsequent experiments by Griffin and a fellow graduate student at Harvard, Robert Galambos (1914–2010), provided crucial experimental evidence in support of Hartridge's hypothesis. Similar to the experiments conducted by Spallanzani, Griffin and Galambos demonstrated that plugging of the bat's ears severely interfered with its ability to avoid obstacles. The same kind of impairment occurred when they taped shut the mouth of the bats, thereby preventing the animals from emitting sound. They concluded their report published in 1941 as follows: '*Flying bats detect obstacles in their path by (1) emitting supersonic notes, (2) hearing these sound waves when reflected back to them by the obstacles, and (3) detecting the position of the obstacle by localizing the source of this reflected sound.'* In 1944, Griffin coined the term **echolocation** for this phenomenon.

Classification of bat ultrasound

Based on their frequency spectrum, the sound produced by bats can be divided into two major categories:

1 Frequency-modulated (**FM**) signals.

2 Constant-frequency (**CF**) signals.

FM signals consist of short pulses, typically lasting for less than 5msec, that quickly sweep downward in frequency during the course of the pulse. As they cover a wide range of frequencies, these sounds are also called **broadband signals**. The North American big brown bat (*Eptesicus fuscus*) is such an 'FM bat.' A sonogram of the chirp-like pulses emitted by this species is shown in Fig. 5.3(a). As this plot demonstrates, the pulses last between 0.5 and 3msec and sweep downward by about one octave. The figure shows, in addition, that, when the bat gets closer to a target, the pulses become shorter, and the frequency range covered by the individual pulses is shifted toward lower frequencies.

CF signals dominate in a rather narrow frequency range, and are thus often referred to as **narrowband signals**. CF signals last much longer than FM signals, typically between 10 and 100msec. They are also less homogenous than FM signals. In particular, they may be followed by a downward frequency-modulated sweep. A well-studied example of this type of signal is the sound of the greater horseshoe bat (*Rhinolophus ferrumequinum*). This signal is comprised of a constant-frequency component of approximately 83kHz, followed by a brief downward frequency-modulated sweep. Such pulses of *Rhinolophus*, recorded during pursuit of an insect, are shown in Fig. 5.3(b). As also revealed by the recording, the calling pattern changes in the course of pursuit. The closer the bat gets to the prey, the more pronounced the FM component, and the shorter the individual pulses, become.

The differences between the different types of call reflect adaptations to specific foraging areas and

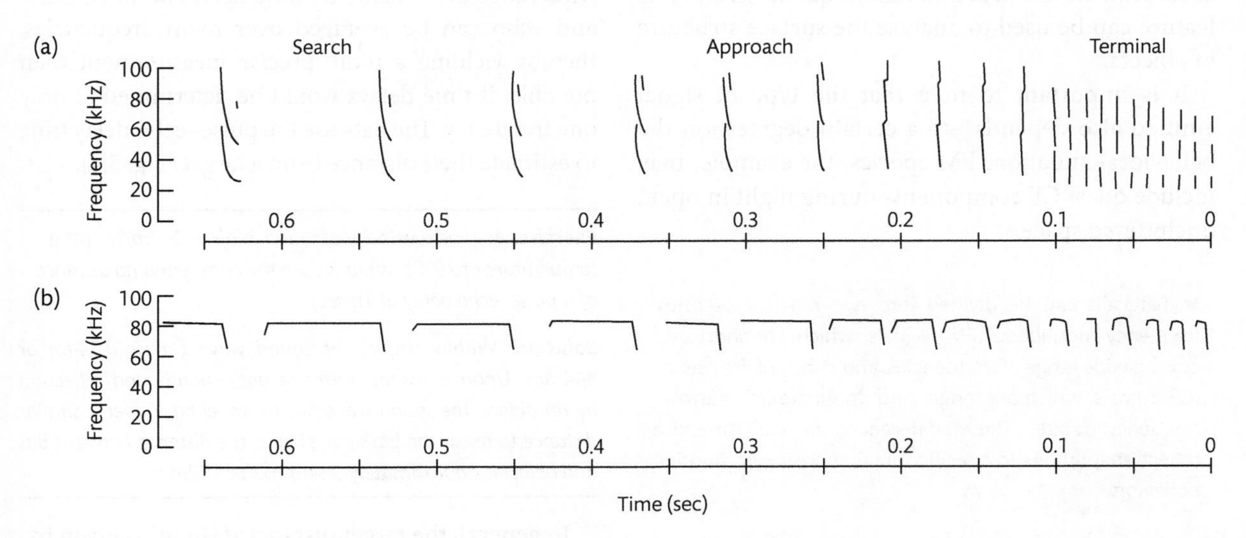

Figure 5.3 Sonograms of echolocation signals emitted by (a) the big brown bat (*Eptesicus fuscus*) and (b) the greater horseshoe bat (*Rhinolophus ferrumequinum*). The signals were recorded while the bats were approaching an insect in flight. Each of the two sequences includes pulses produced during the search, approach, and terminal capture stage. The timescale is plotted in such a way that it counts down to the time of capture at zero. The sonograms illustrate the different types of signals produced and the distinct signal patterns occurring in the different stages of prey hunting. (After: Simmons, J. A., Fenton, M. B., and O'Farrell, M. J. (1979).)

hunting behaviors. Long narrowband signals are typically found in species that forage in open spaces, where the bats use them for long-range echolocation. As will be described in detail below (see 'Doppler shift analysis'), CF signals are also useful for characterization of the target, if the bat analyzes the amplitude and frequency modulations in the echoes caused by the movements of the target. Narrowband signals are, however, less suited to measuring the distance between the bat and the target. The reason is that distance is encoded in the time delay between the emitted signal and the returning echo (see 'Distance estimation'). However, when using narrowband signals, the exact time points of sound emission and echo reception are difficult to determine because such signals stimulate the corresponding auditory receptors for prolonged periods.

Brief broadband signals are predominantly used by species that hunt near the ground or in an environment rich in denser vegetation. The bats employ the latter category of signals to distinguish prey from background clutter, and for measuring echo travel times (see 'Distance estimation'). The echo of broadband signals also encodes information about the properties of the object. For example, the texture of objects has a different effect on the absorption of sound at different frequencies, and this feature can be used to analyze the surface structure of objects.

It is important to note that the type of signal emitted also depends—to a certain degree—on the behavioral situation. FM species, for example, may include quasi CF components during flight in open, unculttered space.

> ➤ Bat calls can be divided into two major categories: frequency-modulated (FM) signals, which are short and cover a wide range of frequencies; and constant-frequency (CF) signals, which are longer and dominate in a narrow-frequency range. These differences in call structure reflect adaptations to specific foraging areas and hunting behaviors.

In the following two sections, we will describe in detail two aspects of the neuroethology of echolocation in bats—the mechanisms underlying distance estimation and the neural substrate of the so-called Doppler shift analysis.

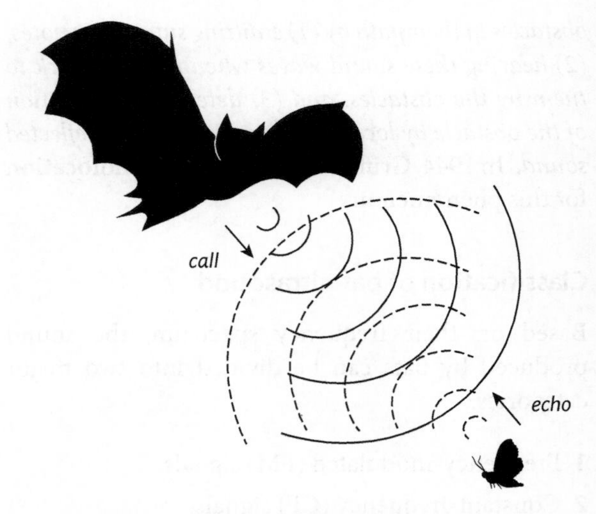

Figure 5.4 Estimation of target distance by an echolocating bat. The bat computes this distance based on the pulse–echo delay. (After: Kössl, M., Hechavarria, J. C., Voss, C., Macias, S., Mora, E. C., and Vater, M., © 2013 Elsevier Ltd.)

Distance estimation

Theoretical considerations have suggested that FM signals are well suited to measuring the time delay between the emitted pulse and the returning echo. First, because of their short duration, FM signals function as well-defined time stamps. Second, since they cover a wide range of frequencies, time delays between pulse and echo can be averaged over many frequencies, thereby yielding a more precise measurement than possible, if time delays would be determined at only one frequency. The bats use the pulse–echo delay time to estimate their distance from a target (Fig. 5.4).

Exercise: *Assume a velocity of sound in air of 343m/sec (at a temperature of 20°C). What is then the corresponding distance of a pulse–echo delay of 1msec?*

Solution: *Within 1msec, the sound wave travels 0.344m or 344mm. Upon emission from the bat's mouth and reflection by an object, the sound wave has to travel back over a similar distance to reach the bat's ear. Hence, the distance between bat and object is approximately 344mm/2 or 172mm.*

In general, the target distance d (in m) is given by

$$d = \frac{tc}{2},$$

where t is the pulse–echo delay time (in sec) and c is the velocity of sound (in m/sec).

Figure 5.5 James A. Simmons. (Courtesy of James Simmons.)

Behavioral experiments by James A. Simmons (now at Brown University in Providence, Rhode Island) (Fig. 5.5) on *Eptesicus fuscus* have shown that the animals, indeed, use signal-to-echo delays to estimate target distance. Simmons initially trained bats to discriminate targets placed at different distances. Then, he replaced the natural echo caused by reflection of the sound from the target by electronically controlled 'phantom echoes.' This experimental manipulation allowed Simmons to vary the delay between the emitted sound and the returning 'echo,' so as to minimize the acoustic distance between the two phantom targets. These experiments demonstrated that the threshold above which the bats could still discriminate two targets based on signal-to-echo delays was 60μsec. This corresponds to a difference in target distance of 10–15mm.

Further experiments have revealed that bats can even discriminate between a target fixed at a certain distance and a second jittering target with a temporal resolution as good as 1μsec. This is equivalent to 1/1000 of the duration of a 'typical' action potential, and corresponds to a forward and backward movement of the second target of just 200μm—a more than astonishing ability! How the central nervous system of the bat achieves this enormous temporal resolution is unknown.

Neurophysiological recordings have revealed neurons within the bat auditory system (particularly the inferior colliculus, the medial geniculate body, and the auditory cortex) whose properties make them well suited to measuring the delay between the emission of a pulse and the arrival of the returning echo. These combination-sensitive neurons respond best to pairs of FM sounds by encoding for the time interval between the emitted pulse and the returning echo. They are, therefore, called **delay-tuned neurons** or **FM–FM neurons.** A characteristic feature of these neuronal cells is that each of them is tuned to a rather narrow range of pulse–echo delays, and to a particular frequency within the FM sweep (and to the same frequency of the returning echo). This property enables different neurons to lock on to the emitted pulse and the returning echo of different frequency components within one FM sweep. As a result, multiple measurements of the time delay can be taken. As mentioned above, this averaging over many individual measurements of the time delays increases the precision with which the distance between bat and target is estimated.

How do the bats know which of the two sounds within a pulse–echo pair is the emitted pulse, and which one is the returning echo? Presumably, they distinguish the two by their relative intensities—the returning echo is predictably far less intense than the emitted pulse.

Since both the pulse and the echo sweep extremely rapidly through a broad range of frequencies, the individual neurons tuned to specific frequencies have just enough time to mark the occurrence of the sweep with a single action potential. Figure 5.6 illustrates this response characteristic based on nine neurons recorded in the inferior colliculus, an important auditory structure in the midbrain. In this experiment, big brown bats were stimulated with a sequence of 54 pulse–echo pairs mimicking the search, approach, and terminal stages of a pursuit (Fig. 5.6(a)). The nine rows shown below in Fig. 5.6(b) indicate the responses of the nine neurons during 32 presentations. The time at which each action potential was generated is indicated by a dot. As evident from the raster plot, the timing of the action potentials produced across all of the nine FM–FM neurons reflects the timing pattern of the pulses and the echoes fairly well. However, as is also obvious, different neurons respond selectively to different phases of the pursuit sequence.

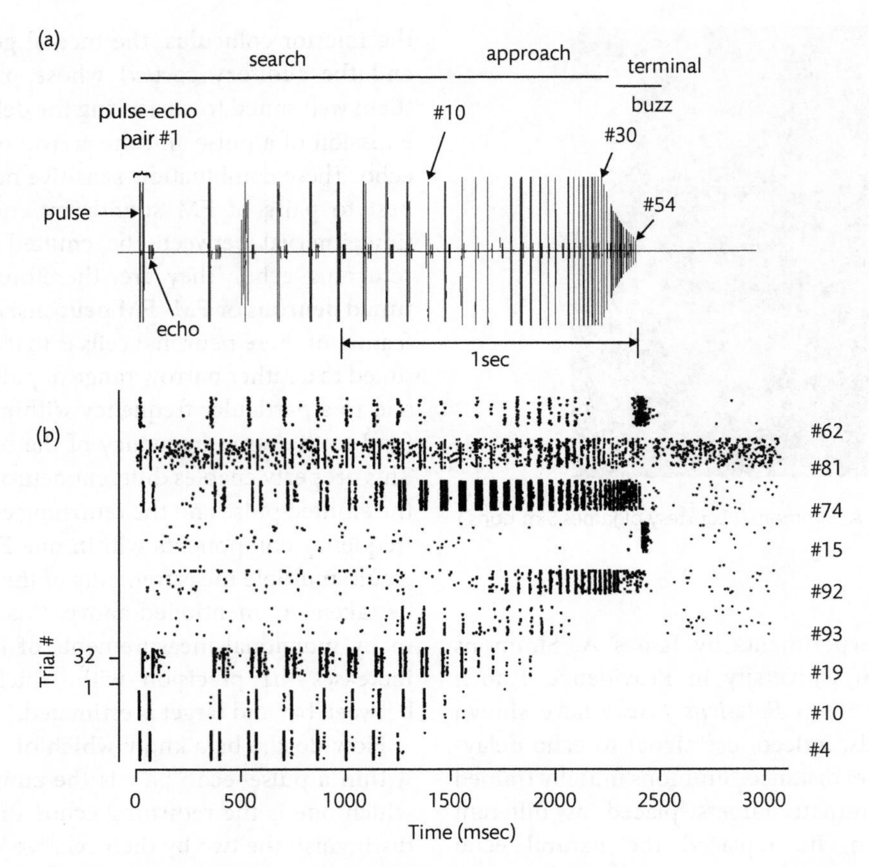

Figure 5.6 Responses of neurons in the inferior colliculus of big brown bats to a sequence of pulse–echo pairs simulating the search, approach, and terminal phases of a pursuit. (a) The sequence consists of a total of 54 pulse–echo pairs. Within each pair, the echo follows the pulse with a short delay. These pulse–echo delay intervals become shorter when the sequence progresses from the search to the approach phase, and from the approach to the terminal phase. The intensity of the echo is much lower than the intensity of the pulse. (b) Raster plots of the responses of nine neurons (numbered to the right) shown during 32 presentations of the stimulation sequence. The generation of each action potential by these neurons is indicated by a dot. (After: Simmons, J. A., © 2012 Elsevier Ltd.)

In the next step, the information encoded by the time-marker neurons in the inferior colliculus is—together with other pieces of information collected in lower auditory sensory structures—conveyed to the auditory cortex. There, different parameters of auditory signals are encoded in different, anatomically separated areas. One of these regions, the FM–FM **area**, processes information related to echo delays. The properties of the neurons in this area have been analyzed in great detail by Nobuo Suga (Fig. 5.7) of Washington University in St. Louis, Missouri. Suga, one of the pioneers of the neuroethology of bats, has carried out most of his studies on the mustached bat, *Pteronotus parnellii*, which emits CF signals with a downward FM component at the end.

Similar to the neurons in the inferior colliculus, neurons in the FM–FM area of the auditory cortex of *Pteronotus parnellii* respond poorly when a pulse, echo, CF signal, or FM sweep is presented alone but fire vigorously when a sound pulse is followed by an echo, at a particular delay time (Fig. 5.8). Detailed analysis has shown that they compare the emitted pulse with the delayed echo. These delay-sensitive neurons are arranged in a topographic fashion, with the delay time increasing along one axis. The range of delays represented along this axis varies from 0.4 to 18msec, which corresponds to target distances of between 7 and 310cm. This is in excellent agreement with the range over which the bats react to prey under natural conditions.

Figure 5.7 Nobuo Suga. (From: *The History of Neuroscience in Autobiography*, Volume 6. edited by Squire (2009). By permission of Oxford University Press, U.S.A.)

➤ FM signals are particularly well suited to estimating the distance between the bat and a target. This is achieved by computation of pulse-echo delay times. Combination-sensitive neurons in several brain areas play a critical role in this task by responding best to pairs of FM sounds encoding specific pulse-echo delay times.

The FM–FM area of the auditory cortex of bats is a special case of a **computational map**. Such maps play a key role in the information processing by the nervous system. In these maps, the values of a computed parameter (in the FM–FM area: the pulse–echo delay) vary systematically across at least one dimension of the neural structure. We will discuss such computational maps in more depth in Chapter 7.

Doppler shift analysis

As mentioned above, the echolocation signals of *Rhinolophus ferrumequinum* consist of a rather long (typically 10–100msec) CF component with a frequency of 83kHz, followed by a short downward-sweeping

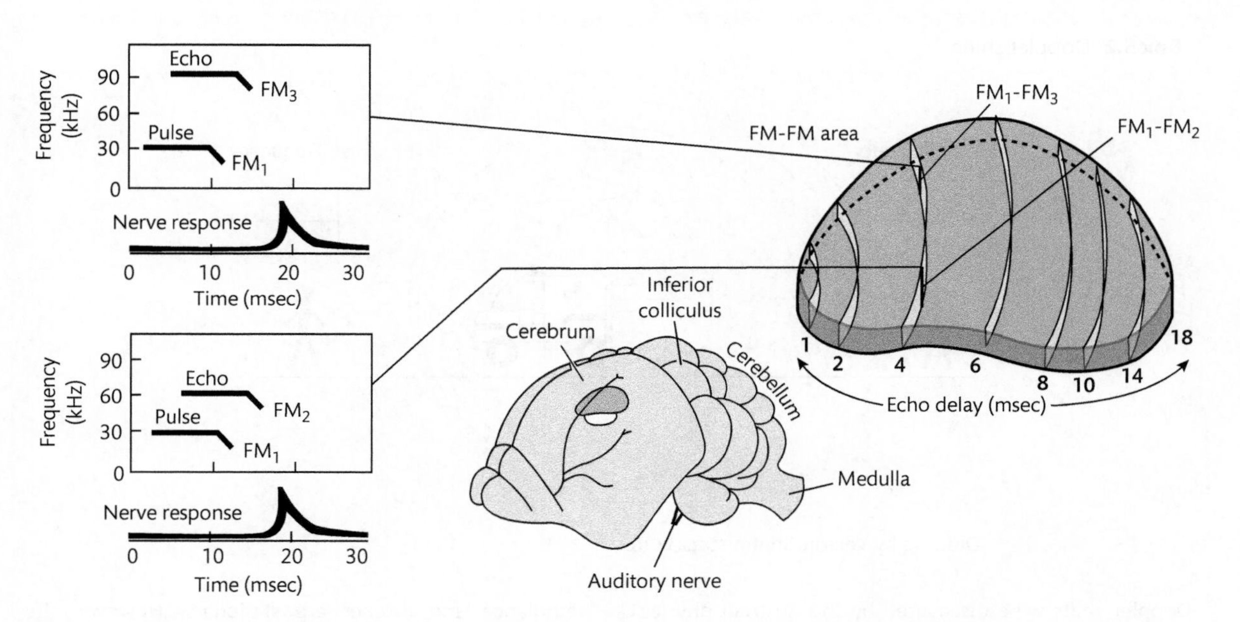

Figure 5.8 Auditory cortex of the mustached bat. The location of the FM–FM area (*dark gray*) is enlarged on the right. Neurons within this computational map are arranged in a topographic fashion, according to their responses to pulse–echo delay times. The echoes contain, in addition to the fundamental frequency (FM_1), higher harmonics, including FM_2 (at two times the fundamental frequency) and FM_3 (at three times the fundamental frequency), which encode information about the target. The importance of these higher harmonics in the echo is reflected by a systematic organization of neurons in the FM–FM area that respond best to combinations of the fundamental frequency of the pulse and the higher harmonics (FM_2, FM_3) of the returning echo. Characteristic responses of such combination-sensitive neurons are shown on the left. (Courtesy of Patricia J. Wynne.)

FM component (typically lasting a few milliseconds). Theoretical considerations, similar to the ones used in the interpretation of radar signals, suggest that the long CF part of the signal could be well suited to a so-called **Doppler shift analysis.** Doppler shifts occur whenever the source of a sound and the receiver of this sound are in relative motion towards one another (Box 5.2). The bat experiences such Doppler shifts under two behaviorally relevant situations.

The first situation arises when the bat emits a pulse of 83kHz while it is flying. The perceived echo returning from an object that the bat is approaching is higher than 83kHz, usually between 83 and 87kHz. This causes a problem, as the bat's acoustic system is most sensitive to sound at 83kHz, and the range of tuning to this frequency is quite narrow. To compensate for this loss in sensitivity, the bat lowers the frequency of the sound emitted as soon as it detects a positive Doppler shift, that is, an increase in frequency of the returning echo compared to the internal reference frequency. The adjustment in emitted frequency is made according to the magnitude of the Doppler shift. This behavioral response, called **Doppler shift compensation**, was discovered in 1968 by Hans-Ulrich Schnitzler (now at the University of Tübingen, Germany). It ensures that the frequency of the echo is kept in the range of most-sensitive hearing. In addition, the lowering of the frequency of the emitted sound helps *Rhinolophus* to protect itself from deafening with its own very intense sounds, as the frequency region just below 83kHz is distinguished by a remarkably high hearing threshold.

➤ CF bats experience Doppler shifts in the returning echo of their calls when flying towards a target. To keep the frequency of the echo in the range of most-sensitive hearing, they perform Doppler shift compensation by lowering the frequency of the emitted pulse.

A second situation in which *Rhinolophus* makes use of Doppler shift analysis occurs during prey detection. *Rhinolophus* usually hunts for insects, mainly flying moths and beetles. The fluttering of

Box 5.2 Doppler shifts

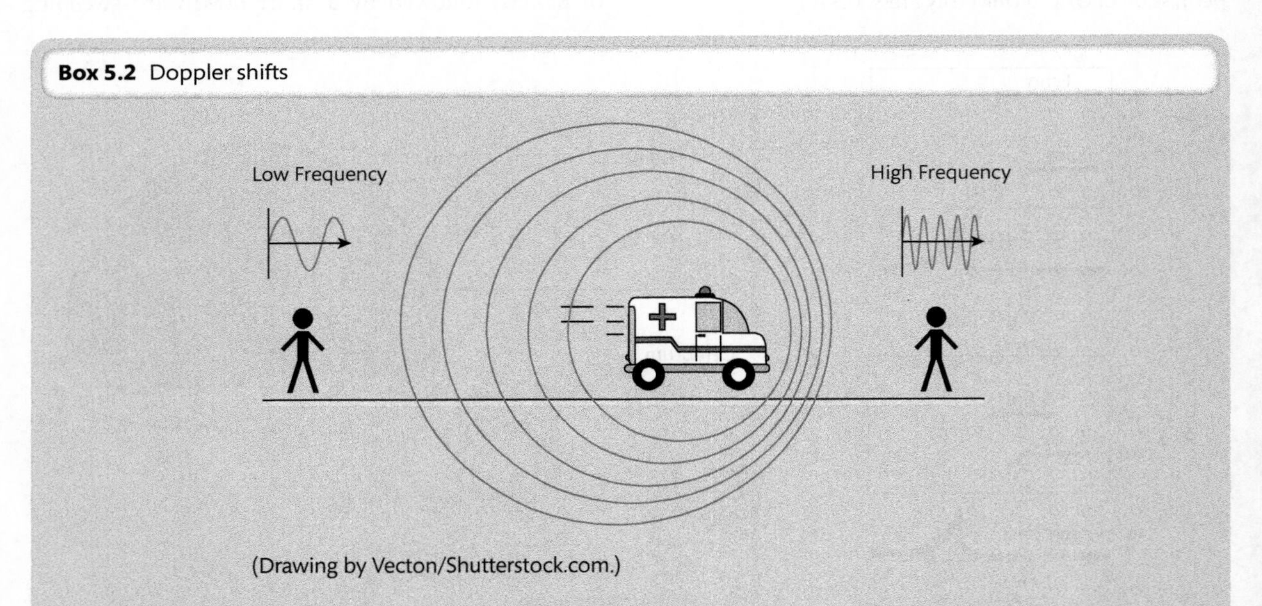

(Drawing by Vecton/Shutterstock.com.)

Doppler shifts were discovered by the Austrian physicist Christian Johann Doppler (1803–1853) in 1842. They involve a shift in frequency and occur whenever the source of any kind of propagating energy, such as light or sound, moves in relation to the receiver of this energy. For example, the siren of an ambulance is perceived at a constant pitch (= frequency) by a listener as long as both ambulance and listener are stationary. However, the pitch will be different when the ambulance moves. As the ambulance approaches, the frequency is higher, because sound waves from the siren are compressed toward the observer. On the other hand, when the ambulance moves away, the frequency is lower as the sound waves are stretched relative to the listener.

the wings of the prey animal produces Doppler shifts in the echo frequency, which, despite their small size, can be detected by the bat. Figure 5.9 shows an oscillogram and real-time spectra of echoes returning from a flying noctuid moth. Comparison of the oscillogram and the spectra with those of a stationary moth demonstrates that the movement of the wings causes strong rhythmical changes in amplitude and spectral composition, including an asymmetrical broadening of the frequency of the echo compared with the emitted sound.

Unlike the Doppler shifts caused by the relative motion of an object in the flight path of the bat, *Rhinolophus* is unable to compensate for the Doppler shifts produced by the fluttering of targets. The latter types of Doppler shift occur at each wing beat of the insect, thus making them too fast for the vocal system of the bat to follow. The frequency of the echo returning from a flying insect is, therefore, continuously oscillating up and down around 83kHz.

Can *Rhinolophus* detect the rather small Doppler shifts originating from the flutter of an insect's

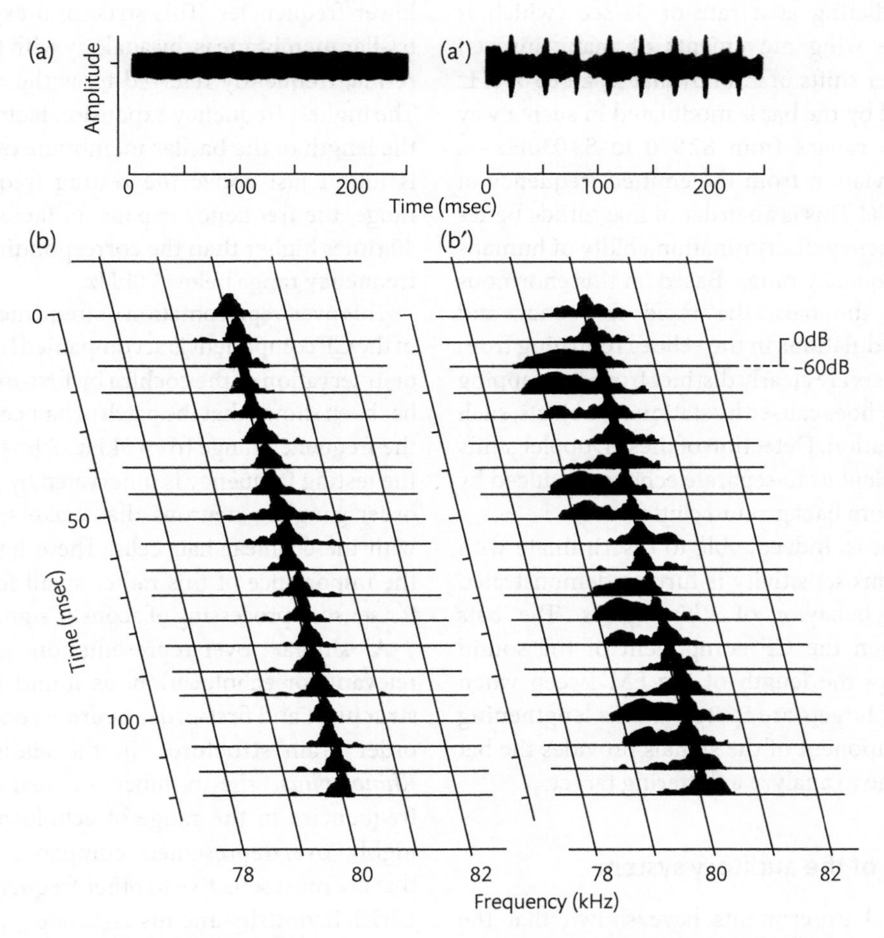

Figure 5.9 Frequency and amplitude modulations of the echolocation signals of the greater horseshoe bat caused by flying moths. For the analysis, a noctuid moth was mounted in the acoustic beam of a loudspeaker transmitting a continuous sinusoidal signal at a frequency of 80kHz. (a) Oscillogram of the echo returning from the moth *Phytographa gamma* at rest. (a′) Oscillogram of the echo returning from the flying moth. The clearly visible amplitude modulations are caused by the fluctuations of the reflection diameter of the moth, which originate from the movements of the wings. (b) Real-time spectra of the echoes returning from the moth at rest. The echoes exhibit single peaks at the transmitted frequency, 80kHz. (b′) Real-time spectra of the echoes returning from the flying moth. The spectra are asymmetrically broadened toward lower frequencies. These broadenings occur in the rhythm of the wing beat of the moth. (After: Schnitzler, H.-U. and Ostwald, J. (1983).)

wings? In order to answer this question, Schnitzler and his group trained bats to discriminate between a sinusoidally oscillating target and a similar but motionless target. By controlling the amplitude and frequency of the target's oscillations, defined modulations of the returning echo could be produced. After a training phase, the bats were able to discriminate between an oscillating and a motionless target. The magnitude of the oscillations was then reduced to determine the discrimination threshold. The results of these experiments demonstrated that the bat's detection system is extremely sensitive. A target oscillating at a rate of 35/sec (which is typical of the wing movements of many insects) causes Doppler shifts of ±30Hz, that is, an 83 000Hz signal emitted by the bat is modulated in such a way that the echo ranges from 82 970 to 83 030Hz—a maximum deviation from the emitted frequency of less than 0.04%! This is an order of magnitude better than the frequency discrimination ability of humans in the best frequency range. Based on this enormous sensitivity of the bats, the rapid frequency and amplitude modulations in the echoes returning from fluttering insects are clearly distinct from overlapping background echoes caused by stationary objects, such as dense vegetation. Detection of these Doppler shifts enables *Rhinolophus* to separate echoes produced by prey targets from background clutter.

That the bat is, indeed, able to discriminate with this tremendous sensitivity is further demonstrated by a second behavior of *Rhinolophus*. The bats always lengthen the CF component of the sound while reducing the length of the FM sweep when oscillations of targets are imitated. This lengthening of the CF component of the signals provides the bat with more time to analyze a fluttering target.

Adaptations of the auditory system

The behavioral experiments have shown that the sound-transmitting structures of *Rhinolophus* are specialized so as to keep the frequency of the returning echo in an 'expectation' window of 83kHz, as long as the Doppler shifts do not occur at too high a rate. Is this specialization of the sound-transmitting structures paralleled by a specialization of the sensory and neural structures receiving and processing the echoes?

In most mammals, the frequency of sound encoded by displacement of the basilar and tectorial membranes at specific places on the cochlea decreases on a logarithmic scale with increasing distance from the oval window (Box 5.3). However, in *Rhinolophus*, similar to other bats that use a long CF component in their echolocation signals, this representation of frequencies on the basilar membrane exhibits a remarkable specialization. Around the place where 83kHz is represented (roughly in the region of 80–86kHz), the basilar membrane is greatly expanded in terms of both length and thickness compared with lower frequencies. This structural expansion of the basilar membrane is, by analogy with the fovea of the retina, frequently referred to as the **acoustic fovea**. The highest frequency expansion factor, expressed as the length of the basilar membrane over one octave, is found just above the resting frequency. In this range, the frequency expansion factor is more than 40 times higher than the corresponding factor in the frequency range below 70kHz.

This over-representation of frequencies in the range of the CF component is accompanied by a high density of innervation of the cochlea by first-order neurons. It has been shown that the patch of hair cells representing the frequency range from 3kHz below to 3kHz above the resting frequency is innervated by 21% of all first-order ganglion neurons that make synaptic contact with the cochlear hair cells. These figures underline the importance of this rather small frequency range for sensory processing of acoustic signals.

A similar over-representation of frequencies relevant for echolocation, as found in the sensory structures and first-order neurons, occurs in higher-order brain structures. In the auditory cortex of *Rhinolophus*, the number of neurons tuned to frequencies in the range of echolocation signals is highly over-represented compared with neurons that are most sensitive to other frequencies. As Hans-Ulrich Schnitzler and his associate Joachim Ostwald found, these neurons can be divided into two groups: one that encodes frequencies within the 'expectation window' of the CF component, and the other with a frequency representation in the range between the resting frequency of the CF component and approximately 70kHz, which covers the bandwidth of the FM sweep. Schnitzler and Ostwald called these regions the 'CF area' and 'FM area,' respectively.

Box 5.3 Encoding sound frequency in the inner ear: The place theory of hearing

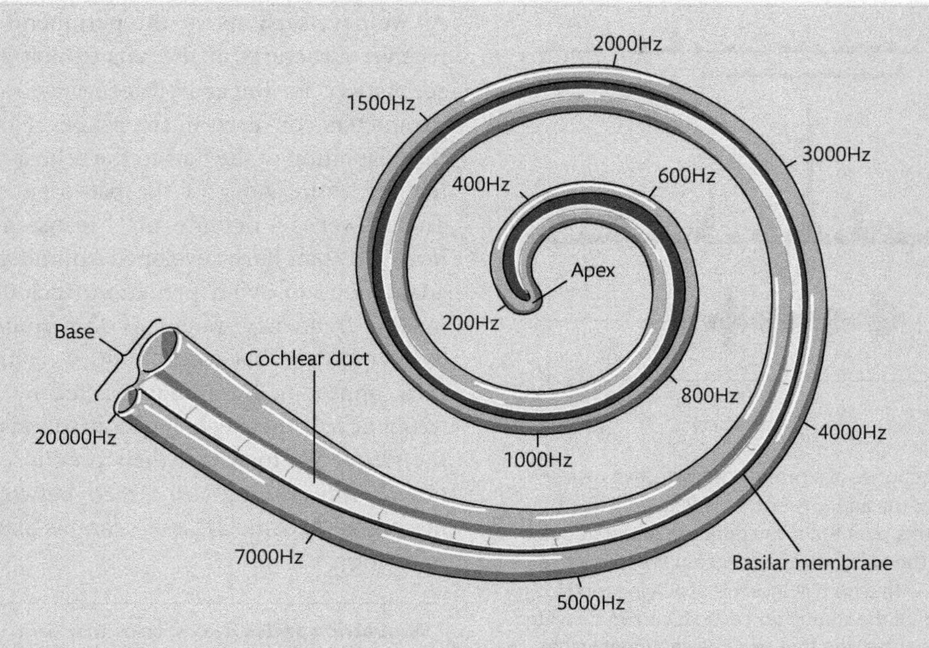

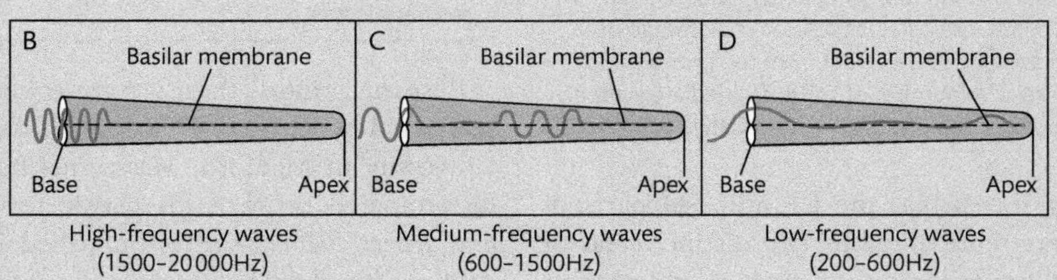

(Drawings by Encyclopaedia Britannica/UIG via Getty Images.)

In mammals, sound pressure waves are transmitted down the external auditory canal of the outer ear to the eardrum or tympanum. There, the sound pressure waves cause the tympanum to vibrate at the same frequency (in case of a pure tone) or frequencies (in case of more complex sound) as the sound wave. These vibrations are transmitted from the outer to the inner ear via the three middle-ear ossicles—the malleus, incus, and stapes. The footplate of the stapes adheres to the oval window, a patch of membrane that vibrates in concert with the vibrations of the tympanum and the three ossicles. Vibrations of the oval window cause vibrations in the cochlear fluid of the scala vestibuli, one of the three chambers of the cochlea. These vibrations travel along the coils of the cochlea in the scala vestibuli to the apex, where they enter the scala tympani. In the latter structure, the vibrational waves are propagated back down the coils of the cochlea. The third cochlear duct, the scala media, is positioned between the scala vestibuli and the scala tympani. It contains the auditory receptor cells—the inner and outer hair cells.

The hair cells are arranged in several rows on the so-called basilar membrane. Their stereocilia project upward, where at least some of them make contact with the tectorial membrane. Vibrations in the cochlear fluid induce displacement of the basilar and tectorial membranes, which, in turn, produces a shear force on the hair cells, including their stereocilia. Like in any other hair cell and as described in detail in Chapter 4, this deflection leads to a systematic change in the membrane potential of the hair cells.

How are the various parameters of sound, particularly frequency, encoded by the cochlea? Studies by the Hungarian scientist and engineer Georg von Békésy (1899–1972), in the 1930s and 1940s, showed that different frequencies of sound exert their maximal effect at different points along the basilar membrane. High frequencies lead to a maximal displacement of the basilar membrane near the oval window, and low frequencies closer to the apex. This encoding of sound frequency, by the position of the hair cell along the basilar membrane, is referred to as the **place theory of hearing**.

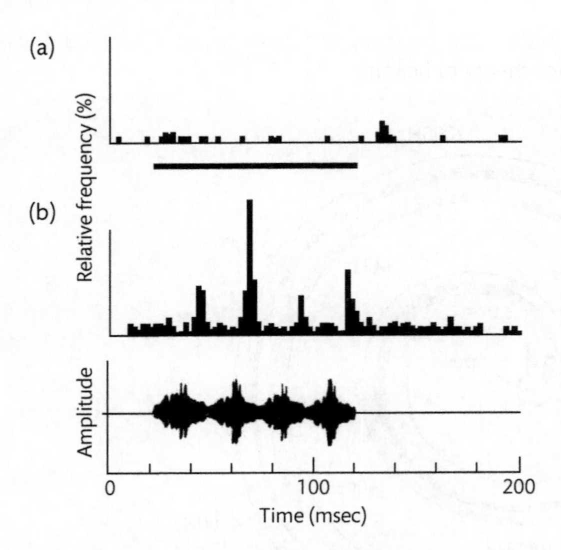

Figure 5.10 Response of a moth-echo-selective neuron in the CF area of the auditory cortex of *Rhinolophus*. (a) Stimulation with a pure tone. The duration of the stimulus is indicated by the horizontal bar beneath the histogram. (b) Stimulation with a mimicked echo of a flying moth. The oscillogram of the stimulation echo is shown beneath the histogram. Each of the four wing beats, evident from the amplitude peaks, causes a phase-locked discharge of the neuron. This response is evident from the pronounced increase in the relative number of spikes immediately after the wing beat. (After: Schnitzler, H.-U. and Ostwald, J. (1983).)

This finding shows that the CF and FM parts of the echoes are represented by two distinct areas of the auditory cortex. Such a distinct representation suggests that these two areas of the auditory cortex analyze different types of information contained in the different parts of the echo.

Further physiological experiments revealed neurons within the CF area of *Rhinolophus* that are selectively responsive to signals mimicking the echo caused by a flying moth. Figure 5.10 shows such a response. As a control signal, a pure sine-wave tone was used. No response of the neurons was evident to control signals (Fig. 5.10a). However, as soon as the pure tone was modulated so as to reproduce moth echoes, a clear response became apparent (Fig. 5.10b). Each wing beat in the moth echo caused a phase-locked discharge of the moth-echo-selective neurons.

➤ Doppler shifts in the echo frequency are also produced by the fluttering of wings of prey insects. Neurons in the CF area of the auditory cortex respond selectively to such modulations.

Counter-sonics: The prey's adaptations

As we have seen above, the peripheral and central receiver structures of the bats exhibit a number of adaptations for the analysis of behaviorally relevant parameters conveyed in the echoes. However, does the adaptation of the bats to the echoes of their prey tell the entire story of the bat–prey relationship? The answer is a definite 'no,' because many insects hunted by bats have developed a number of counter-adaptations to avoid predation, including hearing in the frequency range of the sounds emitted by **sympatric** bat communities. Some bats, in turn, appear to have counteracted by moving the frequency of their echolocation sounds away from the frequency to which their prey insects are best tuned. The ensuing 'arms race' between bats and insects has become a classic example of co-evolution in biology.

Sympatric species Two species that exist in the same geographic area.

Research into these counter-strategies was pioneered by Kenneth D. Roeder (Box 5.4) of Tufts University in Medford, Massachusetts, and widely disseminated through his classic text *Nerve Cells and Insect Behavior*, first published in 1963. The studies carried out by Roeder and other investigators suggest that **tympanic organs**, with sensitivity in the ultrasonic frequency range, have evolved as an anti-bat strategy multiple times in various insect taxa, including lepidopterans, beetles, mantids, lacewings, crickets, mole crickets, katydids, and locusts.

➤ The counter-strategies of various insect groups to avoid predation by bats are based on tympanic organs sensitive in the ultrasonic frequency range.

The high-frequency-sensitive tympanic organs are found in various parts of the body, including thorax, abdomen, and mouthparts, and are thought to have arisen from vibration-sensitive proprioceptors. The latter sensory organs are particularly well pre-adapted for the detection of airborne sound, including ultrasound emitted by echolocating bats. Despite the high diversity of these ears in body location and morphology, their physiological properties are quite

Box 5.4 Kenneth D. Roeder

Kenneth D. Roeder. (From: Dethier, V. G. (1993). Kenneth David Roeder. In *Biographical Memoirs* Vol. 62 (ed. National Academy of Sciences), pp. 350–366. © 2017 National Academy of Sciences.)

In an essay published two years before his death, Donald R. Griffin referred to bat echolocation as a 'magic well'—hiding countless secrets yet to be discovered. These secrets are related, as Griffin stressed, not only to the adaptation of the echolocation of bats for insect hunting but also to the countermeasures against bat predation. It was a contemporary of Griffin, Kenneth David Roeder, who established research in the latter area.

Kenneth D. Roeder was born in Richmond, a suburban town of London, England, in 1908. His father introduced the young boy to collecting insects. Triggered by this early activity, the interest in this animal group stayed with Kenneth for the rest of his life. He studied classical zoology at the University of Cambridge, from which he graduated with Bachelor's and Master's degrees. After working for a brief period as a teaching assistant at the University of Toronto, he was appointed instructor in biology at Tufts College (now Tufts University) in Medford, Massachusetts in 1931.

At about the same time, the zoologist George Howard Parker invited him to become one of his graduate students at Harvard University, but Roeder declined this offer. As unusual as this decision might seem, it reflected Roeder's unorthodox approach to science. As he explained later, he would not have had any fun carrying out experiments that someone else told him to do. To him, freedom to play (an alternate expression he sometimes used for doing research) was more important than pursuing a Ph.D. degree from Harvard. Nevertheless, even without a doctorate, but based upon his accomplishments, he progressed at Tufts through the faculty ranks to become finally Professor of Physiology in 1951. He received wide national and international recognition, including election to membership of the U.S. National Academy of Sciences, honorary member of the Royal Entomological Society of England, and member of the Deutsche Akademie der Naturforscher Leopoldina, Germany. (As a biographical footnote, it should be mentioned that he finally received an advanced graduate degree—a honorary doctorate from Tufts in 1952, the year after he was promoted to Full Professor!)

Roeder's research was characterized by taking observations from the field (many of which he made in the backyard of his home in Concord, Massachusetts) to the lab, where through physiological experiments he related neural activity, even at the single-cell level, to the behavior of the whole animal. He applied this approach to his studies at a time when behavioral science and neurophysiology were completely separate disciplines. He combined this approach with mastership in the design and construction of experimental setups. For example, he modified cameras to take photographs of moths, at night, while they evaded hunting bats. Based on these observations in the field, he examined the physiological properties of sensory receptor cells associated with the tympanic organ of moths, and compared these properties with the ultrasonic sound produced by the hunting bats.

Kenneth Roeder was also among the first physiologists who described spontaneous activity in the nervous system, and related this activity to the normal behavior of the organism. Until the 1960s, the predominant view was that such activity is just physiological 'noise', and behavior is exclusively controlled by reflex chains (see Chapter 6). Roeder, however, discovered, in the last abdominal ganglion of mantids and cockroaches, efferent neurons that fire spontaneously and innervate muscles of the copulatory apparatus. The endogenous discharges of these neurons are modulated by inhibition originating from the subesophageal ganglion. As Roeder showed, removal of this ganglion results in increased efferent activity and sexual behavior.

Roeder's research interests extended also to the study of the role of synaptic transmission in the control of behavior in insects. His expertise in this area led to some of the early investigations on the effect of the then newly introduced insecticide DDT, and to his appointment by President John F. Kennedy to a commission that explored the possibilities of

... continued

eradicating insect pests. Roeder was writing the commission's final report at the White House on the day Kennedy was assassinated.

Besides his research accomplishments, Roeder's influence was important in bridging the then wide gap between behaviorism in the U.S.A. and ethology in Europe. Throughout his life, he maintained personal contact with European ethologists, including Konrad Lorenz, Niko Tinbergen, and Erich von Holst. Through his familiarity with both American and European cultures his role was key in the move toward a more unifying understanding of animal behavior. Equally important, his demonstration of the power of combining ethology and neurophysiology, to explore neural mechanisms of animal behavior, facilitated setting the path for the establishment of neuroethology as a discipline in its own right.

In his last paper, entitled *Joys and Frustrations of Doing Research*, Kenneth Roeder gave a personal account of his approach to research in general and to his investigations of the moth–bat interaction in particular. This essay described not only the joy he had experienced in doing research, but also the recognition of the scientist's limitations in finding the truth. In concluding his essay, he wrote that *'the doing of science is a very human endeavor, and the direction taken by the expanding edge of this logic framework is often influenced by human bias, insight, blindness, and imagination as well as by chance. When he is reporting research the scientist rightly attempts to discount these imponderables . . . But when he is doing research they play a vastly important part both in his success and his failures. Not to recognize and admit to, perhaps even to court, one's subjectivity at this time is to delude oneself; it is also to miss the special joy of scientific discovery and to reduce the adventure to a form of computation.'*

In 1979, three years after publication of this final reflection, Roeder died at his home in Concord.

similar. The high diversity of the tympanic organs suggests that they have evolved independently, yet the similarities of these organs in physiology point to convergence in response to similar selective pressure.

The sensory core structure of high-frequency-sensitive tympanic organs consists of one to four **sound-sensitive receptor cells.** The most sensitive cell, **A1**, is typically tuned to frequencies of 20–60kHz, but sometimes the upper limit can even exceed 100kHz. Although in the best-studied example, moths, hearing is less sensitive than the auditory system of bats, species of the former taxon can detect bats at distances significantly greater than those at which bats can detect moths. The reason for this seeming discrepancy is that moths need to detect 'only' the sound emitted by the bats, whereas the bats have to be able to perceive and process the much lower intensities of the returning echo. As a result, moths can detect bats at approximately 20–100 m, whereas bats most typically can detect moths only at distances of 1–10 m.

The **avoidance responses** of the insects include a number of behaviors. Airborne moths react to bat sound either by turning away from the sound source, or by performing erratic flight maneuvers, such as dives, loops, or spirals. Non-aerial moths respond to ultrasound by freezing all movement; this reaction reduces the risk of predation by gleaning insectivorous bats, which are specialized in catching insects on substrate.

Male praying budwing mantises (*Parasphendale agrionina*) perform different types of evasive behaviors upon stimulation by ultrasound, depending on the distance from the source of sound. This differential response has been elucidated through a collaborative effort by David D. Yager and Michael L. May of Cornell University in Ithaca, New York, and M. Brock Fenton of York University, Ontario (Canada). The three investigators found that, when the mantises are far from the source, their evasion maneuvers consist of simple turns, without changing the elevation. When, on the other hand, the mantises are closer to the sound source, the turns are accompanied by changes in elevation. Finally, when the mantises are very close to the ultrasound source, they frequently go into a steep, power dive (with or without a spiral component), which takes them rapidly to the ground, where they are less prone to detection by the bats.

Green lacewings (*Chrysoperla carnea*) exhibit a similar differential avoidance response as praying mantises do. They fold their wings in a 'V' fashion horizontally above the body and nose dive when they detect the cries of hunting bats. This early avoidance response usually takes them out of the bat's acoustic field. Measurements by Lee A. Miller and Jens Olesen of the University of Odense, Denmark, have shown that during the nose dive the air speed of the green lacewings increases 3–5-fold, compared to normal flying. If, however, the bat continues its attempts to catch the lacewing, the latter may resort to a last-chance maneuver consisting of a momentary extension of the wings, or a wing flip. These evasion strategies are likely to account for the rather low success rate of the bats of only 27%. The critical role of hearing in triggering these evasion strategies is underscored by experiments in which Lee Miller selectively destroyed the ears of green lacewings. The results demonstrated that lacewings that can hear the ultrasound buzzes of the bats have a 47% selective advantage in escaping the attacks of these predators, compared to earless green lacewings.

> ➤ The type of behavioral response of insects to avoid predation by bats frequently depends on the distance from the source of the ultrasonic sound.

Certain tiger moths (family Erebidae, subfamily Arctiinae) have developed an even more sophisticated anti-bat strategy; they emit high-frequency clicks when they hear the buzzes of bats. The peak intensities of these sounds are in the frequency range of 30–75kHz. The sound pressure levels of these clicks are as high as 119dB measured at a distance of 2cm from the sound-producing organs, the **tymbals**.

The tymbals of the tiger moths consist of convex blisters of cuticular plates (**episternites**) on either side of the third segment of the thorax (Fig. 5.11a). These plates are modified by thinning of the cuticule, giving them a translucent appearance. The surfaces of the tymbals are either smooth or possess small ridges referred to as **microtymbals**. Flexion of the muscles beneath the tymbals causes an inward depression of either the entire tymbal surface (in the non-striated type) or each microtymbal (in the striated type). Conversely, relaxation of the steering muscles results in restoration of the resting convex shape of the tymbal due to elastic recoil. The deformation of the tymbal leads to the production of discrete clicks, or in case of striated tymbal surfaces, trains of clicks (Fig. 5.11b). These clicks are generated in two phases, the first when the tymbal collapses, and the second when the tymbal relaxes. The two phases are separated by a brief interval of silence. The clicks are broad-banded (i.e., they cover a broad range of frequencies), with peak intensities typically in the ultrasonic frequency range, thus being reminiscent of the frequency sweeps of the bat calls.

A number of observations suggest that tiger moths use the ultrasonic clicks as a defensive strategy against predation by insectivorous bats. Three behavioral mechanisms have been hypothesized to mediate this strategy: startling of the bat; acoustic **aposematism**; and jamming of the bat's echolocation system.

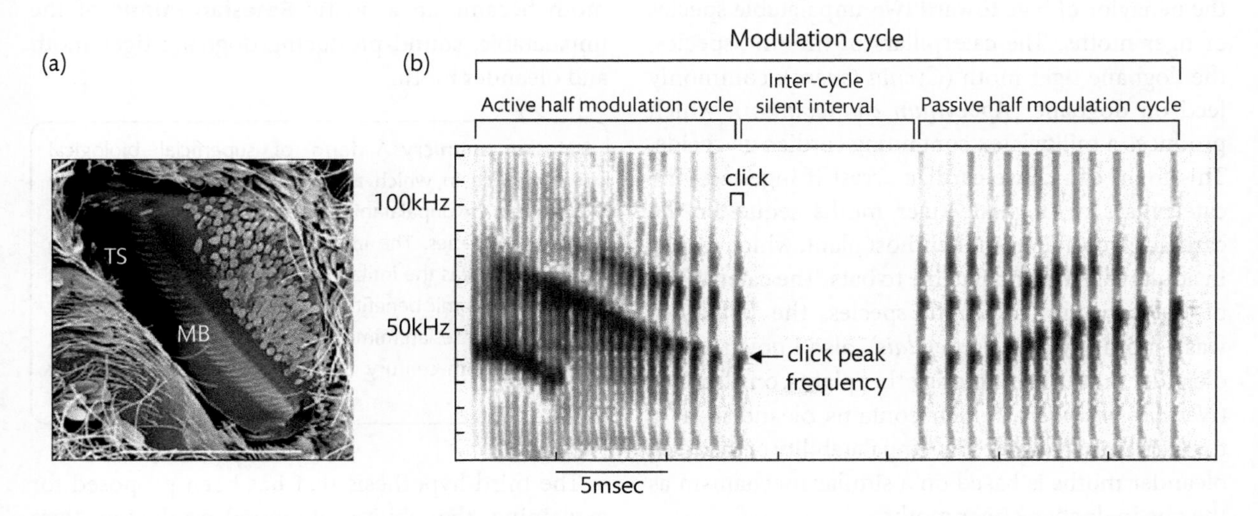

Figure 5.11 Ultrasound production by the tiger moth *Bertholdia trigona*. (a) Scanning electron micrograph of the tymbal organ. Part of the tymbal surface (TS) is adorned with small ridges, forming a microtymbal band (MB). (b) Spectrogram of the tymbal sound produced. The sequential buckling of the microtymbals results in a train of broad-band clicks. The peak intensities of these clicks are well in the ultrasonic frequency range. The sequences of clicks generated during the inward depression and the outward restoration of the tymbal surface are separated by a brief interval of silence. (After: Corcoran, A. J., Conner, W. E., and Barber, J. R. (2010). Anti-bat tiger moth sounds: Form and function. *Current Zoology* **56**: 358–369.)

Aposematism The use of a conspicuous signal, such as bright coloration, by an animal to advertise to potential predators that it is toxic or distasteful.

➤ Certain species of tiger moths generate with their tymbals ultrasonic clicks as a defensive measure against predation by sympatric insectivorous bats.

The first hypothesis suggests that the clicks produced by the tiger moths evoke a **startle reflex** in bats, thereby providing a momentary advantage to the tiger moths so that they can escape ('**startle hypothesis**'). However, if this mechanism plays, indeed, a role in the defense strategy of tiger moths, then it is likely restricted to naïve bats, because experiments have shown that bats quickly habituate to the tiger moth-produced clicks.

The second hypothesis states that bats learn to associate the clicks produced by tiger moths with defensive chemicals they produce *de novo* or sequester from their host plants ('**acoustic aposematism hypothesis**'). Several lines of evidence have supported the acoustic aposematism hypothesis. In one of these studies, Jesse R. Barber and William E. Conner of Wake Forest University, Winston-Salem, North Carolina (USA) examined the behavior of bats toward two unpalatable species of tiger moths. The caterpillars of the first species, the dogbane tiger moth (*Cycnia tenera*), commonly feed on dogbane (*Apocynum cannabinum*), which produces a milky latex containing cardiac glycoside. This toxin can cause cardiac arrest if ingested. The caterpillars of dogbane tiger moths sequester the cardiac glycoside from their host plant, which results in adults that are unpalatable to bats. The caterpillars of the second tiger moth species, the polka-dot wasp moth (*Syntomeida epilais*), also known as the oleander moth, feed, among other plants, on oleander (*Nerium oleander*), which contains oleandrin as a toxic cardiac glycoside. The unpalatability of the adult oleander moths is based on a similar mechanism as the one in dogbane tiger moths.

For their study, Barber and Conner presented adults of either dogbane tiger moths or oleander moths for five nights to naïve individuals of two bat species, the big brown bat (*Eptesicus fuscus*) and the red bat (*Lasiurus borealis*). The results for the presentation of each of the two unpalatable, sound-producing tiger moth species were virtually identical. Within a short time, the bats learned to associate the sounds produced by the unpalatable tiger moths with their bad taste and stopped hunting them. This observation provided strong evidence in favor of the acoustic aposematism hypothesis.

Based on these findings, Barber and Conner took the idea that the sound produced by the tiger moths warns the bats of their distastefulness one step further. Following the conditioning during the first five nights, they presented to the bats on nights 6 through 10 a novel moth species, the milkweed tiger moth (*Euchaetes egle*). Like the dogbane tiger moth and the oleander moth, the milkweed tiger moth produces click-like sounds, but unlike the former it is totally palatable. The bats captured the milkweed tiger moths less frequently than the control moths, suggesting that the bats were deceived by the sound produced by the palatable milkweed tiger moths. This notion was further substantiated when the investigators removed the tymbals from the milkweed tiger moths and presented them on night 11 to the bats. Now, all of the silenced moths were captured and eaten. Taken together, these results indicate that through quick learning by the bats in the course of the experiments the milkweed tiger moth became an **acoustic Batesian mimic** of the unpalatable, sound-producing dogbane tiger moth and oleander moth.

Batesian mimicry A form of superficial biological resemblance in which a noxious species, protected by its toxicity or unpalatability, is mimicked by a harmless, palatable species. The imitated species is known as the model, whereas the imitating species is referred to as the mimic. The mimic benefits because predators mistake it for the model. Batesian mimicry is named after its discoverer, the nineteenth-century English naturalist Henry Walter Bates.

The third hypothesis that has been proposed for explaining the ability of sound-producing tiger moths to reduce the risk of predation has become known as the **sonar jamming hypothesis**. The core idea of this hypothesis is that the click-like sound of the moths interferes with the bat's perception of the returning echoes from its own sound emitted. If this

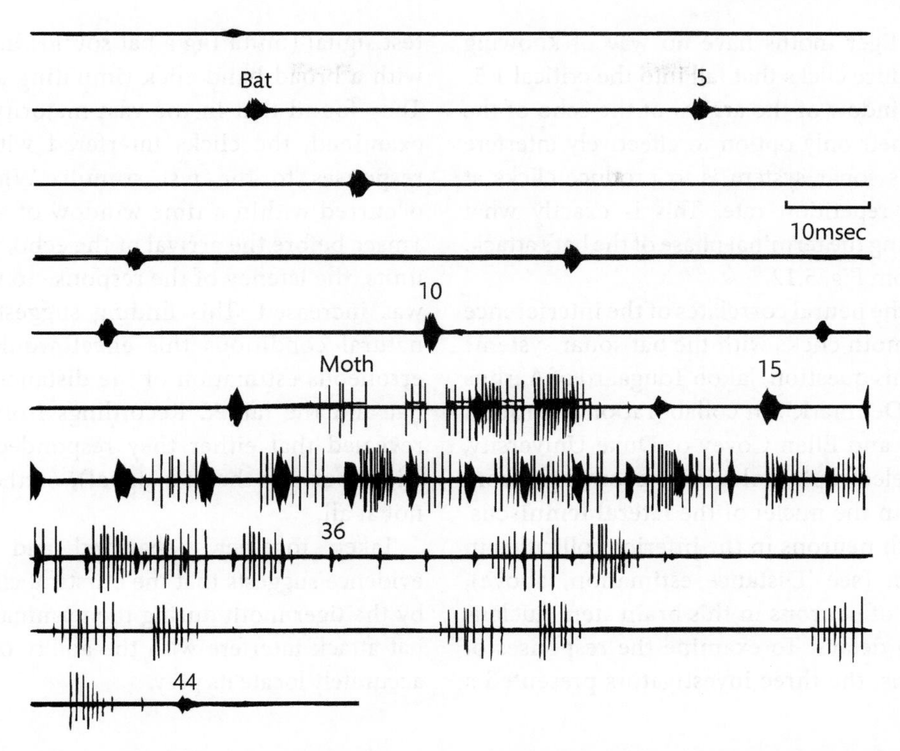

Figure 5.12 Acoustic response of a dogbane tiger moth to the attack echolocation response of a big brown bat. The moth was stationary and stimulated in the laboratory with the playback of a recorded sound sequence produced by a big brown bat during attack. This experimental design simulates an encounter of the moth with a bat. The oscillogram shows both the calls of the bat ('Bat') and the click responses of the moth ('Moth'). The numbers above some of the individual bat calls indicate the number of these calls in the sequence. The experiment shows that the moth produced its clicks during the terminal phase of the simulated bat's attack. (Fullard, J. H., Simmons, P. A., and Saillant, P. A. © The Company of Biologists Limited. (1994).)

hypothesis is correct, one would expect that the moth clicks are primarily generated during the later stages of the bat attack so as to maximize confusion—by simply not allowing the bat enough time to adjust its behavior.

Observations by James H. Fullard of the University of Toronto at Mississauga, Ontario (Canada) and James A. Simmons and Prestor A. Saillant of Brown University, Providence, Rhode Island have, indeed, confirmed this prediction. Dogbane tiger moths were presented with a recording of the echolocation sequence emitted by a big brown bat. The tiger moths did not produce sound in response to search or approach calls but waited to emit their clicks until the terminal phase of the attach (Fig. 5.12).

Psychophysical experiments conducted by Lee A. Miller (also using big brown bats, but a different tiger moth species, the ruby tiger moth *Phragmatobia fuliginosa*), have shown that the clicks emitted by the tiger moths greatly diminished the bat's precision to determine the distance to targets. When the burst of clicks produced by the tiger moths started within a time window of about 1.5msec before the echo of the bat sound returned, the bat's performance to discriminate differences in target distance dropped by a factor of as much as 40!

Behavioral observations, including three-dimensional flight-track analysis, indicated that bats exposed to a sufficiently large number of jamming tiger-moth clicks frequently missed their prey. Notably, their mean error was approximately 16cm—a value similar to the one observed in the target distance discrimination tests. The relatively low number of successful attacks occurred typically on moths that produced rather few clicks.

Since the tiger moths have no way of knowing when to produce clicks that fall into the critical 1.5-msec time window of the arrival of the echo of the bat sound, their only option to effectively interfere with the bat's sonar system is to produce clicks at a very high repetition rate. This is exactly what happens during the terminal phase of the bat's attack, as evident from Fig. 5.12.

What are the neural correlates of the interference of the tiger moth clicks with the bat sonar system? To address this question, Jakob Tougaard of Aarhus University (Denmark), in collaboration with John H. Casseday and Ellen Covey of Duke University, performed electrophysiological recordings from single units in the nuclei of the **lateral lemniscus**. Together with neurons in the inferior colliculus in the midbrain (see 'Distance estimation,' above), the majority of neurons in this brain stem nucleus encode echo delays. To examine the responses of these neurons, the three investigators presented a test signal (imitating a bat sound) in combination with a broad-band click (imitating a moth click). They found that in the vast majority of the units examined, the clicks interfered with the neural responses to the test stimuli. When the click occurred within a time window of approximately 1 msec before the arrival of the echo, then, in some units, the latency of the response to the test signal was increased. This finding suggests that under natural conditions this effect would lead to an erroneous estimation of the distance between the bat and the target. Recordings from other units revealed that either they responded only to the click (but not the test signal), or they responded not at all.

Taken together, behavioral and physiological evidence suggests that the bursts of clicks produced by the tiger moth during the terminal stages of the bat attack interfere with the ability of the latter to accurately locate its prey.

Summary

- Echolocation as an active orientation mechanism involves the emission of high-intensity sound, followed by perception and analysis of the returning echo, to negotiate the animal's surroundings.
- Although some animals use sound in the frequency range of human hearing, most bats produce ultrasound for echolocation.
- Due to its short wavelength, ultrasound is better suited than lower-frequency sound for the detection of small objects, although its disadvantage is stronger atmospheric attenuation.
- Based on the frequency spectrum, the sound produced by bats can be divided into two major categories—frequency-modulated (FM) signals and constant-frequency (CF) signals.
- FM signals are, among others, used to estimate the distance of the bat from an object. This is achieved by measuring the time delay between the emitted signal and the returning echo.
- In the inferior colliculus of bats, neurons have been identified that serve as time markers by encoding the time interval between the emitted pulse and the returning echo. In the auditory cortex, neurons exist that respond best to a particular pulse–echo delay time.
- The CF signal of bat ultrasound is particularly well suited for the so-called Doppler shift analysis, which is performed during flight and when localizing flying insects. During flight movements of the bat, this analysis forms the basis for Doppler shift compensation to keep the frequency of the returning echo in the range of maximal sensitivity of the auditory system. During prey hunting, Doppler shift analysis is used to detect the fluttering of wings of the flying insects.
- Sensory structures in the ear and neural structures in the brain of the bat are both specialized to maximize sensitivity to frequencies in the range of the CF component of the returning echo. This includes an

over-representation of such frequencies in the cochlea and auditory cortex.

- To counteract bat echolocation, several groups of insects have developed a number of adaptations. These include the development of ears sensitive in the ultrasound range and, in some tiger moths, the production of high-frequency clicks.

The bigger picture

In the last chapter, orienting behaviors were discussed that rely on passively perceived sensory information from the animal's environment. Although all animals use such information for orientation, some have in addition developed mechanisms through which they can actively probe the environment by self-generated signals. After interaction of these signals with objects in the environment, these signals serve as carriers of information about these objects. To use this information for short-distance navigation and other behavioral decisions, the animal must be able to extract information relevant to the behavioral task from the signal. The most intensively studied of such active orientation mechanisms is echolocation of bats. It has attracted substantial interest by neuroethologists, mainly due to the compelling demonstration of its behavioral functions, foremost the localization of insects that many bats prey on. A similar exemplary behavioral analysis has been applied to the counterpart of bat echolocation—the behavioral strategies used by many insects to reduce the risk of predation by bats. The comprehensive body of behavioral information about these two closely interrelated systems has greatly aided the identification of adaptations in the underlying sensory and motor pathways, making the echolocation of bats and the behavioral counter-strategies of insects the best characterized prey–predator model system of neuroethology.

Recommended reading

Brinkløv, S., Fenton, M. B., and Ratcliffe, J. M. (2013). Echolocation in oilbirds and swiftlets. *Frontiers in Physiology* **4**:123.

Outside the mammalian class, echolocation has been found, to date, only in two groups of birds—the oilbirds and several species of swiftlets. Although echolocation in these birds has by far not been studied as intensively as in bats, this review article provides some fascinating insights into the biology of oilbirds and swiftlets, the design principles of the acoustic signals used for echolocation, the role of this behavior for orientation and food detection, and its evolution.

Griffin, D. R. (1958). *Listening in the Dark. The Acoustic Orientation of Bats and Men.* Yale University Press, New Haven.

Although published more than half a century ago, this is still a superb book to read. Written by the discoverer of bats' ultrasonic sounds who was been not only an outstanding scientist, but also a stimulating writer.

Madsen, P. T. and Surlykke, A. (2013). Functional convergence in bats and toothed whale biosonars. *Physiology* **28**:276–283.

Echolocating bats and toothed whales have independently evolved biosonar systems for navigation and hunting. Despite the many differences between these two mammalian taxa and the media in which they live, the authors demonstrate a striking convergence in the way these animals use ultrasound to achieve these tasks.

Short-answer questions

5.1 What is the advantage of using ultrasound, compared to sound of lower frequencies, as an echolocation signal? What is the disadvantage of high-frequency sound for biosonar?

5.2 Contrast FM signals with CF components of echolocation signals in terms of their duration and range of frequencies covered.

5.3 Bats use measurements of the time delay between the emitted pulse and the return of the echo to estimate their distance from a target. Assuming a velocity of sound of 300m/sec, to what distance does a pulse–echo delay of 1msec correspond at night?

5.4 Why are FM signals particularly well suited for measuring the time delay between the emitted pulse and the returning echo?

5.5 To what physical properties of a sensory signal are FM–FM neurons in the inferior colliculus of bats tuned? How do these neurons respond to the occurrence of signals with such features?

5.6 A greater horseshoe bat emits echolocation signals while it is flying toward a stationary object. The CF component of the echolocation sound has a frequency of 83kHz. How does the motion of the bat in relation to the object affect the perceived echo frequency?

5.7 How does Doppler shift analysis help greater horseshoe bats to separate echoes produced by fluttering insects from background clutter?

5.8 Studies by Georg von Békésy, in the 1930s and 1940s, showed that different frequencies of sound cause maximal displacement at specific positions along the basilar membrane. What is the name of the theory that describes this effect?

5.9 What is the acoustic fovea in the inner ear of greater horseshoe bats? What is the functional adaptation of this structure for the perception of echolocation signals?

5.10 What sensory adaptations have various taxonomic groups of insects developed as a counter-strategy to avoid predation by echolocating insectivorous bats?

5.11 What strategy do certain tiger moths use to interfere with the ability of bats to accurately localize prey?

5.12 **Challenge question** The diet of bats is remarkably diverse and includes not only insects but, among others, in some species also fish. Laboratory experiments have, however, shown that these bats are unable to detect pieces of fish tissue or other objects in water, even if they are just a few millimeters below the surface of the water. Formulate a hypothesis how, then, piscivorous bats could detect fish.

Essay questions

5.1 How do bats that emit FM signals gauge distance to a prey animal? What neural substrate underlies this behavioral ability?

5.2 How does Doppler shift analysis help bats to detect prey? What adaptations of the bat's auditory system make this analysis possible?

5.3 Certain tiger moths emit high-frequency clicks as an anti-bat strategy. How are these signals produced? Describe the three hypotheses that have been advanced to explain how the ultrasonic clicks help the tiger moths to reduce the risk of predation by echolocating insectivorous bats. What experimental evidence has been obtained in support of these hypotheses?

Advanced topic Do bat calls play a role in communication?

Background information

The echolocation calls produced by bats are very loud, with sound pressure levels typically greater than 120dB at a distance of 10cm in front of the bat's head. As a consequence, these calls can be heard by other bats over large ranges—much larger than the significantly less intense echoes. It has been estimated that a pulse at a frequency of 20kHz could be perceived by other bats over 128m (for comparison, the bat would receive an audible echo of the same pulse reflected by a large insect over only 11m). As we have seen above, several groups of insects have developed hearing in the ultrasonic range to detect and evade marauding bats. The information extracted from the bat calls by the tympanic ears of the insects appears to be limited to the presence of the bat sound, its intensity, and, in some cases, the rates at which the individual pulses are generated. However, recordings of bat sounds in both the laboratory and the field have revealed that these vocalizations are a much richer source of information. Analysis has shown that they frequently vary among individuals of a given species, and that these variations may correlate with sex, age, body size, social group, and geographic location. These findings have been used as evidence to propose a communicatory function of the bat calls, in addition to their echolocation role.

Essay topic: Do the sounds produced by bats serve in communication?

A potential role of the vocalization of bats in communication, although an attractive hypothesis, is by far not as well documented as auditory communication in some other vertebrate taxa. When reading the two starter references, and possibly some of the original research papers cited in these review articles, consider first the stringency of the definition of communication used in these publications: Is the underlying assumption that certain calls, or certain properties of calls, have specifically evolved to optimize intentional intraspecific transmission of information from a sender to a receiver? Or does empirical evidence suggest that the bat echolocation calls, evolved for orientation and foraging, are transmitted inadvertently, and as such received by eavesdroppers? Then, in a major section of your essay, summarize some of the key findings that have demonstrated correlations of variations in bat vocalizations with individual- and colony-specific properties. While such correlative evidence is consistent with the idea of a role of bat calls in communication, playback experiments are required to verify a communicatory function. Describe the experimental design and the outcome of the rather few studies that have used this approach. Propose additional experiments to test the communication hypothesis of bat vocalization.

Starter references

Fenton, M. B. (2003) Eavesdropping on the echolocation and social calls of bats. *Mammal Review* **33:**193–204.

Jones, G. and Siemers, B. M. (2011). The communicative potential of bat echolocation pulses. *Journal of Comparative Physiology A* **197:**447–457.

Smotherman, M., Knörnschild, M., Smarsh, G., and Bohn, K. (2016) The origins and diversity of bat songs. *Journal of Comparative Physiology A* **292:**535–554.

To find answers to the short-answer questions and the essay questions, as well as interactive multiple choice questions and an accompanying Journal Club for this chapter, visit **www.oup.com/uk/zupanc3e**.

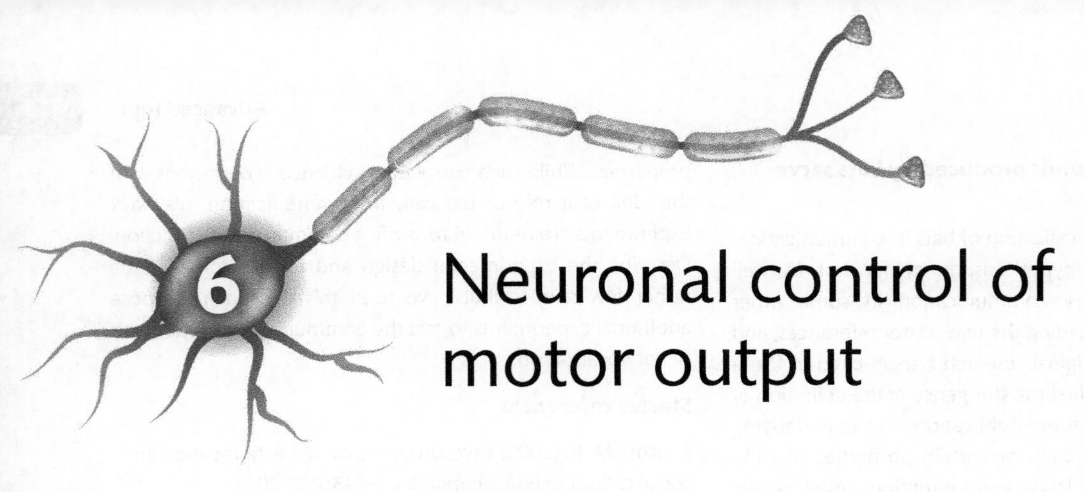

6 Neuronal control of motor output

Introduction

Although simple **reflexes** in adult mammals were described as early as 1906 by Charles Scott Sherrington, we still lack a comprehensive mechanistic understanding of how the contractions of the muscles are achieved in response to a tactile stimulus, or how simple rhythmic motor patterns, such as walking, are generated in mammalian species, including humans.

To avoid the difficulties associated with the complexity of the adult mammalian nervous system, behavioral neurobiologists have searched for model systems in which the neural circuits controlling motor patterns have a simple structure and function and are readily accessible to experimental analysis. This strategy has led to some remarkable success. One of the model systems established by using this approach centers around a simple behavior, escape swimming in hatchling *Xenopus* tadpoles. This behavioral response can be readily evoked through sensory stimulation and is controlled by a rather small number of defined neurons in the spinal cord. The research involving this amphibian organism will be presented in detail in the 'Model system' section of this chapter.

Key concepts

At the beginning of the twentieth century, by studying locomotion in a number of mammalian species (including dogs, cats, rabbits, and guinea pigs), several investigators addressed the problem of how the coordinated contractions of muscles are controlled at the neural level. The most prominent among them was Charles Scott Sherrington (for a biographical sketch, see Box 6.1), who found that **spinal preparations** of the animals were capable of executing movements with their hind legs, and that the behavioral pattern generated closely resembled that of natural stepping. This finding demonstrates that neural control by the brain is not required to perform this behavior. In the experiments, the so-called **reflex stepping** is elicited from the preparation by lifting it from the ground, with its longitudinal axis parallel to the ground and the hind limbs pendent. The stepping

Spinal preparation Animal with its spinal cord isolated from higher regions of the central nervous system by transection of the cord at its upper segments.

movement can be stopped by passively flexing one limb at the joint of the hip. Importantly, Sherrington was able to evoke this locomotory behavior even after severing all the afferent connections from the skin of the spinal preparation. Based on these observations, he concluded that the reflex stepping is triggered by input of **proprioceptors** in the muscles, including the hip flexor muscle, to the spinal cord.

Proprioception The perception of spatial orientation and movement by sensory receptors within the body, which are referred to as proprioceptors. Example: Muscle spindles in limbs are proprioceptors that provide information about changes in muscle length.

Although Sherrington did consider the possibility that the rhythm of stepping is under control of

Box 6.1 Charles Scott Sherrington

Charles Scott Sherrington. (Courtesy: Wellcome Library, London.)

To neuroscientists, Charles Scott Sherrington is known as the neurophysiologist whose pioneering work has shaped our understanding of functional integration of neurons within the central nervous system. Yet, his curiosity and ingenuity spanned far beyond neurobiology and neurology, and even the sciences.

Sherrington was born in London in 1857. The family and the school environment in which he grew up instilled in him a lifelong love of art and the classics. He studied medicine at St. Thomas' Hospital Medical School, London, and physiology at Cambridge University. During that time, he became interested in the structure and function of the brain and spinal cord, and together with the renowned Professor of Physiology at Cambridge, John Newport Langley, in 1884, he published his first paper. Nevertheless, Sherrington considered becoming a pathologist, and when cholera broke out, first in Spain in 1885, then in Italy in 1886, he went to those countries to perform autopsies and collect tissue specimens for pathological examination. Part of these examinations was done in the Berlin laboratories of Rudolf Virchow and Robert Koch—then leading authorities in bacteriology.

Despite his initial interest in pathology, Sherrington finally turned to neuroanatomy and neurophysiology. His academic appointments included professorships at the University of London, the University of Liverpool, and Oxford University. His research covered a wide range of topics, but, time and again, centered around nervous system functions. Contrary to the then prevailing notion that reflexes are the result of isolated reflex arcs, Sherrington showed that reflexes must be regarded as activities integrated within the central nervous system. He also demonstrated reciprocal innervation of muscles—when one set of muscles is stimulated, then the antagonistic muscles are simultaneously inhibited. This phenomenon has become known as Sherrington's law of reciprocal innervation. Sherrington is also credited with introducing several new terms, such as 'synapse,' 'proprioception,' and 'motor unit'. His book, *The Integrative Action of the Nervous System*, published in 1906, became a classic in the field of physiology. For his achievements, Charles Sherrington received numerous awards, including the Nobel Prize in Physiology or Medicine in 1932.

Throughout his academic career, Sherrington retained his interest in the classics, poetry, art, philosophy, and history. After he retired from Oxford, in 1936, he devoted his time and energy to the study of these disciplines. He published a book with a collection of his poems (*The Assaying of Brabantius and other Verse*) and several books with his prose writings, including *Goethe on Nature and on Science*, *The Endeavour of Jean Fernel*, and *Man on his Nature*. In the latter, he explored the nature of mind, essentially adopting a dualistic position (on the problem of mind-body dualism, see Chapter 3). This puzzled many of his fellow scientists who felt that it was his neurophysiological work that had paved the path to ultimately explaining how the brain functions as the organ of the mind. Despite this criticism, Sherrington held, up to his death in 1952, that science might be unable to answer the question that Aristotle had already raised two thousand years earlier: 'How is the mind attached to the body?'

neural activity in the spinal cord, overall he favored a model that explains the rhythm and pattern of hind limb stepping movements primarily as the result of sequential hind limb reflexes. As an irony of history, it was one of his associates, Thomas Graham Brown (for a biographical sketch, see Box 6.2) who, while working in Sherrington's laboratory, provided evidence for the potential existence of a spinal neural network whose autonomous activity generates the basic pattern of stepping. Brown found that stepping movements in the hind limb are spontaneously evoked in animals, while they are lying on one side, under general anesthesia after transecting the spinal cord and in the absence of any peripheral afferent input. Since the level of anesthesia used in these experiments was so strong that it abolished proprioceptive reflexes, this finding was difficult to reconcile with Sherrington's model but could readily be explained by assuming that **autonomous center(s)** do exist in the spinal cord and that their spontaneous activity determines the basic rhythm and pattern of the hind limb stepping movement.

➤ Two models have been central to our understanding of neural control of locomotor activity: the model of Charles Scott Sherrington that explains the rhythm and pattern of such movements primarily as the result of sequential reflexes; and the model of Thomas Graham Brown that emphasizes the significance of autonomous neural activity in the spinal cord.

In other experiments, Brown discovered that the stepping movements of one limb persisted after he had removed the opposite half of the lumbar **spinal cord segments**. This finding led Brown to propose a **half-center** model. According to this model, the half-centers are composed of two groups of neurons in the spinal cord that **mutually inhibit each other** and, together, produce the basic stepping rhythm. One half-center controls the contraction of extensors, whereas the opposing half-center is involved in the contraction of flexors. While, for example, extension is occurring through excitation of the extensor half-center neurons, this half-center simultaneously inhibits the flexor half-center, thus preventing the flexors from contraction. After a period of 'fatigue' of the extensor half-center, the flexor half-center takes the lead in driving the rhythm by exciting the flexors, while inhibiting the extensor half-center.

Spinal cord segments The spinal cord is divided into four regions: cervical (which is closest to the head), thoracic, lumbar, and sacral. Each of these regions consists of several segments.

Figure 6.1, reproduced from a paper published by Graham Brown in 1916, succinctly summarizes his concept of how stepping is controlled at the neural level. This model does not necessarily exclude the

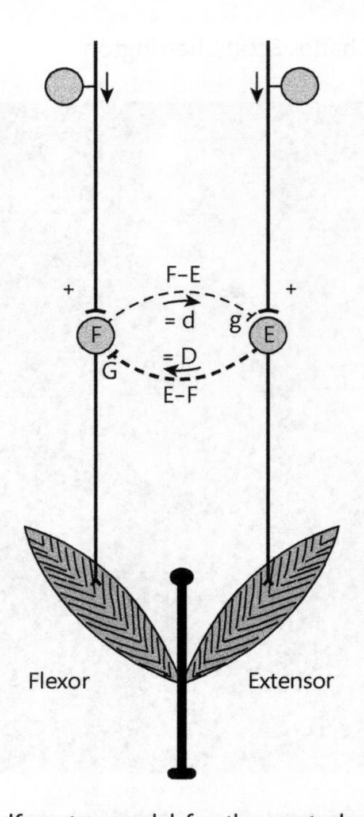

Figure 6.1 Half-center model for the control of stepping, as originally proposed by Graham Brown in 1916. A key element of this model is the reciprocal innervation between the flexor half-center (F) and the extensor half-center (E) in the spinal cord. In the phase of the step shown here, extension takes place. During this phase, the activity of E is high, and this half-center exerts a strong inhibition (symbolized by 'D') upon F. At the same time, the activity of F is low, and the inhibition exerted by this half-center upon E is weak (symbolized by 'd'). However, the activity of E progressively decreases as a consequence of 'fatigue', the level of which is inversely related to the activity of a half-center ('G' denotes strong fatigue, 'g' weak fatigue). The increasing amount of fatigue of a half-center is accompanied by a reduced inhibition that this half-center exerts on the other one. This leads to a point when, for example, F is no longer effectively inhibited, and its activity becomes so strong (due to low levels of fatigue) that balance is not only regained but overshot. At this point a phase switch occurs—now the flexor pathway is activated, and the extensor half-center is inhibited. (From: T. Graham Brown (1916). © Springer.)

Box 6.2 Thomas Graham Brown

Thomas Graham Brown at the age of approximately 40 years. (Photograph from 'Thomas Graham Brown 1882–1965' by Lord Adrian in *Biographical Memoirs of Fellows of the Royal Society* Vol. 12 (November, 1966).

As a scientist, Thomas Graham Brown made a discovery of fundamental importance. Yet, as a person, he remained somewhat of an enigma to most of his contemporaries.

Brown was born in 1882, in Edinburgh, into a family of distinguished Scottish physicians. After graduating with a B.Sc. and a medical degree from the University of Edinburgh, he went to the University of Strasbourg to work with the renowned German physiologist Julius Ewald. He then returned to Scotland to become a research assistant in the Physiological Laboratory of the University of Glasgow. In 1910, he was awarded a Carnegie Fellowship to do research in the laboratory of the then already internationally renowned Charles Sherrington (see Box 6.1) at the University of Liverpool. It was during the three years in Sherrington's lab that Graham Brown began to collect experimental evidence in support of the then revolutionary idea that locomotion in mammals is controlled by autonomous neural activity generated in half-centers of the spinal cord. He continued these studies at the University of Manchester, where he held an appointment as Lecturer in Experimental Physiology, from 1913 to 1915. Although still in an early stage of his career, the years in Liverpool and Manchester turned out to be scientifically the most productive of his life, resulting in a remarkably large number of papers, most of which he published as sole author. After serving in the Royal Army Medical Corps during the First World War, he was appointed to the Chair of the Cardiff Institute of Physiology in 1920, and elected Fellow of the Royal Society, in 1927. However, after he had become Fellow, Brown virtually stopped publishing within the sciences, and his interest shifted to a completely different area–mountaineering–although he remained Chair at Cardiff until his compulsory retirement in 1947.

In 1927, Brown, together with another English climber, Frank Smythe, succeeded in the first ascent of Mont Blanc via the Sentinelle Route. This was the first of many first ascents in the Alps and other parts of the world, including Alaska and the Himalayas. Despite the fact that Brown was already in his forties when he started his career as a mountaineer, he became one of Britain's most accomplished climbers. During the winters, while in Cardiff, he devoted most of his time to exploring new routes, which he tried out during the summer holidays. He published detailed accounts of these expeditions in journal articles and in two books.

It seems that mountaineering provided a somewhat better match to Brown's reserved and often difficult character than the duties that came with his positions as department head and university professor. In fact, Graham Brown never mentored any graduate student or postdoc, clashed with many people, remained a lifelong bachelor, and lived in a hotel near the institute in Cardiff throughout his 27-year tenure as Chair. For his accomplishments as a climber, credit usually followed immediately. However, for his ideas on the neural control of locomotion, he had to wait until shortly before his death to receive recognition. Only then, in the late 1950s and early 1960s, did it become possible to demonstrate intrinsic neural oscillations–the decisive feature of central pattern generators–by performing intracellular recordings.

Thomas Graham Brown died in Edinburgh in 1965 after a stroke.

possibility that more than one pair of half-centers do exist in the lumbar spinal cord. The latter aspect was, however, only of minor importance to Brown.

Although Brown's spinal half-center model provided a better explanation than the reflex model for the experimental data collected by himself and other investigators, it received little attention during his lifetime. This attitude of his contemporaries is quite well summarized in an obituary published in 1966 by Edgar D. Adrian (who shared the Nobel Prize in Physiology or Medicine with Charles Sherrington in 1932). In this article, he characterized the outcome of Brown's work as 'disappointing both to him and to those who realized its merit . . . it added to the general store of information on the spinal reflexes, yet apart from that it cannot be said to have much influence on the progress of physiology'. Indeed, until the 1960s, the predominant viewpoint was that rhythmic activities of organisms, including the basic pattern of locomotion, are determined by sensory input, and that each cycle of a movement is triggered by the previous one through a chain of reflexes.

This viewpoint changed only gradually after an increasing number of investigators had obtained experimental evidence that suggested the existence of rhythmic neural activity in the central nervous system in the absence of input from higher brain centers or from the peripheral nervous system. Among these investigators was Edgar D. Adrian of the University of Cambridge who reported in 1931 that isolated thoracic and abdominal ganglia of the great diving beetle (*Dytiscus marginalis*) display considerable spontaneous activity with frequencies similar to respiratory movements. Other scientists, including Donald M. Wilson and Robert J. Wyman of the University of California, Berkeley, and George Morgan Hughes and Cornelis Adrianus Gerrit Wiersma of the California Institute of Technology, confirmed as part of their studies, at the beginning of the 1960s, that the central nervous system is capable of spontaneously generating rhythmic activity, while highlighting another important aspect: That, in addition to the intrinsic activity of the respective part of the central nervous system, input from peripheral proprioceptors is required to trigger the rhythmic activity and/or to attain a frequency of the oscillations similar to the one in the intact system. For example, in grasshoppers stimulation of the thoracic ganglia is necessary to elicit a rhythmic motor output pattern. This activity can be observed even in the absence of any input from higher 'brain' centers and after ablating the wing proprioceptors. Nevertheless, the output frequency is much lower than the normal wing stroke frequency, and largely independent from the frequency with which the thoracic ganglia are stimulated. This deficiency can be overcome by leaving the input from the stretch receptors associated with the wings intact. Then, the thoracic ganglia are capable of attaining normal wing stroke frequency. With these and other findings accumulated by the early 1960s, the idea of the **central pattern generator** was born (for a detailed description, see Box 6.3).

Box 6.3 Central pattern generator

The production of rhythmic motor patterns is a universal phenomenon in animals. Examples are not only basic physiological activities, such as ventilation and chewing, but also more obvious behaviors, such as swimming and walking.

At the beginning of the twentieth century, two models were proposed to explain the generation of such rhythmic movements. One model, the main proponent of which was Charles Sherrington (see Box 6.1), is based on the assumption that sensory receptors continuously monitor the activity of the muscles. This information is then fed back into the motor system to activate reflexes that trigger the next step in the sequence of the individual repeated actions. Thus, this model proposes that rhythmic motor patterns are the result of a **chain of reflexes**.

The alternative model assumes the existence of an **oscillator** within the central nervous system. This oscillator is based on a network of neurons that provide the timing to rhythmically activate the muscles. This concept was first proposed by Graham Brown (see Box 6.2) of the University of Liverpool (U.K.). The neuronal oscillator is nowadays commonly referred to as the **central pattern generator**.

A central pattern generator may thus be defined as a neuronal network that can produce a rhythmic motor pattern in the absence of any sensory input or input arising from a higher center within the central nervous system and conveying timing cues. The number of neurons encompassing central pattern generators ranges between ten and thousands, depending on the system and the animal species.

The requirement that the rhythmicity be generated in the absence of sensory input does not exclude the possibility that such input may shape the rhythmically generated motor pattern to adjust the output produced at a given time to the other behaviors performed by the animal. Moreover, the neuronal network underlying the central pattern generator is subject to modulation by neuromodulators, such as catecholamines or neuropeptides (see Chapter 9). This modulation may lead to alterations in the frequency or intensity of the rhythmic motor output.

Two major mechanisms of rhythm generation have been identified. In some central pattern generators, the individual

... continued

neurons themselves exhibit oscillatory activity and thus act as **pacemakers**. In other types of central pattern generators, the rhythmicity is the result of synaptic interactions among neurons that are themselves not rhythmically active. In other words, in the latter types, the rhythmicity is a property of the circuit, but not of the individual neuron. These two mechanisms of rhythm generation are not necessarily mutually exclusive. In many central pattern generators, both neurons with pacemaker properties and rhythmicity caused by synaptic interactions have been demonstrated.

Pacemaker cells are characterized by their ability to generate action potentials spontaneously and cyclically, independent of synaptic input. Although the precise mechanism that underlies the intrinsic pacemaker activity is still debated, studies of rhythmogenesis in heart and brain have provided support for two major models. According to the **membrane or voltage model**, membrane hyperpolarization at the end of an action potential activates a specific type of channel. The resulting current is sometimes referred to as 'funny' (I_f) because the majority of voltage-sensitive currents are activated by depolarization, rather than hyperpolarization. The hyperpolarization-activated channel is permeable to both sodium and potassium ions, but at physiological voltages the Na^+ influx into the cell via this channel is larger than the K^+ efflux. This leads to depolarization of the cell until threshold is reached and an action potential is generated.

The **calcium clock model** assumes that spontaneous release of Ca^{2+} from intracellular calcium stores increases the calcium concentration in the cytoplasm, thereby activating the sodium/calcium exchanger. As a result, for the removal of each Ca^{2+} ion from the cell, 3 Na^+ ions are imported. This net addition of positive charge to the interior or the cell leads to membrane depolarization and, upon reaching threshold, firing of an action potential.

The rhythmic activity of central pattern generators can be maintained even in the absence of real motor activity, for example after isolation of neuronal tissue containing the central pattern generator or in immobilized animals. This allows the researcher to study the physiological properties of the central pattern generator circuit while it produces fictive behaviors, instead of real motor activities.

Central pattern generators, or **CPGs** as they are commonly called, are characterized by their intrinsic ability to generate a rhythmic output signal. Yet, another property of the neurons that constitute a given CPG is that their output pattern, including the frequency, can be modulated by various inputs. We will explore this neuromodulatory property in Chapter 9.

Research in the area of CPGs and neural control of locomotion was further advanced by the application of **intracellular recording** techniques (see Chapter 2) to study the activity of individual neurons. Such experiments, conducted by Anders Lundberg at Gothenburg University in Sweden in the 1960s, provided the first direct evidence for the existence of a network of neurons in the lumbar segments of the spinal cord reminiscent of the half-center proposed by Graham Brown four decades earlier.

Since this initial experimental demonstration of a half-center-like organization in the spinal cord, the original model has been refined in several aspects. Nevertheless, the underlying idea has stimulated numerous investigators in the field of neural control of locomotion, including those studying the escape behavior in hatchling tadpoles of *Xenopus*. Research in the latter area has been pioneered by Alan Roberts (for a biographical sketch, see Box 6.4) and his group at the University of Bristol (U.K.). Their key findings are described in the following section.

Model system: Escape swimming in toad tadpoles

The behavior

Embryos of the clawed toad (*Xenopus laevis*) hatch after two days of development. The first day out of the egg, the 5–6mm-long tadpoles spend most of their time hanging from a strand of mucus secreted by a **cement gland** located on the head (Fig. 6.2(a, b)). During this developmental stage, which lasts one day or so, the digestive system matures and the mouth opens. The next phase of development consists primarily of constant swimming and filter feeding of the tadpole.

Although the tadpole is at rest most of the time during the first day of free life, it exhibits certain behavioral responses if it becomes detached, or when it is stimulated by dimming of the light, or by touching any part of its body. In the simplest case, it flexes to the opposite side where the stimulus was applied, a behavior referred to as **flexion reflex**. In

Box 6.4 Alan Roberts

Alan Roberts in his laboratory at the University of Bristol. (Courtesy: A. Roberts.)

Alan Roberts, born 1941 in Rugby (U.K.), is Professor Emeritus at the School of Biological Sciences of the University of Bristol (U.K.). From 1960 to 1963, he studied zoology at the University of Cambridge, where his scientific hero was Hans Lissmann (see Chapter 8). Roberts then went to the U.S.A. to do his Ph.D. on crayfish escape behavior with Theodore Bullock (see Chapter 3), first at the University of California, Los Angeles, and later at the University of California, San Diego. In 1967, Roberts returned to the U.K. to become a Research Fellow in the Department of Zoology of the University of Bristol, where he subsequently progressed through the faculty ranks, until he was appointed to a Personal Chair in 1991.

When commencing his first position at the University of Bristol, Alan Roberts was looking for an easily available animal species with simple behavior and a nervous system composed of only a limited number of neurons. Although he had done his Ph.D. on crayfish, Roberts nevertheless decided that the central nervous system of most adult invertebrates is still too complex and causes intractable problems. He therefore turned to the developing nervous system of 'simple' vertebrates. His final decision to use hatchling *Xenopus* embryos as a new model system was, as he says, 'inspired by George E. Coghill's book on *Anatomy and the Problem of Behavior*, published in 1929, and by the fact that a breeding colony of *Xenopus* was already established by other colleagues at Bristol at the time of [his] arrival.' Since then, Alan Roberts' choice has, indeed, proven to be an excellent one, having led to important breakthroughs in the analysis of the origin of behavior at the level of individual neurons.

most cases, however, the tadpole swims off away from the side stimulated. This behavior is known as **escape swimming.** The sequences of these two behaviors are shown at 5msec intervals in Fig. 6.3(a,b).

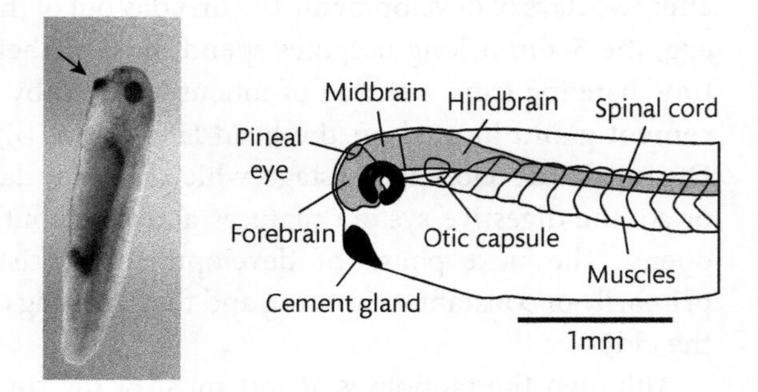

Figure 6.2 Hatchling *Xenopus* tadpole. (a) Photograph of an approximately 5mm-long tadpole attached to the glass side of an aquarium by mucus secreted from its cement gland (arrow). (b) Schematic drawing of the head end, showing the location of the cement gland and the major parts of the central nervous system present at this stage of development. (© 2010 Roberts, Li, and Soffe.)

Analysis of the escape swimming reveals that, when the tadpole is touched, it bends to the opposite side and moves away by producing **lateral undulations** of the body at a frequency of 10–25Hz. These undulating movements are the result of alternate contractions of the antagonistic segmented trunk muscles on the left and right sides. The contractions produce waves of bending spreading from the head to the tail at a speed of approximately 15cm/sec, which drive the tadpole forward at about 5cm/sec. Under natural conditions, swimming ceases when the tadpole bumps into an object, at which point it attaches itself to a substrate using secreted mucus. Under experimental conditions, swimming can be stopped by applying gentle pressure to the head or the cement gland. In the free-swimming tadpole, the stimulus-induced swimming helps the animal to escape from predators and can thus be regarded as an **escape behavior.**

➤ Swimming in the hatchling tadpole is performed by lateral undulations of the body and forms part of the escape behavior.

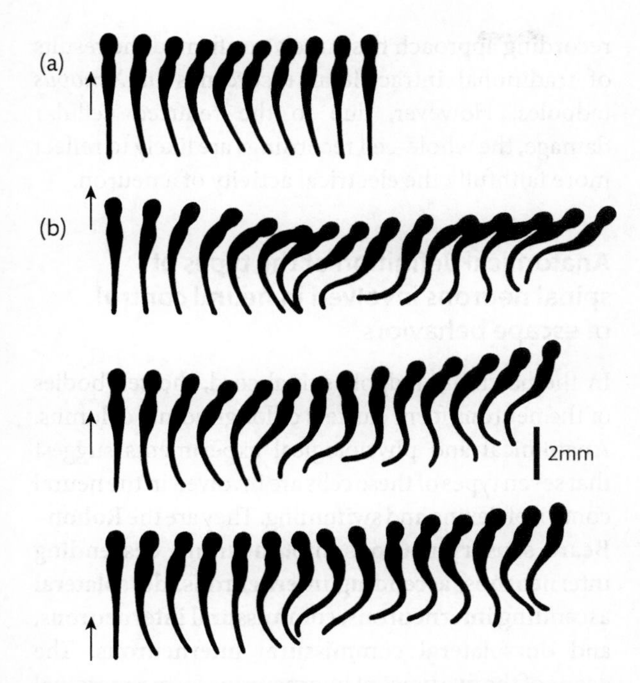

Figure 6.3 Withdrawal responses in hatchling tadpoles, seen from above and based on analysis of movies taken at 200 frames per second. (a) Flexion reflex. If a tadpole at rest is touched on the trunk or tail with a hair, it may show flexion behavior, a bending away from the stimulus. (b) Escape swimming. Upon touch stimulation with a hair, the tadpole swims off, away from the side stimulated. The arrows indicated the site of tactile stimulation with the hair. (© 2010 Roberts, Li, and Soffe.)

The hatchling *Xenopus* tadpole: A powerful model system to study motor control

The small size of the hatchling *Xenopus* tadpoles has both advantages and disadvantages for experimentation. Figure 6.2(b) provides a side view of the head end. At this stage, the central nervous system can be divided into fore-, mid-, and hindbrain, and spinal cord. The spinal cord is small, 100μm in diameter and a few millimeters in length, and there are only approximately 1000 neurons on each side of the cord. As described in detail below, these spinal neurons form eight anatomical categories. Seven of these neuronal types play a role in the neural control of the flexion reflex and escape swimming.

These few neural categories and the low total number of neuronal cells simplify relating neuronal activity to certain behavioral patterns, compared to more complex systems, such as the spinal cord of adult tetrapods. Another benefit of using hatchling *Xenopus* tadpoles for the study of behavioral neurobiology is that their small size eases pharmacological manipulations, as there is rapid diffusion of drugs to neurons of the spinal cord. Also, in whole-mount preparations, all neurons can be seen from the surface, thus making it unnecessary to cut sections of the central nervous system to study the morphology of the neurons.

On the other hand, the small size makes electrophysiological recordings from the entire free-swimming animal impossible. However, Roberts and associates discovered that this limitation can be overcome by studying **fictive swimming**, instead of real swimming behavior. This is achieved through immobilization of the animal by blocking synaptic transmission at the neuromuscular junction with **α-bungarotoxin**. The tadpole can then be held on its side in a small dish, perfused with a physiological saline solution, and dissected to expose the muscles and spinal cord. **Extracellular recordings** can, for example, be made by placing electrodes near motoneuron axons innervating the swimming muscles. **Intracellular recordings** are possible from individual neurons of the spinal cord using glass capillary microelectrodes inserted across the cell membrane (for technical details of these recording techniques, see Chapter 2). Figure 6.4(a) outlines the preparation for such physiological experiments. The top trace of Fig. 6.4(b) shows the neural activity of a single motoneuron recorded with an intracellular microelectrode during fictive swimming, which can be evoked by similar stimuli to actual swimming in the intact animal. The two traces below are extracellular recordings of the alternating motor discharge traveling to the swimming muscles on either side of the body. They demonstrate the alternation of firing of the motoneurons on the left and right side of the spinal cord.

Fictive behaviors Sequences of motoneuron activity occurring without the production of actual movement or muscular contraction. Such behaviors can be observed in immobilized whole animals or in isolated preparations of the nervous system in which muscles are removed.

α-Bungarotoxin A constituent protein of the venom of the Southeast Asian many-banded krait (*Bungarus multicinctus*). By binding to nicotinic postsynaptic receptor sites, this neurotoxin irreversibly blocks cholinergic transmission at the neuromuscular junction, thus producing muscle paralysis.

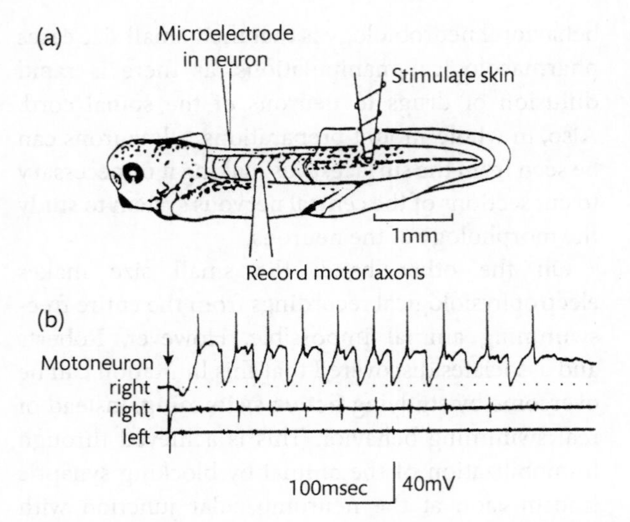

Figure 6.4 (a) Preparation for monitoring physiological activity of spinal neurons during fictive swimming in a tadpole with the brain removed. Fictive swimming is evoked by stimulation of the skin. (b) Activity of motoneurons during fictive swimming. The lower two traces are the results of extracellular recordings and show the alternating motor discharge traveling to the swimming muscles, on either side of the body. Note the alternation of discharges of the motoneurons on the left and right side. The top trace, obtained by using intracellular techniques, shows the activity of a single motoneuron. (After: Roberts, A. (1990).)

In addition to extracellular and intracellular recordings, Alan Roberts and associates have employed **whole-cell recordings**. After opening the spinal cord dorsally along the midline under microscopic control, a 'patch pipette' is advanced to the neuron of interest. This pipette is filled with a solution mimicking the composition of the cytosol; in addition, it can contain tracer substances for later intracellular labeling of the cell. Typically, the tip diameter of the pipette is 0.5–2μm. The pipette is placed in such a way that it touches a patch of cell membrane (hence, the name 'patch pipette') and a weak suction is applied. This results in the formation of a seal between the cell membrane and the glass pipette with an extremely high resistance (a so-called giga-seal). The cell membrane is then ruptured, and recordings from the whole cell are made. The whole-cell recording technique is superior to the traditional intracellular recording technique because less damage to the recorded neuron is caused. Moreover, as the tight seal is not compromised, even extremely low currents can be measured. The whole-cell recording approach has largely confirmed the results of traditional intracellular recordings in *Xenopus* tadpoles. However, due to the reduced cellular damage, the whole-cell recordings are likely to reflect more faithfully the electrical activity of a neuron.

Anatomical definition of the types of spinal neurons involved in neural control of escape behaviors

In the hatchling tadpole spinal cord, the cell bodies of the neurons form clusters of longitudinal columns. Anatomical and physiological experiments suggest that seven types of these cells are involved in the neural control of flexing and swimming. They are the **Rohon–Beard sensory neurons, motoneurons, descending interneurons, ascending interneurons, dorsolateral ascending interneurons, commissural interneurons,** and **dorsolateral commissural interneurons.** The axons of the five types of interneurons form a marginal zone outside the area defined by the cell body clusters. They make synaptic contact with dendrites that grow into this layer.

Table 6.1 lists the seven neuronal types, together with the abbreviations used in the figures and with a summary of their morphological properties. Figure 6.5 illustrates some of these features. To be able to better navigate through the complex anatomy of this neural circuitry in the following text, it might be helpful to consult the table and figure as a reference.

The neural circuitry controlling flexion reflex

As mentioned above, the flexion reflex is a very simple behavior consisting of a brief flexion upon a touch stimulus applied to the skin. To elucidate the neural path underlying this behavior, Roberts and colleagues conducted whole cell recordings from pairs of neurons in tadpoles immobilized with α-bungarotoxin. When the nerve endings in the skin of the **Rohon–Beard neurons** are stimulated with a sufficiently strong current pulse, a brief spike is produced in the motor nerves on the opposite side. In the non-immobilized tadpole, this physiological response would result in a flexion reflex by contraction of the corresponding trunk muscles. Figure 6.6 shows the pathway that mediates this response.

Table 6.1 Types of spinal neurons involved in control of flexing and swimming in hatchling *Xenopus* tadpoles.

Name	Abbreviation	Morphological characterization
Rohon–Beard neurons	RB	Sensory neurons located along the dorsal midline; they innervate the skin of the trunk and mediate sensory perception of touch stimuli
Motoneurons	mn	Motoneurons located ventrolaterally; they innervate swimming muscles of the trunk
Descending interneurons	dIN	Interneurons that project in caudal direction (i.e., towards the tail)
Ascending interneurons	aIN	Interneurons whose primary axons project rostrally (i.e., towards the head); axonal branches form additional descending projection
Dorsolateral ascending interneurons	dla	Interneurons whose somata are located dorsolaterally; axons project rostrally
Commissural interneurons	cIN	Interneurons that project contralaterally (i.e., to the other half of the spinal cord); there, they branch to form both descending and ascending projections
Dorsolateral commissural interneurons	dlc	Interneurons whose somata are located in a dorsolateral position; their primary axons travel contralaterally where they branch to form both ascending and descending projections

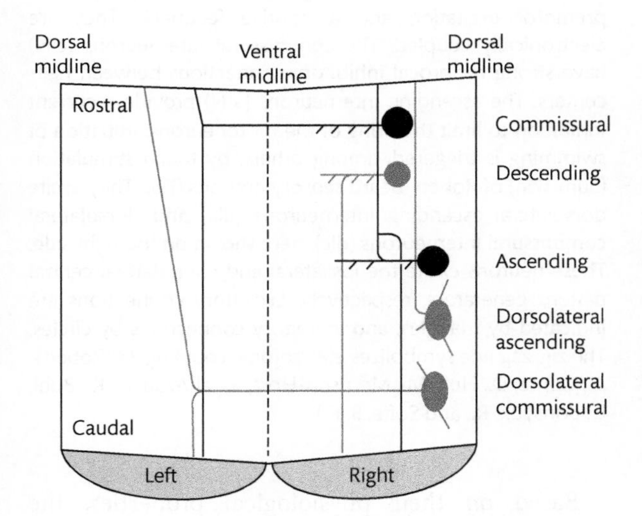

Figure 6.5 Basic features of the five types of spinal interneurons involved in escape swimming in the hatchling *Xenopus* tadpole. The schematic diagram is a dorsal view of the spinal cord as if it were divided along the dorsal midline and then opened like a book. As a result, the dorsal midline is represented by the left and right edges of the drawing, and the ventral midline by the dashed line in the middle. The longer lines indicate the branching of the axons and the projection patterns of interneurons. Descending projections are symbolized by axons running in the caudal direction; ascending projections by axons travelling rostrally. The shorter oblique lines represent the dendrites; they arise either directly from the somata or from the primary processes. Excitatory neurons are gray and inhibitory neurons are black. (After: Li, W.-C., Perrins, R., Soffe, S.R., Yoshida, M., Walford, A., Roberts, A. (2001). © 2001 Wiley-Liss, Inc.)

Upon firing of an action potential by the Rohon–Beard neurons, **dorsolateral commissural interneurons** are stimulated so that they produce an EPSP, and often an action potential. This stimulation is mediated by the release of glutamate that binds to **AMPA receptors**. The dorsolateral commissural interneurons, in turn, excite **motoneurons** on the opposite side, which also involves release of glutamate, but at this junction activation not of AMPA receptors but of **NMDA receptors.** Since the latter excitation is rather weak, activation of many dorsolateral commissural interneurons is required to result in a motoneuron firing and flexion on the

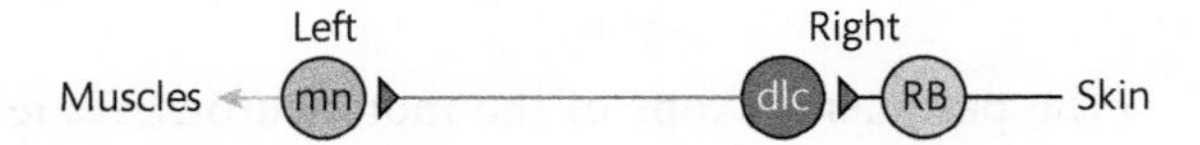

Figure 6.6 Pathway for the flexion reflex. Rohon–Beard sensory neurons (RB) respond to a touch stimulus in the skin (here, on the right side) by firing of an action potential. This leads to excitation of the dorsolateral commissural interneurons (dlc), which may result in the generation of an action potential. The action potentials produced by many dorsolateral commissural interneurons are conveyed via their axons to the opposite side (here, the left side) of the spinal cord, where they cause excitation of motoneurons (mn), which, in turn, leads to the firing of these neurons and the contraction of the innervated muscles, and, ultimately, to the flexion response. (© 2010 Roberts, Li, and Soffe.)

opposite side. Thus, overall three types of neurons—Rohon–Beard neurons, dorsolateral commissural interneurons, and motoneurons—are involved in neural control of the flexion reflex response upon tactile stimulation of the skin.

AMPA receptor An ionotropic glutamate receptor that can be activated by the glutamate analogue α-amino-3-hydroxy-5-methyl-4-isoxazolepropionic acid, commonly known as AMPA.

NMDA receptor An ionotropic glutamate receptor. The agonist *N*-methyl-D-aspartate (NMDA) binds selectively to this type of glutamate receptor but not to other glutamate receptors.

The neural circuitry controlling escape swimming

Out of the seven types of neurons in the spinal cord of hatching tadpoles involved in neural control of swimming, four form the core of the underlying CPG. They are the motoneurons, descending interneurons, ascending interneurons, and commissural interneurons. The CPG in each segment of the spinal cord consists of two parts, one on the left side and the other on the right side. They are called **half-centers.** The pattern of connectivity within the underlying neural network is shown in Fig. 6.7.

➤ The central pattern generator controlling the rhythmic muscle contractions during swimming in hatchling tadpoles consists of two half-centers in the spinal cord and four types of neurons—motoneurons, descending interneurons, ascending interneurons, and commissural interneurons.

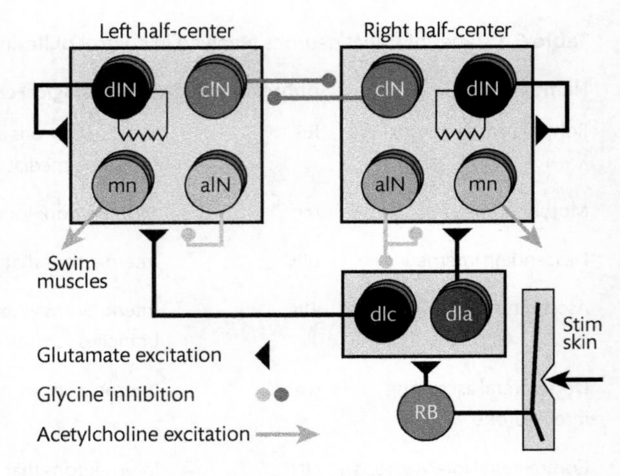

Figure 6.7 Basic circuitry of neurons in the spinal cord controlling escape swimming in the *Xenopus* tadpole. As this wiring diagram shows, the central pattern generator consists of two half-centers, one on each side of the spinal cord. Each circle represents a small population of neurons. The motoneurons (mn) innervate the swim muscles on the ipsilateral side. The descending interneurons (dIN) provide premotor excitation and a positive feedback. They are electronically coupled. The commissural interneurons (cIN) have strong reciprocal inhibitory connections between half-centers. The ascending interneurons (aIN) provide recurrent inhibition to limit the firing of the motoneurons. Initiation of swimming is triggered, among others, by touch stimulation (Stim skin) of Rohon–Beard sensory neurons (RB). They excite dorsolateral ascending interneurons (dla) and dorsolateral commissural interneurons (dlc), here shown on the right side. These neurons excite the ipsilateral and contralateral central pattern generators, respectively. Excitatory connections are indicated by triangles, and inhibitory connections by circles. The zig-zag line symbolizes electrotonic coupling. (© Roberts, A., Conte, D., Hull, M., Merrison-Hort, R., al Azad, A. K., Buhl, E., Borisyuk, R., and Soffe, S. R.)

The peripheral axons of the motoneurons excite the swimming muscles in the trunk, which are segmentally organized, by releasing the transmitter substance **acetylcholine.** Descending longitudinal axons originating from these neurons make excitatory synaptic contact with other, more caudally located motoneurons. This excitation is mediated by **nicotinic acetylcholine receptors** and provides positive feedback to motoneurons in other segments. In addition to these cholinergic synaptic contacts, the motoneurons within one segment also are **electrotonically coupled.**

Based on their physiological properties, the descending interneurons are classified as **excitatory.** There is also some evidence that they have **pacemaker** properties. The descending interneurons excite not only the motoneurons but also the commissural interneurons and the ascending interneurons. Most importantly, by employing whole-cell recordings, Roberts and associates were able to demonstrate that, in a small region of caudal hindbrain, there is a population of these excitatory descending interneurons that make reciprocal excitatory connections with each other. It is thought that this **feedback excitation** provides a key mechanism to generate persistent responses. Such a prolonged response occurs when a skin stimulus, in spite of its short duration, triggers motor nerve activity to the swimming muscles lasting for seconds or minutes. The notion that neural activity persists because neurons in a neural network feed excitation back onto each other is sometimes referred to as **reverberation.**

The excitatory descending interneurons release **glutamate** as a transmitter at their synapses, which then binds to postsynaptic receptors. The EPSPs produced in the target cells consist of two components: a **slow component mediated by NMDA receptors**, and a **fast component mediated by AMPA receptors**. The fast AMPA excitation helps more caudal neurons to reach the firing threshold, whereas the slow NMDA excitation sustains the next cycle of activity on the same side of the spinal cord.

In contrast to the descending interneurons, the **commissural interneurons** exert an **inhibitory action** upon their target cells. They project to the opposite side of the cord, where their release of the amino acid glycine leads to the production of IPSPs (see Chapter 2). Upon binding of this transmitter to receptor molecules in the postsynaptic neuron, channels are opened that result in an influx of chloride ions into the cell. This current drives the potential in a negative direction.

During fictive swimming, the descending interneurons, the commissural interneurons, and the motoneurons in one given half-center of the central pattern generator produce only one spike per swim cycle. Due to the electrical coupling of the motoneurons, the spikes generated by the neurons on one side appear almost simultaneously. However, the spikes produced by the left and the right half-centers alternate because of the inhibitory commissural interneurons. The ascending interneurons are the only type of neurons in the CPG that can fire more than once at each cycle. They release the inhibitory amino acid transmitter glycine, and their function is to **limit the firing of the motoneurons.**

The production of spikes by the motoneurons is caused by the EPSPs evoked by input originating from the ipsilateral excitatory descending interneurons. After firing, the motoneurons remain depolarized relative to the resting potential before swimming is initiated. When neurons on one side of the central pattern generator fire, all of the neurons on the opposite side receive, in the middle of their activity cycle, IPSPs mediated by the contralateral projection of the inhibitory interneurons. Thus, overall the two half-centers of the swimming central pattern generator exert an inhibitory action on each other.

How is the rhythmic activity of the central pattern generator produced? In addition to the feedback excitation exerted by the descending interneurons, experimental evidence suggests that inhibition of the spinal neurons also plays a key role in this oscillation. As illustrated in Fig. 6.8, in the absence of rhythmical drive from the central pattern generator, injection of a suprathreshold depolarizing current into a motoneuron elicits an action potential. This is a result to be expected. Quite surprisingly, however, injection of a pulse of hyperpolarizing current into the same motoneuron can also evoke firing, so-called **rebound firing.** This effect is observed only if the cell is steadily depolarized, as is the case in swimming.

> **Rebound firing** The production of action potential(s) by a neuron after it has been hyperpolarized.

Rebound firing also occurs in the descending interneurons. At rest, IPSPs or negative current injection rarely lead to rebound firing in these cells. However, when they are depolarized, the same stimuli cause them to exhibit post-inhibitory rebound firing. It is thought that the descending interneurons are kept in a state of excitation by feeding back onto themselves long-duration glutamate excitation, mediated by NMDA receptors. Although they are strongly depolarized through this mechanism, this

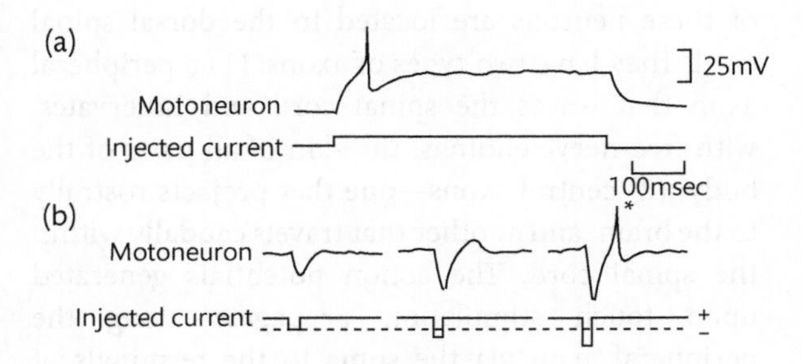

Figure 6.8 Effect of depolarization and hyperpolarization of spinal motoneurons on spike generation. (a) Injection of a suprathreshold depolarizing current elicits a single spike, but no repeated firing. (b) Injection of pulses of hyperpolarizing current into a steadily depolarized cell elicits rebound firing (indicated by an asterisk). (After: Arshavsky, Y. I., Orlovsky, G. N., Panchin, Y. V., Roberts, A., and Soffe, S. R. (1993).)

depolarizing stimulus does not lead to firing shortly after they produce an action potential. However, when an action potential from the commissural interneuron leads to the generation of a short IPSP in the depolarized descending interneuron, then the latter fires a delayed action potential on rebound from the inhibition.

The hypothesis that the rebound phenomenon plays an important role in the production of the oscillations (in addition to the oscillations generated by feedback excitation) is consistent with the observation that the cycle period of the swimming rhythm is the duration of the reciprocal IPSPs. Indeed, the cycle period during swimming (40–100msec) is effectively the sum of the IPSP durations in each half-center, namely two times 25–40msec.

Neural mechanism for initiation of swimming

Although CPGs are defined by their ability to produce a rhythmic output even in the absence of any input, the operation of most CPGs requires some form of input, for example to initiate or modulate their activities. As behavioral observations in hatchling *Xenopus* tadpoles have shown, escape swimming is initiated by a number of sensory stimulations, including dimming of the light or touching of the body. How are such sensory stimuli translated into the initiation of swimming?

When the tail skin is stimulated with a touch (or, in experiments, a brief current pulse), a few **Rohon–Beard sensory neurons** are excited. The cell bodies of these neurons are located in the dorsal spinal cord. They have two types of axons: (1) a peripheral axon that leaves the spinal cord and innervates, with free-nerve endings, the skin of all parts of the body; (2) central axons—one that projects rostrally to the brain, and another that travels caudally within the spinal cord. The action potentials generated upon touch stimulation propagate along the peripheral axon via the soma to the terminals of the central axons, where they excite, mediated by AMPA receptors, two types of spinal interneurons ipsilaterally (Fig. 6.7): **dorsolateral ascending interneurons** and **dorsolateral commissural interneurons**. Each of these two neuronal types makes excitatory glutamatergic synaptic contacts with all neurons of the CPG that controls escape swimming. The major difference between them is that the dorsolateral ascending interneurons project to the ipsilateral CPG, whereas the postsynaptic targets of the dorsolateral commissural interneurons are cells of the contralateral CPG.

Experiments have indicated that excitation of the CPGs in rostral regions, and in both hemispheres of the spinal cord, is required to initiate swimming. This is achieved through rostral projections of the dorsolateral ascending interneurons and the dorsolateral commissural interneurons, and through their ipsilateral and contralateral projections, respectively. Typically, swimming starts some tens of milliseconds after EPSPs are generated by the CPG neurons in response to receiving excitation from the dorsolateral ascending interneurons and the dorsolateral commissural interneurons. As part of this process, the skin stimulation signal is amplified—a few action potentials produced by Rohon–Beard neurons result in the excitation of many dorsolateral ascending interneurons and dorsolateral commissural interneurons. The critical role that the latter two types play for the initiation of swimming appears to be their sole function. During swimming, both receive glycinergic inhibition from ascending interneurons, thus making it unlikely that they are involved, for example, in the maintenance of the swimming rhythm.

Neural mechanism for stopping of swimming

Tadpoles stop swimming when their heads bump into an object, such as vegetation. As for the flexion reflex and escape swimming, the neural mechanism mediating the stopping of swimming can be best studied in animals immobilized by α-bungarotoxin. In these experiments, fictive swimming is induced by electrical stimulation of the trunk skin. An example of the neural activity that accompanies this fictive behavior is shown in the upper trace of Fig. 6. 9, recorded from the ventral roots (which are formed by the axons of motoneurons). Fictive swimming can be stopped by pressing the skin or the cement gland on the head slowly with a blunt probe. This experimental manipulation mimics, for example, a situation when a swimming tadpole bumps into an object.

Several lines of evidence suggest the following neural mechanism that underlies the stopping of swimming. Upon mechanical stimulation, glutamatergic **sensory neurons in the trigeminal ganglion**, which innervate the head region, fire and excite GABAergic **mid-hindbrain reticulospinal neurons** (lower trace in Fig. 6.9). The latter, in turn, inhibit rhythmic spinal neurons and cause swimming to stop.

Intriguingly, the trigeminal sensory neurons that innervate the cement gland produce action potentials continuously, at a frequency of approximately 1Hz, when the tadpole is at rest, hanging from a strand of mucus. Modeling was done, based on the number of axons of the trigeminal sensory neurons that innervate the cement gland, and on the data obtained through recordings of the activity of the mid-hindbrain reticulospinal neurons. Results of these experiments have suggested that the excitation in the mid-hindbrain reticulospinal neurons, resulting from the summation of the inputs from the individual axons of the trigeminal sensory neurons, might be sufficient to produce, for most of the time when the tadpole is at rest, an effective inhibition of the spinal swimming network.

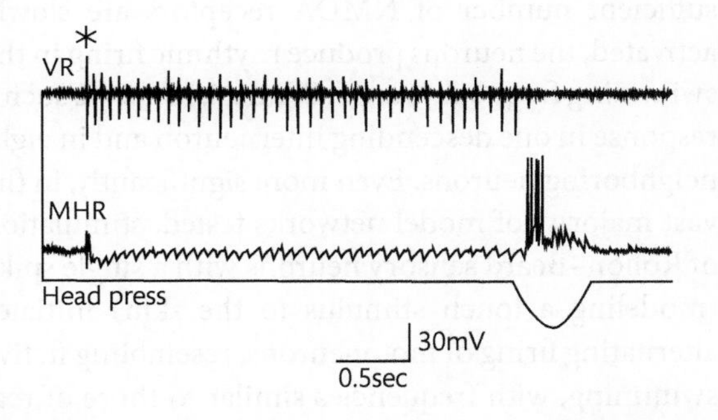

Figure 6.9 Activity of a mid-hindbrain reticulospinal neuron during swimming and head press. Fictive swimming in the immobilized tadpole was triggered by electrical stimulation of the trunk skin for 1msec (indicated by *asterisk*). Extracellular recording of motor activity from the ventral roots (VR) demonstrates that fictive swimming was reliably elicited. Intracellular recording from a mid-hindbrain reticulospinal neuron (MHR) shows that, during fictive swimming, this neuron received rhythmic inhibitory postsynaptic potentials. However, upon pressing the head skin in the region of the cement gland, the same neuron fired action potentials and the fictive swimming behavior stopped. The duration of the head press is indicated by the downward deflection of the respective trace. (After: Perrins, R., Walford, A., and Roberts, A. © 2002 Society for Neuroscience.)

Coordination of oscillator activity along the spinal cord

During swimming, motor activity produced in the half-centers of each central pattern generator starts at the head and progresses toward the tail. The resulting rostrocaudal delay in motor output, which assumes values between 1.5 and 5.5msec/mm, appears to be controlled by a combination of several mechanisms. One mechanism is based on the projection pattern of the excitatory descending interneurons. These neurons not only make synaptic contact with motoneurons and commissural interneurons within their own half-center, but also activate, via descending axons, half-centers in more caudal segments. This pattern of connectivity promotes a progression of activation of the swimming central pattern generators in the rostrocaudal direction.

Another factor causally involved in controlling the longitudinal progression of motor activity is a reduction in the number of excitatory interneurons and inhibitory interneurons found per segment in the rostrocaudal direction, as regions located closer to the head have more interneurons than those closer to the tail. Related to this rostrocaudal reduction in the number of interneurons is a rostrocaudal gradient in both the excitatory and the inhibitory synaptic input received by the motoneurons. The importance of this gradient in setting up the longitudinal delay in motor activity can be demonstrated by pharmacological manipulations: Reduction of the gradient in excitatory input by application of NMDA to the caudal spinal cord reduces the delay time, whereas application of an NMDA antagonist to caudal segments increases the delay by increasing the rostrocaudal excitatory gradient.

Taken together, these mechanisms indicate that a **'leading' oscillator** is located in spinal segments near the head.

Mathematical and computational modeling of the neural network underlying escape swimming

The neural network that controls escape swimming in the hatchling tadpole exhibits an easily overseen but nevertheless remarkable feature—it is fully functional so that the larvae can avoid predation,

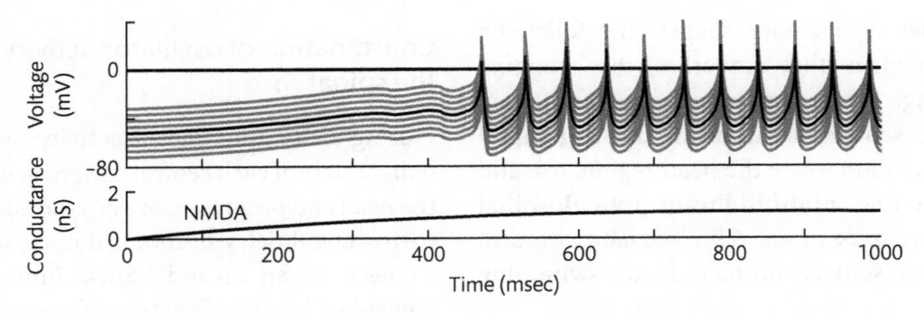

Figure 6.10 Responses to NMDA receptor activation in an electrically coupled population of 30 model descending interneurons. To simulate bench-type experiments in which real descending interneurons are perfused with NMDA, in this modeling experiment an NMDA conductance is slowly activated (lower panel). As soon as sufficient NMDA receptors in the descending interneurons are activated, and this activation is maintained, the depolarized descending interneurons exhibit pacemaker-like firing in the swimming frequency range (upper panel). Shown is a single neuron (black line) and its eight neighboring electrically coupled neurons (paler lines). For the sake of clarity, the traces of the neighboring neurons are displaced vertically. (© Roberts, A., Conte, D., Hull, M., Merrison-Hort, R., al Azad, A. K., Buhl, E., Borisyuk, R., and Soffe, S. R.)

despite the fact that it still is in an early stage of the organism's development. This simple observation suggests that the neural network in the *Xenopus* larvae is not simply a precursor of a more complex network that starts to work properly only once its synaptic connections are refined by developmental maturation. Rather, it is a fully functional neural network in its own right.

To decipher some of the rules that govern the establishment of this neural network, Alan Roberts and co-workers combined anatomical and physiological approaches with mathematical and computational modeling. Based on a wealth of biological measurements obtained over several decades of research, they generated a model tadpole network composed of up to 2000 cells of the seven brainstem and spinal cord neuronal types involved in the control of swimming, as well as of their axons, dendrites, and synapses. During the self-assembly of this model network, axons are allowed to 'grow' controlled by certain physical barriers (such as parameters reflecting the edges of the brainstem and spinal cord) and by some simple chemical gradient cues (such as information representing diffuse chemical gradients that define a head-to-tail polarity). During their outgrowth, axons are also permitted to branch and to make synaptic contacts with dendrites according to some predefined rules.

When the resulting **'anatomical' network** is mapped onto a **'physiological' model**, and simulations are run in which current is injected into a single descending interneuron, these neurons respond with single spikes; sometimes, they also show post-inhibitory rebound firing, as can be observed in experiments using immobilized tadpoles. After electrical coupling of the population of descending interneurons, they behave like a CPG—when a sufficient number of NMDA receptors are slowly activated, the neurons produce rhythmic firing in the swimming frequency range. Figure 6.10 shows such a response in one descending interneuron and in eight neighboring neurons. Even more significantly, in the vast majority of model networks tested, stimulation of Rohon–Beard sensory neurons with a single spike (modeling a touch stimulus to the skin) initiates alternating firing of motoneurons, resembling fictive swimming, with frequencies similar to those of real escape swimming. Thus, the reconstructed network built by Alan Roberts and his collaborators is capable of reproducing the complex rhythmic behavior of the whole animal, a finding that further validates the biological adequacy of the model.

How robust is the network model against deviations? After 500 runs, the self-assembly of the growth model resulted, on average, in 86 655 synapses between all the neurons in the model network. This number could be altered in experimental runs by modifying the probability with which a synapse is formed when an axon crosses a dendrite. Surprisingly, modifications of this probability over a wide range did not affect the reliability of the production of a swimming rhythm. Likewise, if the details of the axon trajectories and dendritic arborizations of each neuron type were discarded, the resulting network still generated reliable swimming activity, as long as the total synapse number of the network was conserved.

> ➤ Mathematical and computational modeling suggests that the neural network controlling escape swimming in hatchling tadpoles develops using surprisingly simple rules, without the need to recognize 'correct' target neurons.

These findings lead to an important conclusion—surprisingly simple rules seem to be required for the functional networks to form in the brainstem and spinal cord during early development. During these early developmental stages, it appears to be sufficient that axons grow into an adequate region and in a broad direction, and that they make synaptic contact with dendrites (but not, for example, glial cells). On the other hand, recognition of the 'correct' type of neuron(s) is probably rather irrelevant. However, it is likely that in later stages of development **refinement of synaptic connections** plays a more important role.

Summary

- Intensive research has been undertaken over the last one hundred years on how coordinated contractions of muscles are controlled at the neural level, but our understanding of this phenomenon is still incomplete. These muscle activities are associated with simple reflexes, and rhythmic behaviors that often persist over long periods.
- One model, originally proposed by Charles Scott Sherrington at the beginning of the twentieth century, explains such coordinated activities as the result of a chain of reflexes, coordinated by proprioceptive information, in the spinal cord.
- A second major model, originally proposed by Thomas Graham Brown, also in the early twentieth century, assumes the existence of an intrinsic oscillator in the spinal cord that drives the coordinate activity of the muscles. The neurons of this central pattern generator, or CPG, can produce the basic rhythmic pattern even in the absence of any input, although most commonly their oscillatory activity is initiated or modulated by afferent information, or information from higher centers of the central nervous system.
- One of the premier model systems to explore the anatomical structure and the physiological function of the neural circuitries that control reflexive motor actions in response to sensory stimulation is the hatchling *Xenopus* tadpole. Among others, this animal performs flexion reflex and escape swimming in response to sensory stimuli.
- Although the physiology of these behaviors cannot be studied in intact tadpoles, the neural activity can be explored in immobilized animals exhibiting fictive behaviors.
- Such studies have revealed the neural pathway that controls the flexion reflex. Upon tactile stimulation, Rohon–Beard sensory neurons excite dorsolateral commissural interneurons, which in turn excite motoneurons that control the contraction of the trunk muscles on the opposite side where stimulation occurred.
- The neural circuitry that controls swimming consists of seven types of neurons, four of which (motoneurons, descending interneurons, ascending interneurons, and commissural interneurons) form the core of the CPG. The remaining three neuronal types (Rohon–Beard sensory neurons, dorsolateral ascending interneurons, and dorsolateral commissural interneurons) are involved in the initiation of swimming behavior upon sensory stimulation. This swim circuitry is present bilaterally in each segment of the spinal cord, thus being organized as half-centers.
- The descending interneurons excite all three other types of neurons of the CPG. The excitation of the motoneurons leads ultimately to the contraction of the swimming muscles in the

trunk. Some of the descending interneurons make reciprocal excitatory connections with each other. This feedback excitation is thought to contribute to the generation of persistent motor nerve activity.

- The commissural interneurons project to the contralateral CPG, where they cause inhibition of all the neurons in the middle of the activity cycle. This inhibitory input to the depolarized descending interneurons causes the latter to exhibit post-inhibitory rebound firing. It is thought that both the feedback oscillation and the rebound firing are part of key mechanisms mediating the oscillations generated by the half-centers.
- During swimming, the rhythmic activity produced in the half-centers of the CPGs progresses from head to tail along the spinal cord. The resulting rostrocaudal delay in motor output is presumably caused by: (i) synaptic contact of descending interneurons with half-centers in more caudal segments of the spinal cord; (ii) a rostrocaudal reduction in the number of descending interneurons and commissural interneurons; (iii) a rostrocaudal gradient in both excitatory and inhibitory input received by motoneurons.
- Mathematical and computational modeling of the neural network underlying escape swimming indicate that modifications, over a wide range, of the connectivity of this network do not affect the reliability of the swimming rhythm. Refinement of synaptic connections is likely to play a significant role in later stages of development.

The bigger picture

In this chapter, we have reviewed neural mechanisms involved in generating locomotor behavior. While the overwhelming majority of scientists in the first half of the twentieth century were convinced that these mechanisms operate solely on the basis of reflexes, in the second half of the century the alternate notion, formulated several decades earlier by Graham Brown, was gradually accepted: That central pattern generators play a critical role in the neural control of locomotory activity in particular, and of rhythmic behavioral patterns in general.

However, the dispute over the relative contribution of reflexes to the function of the system extends beyond their involvement in locomotion. To be able to fully appreciate the significance of this dispute, it is helpful to take a look back at the beginning of the twentieth century. Then, scientists witnessed what turned out to be a true revolution in the study of behavior—the abandoning of subjective, anthropomorphic approaches, and their replacement by objective, scientific approaches. The reflex concept as a child of this paradigm shift was celebrated as an achievement equivalent to the concept of the atom in the physical sciences. Ivan Pavlov, in a lecture delivered in 1927, expressed his trust that the mechanism of 'machine-like activities,' as he referred to the behavior of organisms, would be fully unraveled at some point in the future. He viewed any type of nervous activity (at the neural level) or behavior (at the organismal level) as the result of a chain of reflexes. This included 'instincts,' which he regarded as innate reflexes, and even thoughts, which were, according to this model, reflexes in which the efferent path is inhibited. Today, we know that Pavlov's attempt to explain *any* behavior, emotion, or thought as a chain of reflexes is far too simplistic. Nevertheless, when applied with a less broadly defined scope, the reflex concept has stood the test of time—more so than most other concepts of the last one hundred years.

Recommended reading

Grillner, S., El Manira, A., Kiehn, O., Rossignol, S., and Stein, P. S. G. (eds) (2008). Networks in Motion. *Brain Research Reviews* 57:1–269.

This special issue of the journal Brain Research Reviews contains 29 articles on various aspects of neural control of locomotion, written by some of the leading experts in the field.

Ijspeert, A.J. (2008). Central pattern generators for locomotion control in animals and robots: A review. *Neural Networks* 21:642–653.

The problem of controlling locomotion is an area in which neurobiology and robotics can fruitfully interact. This article reviews both how CPG models can be used in robotics, and how robots can be used as a scientific tool to achieve a better understanding of the functioning of biological CPGs.

Rossignol, S. and Frigon, A. (2011). Recovery of locomotion after spinal cord injury: Some facts and mechanisms. *Annual Review of Neuroscience* 34:413–440.

This comprehensive review focuses on locomotor recovery after spinal cord injury. The authors argue that plasticity within intrinsic spinal circuits is a critical component of hindlimb locomotor recovery after such injury.

Short-answer questions

6.1 What pioneering discovery did Thomas Graham Brown make when he studied the neural basis of reflex stepping in mammalian model systems?

6.2 What are proprioceptors? Define the term in one sentence and give one example of a specific type of proprioceptor.

6.3 Describe, in general terms, the key features of the half-center model.

6.4 What is the defining property of central pattern generators?

6.5 What are fictive behaviors?

6.6 Complete the following sentence: Alpha-bungarotoxin is a constituent protein of the venom of the Southeast Asian many-banded krait. By binding to .... [specify sub-type] postsynaptic receptor sites, this neurotoxin irreversibly blocks .... [specify type] transmission at the neuromuscular junction.

6.7 Define, in one sentence, the term 'rebound phenomenon'.

6.8 Specify the two types of glutamate receptor. How can they be distinguished pharmacologically?

6.9 What is escape swimming in hatchling *Xenopus* tadpoles? What sensory stimuli can trigger this behavior?

6.10 Briefly outline the neural pathway controlling the flexion reflex in hatchling *Xenopus* tadpoles.

6.11 Define the anatomical structure of the central pattern generator that controls the rhythmic muscle contractions during escape swimming in hatchling *Xenopus* tadpoles.

6.12 Define the neural correlates that initiate escape swimming in hatchling *Xenopus* tadpoles.

6.13 **Challenge question** Explain the peristaltic movements of an earthworm based on the reflex-chain theory. What role does sensation play in this process?

6.14 **Challenge question** Two alternate theories have been proposed to explain the peristaltic movements of earthworms—one based on a reflex chain, the other on autonomous neural control. How could one experimentally interfere with the underlying physiological process to decide between these two hypotheses?

Essay questions

6.1 Provide a historical sketch of the development of the half-center model of locomotion. What were the major contributions of Charles Scott Sherrington and Thomas Graham Brown to this development?

6.2 What is 'fictive behavior'? How can it be studied in an animal? What are the advantages and disadvantages of analyzing fictive behavior, instead of real behavior?

6.3 What are central pattern generators? How do they function? What experimental approaches are available to study their physiological properties?

6.4 Describe the structural organization and physiological operation of the neural circuitry controlling escape swimming in hatching *Xenopus* tadpoles. Include in your essay a description of how, at the cellular level, swimming is initiated and stopped, and how the rhythmic pattern is generated and maintained.

Advanced topic Escape swimming in *Tritonia*: from behavior to central pattern generator

Background information

An important but largely unresolved issue related to the neural control of rhythmic motor behavior is **whether the underlying circuit consists of a fixed number of 'hard-wired' neurons that participate in every cycle of the motor program, or whether this circuit is more dynamic, with variably participating neurons**. To address this issue, it is necessary to monitor the activity of all cells in the neural assembly simultaneously while the motor program is running. This is difficult to do in vertebrate species where often hundreds, or even thousands, of cells in the spinal cord are involved in the execution of such programs. Moreover, these cells are typically not well accessible for recording of their activities.

As demonstrated in the paper published by Hill and co-authors (see 'Starter references,' below), these difficulties can be overcome by using a suitable invertebrate organism, and combining such a model system with **large-scale voltage-sensitive dye imaging**. These authors studied *Tritonia diomedea*, a well-established molluscan model system that exhibits rhythmic escape swimming in response to an aversive skin stimulus. This highly stereotypic behavior is controlled by a relatively small number of large neurons located on the dorsal surface of the central ganglion, where they are accessible to large-scale optical recording. These experiments revealed that the network underlying escape swimming in this species is comprised of two populations of neurons: A core of neurons dedicated to the control of the swim behavior, and an additional assembly of neurons that, surprisingly, exhibit moment-to-moment flexibility in their participation in the network operation.

Essay topic: Escape swimming in *Tritonia*: how rigidly hard-wired is the underlying motor network?

In the first part of your essay describe the escape swimming of *Tritonia* in its biological context, and summarize what is known about the anatomical structure and physiological operation of the neural network underlying this behavior. In the second part, explain the principles of the large-scale voltage-sensitive dye imaging technique and the Independent Component Analysis that were used by Hill and co-authors (2012) to collect and analyze the recordings of multiple neurons. Summarize the major findings reported in this paper, and discuss them in a broader context: How do these results affect the traditional view of hard-wired neural networks composed of dedicated neurons? What might be the advantage of a dynamic structural organization of motor networks?

Starter references

Frost, W. N. and Katz, P. S. (1996). Single neuron control over a complex motor program. *Proceedings of the National Academy of Sciences U.S.A.* **93**:422–426.

Getting, P. A. (1977). Neuronal organization of escape swimming in *Tritonia*. *Journal of Comparative Physiology A* **121**:325–342.

Hill, E.S., Vasireddi, S.K., Bruno, A.M., Wang, J., and Frost, W.N. (2012). Variable neuronal participation in stereotypic motor programs. *PLoS ONE* **7**:e40579.

Willows, A. O. D. and Hoyle, G. (1969). Neuronal network triggering a fixed action pattern. *Science* **166**:1549–1551.

Wyeth, R. C. and Willows, A. O. D. (2006). Field behavior of the nudibranch mollusc *Tritonia diomedea*. *Biological Bulletin* **210**:81–96.

To find answers to the short-answer questions and the essay questions, as well as interactive multiple choice questions and an accompanying Journal Club for this chapter, visit **www.oup.com/uk/zupanc3e**.

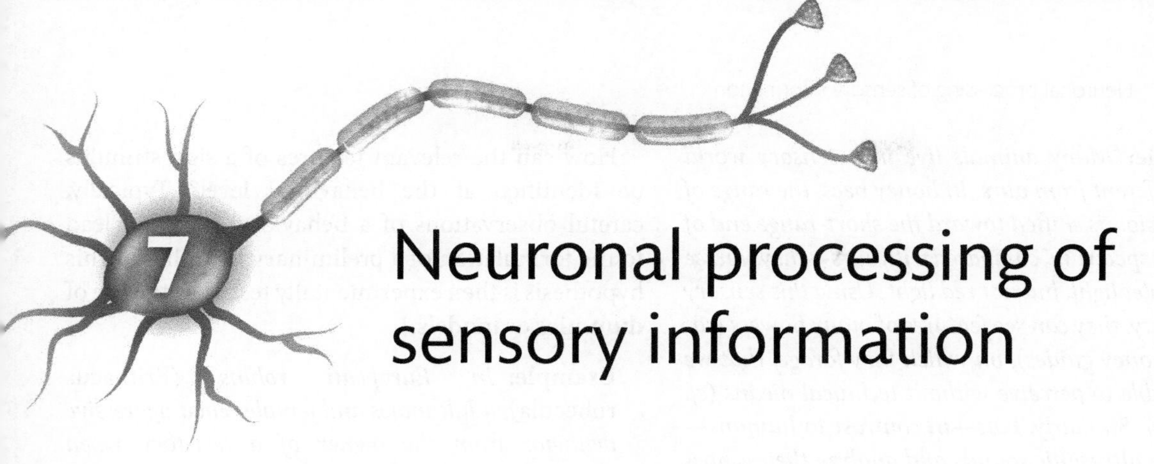

7 Neuronal processing of sensory information

Introduction

Our sensory organs are confronted every second with an enormous amount of information from the environment. However, only a tiny fraction of this information flow reaches our brain, and even less is consciously perceived. As a consequence of this tremendous reduction in the information flow, our perception of the world around us does not, by any means, reflect the 'true' and complete information present in the environment. Rather, what we perceive is a reflection of both the stimuli received and the sensory and neural structures involved in the process of perception.

The aim of this chapter is to discuss how organisms distinguish behaviorally relevant information from behaviorally irrelevant background noise, and how recognition of such behaviorally important information is achieved at the neural level. We will use two particularly well-examined model systems to illustrate this. Before describing these two model systems in detail, we will review some of the ethological key concepts that are critical to understanding how behaviorally relevant sensory stimuli are processed by sensory organs and the brain.

Key concepts

Sign stimuli, releasing mechanisms, and the Umwelt of organisms

The observation that only a tiny fraction of the information from an organism's environment is perceived by the sensory receptors and further processed in the brain led biologist Jakob von Uexküll (see Chapter 3), at the beginning of the twentieth century, to the formulation of the **Umwelt** concept. Von Uexküll used this term not in the sense of its literal translation from German, namely 'environment,' but as 'the environment we perceive'.

Umwelt That part of the environment which is perceived after sensory and central filtering.

Example: *Many animals live in a sensory world very different from ours. In honey bees, the range of color vision is shifted toward the short-range end of the light spectrum, compared with ours—they can see ultraviolet light, but not red light. Using this sensory capability, they can see features of many flowers (so-called honey guides), on which they forage, that we are unable to perceive without technical means (cf. Fig. 3.9). Similarly, bats—in contrast to humans—produce ultrasonic sounds and analyze their echoes for the purpose of orientation. Weakly electric fish are able to perceive electric signals of low amplitude that we cannot sense. Each of these three examples demonstrates that the Umwelt of these animals is very different from ours, although the absolute physical environment is identical.*

Reduction in the flow of information originating from the animal's environment has prompted ethologists to propose the existence of sensory and central filter mechanisms that select those stimuli from the environment that are biologically relevant, but ignore the others. These combined sensory and central filter mechanisms are called **releasing mechanisms.** They determine the kind of stimuli to which the animal responds by producing an associated behavior. Thus, releasing mechanisms can be viewed as a sensory/central link between the stimulus and the resultant behavior. The component of the environment that triggers a given behavior is termed a **sign stimulus.** If the sign stimulus occurs in the context of social communication, it is often referred to as a **releaser.**

Sign stimulus The component of the environment that triggers a specific behavior.

When early ethologists developed the concepts of releasing mechanisms and sign stimuli, they knew very little about the actual sensory and neural mechanisms responsible for stimulus filtering. What they studied was the relationship between stimuli and motor patterns purely from a behavioral point of view. The structural correlate of a releasing mechanism was treated as a 'black box.' Elucidation of the physiological nature of this black box remains the primary goal of sensory physiology and neurobiology in general, and neuroethology in particular.

How can the relevant features of a sign stimulus be identified at the behavioral level? Typically, careful observations of a behavioral situation lead to the formulation of a preliminary hypothesis. This hypothesis is then experimentally tested by the use of **dummies** or **models.**

Example: *In European robins* (Erithacus rubecula)*, adult males and females elicit aggressive behavior from the owner of a territory when trespassing on its territory, while juveniles are not attacked. Adults have a red breast, whereas juveniles do not. This observation makes plausible the hypothesis that the red breast feathers are a key component of the stimulation regime triggering attacks by the owner of a territory. In his classic experiments, published in 1939, David Lack confirmed this hypothesis by placing various types of dummy into the territory of wild male robins, such as a stuffed adult robin, a stuffed juvenile robin (which has no red breast), a stuffed adult robin with the underparts painted over with brown ink, and a bunch of red breast feathers without the rest of the body (Fig. 7.1). The male robins threatened the bunch of breast feathers almost as heavily as the stuffed adult robin. By contrast, the stuffed juvenile robin and the adult robin with the underparts stained brown elicited almost no threat responses and attacks, although these dummies look (to us!) more robin-like than the bunch of red feathers.*

Supernormal stimuli

Frequently, it is possible to make models that produce a greater response from an animal than the natural object does. Such a model provides **supernormal stimuli.**

Experiments in several avian species have shown that, within certain limits, the parental bird prefers clutches with more eggs or giant supernormal eggs over the smaller natural eggs for incubation. Niko Tinbergen (see Box 3.2) demonstrated this in a classic series of experiments using oystercatchers (*Haematopus ostralegus*). If they have the choice, these birds prefer clutches with five eggs over normal clutches, which consist of only three eggs (Fig. 7.2). Similarly, when offered eggs of different sizes, they prefer supernormal eggs (Fig. 7.3).

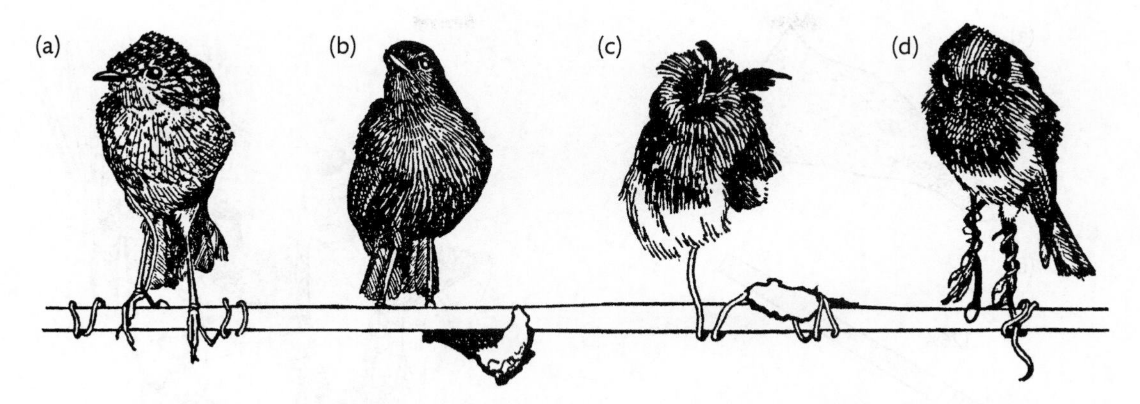

Figure 7.1 David Lack's classic experiment to examine the attack response of European robins. Among others, he placed one of these four dummies into the territories of male robins and monitored their behavior. While a mounted, naturally looking juvenile robin with its dull-brown breast (a) or an adult with the red breast and white abdomen stained brown (b) elicited either no attacks at all or only a few threat responses, a simple bunch of red breast feathers (c) triggered similar attacks from the territorial owner as did a stuffed adult (d). (After: Lack, D. (1953).)

➤ A supernormal stimulus produces a greater response from an animal than does the natural sign stimulus.

A second classic example is provided by the pecking response of infant herring gulls (*Larus argentatus*) toward the parents' bill. The adult herring gull has a red spot near the end of the lower mandible, which contrasts with the yellow color of the bill. When the parent returns to the nest, the chick pecks at this spot, which in turn causes the parent to regurgitate food. In a detailed study, Niko Tinbergen and Albert Perdeck examined the importance of the various features (such as color of the spot and color of the bill) of this stimulation regime for evoking the pecking response. While most variations elicited fewer responses than the natural stimulus, a thin red rod adorned at its tip with three sharply edged white bands had the opposite effect (Fig. 7.4). Now, the chicks pecked more vigorously than toward dummies imitating the parent's natural bill.

How might the preference for supernormal stimuli over normal stimuli have evolved? This question is particularly intriguing given that it is unlikely that oystercatchers, for example, experience in the wild

Figure 7.2 An oystercatcher incubates a 'supernormal' clutch of five eggs instead of the normal clutch consisting of only three eggs. (After: Tinbergen, N. (1969). *The Study of Instinct*. Oxford University Press, London. Reprinted by permission of Oxford University Press.)

Figure 7.3 An oystercatcher prefers a supernormal egg for incubation over her own egg (foreground) or a herring gull's egg (left). (After: Tinbergen, N. (1969). *The Study of Instinct*. Oxford University Press, London. Reprinted by permission of Oxford University Press.)

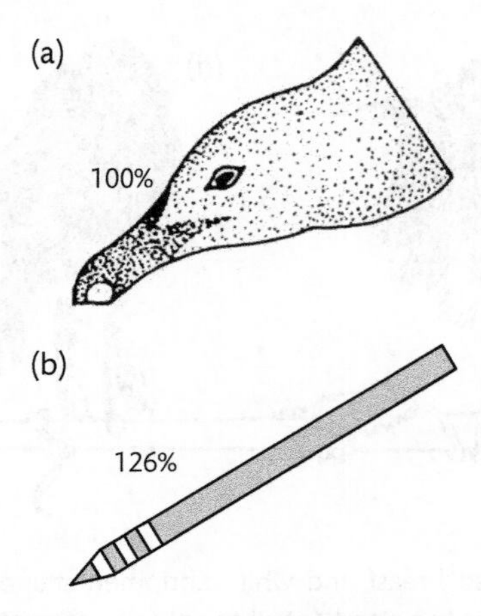

Figure 7.4 Pecking response toward a dummy imitating the natural appearance of an adult herring gull (a) and toward a thin red rod with three white bands near its tip (b). The percentage indicates the normalized responses elicited by each of the two models. (After: Tinbergen, N. and Perdeck, A. C. (1950). On the stimulus situation releasing the begging response in the newly hatched herring gull chick (*Larus argentatus argentatus* Pont.). *Behaviour* **3**:1–39. Reprinted by permission of Brill Academic Publishers.)

Figure 7.5 Human lips made supernormal by the use of lipstick. (Courtesy: Jose Luis Pelaez, Inc./CORBIS.)

eggs as large as Tinbergen used in his experiments (because the size and the anatomy of the bird sets an upper limit of egg size). In case of the oystercatcher egg-size preference, the psychobiologist John E. R. Staddon of Duke University in Durham, North Carolina, hypothesized that there is an inverse relationship between egg size and viability—smaller-than-normal eggs are less viable than average-sized eggs. As a result, selection pressure should be asymmetrical, favoring the development of preference by the parental bird for supernormal eggs over normal eggs. However, it is unclear whether asymmetrical selection pressure can also explain the evolutionary cause of supernormality in all other instances.

Figure 7.6 Enlargement of men's shoulders in various societies to emphasize the 'male' appearance. Top: Waika Indian from South America. Middle: Kabuki actor from Japan. Bottom: Alexander the Second of Russia. (After: Eibl-Eibesfeldt, I. (1974). *Grundriß der vergleichenden Verhaltensforschung*, 4th edn. R. Piper & Co Verlag, München/Zürich. Reprinted by permission of Irenäus Eibl-Eibesfeldt.)

In humans, socially relevant features are often made supernormal by exaggerating them. One example is the use of lipsticks to create supernormal lips in women (Fig. 7.5). Another example is the enlargement of men's shoulders, in many societies, to emphasize the 'male' appearance (Fig. 7.6). The creation of supernormal shoulders is likely to have its roots in the behavior of human ancestors. When they ruffled up their hairs, for example, in aggressive situations, their shoulders seemingly increased in size.

Law of heterogeneous summation

A stimulation regime that elicits a specific behavioral response often consists not just of one stimulus feature. Rather, several independent and heterogeneous features may be present. How do they influence the outcome of the behavioral response?

One possibility is that such independent and heterogeneous features of a stimulus are additive in their effect upon the behavior. This is known as the **law of heterogeneous summation** formulated in 1940 by Alfred Seitz, a student of Konrad Lorenz (see Box 3.1).

> ➤ The law of heterogeneous summation defines the additive effect of the independent and heterogeneous features of a stimulus upon a behavior.

An intriguing quantitative demonstration of this 'law' was performed by Daisy Leong, a student of Walter Heiligenberg (see Box 8.1), at the Max Planck Institute for Behavioral Physiology in Seewiesen, Germany. For her experiments, Leong used Burton's mouthbrooder (*Haplochromis burtoni*), a cichlid fish from Lake Victoria in East Africa. Males of this species are highly territorial and extremely aggressive toward other fish, especially males. As Leong's experiments revealed, the following two features of male coloration are crucial in controlling aggressive responses from other males:

- A black eye-bar that runs from the posterior end of the mouth to the eye.
- Bright orange spots in the pectoral region, as well as on the dorsal, caudal, and anal fins.

To test the effect of these two stimulation features quantitatively, Leong divided an aquarium into two parts using a glass partition. In one part, an adult male was placed together with a group of 10–15 young fish. In the second part, dummies were presented whose head pattern and pectoral coloration were systematically changed as follows (Fig. 7.7):

1. Dummy 1 had a black eye-bar, but no orange pectoral spots. Presentation of this dummy behind the glass partition increased the average bite rate toward the young fish by 2.81 bites/min, relative to baseline bite rate when no dummy was present.
2. Dummy 2 had orange spots in the pectoral region, as well as on the dorsal, caudal, and anal fins, but no black eye-bar. Its presentation lowered the number of attacks by 1.77 bites/min, relative to baseline.
3. Dummy 3 combined both features, having the black eye-bar and the orange spots. This dummy elicited an average increase in the attack rate by 1.12 bites/min, relative to baseline.

Application of the law of heterogeneous summation predicts the following effect when dummy 1 and dummy 2 are combined: 2.81 attacks/min + (−1.77 attacks/min) = 1.04 attacks/min. Dummy 3, which incorporated the features of both dummy 1 and dummy 2 in one model, elicited 1.12 attacks/min. Thus, the predicted value (1.04 attacks/min) and the value observed in the experiments (1.12 attacks/min) are in remarkable agreement.

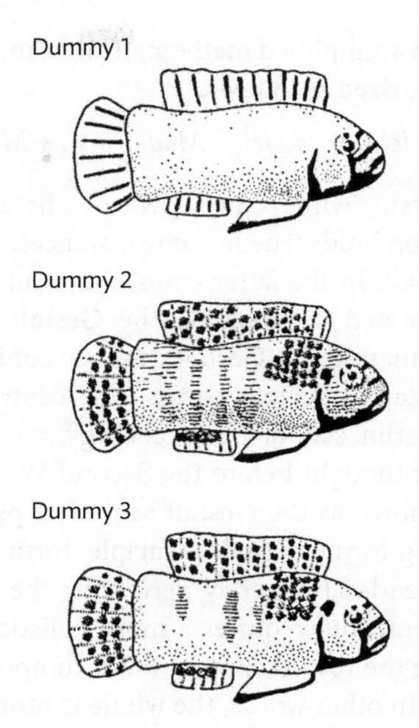

Figure 7.7 Dummies used in the experiment by Leong. Dummy 1 had a black eye-bar, but lacked an orange pectoral patch, dummy 2 had an orange patch, but no black eye-bar, and dummy 3 combined these two features, incorporating both the black eye-bar and the orange patch. (After: Leong, C.-Y. (1969). The quantitative effect of releasers on the attack readiness of the fish *Haplochromis burtoni* (Cichlidae, Pisces). *Zeitschrift für Vergleichende Physiologie* **65**:29–50. Reprinted by permission of Springer-Verlag. © 1969 Springer-Verlag.)

Gestalt principle

The effect of the orange spots and black eye-bar on the aggressive behavior of *Haplochromis burtoni* provides an excellent example that obviously follows the law of heterogeneous summation. The total number of responses (here, attack rate toward the young fish) released by these two features of the stimulation regime is the same when presented successively as when released by the whole. In general terms, the law of heterogeneous summation states that the successive presentation of the individual components of a model elicits the same response as their simultaneous presentation in a combined model. In a simplified mathematical form, this could be summarized as follows:

$$Model_{sum} = Model_1 + Model_2 + \ldots + Model_n$$

However, while the law of heterogeneous summation holds true in some instances, it does not do so in all. In the latter cases, the results can best be interpreted by applying the **Gestalt principle.** The German term Gestalt means 'configuration.' The Gestalt principle forms the central dogma of the Berlin school of psychology, an influential school of thought before the Second World War. It is also known as the **Gestalt school of psychology.** According to the Gestalt principle, form cannot be comprehended by merely perceiving the individual components, but requires a more holistic approach by taking the relationship of these components into account. In other words, **the whole is more than the sum of its parts:**

$$Model_{sum} >> Model_1 + Model_2 + \ldots + Model_n$$

In the most extreme case, the presentation of the individual components may not elicit any response. Rather, only the complete model that incorporates all of the individual components will lead to a behavioral response.

> ➤ The Gestalt principle states that the effect of a stimulus upon a behavior is greater when presented as a whole than when presented sequentially in its individual components.

Arguably, it has been suggested that the law of heterogeneous summation applies predominantly to taxonomically 'lower' animals, whereas the Gestalt principle appears to underlie the mechanism of perception in more highly developed organisms. Moreover, there is some evidence that the mechanisms mediating the perception of behaviorally relevant signals change in the course of ontogenetic development. An intriguing study demonstrating such an alteration of the applicability of the law of heterogeneous summation with age was published in 1966 by Thomas ('Tom') G. R. Bower, then at Harvard University.

In this investigation, Bower trained human infants to turn their heads when a stimulus consisting of three individual components (resembling the face of an adult) was presented. The reinforcement consisted of a 15-second-long 'peek-a-boo' from the experimenter. Then, in the test situation, the combined stimulus and its individual components were presented in a counterbalanced order.

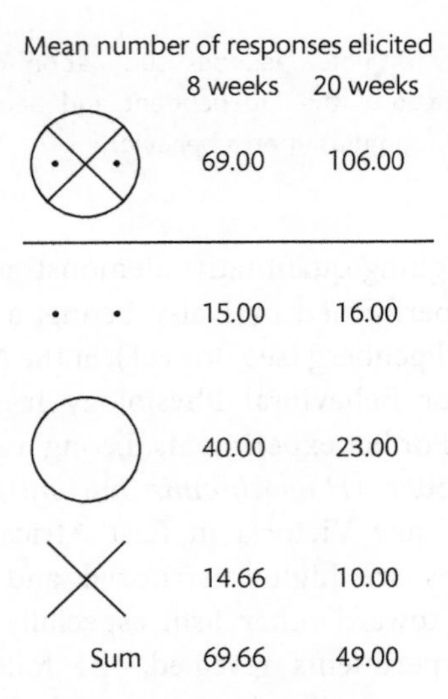

Figure 7.8 Law of heterogeneous summation versus the Gestalt principle. Human infants were trained to make a head movement as a conditioned response when stimulated with a rough sketch of a human face (top row) or the three components of this face (lower three rows). In eight-week-old infants, the sum of the mean number of responses elicited by the individual components is very similar to the mean number of responses evoked by the whole (69.66 versus 69.00). By contrast, in 20-week-old infants, the whole stimulus elicits more responses than the sum of its parts (106.00 versus 49.00). (Courtesy: G. K. H. Zupanc, based on data by: Bower, T. G. R. (1966). Heterogeneous summation in human infants. *Animal Behaviour* **14**:395–398.)

When these experiments were performed on eight-week-old infants, their response confirmed the applicability of the law of heterogeneous summation (Fig. 7.8). In contrast, when identical experiments were conducted on 20-week-old infants, they produced quite a different outcome. The whole was found to elicit more responses than the sum of the responses elicited by its individual components.

Importance of motivation of recipient

It is a well-known fact that the effect of a sign stimulus depends not only on the stimulus presented, but also on the motivation of the recipient. For example, female frogs are strongly attracted by the calls of territorial males if they are physiologically ready to lay eggs. If they have already laid eggs, or if they have not yet developed ripe eggs, the vocalizations of males elicit little approach behavior in females. We will discuss the significance of the motivation of an animal on its behavior, and the neural mechanisms underlying motivational changes in behavior, in detail in Chapter 9.

A quantitative demonstration of how the motivation of the recipient of a stimulus can determine the type of behavior elicited was published by Oskar Drees of the University of Kiel, Germany, in 1952. He used jumping spiders, a family (Salticidae) comprising approximately 2800 species. These spiders do not build webs, but capture prey by jumping onto them. Among their prey animals are insects, such as flies, that resemble, to a certain extent, potential sexual mates. It is, therefore, possible to construct dummies that are capable of evoking, in males, either attempted prey capture or courtship behavior. Which of the two types of behavior is elicited depends on the motivational state of the male. Satiated males respond even to rather crude dummies—black filled circles with one pair of legs—largely with courtship behavior (Fig. 7.9). If the males are left without food, the proportion of courtship behavior relative to attempted prey capture gradually decreases proportionally to the length of time the male has gone without food. However, it is possible again to increase the relative number of courtship behaviors displayed by presenting models with three pairs of legs, thus making the dummies more similar to potential mating partners.

In the next section, we will have a look at another important aspect of the stimulus–motivation relationship—how releasers can influence the motivation of its recipient.

➤ The effectiveness of a sign stimulus on a behavior depends on both the stimulus presented and the motivation of the recipient.

Motivational effect of releaser

The above examples may give the impression that stimuli exert only direct releasing effects. Although this is the most studied aspect of stimulation, a second behavioral consequence is probably of equal importance. Studies have shown that stimuli may also influence the **motivation** of the receiver. An elegant demonstration of this effect was achieved by Walter Heiligenberg together with his assistant Ursula Kramer. They used the same experimental setup as employed by Daisy Leong in the experiments described above (Fig. 7.10). Initially, a male *Haplochromis burtoni* was

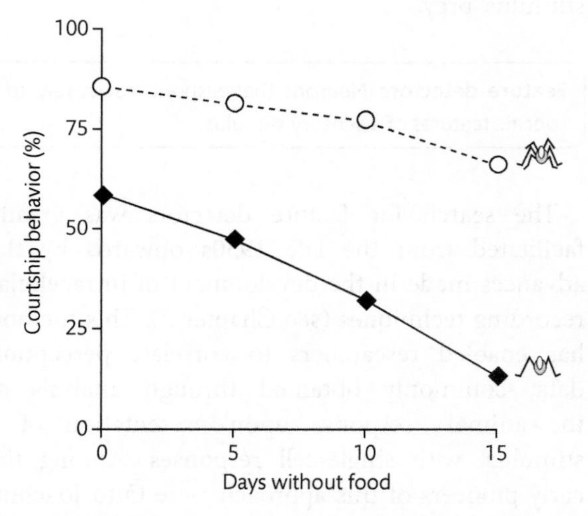

Figure 7.9 Effect of motivation of recipient on courtship behavior in male jumping spiders. The number of courtship displays relative to the number of prey-capture activities depends on both the time the male is left without food and the complexity (three pairs of legs versus one pair of legs) of the model presented. (Courtesy: G. K. H. Zupanc, based on data by: Drees, O. (1952). Untersuchungen über die angeborenen Verhaltensweisen bei Springspinnen (Salticidae). *Zeitschrift für Tierpsychologie* 9:169–207.)

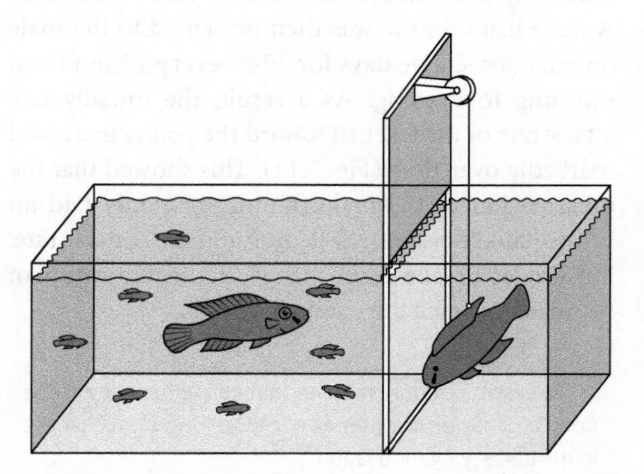

Figure 7.10 Test aquarium to examine the long-term effect of stimulation on the aggressiveness in the cichlid fish *Haplochromis burtoni*. An adult male of this species was kept isolated in an aquarium with ten small fish of the species *Tilapia mariae*. A dummy hanging on strings above the water, behind a screen, was invisible to the fish, until it was lowered, by remote control, into the water behind a glass partition. As an indication of its aggressiveness, the attack rate of the adult male toward the young fish was measured. (Courtesy: G. K. H. Zupanc.)

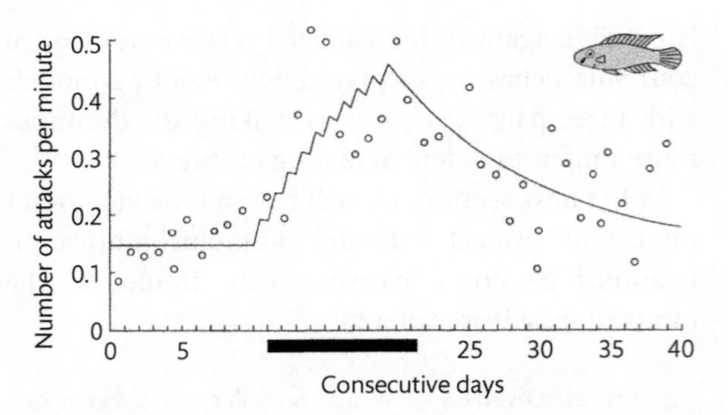

Figure 7.11 Long-term effect of repeated presentation of a male fish dummy (inset) on the attack rate of the cichlid fish *Haplochromis burtoni*. To establish the baseline level of aggression, the fish's attack rate was measured on the first ten consecutive days. Then, starting with day 11, a model was presented every 15min for 30sec over 8h every day for ten days (indicated by the bar underlying the *x*-axis). Upon completion of the stimulation, the attacks exhibited toward the young fish continued to be measured for another 20 days. The data points are the mean of six experiments. The line indicates the theoretical build-up and decay of the attack rate. (Courtesy: G. K. H. Zupanc. Based on data by: Heiligenberg, W. and Kramer, U. (1972). Aggressiveness as a function of external stimulation. *Journal of Comparative Physiology* **77**:332–340. Reprinted by permission of Springer-Verlag. © 1972 Springer-Verlag.)

kept with young fish in its part of the tank, without a dummy presented in the second part of the tank. A male fish dummy was then presented to the male on ten consecutive days for 30sec every 15min from morning to evening. As a result, the initially low attack rate of the test fish toward the young increased markedly over time (Fig. 7.11). This showed that the presentation of the male dummy not only had an immediate releasing effect on the male's attack rate, but also led to long-term changes in the motivation of the male to attack the young fish.

➤ Releasers not only have an immediate releasing effect upon a behavior, but may also lead to long-term changes in the underlying motivation.

Model system: Recognition of prey and predators in the toad

We will now apply the concept of the sign stimulus to explore how the brain processes the information associated with such behaviorally important sensory stimuli. The first model system described here centers around the toad's ability to visually recognize prey objects and distinguish them from predators. To better understand this ability, it is key to address the following two questions: First, what are the prey/predator sign stimuli? Second, how are these sign stimuli extracted by the eye and the brain from the wealth of sensory information received? In 1959, Jerome ('Jerry') Y. Lettvin, Humberto R. Maturana, Warren S. McCulloch, and Walter H. Pitts from the Massachusetts Institute of Technology in Cambridge, Massachusetts, proposed, in a highly cited paper entitled '*What the frog's eye tells the frog's brain*,' that much of this recognition process takes place at quite an early stage of sensory processing. They described 'bug detectors' in the retina of the frog that respond best '*when a dark object, smaller than a receptive field, enters that field, stops, and moves about intermittently thereafter.*' Although, as we will see later in this section, subsequent investigations have shown that the real situation is far more complex, this work has stimulated an intensive search for **feature detectors**, neurons that respond selectively to rather specific features of a sensory stimulus—namely those that define the sign stimulus 'prey.'

Feature detectors Neurons that respond selectively to specific features of a sensory stimulus.

The search for feature detectors was greatly facilitated from the late 1950s onwards by the advances made in the development of intracellular recording techniques (see Chapter 2). This method has enabled researchers to correlate perception data, commonly obtained through analysis of the animal's response upon presentation of a stimulus, with single-cell responses. Among the early pioneers of this approach were Otto-Joachim Grüsser, Otto Creutzfeldt, and Günter Baumgartner in the laboratory of Richard Jung at the University of Freiburg (Germany). This trio of German neurobiologists performed a quantitative analysis of the properties of single neurons in the cortex of cats. One of the first scientists who applied this approach to more neuroethologically oriented studies was Ursula ('Ulla') Grüsser-Cornehls, the wife of Otto-Joachim

Grüsser. She initiated this work using the frog retina during a research visit to the laboratory of Ted Bullock at the University of California, Los Angeles (see Chapter 3). Based on these investigations, Jörg-Peter Ewert (Box 7.1) and his group at the University of Kassel in Germany started in the mid 1960s to study the ability of toads to **recognize prey and enemies.** They used these investigations for the exploration of the behavioral and neural mechanisms of **feature detection.** This research led to the establishment of one of the first model systems in neuroethology, an achievement that had a major impact on the development of this scientific discipline.

The biology and natural behavior of toads

Before we discuss some of the key experiments that led to the identification of feature detectors in toads, we will first look at the biology and natural behavior of these amphibians. The true toads form the family Bufonidae, within which *Bufo* comprises more species than any other genus—approximately 250. They are widely distributed all over the world, with the exception of a few major regions including Australia, Greenland, New Guinea, New Zealand, and Madagascar. Within central Europe, the most abundant species is the common toad (*Bufo bufo*).

When a common toad is motivated to catch prey, and a small moving prey object appears in its visual field, the toad responds with characteristic behaviors, which are shown in Fig. 7.12. One of the first responses is an **orienting movement toward the prey.** The following sequence is variable and depends on the exact stimulus situation, but typically includes (if the prey is close enough) stalking the prey, binocular fixation, snapping, swallowing, and finally wiping of the mouth with the forelimbs.

➤ The natural prey-catching behavior of the toad consists of a series of well-defined individual behavioral patterns.

Box 7.1 Jörg-Peter Ewert

Jörg-Peter Ewert in 2009. (Courtesy: Jörg-Peter Ewert.)

Jörg-Peter Ewert was among the pioneers at the time when neuroethology came into being. His research focused on the identification of neural correlates of the classical ethological concepts of sign stimuli and releasing mechanisms. By employing various experimental approaches in common toads, he and his co-workers explored how diencephalic, mesencephalic, and telencephalic visual structures interact in the detection of prey-like and threatening visual stimuli. This work not only led to the discovery of prey-feature-recognition cells in the toad's optic tectum, but also established one of the first neuroethological model systems.

Ewert was born in the Free City of Danzig in 1938. From 1958 to 1965, he studied biology, chemistry, and geography at the University of Göttingen, West Germany. After receiving his Ph.D. from the University of Göttingen in 1965, and his *Habilitation* from the Technical University of Darmstadt, West Germany, in 1969, Ewert worked as a Fellow of the 'Foundations' Fund for Research in Psychiatry' at McLean Hospital of Harvard Medical School and as a faculty member at the Technical University of Darmstadt. In 1973, he was appointed to a Full Professorship of Zoology and Physiology at the University of Kassel (Germany), where he served on the faculty until his retirement in 2006.

Ewert's contributions to the development of neuroethology are not limited to his accomplishments in research. He authored the first textbook of neuroethology, published in German in 1976 and subsequently translated into several other languages, including English. Ewert also played a key role in the establishment of the International Society for Neuroethology. The decision to found this professional organization of neuroethologists was made during the NATO Advanced Study Institute on *Advances in Vertebrate Neuroethology*, an international meeting organized by Ewert and held at the University of Kassel in 1981.

A first indication of which features may be used to distinguish between prey and enemy objects is given by observations of the toad's natural behavior. Among the enemies that prey on toads are snakes. Therefore, whenever possible, a toad tries to avoid an encounter with a snake. If this is not possible, the toad shows a characteristic avoidance response. It blows itself up, assumes a stiff-legged posture, and exhibits its flank, as shown in Fig. 7.13(a). A similar behavior can be evoked through presentation of a head–rump dummy, as illustrated in Fig. 7.13(b). On the other hand, there are also animals such as leeches that look somewhat similar to snakes, but are not dangerous to toads. Nevertheless, depending on the leech's posture and movement, the toad may confuse them with a snake. This happens if a leech lifts its sucker in the air (Fig. 7.13c). In this case, the toad takes it for an enemy. If, on the other hand, a leech moves jerkily along with its frontal sucker on the ground, it is regarded as prey (Fig. 7.13d).

Figure 7.12 Prey-catching behavior of the common toad, evoked by a worm-like stimulus. When a rectangular stripe is moved in the direction of its long axis within the lateral visual field of the toad, leading to a monocular perception of the object, one of the first responses is a turning movement toward the prey (a). The subsequent behaviors include a binocular fixation of the prey (b), snapping (c), swallowing (d), and wiping of the mouth (e). (After: Ewert, J.-P. (1980).)

Dummy experiments

To examine which features are used to distinguish prey and enemy, toads are placed in a glass cylinder and presented with several models. This experimental setup is shown in Fig. 7.14. The toad's behavior in response to these dummies is then measured by the rate of turning of the head toward the prey.

Dummy experiments in which the toad is stimulated with a small, narrow, dark stripe have

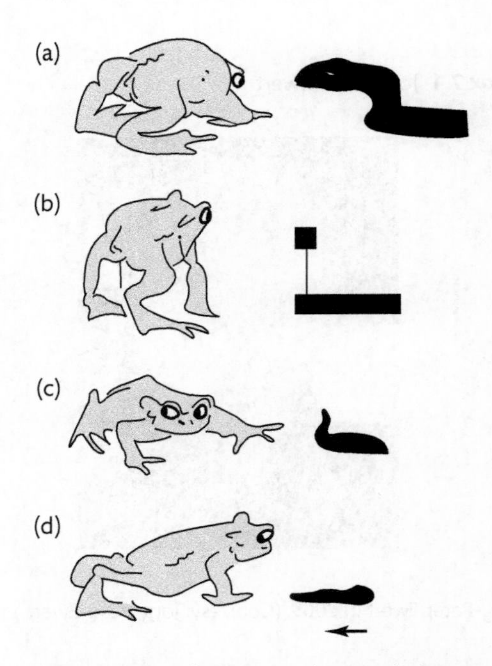

Figure 7.13 Images of enemy and prey of the common toad. The perception of a snake elicits a characteristic avoidance response (a). Presentation of a head–rump dummy results in the same type of defense posture (b). Such a behavior can also be evoked by a leech with a raised frontal sucker (c), implying that it fits the enemy image. On the other hand, as soon as the sucker is on the ground, and the leech moves along its long axis, the toad responds with an orienting movement typically exhibited toward prey (d). (After: Ewert, J.-P. (1980).).

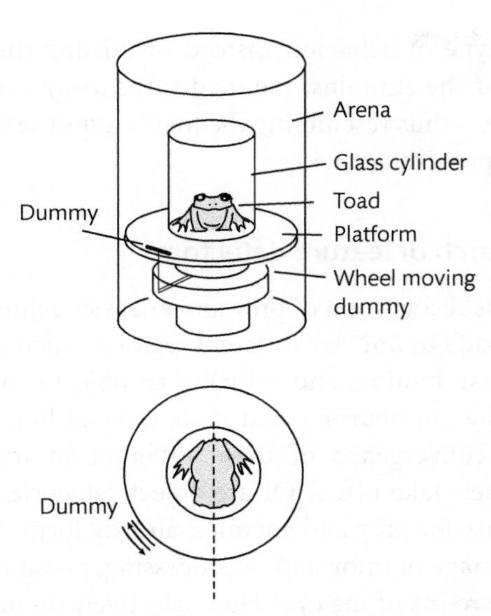

Figure 7.14 Experimental setup to examine the toad's response toward prey objects. The toad is placed on a circular stage within a glass cylinder. Dummies, such as a black stripe, can be moved around the toad at various velocities by means of an electric drive. The wall of the arena, which in reality is opaque and not transparent as shown here, provides the background against which the dummy is viewed. (After: Ewert, J.-P. and Ewert, S. B. (1981). *Warnehmung*. Quelle & Meyer, Heidelberg. Reprinted by permission of Quelle & Meyer Verlag GmbH & Co.)

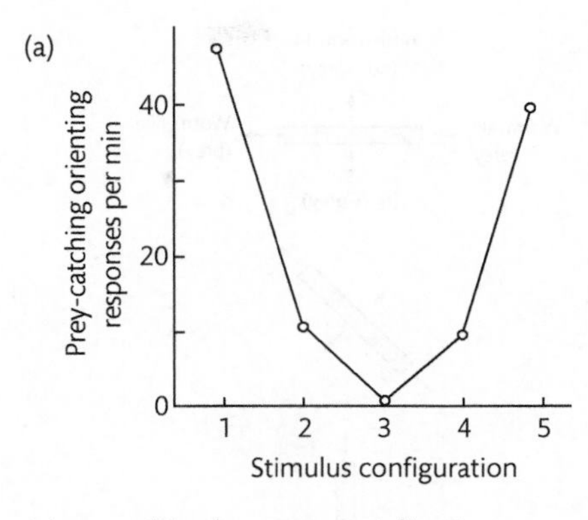

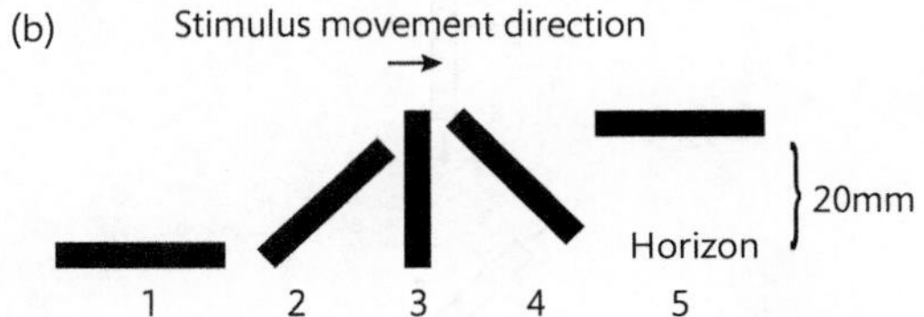

Figure 7.15 Importance of direction of movement of stripe dummies for the efficacy of releasing the prey-catching response in the common toad. (a) The stripes are moved in different directions relative to their longitudinal axis, as indicated in (b). The rate of prey-catching activity is maximal when the stripe is moved parallel to its long axis ('worm configuration'), as done with dummies 1 and 5. The prey-catching response is minimal when the stripe is moved in the direction of its short axis ('anti-worm configuration'), as revealed by dummy 3. Movement in directions intermediate between these two extreme configurations results in intermediate responses, as demonstrated by experiments with dummies 2 and 4. (After: Ewert, J.-P. (1980).)

shown that the following features characterize the key stimulus 'prey':

- Direction of movement.
- Area dimensions (short side vs. long side of the rectangular-shaped stripe) relative to the direction of movement.

The importance of the direction of movement is illustrated by the results of the following experiments (Fig. 7.15). When the stripe moves in a worm-like fashion, that is, parallel to its long axis, it signals 'prey.' When, on the other hand, the long axis of the same stripe is oriented perpendicular to the direction of movement, the stimulus loses its key feature. Such models appear to be perceived as 'threat,' rather than as 'prey,' and elicit a threat reaction or no response at all.

These results have led to the concept of 'worm' and 'anti-worm' configuration of dummies, which are illustrated in Fig. 7.16 and can be summarized as follows:

- A **worm configuration** is achieved by movement of a rectangle in the direction of its long axis.
- An **anti-worm configuration** results when a rectangle moves in the direction of its short axis.

Note that movement along the long axis does not necessarily imply movement in the horizontal direction. Rather, a rectangle presented in an upright position could also move in a vertical direction and still elicit the same kind of response as rectangles moving along their long axis in the horizontal direction. This configuration may, for example, imitate a situation when a caterpillar climbs up a stalk of grass.

> ➤ In the toad's sensory world, 'prey' are elongated objects that move in the direction of their long axis. By contrast, 'enemies' are objects that move in the direction of their short axis.

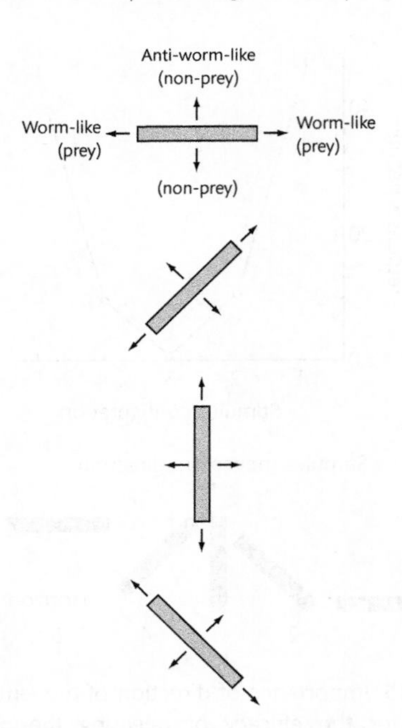

Figure 7.16 Illustration of the concept of worm and anti-worm configurations. Movement of an elongated object, such as a rectangle, along its longitudinal axis signals 'prey.' Movement perpendicular to the long axis elicits no prey-catching response. Rather, this stimulus configuration commonly triggers enemy-avoidance behavior. Note that the 'worm configuration' and 'anti-worm configuration' are solely defined by movement of the object relative to its long axis. Thus, these configurations are invariant to the actual direction of movement, as illustrated here in the two-dimensional plane. (After: Ewert, J.-P. (1980).)

The importance of the area dimension relative to the direction of movement is demonstrated by experiments in which rectangles of various lengths but constant width are used. These rectangles are moved either in the direction of their long axis or in the direction of their short axis. As anticipated, based on the results of the previous experiments, the best response is evoked by movement of the rectangles along their long axis. Moreover, within certain limits, the greater the length of the rectangle (the 'wormier' the object looks like), the greater the toad's response (Fig. 7.17).

By contrast, rectangles moving along the short axis have low releasing values. These values decrease even further if the length of the long axis increases.

Small moving squares have releasing values similar to those of rectangles moved in a worm-like fashion. For longer squares, the releasing value decreases, until it reaches zero. Still longer squares evoke a new type of behavior. Instead of turning the head *toward* the stimulus, the toad turns *away* from the square—thus resembling the toad's response toward an approaching enemy.

In search of feature detectors

How is recognition of prey and enemies achieved in the toad's brain? Are different features (such as size, contrast, motion, and color) of an object processed by different neurons, and if so, at what brain level does convergence of these different information channels take place? Or are object categories, such as those for prey and enemies, already formed at an early stage of information processing, possibly even in the retina of the eye? How selectively do neurons involved in the processing of visual information respond to specific configurations of an object?

To explore the ability of neurons to detect features relevant to object recognition, **recording experiments** are performed. During the recordings, the toad is immobilized by injection of a neuromuscular blocking agent. This treatment is necessary to avoid displacement of the electrode within the brain due to possible movement of the animal. To get access to the brain, a small piece of skull is removed under local anesthesia. Then, by means of a three-dimensionally movable micromanipulator, a recording electrode is advanced to the brain region of interest.

As in the behavioral experiments, the toad is stimulated during the recording experiments with black elongated stripes or squares moving in a worm-like or anti-worm-like fashion. To qualify as a recognition neuron, one expects a cell to exhibit a similar selectivity in discriminating worm- and anti-worm-like stimuli, as shown by the whole animal in the behavioral tests. For example, presentation of worm-like stimuli with increasing length of the long axis should result in similar increases in neuronal activity by prey-recognition neurons. Moreover, only the respective recognition neuron should be responsible for initiating the corresponding motor response, and no other neuron should activate this motor pattern.

> ➤ Feature detectors can be identified by recording from the brain, while exposing the animal to behaviorally relevant stimuli.

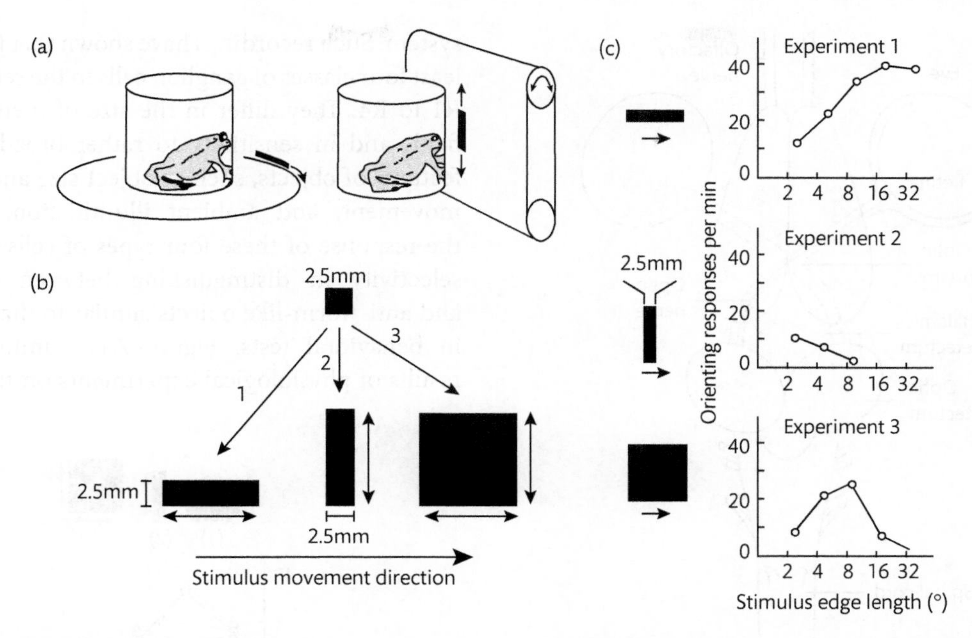

Figure 7.17 Importance of area dimensions relative to direction of movement. (a) The toad is placed in a glass cylinder and stimulated, at a constant distance of approximately 7cm, with models moved in the horizontal or vertical plane. (b) Starting with a square of 2.5 × 2.5mm, the area of the dummy is systematically increased. In the first set of experiments (1), the vertical edge is kept constant at 2.5mm, while the horizontal edge is elongated stepwise in the direction of movement ('the worm becomes longer'). In the second set of experiments (2), the horizontal edge is kept constant at 2.5mm, while the length of the vertical edge is increased stepwise ('the snake becomes taller'). In the third set of experiments (3), squares of increasing size are presented. (c) The number of orienting responses is plotted as a function of stimulus edge length, expressed as the size of the visual angle (in degrees). Using worm-like objects (experiment 1), an increase in the long axis evokes–within certain limits–an increasing number of prey-catching responses. When employing the anti-worm configuration (experiment 2), an increase in the long axis leads to a reduction of the already quite low rate of prey-catching activity. Presentation of squares of increasing size (experiment 3) leads initially to an increase in prey catching. However, further increases in size reduce the prey-catching activity down to zero. (After: Ewert, J.-P. (1980).)

The visual system

The first processing station of the toad's visual system is the retina. The receptor cells are linked, via amacrine and bipolar cells, to the ganglion cells. Typically, one ganglion cell receives input from a number of receptor cells. The corresponding visual area defines the **receptive field** of a particular ganglion cell. The axons of the retinal ganglion cells form the optic nerve. This nerve innervates various locations of the brain. As we will see in detail in the next subsection, two of these areas are particularly important for the recognition of prey and enemies: (i) the **optic tectum**, and (ii) the posterior thalamus and pretectum, commonly called the **thalamic–pretectal area**. Figure 7.18 summarizes the major neural pathways involved in the prey-catching and escape responses in the toad.

The optic tectum forms the roof of the midbrain. Its name comes from the Latin word for roof, *tectum*. The homologous structure in mammals is the superior colliculus.

The link between the eye and the tectum, mediated by the axons of the retinal ganglion cells, is referred to as the **retinotectal projection**. In amphibians, it is entirely contralateral—the left optic nerve terminates in the right optic tectum, and the right optic nerve travels to the left optic tectum.

Receptive field That part of the sensory space (e.g. skin surface) that can elicit neuronal responses (e.g. change in spike frequency) when stimulated.

➤ The major neural parts of the visual system of the toad are: the receptor cells and ganglion cells in the retina; the optic nerve formed by axons of the ganglion cells; and the optic tectum and the thalamic-pretectal area.

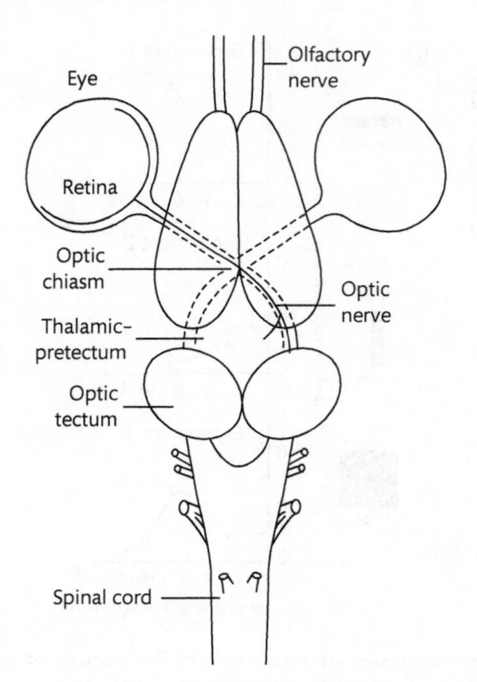

Figure 7.18 Diagram of the basic pattern of connectivity between the retina on the one side, and diencephalic and mesencephalic brain regions on the other, in the common toad. The axons of the ganglion cells form the optic nerve. These cells project contralaterally over wide but specific areas within the thalamic–pretectum in the diencephalon and the optic tectum in the mesencephalon. For sake of simplicity, only one such projection is shown. Also, reciprocal connections between the thalamic–pretectum and telencephalic nuclei are not included in this simplified diagram. (After: Ewert, J.-P. (1974). The neural basis of visually guided behavior. *Scientific American* **230**:34–42.)

As a result of this contralateral projection, neurons located in the right tectum respond to stimulation within the receptive field of the corresponding receptor and ganglion cells in the left retina. Moreover, retinal ganglion cells project to the optic tectum in a systematic point-to-point fashion. Adjacent receptive fields are therefore represented by adjacently located tectal neurons. This leads to the establishment of a topographic map of the visual world on the surface of the optic tectum. This map is specifically referred to as a **retinotopic map.**

Recording experiments

By employing the above approach, Ewert and his group conducted a long series of recording experiments at the different levels of the visual system. Such recordings have shown that there are at least four classes of ganglion cells in the retina, called R1 to R4. They differ in the size of their receptive fields and in sensitivity to rather broadly defined features of objects, such as object size and contrast, movement, and ambient illumination. However, the response of these four types of cells reveals no selectivity in distinguishing between worm-like and anti-worm-like objects similar to that observed in behavioral tests. Figure 7.19 summarizes the results of physiological experiments on three of the

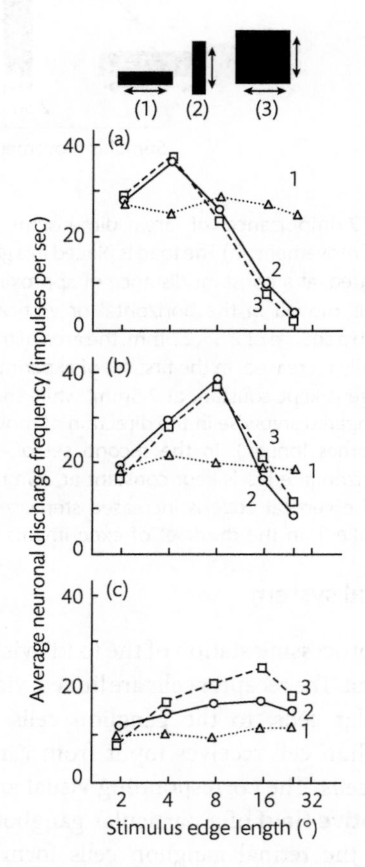

Figure 7.19 Stimulus–response profile of three classes of retinal ganglion cells in the common toad. For the stimulation, the same types of configurations were used as in the behavioral experiments (see Fig. 7.17): '1' denotes a rectangle moved in a worm-like fashion, '2' a rectangle moved in an anti-worm-like fashion, and '3' a square moved along one of the stimulus edges. The length of the stimulus edge is expressed as the size of the visual angle (in degrees). Each data point indicates the mean response of ten different neurons of the same class: retinal ganglion cells of class R2 (a), class R3 (b), and class R4 (c). None of the response profiles resembles those obtained in behavioral experiments. (After: Ewert, J.-P. (1980).)

classes of these retinal cells, R2, R3, and R4 neurons. Obviously, if more selective feature detectors exist, they must be located deeper in the brain.

At the next level, the thalamic–pretectal area, there are a large number of different types of neuron. In their initial work, Ewert and his group identified at least ten different classes, referred to as TP1 to TP11. These different types of neuron become activated by various visual stimuli, some of which are quite complex. One of these neuronal cell types, TP3, responds to large moving visual stimuli. Neurons of this class show little response to stimulation when elongated objects move in a worm-like fashion. Rather, they respond best when such objects are moved in an anti-worm-like fashion, as shown in Fig. 7.20(a). A similar response profile is displayed by TP4 neurons.

Different categories of cell also exist in the optic tectum. Ewert and colleagues grouped them into eight classes, T1 to T8. Each is maximally activated by different visual stimulation regimes. One class, T5, gives strong responses to moving stimuli. Recordings from these cells have revealed two cellular populations defining the subclasses T5(1) and T5(2). T5(1) neurons encode elongation of an object parallel to the direction of movement, but their response to anti-worm configurations does not correspond to the responses shown by the animal (Fig. 7.20b). T5(2) cells show a similar response profile, but are, in addition, able to differentiate between object elongation parallel to and across the direction of movement—their response is strongest to worm-like stimuli and weakest to anti-worm-like stimuli. Moreover, comparison of the stimulus–response profile obtained through behavioral experiments and that defined by the spike frequency of the T5(2) neurons has demonstrated a nearly identical shape to the curves (Fig. 7.20c): elongation of the long axis of a rectangle moved in a worm-like fashion results, initially, in a gradual increase in the response, but in a decline after further elongation. By contrast, elongation of the long axis of the anti-worm-like stimulus leads to a steady decrease in the stimulus–response curve.

➤ T5(2) neurons in the optic tectum show many properties that qualify them as feature detectors responding best to worm-like stimuli.

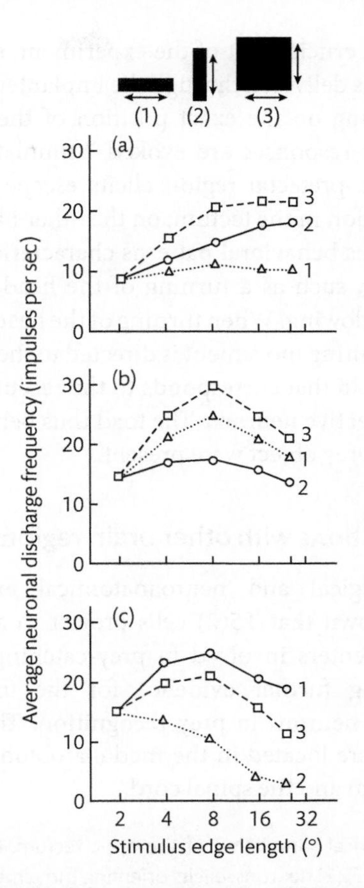

Figure 7.20 Stimulus-response profile of TP3 cells (a) in the thalamic-pretectum and of T5(1) and T5(2) neurons (b and c, respectively) in the optic tectum of the common toad. For the stimulation, rectangles moved in a worm-like fashion ('1') or an anti-worm-like fashion ('2'), and squares moved along one stimulus edge ('3') were employed. Each data point is the mean of 20 recording experiments. For interpretation of the diagrams, see the text. (After: Ewert, J.-P. (1980).)

Stimulation experiments

Compelling evidence for the involvement of T5(2) neurons in prey recognition comes from stimulation experiments carried out in freely moving toads. For electrical stimulation, an electrode is permanently implanted in a small region of the tectum. Before stimulation, the same electrode is used to record the activity of cells located near its tip. By moving a small object across the contralateral visual field, the receptive field of this group of neurons is determined. If the object is moved within the receptive field, neuronal activity is recorded. Outside the receptive field, no activity, or highly reduced activity, is observed.

In the crucial part of the experiment, stimulating current is delivered through the implanted electrode. Depending on the exact position of the electrode, different responses are evoked. Stimulation in the thalamic–pretectal region elicits escape responses. Stimulation in the tectum, on the other hand, results in various behavioral patterns characteristic of prey catching, such as a turning of the head, snapping, and swallowing. When turning of the head is elicited, this orienting movement is directed to the part of the visual field that corresponds to the receptive field of the respective neurons. The toad thus behaves as if a natural prey object were present.

Connections with other brain regions

Physiological and neuroanatomical experiments have shown that T5(2) cells project to appropriate motor centers involved in prey-catching behavior, providing further evidence for the involvement of these neurons in prey recognition. These motor centers are located in the medulla oblongata in the hindbrain and the spinal cord.

> ➤ Electrical stimulation of the optic tectum, believed to activate T5(2) neurons, elicits orienting movement in toads characteristic of prey-catching behavior.

Moreover, and most significantly, T5(2) cells receive pronounced inhibitory input from the thalamic–pretectal area, especially TP3 neurons. It is thought that this inhibition plays an important role in the toad's selectivity for certain configurational features of objects, including those relevant for prey–enemy discrimination.

This hypothesis is supported by both lesioning and stimulation experiments. Lesioning of the thalamic–pretectal/tectal connection produces a number of altered responses of tectal neurons (Fig. 7.21a): (i) a strong increase in the visual responses of tectal neurons; (ii) an impairment of the configurational selectivity of T5(2) neurons; (iii) an increased sensitivity of T5(1) and T5(2) neurons to moving large objects—even to those that were ineffective prior to lesioning; (iv) an inability to distinguish between moving objects and (through movement of the animal) self-induced moving retinal images; and (v) a failure to estimate object distance.

> ➤ T5(2) cells receive inhibitory input from the thalamic–pretectal area. This connection is thought to contribute to the enhancement of selectivity for features relevant for prey–enemy discrimination.

Similar deficiencies are exhibited by toads with thalamic–pretectal/tectal lesions when tested in

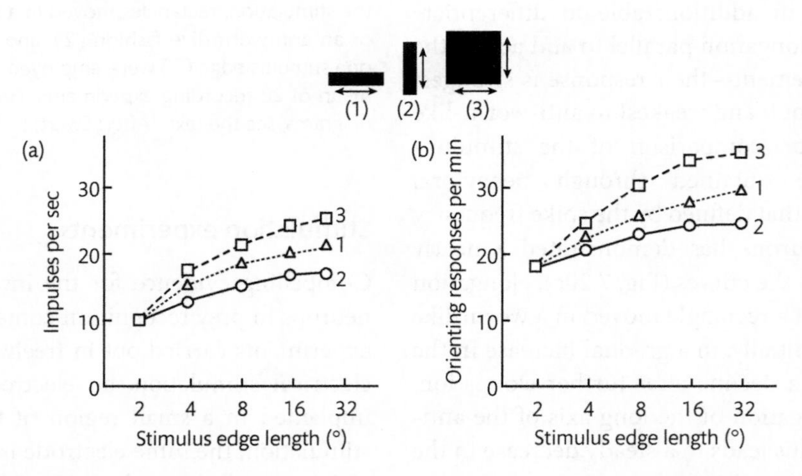

Figure 7.21 (a) Effect of thalamic–pretectal lesions on the response of T5(1) and T5(2) neurons to stimuli of different configurations: '1' denotes a rectangle moving in worm-like configuration, '2' a rectangle moved in an anti-worm-like configuration, and '3' a square moved along one of the stimulus edges. The neurons respond to a size increase in each of the three configurational objects with an increase in firing rate. Their ability to distinguish between the three configurations, normally observed in the intact animal, is largely lost in the lesioned animal. (b) Effect of thalamic–pretectal lesions on prey-catching behavior, as measured by the head-turning response in the common toad. The animals respond to any moving object, independent of its configuration ('1', '2', and '3' as in (a)), with prey-catching activity. The data points are the means of recordings from 20 individual T5 neurons and 20 lesioned toads, respectively. (After: Ewert, J.-P. (1980).)

behavioral experiments. Such toads are unable to discriminate between prey and enemy—they respond to *any* moving object with prey-catching behavior, but lack an enemy-avoidance response (Fig. 7.21b).

Model system: Directional localization of sound in the barn owl

The second model system discussed in detail in this chapter is based on the ability of barn owls (*Tyto alba*) to localize prey solely on the basis of sound generated by the prey animal. As in the toad, elucidation of the underlying neural mechanism in the barn owl has been greatly advanced through focusing on a relatively simple behavioral pattern. Research in this area has been pioneered by Masakazu Konishi and his associates at the California Institute for Technology in Pasadena since the 1970s (see Box 7.2).

The behavior

For his investigations, Konishi has capitalized on an extraordinary ability of barn owls to precisely

Box 7.2 Masakazu Konishi

Mark Konishi in 2003. (Courtesy: M. Konishi.)

Today recognized as one of the leaders in neuroethology, Masakazu ('Mark') Konishi succeeded in combining an interesting ethological question—how owls localize prey using sound—with a rigorous neurobiological approach. In addition to providing an answer to this specific question, the research of his group has also revealed important general computational rules and neural principles of how the brain processes sensory information to produce behavior.

Konishi was born in Kyoto (Japan) in 1933. While still at school, he developed a keen interest in biology. As he puts it, '*As a child, I enjoyed fooling animals. When I read* The Study of Instinct *by Niko Tinbergen in 1953, my junior year in college, I thought I found a perfect field for me. You get paid and praised for fooling animals with dummies.*' As an entry to such a career, he studied zoology at Hokkaido University in Sapporo (Japan). As he felt greatly attracted by the American academic system, Konishi moved to the U.S.A. in 1958 to do his Ph.D. on birdsong under the renowned ethologist Peter Marler at the University of California, Berkeley. Upon receiving his doctorate in 1963, he went to Germany for his postdoctoral training. After a few months in the laboratory of the sensory physiologist Johann Schwartzkopff at the University of Tübingen, he joined the group of Otto Creutzfeldt at the Max Planck Institute for Psychiatry in Munich, where he was involved in one of the first attempts to analyze visual receptive fields of neurons in the visual cortex of cats by single-unit recordings. Although the project was only of limited success, due to the enormous difficulties in recording long enough from neurons to map their receptive field, Konishi greatly benefited from the expertise gained, as well as from the close vicinity of the Max Planck Institute for Behavioral Physiology in Seewiesen, where he met many of the leading ethologists.

After two years in Germany, Mark Konishi assumed an assistant professorship at the University of Wisconsin, Madison, but shortly thereafter moved to Princeton University, New Jersey. Fascinated by the behavioral studies of Roger Payne, he decided to work on sound localization in barn owls. A systematic physiological investigation of this problem began in 1976, a year after Konishi joined the faculty of the California Institute of Technology in Pasadena, to which he still belongs. Following an intensive search, a breakthrough was achieved when he found, together with his postdoc Eric Knudsen, a map of auditory space in the owl's midbrain. This contradicted the then widely held belief that the auditory system should have neither spatial receptive fields nor maps, because the first processing station of the auditory pathway, the inner ear, maps sound frequency instead of space. However, the discovery of Konishi and Knudsen is not only now generally accepted, but also marked the beginning of a systematic and very successful analysis by the laboratory of Mark Konishi of the neural steps involved in the processing of auditory information.

In 2013, Konishi retired and became Professor Emeritus at Caltech.

localize prey animals solely based on the noise produced by them. This sensory capability is likely to be an adaptation to the behavior of field mice, the main food source of barn owls. Field mice forage predominantly at night when they are barely visible. Moreover, they tend to move through tunnels in grass or snow, rather than across open areas. Thus, the visual system is of rather limited use in prey hunting.

While hunting in the dark, the barn owl visits a number of observation perches within its territory to survey the ground. Upon hearing the noise of a prey animal, the barn owl turns its head in a rapid flick so that it directly faces the source of the sound. Then, it swoops down to finally strike the mouse.

The fact that barn owls can find their prey based solely on the **noise** generated by the moving mouse was demonstrated in a series of simple but elegant experiments by Roger Payne while he was a graduate student at Cornell University in Ithaca, New York. In one of these experiments, a barn owl was trained to strike mice released in leaves on the floor of a room in complete darkness. Then, with the lights turned off, a mouse-sized wad of paper was dragged through the leaves on the floor. The owl successfully struck the paper wad. This experiment demonstrates that the owl is not using olfaction, because the paper wad did not release a mouse-like odor. Also, the owl's ability to precisely catch the mouse could not be explained based on infrared sensitivity, as the paper wad and the leaves through which it was dragged were at the same temperature.

> ➤ Barn owls can localize prey solely based on acoustic cues.

In a different experiment, also conducted in complete darkness, the floor of the experimental room was covered not with a layer of leaves, but with sand instead. This enabled the mouse to walk silently. To make the mouse's movements audible nevertheless, the mouse was forced to tow a rustling leaf several centimeters behind its tail. In this experiment, the barn owl struck the leaf, but ignored the mouse.

As the barn owl hunts from points above ground, it is not sufficient for it to locate the prey only in the horizontal plane (which is referred to as the **azimuth**). Rather, it must also determine the angle between itself and the prey in the vertical dimension (which is referred to as the **elevation**). How is this achieved?

> **Azimuth** Horizontal plane.
>
> **Elevation** Vertical plane.

Experimental approach

To answer this question, it was first necessary to obtain detailed information about the accuracy of prey localization. However, owls in free flight are of only limited use for this purpose. When the owl swoops down, it aligns its talons with the long axis of the mouse's body, just before striking the mouse. When the mouse turns to run in a different direction, the owl realigns its talons. This flexibility in the behavior of the owl during the free-flight phase of prey catching makes it difficult to employ a standardized behavioral paradigm. Therefore, owls are trained to remain on their perch during the experiment. Each time they turn their head in response to a sound stimulus, they are rewarded with a small piece of meat.

The experimental setup is shown in Fig. 7.22. At the beginning of the experiments, the head is aligned by attracting the owl's attention with a sound from a fixed source. This sound source is referred to as the **zeroing loudspeaker**. Stimulation takes place with sound from a second loudspeaker called the **target speaker**. This loudspeaker can be changed in both the horizontal and vertical plane, thus imitating the natural situation in which the angle between the prey and the owl also varies in these two dimensions.

The orientation of the response relative to the target speaker is monitored by means of an **electromagnetic angle-detector system** composed of two major components. First, two small coils of copper wire are mounted on top of the owl's head and arranged perpendicularly to each other. These are called the **search coils**. The second part of the electromagnetic angle-detector system consists of two bigger coils between which the owl is positioned. These two coils carry electric current and are called the **induction coils**, as they induce current flow in the search coils. Turns of the head cause predictable

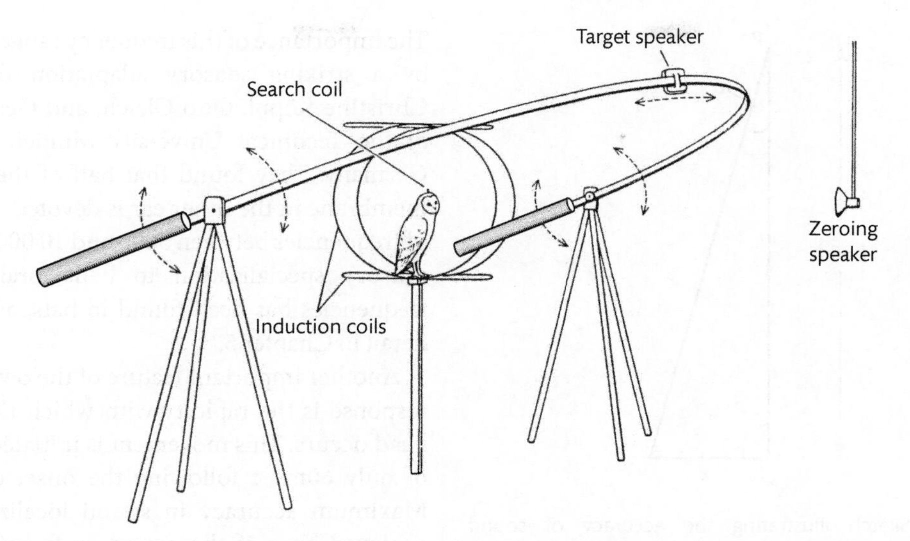

Figure 7.22 Simplified drawing of the electromagnetic angle-detector system. This setup is based on the natural response of hunting owls when hearing a noise—the bird turns its head to face the source of sound. In the experiments, the owl is trained to remain on the perch, instead of approaching the presumptive prey. The movements of the owl's head are monitored by search coils mounted on top of the bird's head and by induction coils between which the owl is positioned. Through this arrangement, any movement of the head induces a current change in the search coils, which is recorded and analyzed by a computer. At the beginning of the experiment, the head is aligned by directing the owl's attention to a zeroing loudspeaker. Sound imitating the noise caused by a prey animal comes from a target speaker. This speaker can be moved around in both the horizontal and the vertical plane, at a constant distance from the owl's head. (After: Knudsen, E. I. (1981).)

changes in current flow within the search coils. Analysis of these variations therefore enables the experimenter to determine both the azimuth and the elevation of head movements.

➤ The accuracy of sound localization, as indicated by turning of the owl's head toward the source of the stimulus, can be monitored by means of an electromagnetic angle-detector system.

Accuracy of orientation response

Experiments employing the electromagnetic angle-detector system have revealed several important features of the barn owl's orientation system. Under optimal conditions (e.g. when the sound source is directly in front of the face), **the barn owl can locate sound within one or two degrees in both the azimuth and elevation.** For comparison, humans are as accurate as owls in the azimuth, but three times worse in the elevation. The biological relevance of this accuracy will be demonstrated in the following exercise.

***Exercise**: A barn owl sits on its perch 5m above ground. It hears a mouse directly underneath on the ground. Within what error range can it locate the mouse on the ground? How does this error range relate to the size of an adult field mouse?*

***Solution**: The situation is sketched in Fig. 7.23. The angle α (1°) denotes the accuracy in the azimuth, h (5m) the height of the perch, c the hypotenuse, and x the error range. Therefore:*

$$\tan \alpha = x/h$$

or

$$x = h \tan \alpha$$

By using the above values, we get:

$$x = 5 \cdot 0.018\text{m}$$

Therefore:

$$x = 0.09\text{m}$$

➤ The barn owl can locate sound with high accuracy in both the azimuth and elevation.

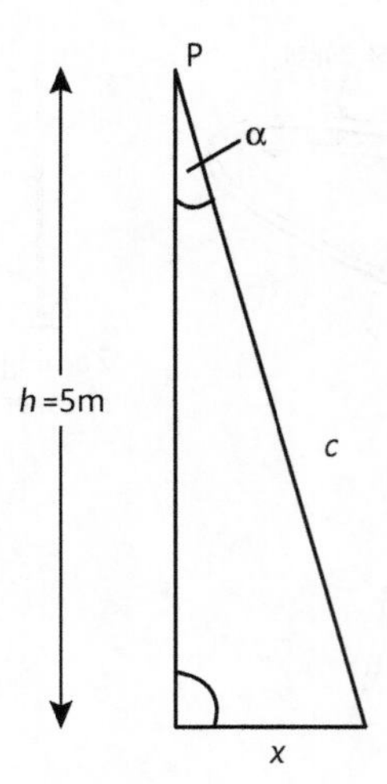

Figure 7.23 Sketch illustrating the accuracy of sound localization in the barn owl. The owl sits on its perch (P) 5m above ground. In the right-angle triangle drawn, α denotes the accuracy in azimuth, h the height of the perch relative to the ground, c the hypotenuse, and x the error range. The angle α is exaggerated for illustrative purposes. For an explanation of calculation of the error range, see the text. (Courtesy: G. K. H. Zupanc.)

The error range of 0.09m (or 9cm) roughly corresponds to the length of a mouse. Thus, within the limits of its sensory accuracy, the barn owl is able to locate the mouse with sufficient precision. Moreover, after having started to swoop down to the mouse, the owl can correct the course of its approach (e.g. if the mouse makes a turn), before it finally strikes the target. This improves the precision of localization even further.

The success rate with which an owl strikes a mouse is illustrated by results of the behavioral experiments conducted by Roger Payne. A trained owl was kept in a completely darkened room, its floor covered with leaves. Single mice were then released. In 16 trials, the owl made 16 strikes at the mouse, which was at least 4m away. It missed the mouse only four times and never by more than 5cm!

The accuracy of prey localization in both the azimuth and elevation varies with several factors, among which the frequency range of the sound plays an important role. Although owls can hear over a broad range of frequencies (roughly from 100 to 12 000Hz), experiments have shown that it is essential for accurate localization that the sound contains relatively high frequencies between 5000 and 9000Hz. The importance of this frequency range is underlined by a striking sensory adaptation discovered by Christine Köppl, Otto Gleich, and Geoffrey Manley of the Technical University Munich in Garching, Germany. They found that half of the owl's basilar membrane in the inner ear is devoted to the analysis of frequencies between 5000 and 10 000Hz. A similar sensory specialization to behaviorally important frequencies has been found in bats, as discussed in detail in Chapter 5.

Another important feature of the owl's orientation response is the rapidity with which the flick of the head occurs. This movement is initiated after a delay of only 60msec following the onset of the sound. Maximum accuracy in sound localization can be achieved even if the sound ends before the head movement begins. This indicates that the owl does *not* use any feedback information while moving the head, and thus *no* iterative adjustment of the head's position relative to the source of sound occurs. Rather, the entire program controlling head movement is laid out before its initiation. In other words, the owl makes the computational decisions in regard to the head movement under **open-loop conditions**. However, mid-course corrections may be made when the owl flies toward a mouse in the dark, by listening to the noise generated by the moving mouse.

Physical parameters of sound involved in head orientation

Indications of which physical parameters are used by the owl in localizing sound are provided by head-orientation experiments in which one ear is blocked. The results of these experiments and their interpretations can be summarized as follows:

1. If one ear is completely blocked, the owl makes large errors in localizing the source of the sound. This demonstrates that the owl's ability to successfully locate prey depends on a comparison of the signals in the two ears.
2. If one ear is partially blocked, this results in significant errors in determining the elevation (Fig. 7.24). However, only slight errors occur in the accuracy of determining the azimuth. Partial blocking of one ear effectively reduces the intensity of the sound reaching the ear, but does not

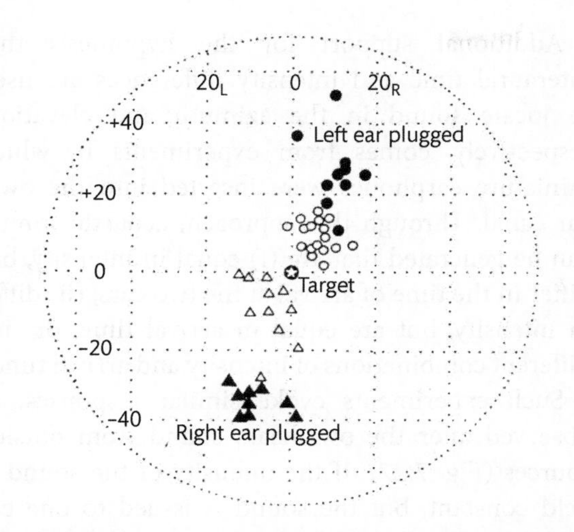

Figure 7.24 The functional significance of interaural intensity differences for localization of sound in the elevation. A barn owl's auditory space is drawn in both the azimuth and elevation. The target is presented in the origin of the bicoordinate system defined by these two parameters. Insertion of a loose ear plug (open symbols) into the left ear (circles) or the right ear (triangles) results in modest errors in the elevation. Occlusion of the ears with tighter ear plugs (closed symbols), on the other hand, causes severe errors in the elevation, while still affecting the accuracy of sound localization in the azimuth, but only to a limited extent. (After: Knudsen, E. I. and Konishi, M. (1979).)

significantly alter the time of arrival of the sound at the two ears. These results therefore indicate that differences in intensity between the ears (referred to as **interaural intensity differences**) are the principal cues for locating sound elevation.

3 If the plug is inserted into the left ear, it causes the owl to direct its head above and a little to the right of the target (Fig. 7.24). Conversely, a plug in the right ear results in the owl facing below and a little to the left of the target. In both cases, the tighter the ear plug, the greater the degree of error.

The latter result reflects a vertical **asymmetry in the directional sensitivity of the two ears.** This can be attributed to a vertical displacement of the two ears. While the left ear is above the midpoint of the eyes and points downward, the right ear is below the midpoint of the eyes and points upward.

Moreover, there is a slight **asymmetry in the arrangement of the facial ruff,** which can be seen in Fig. 7.25. This structure is composed of stiff, dense feathers that are tightly packed. The surface of the facial ruff very efficiently reflects high-

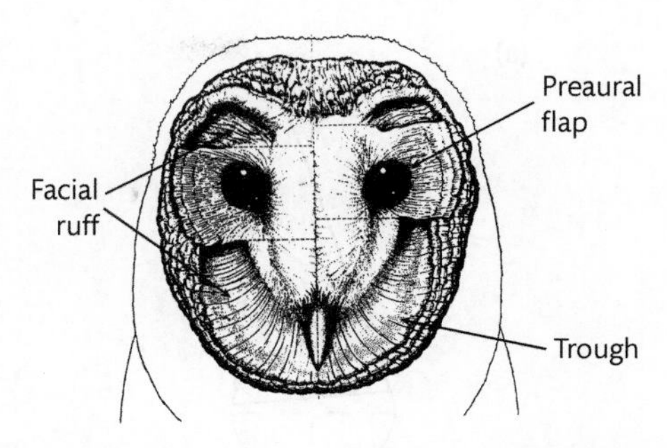

Figure 7.25 Facial ruff of the barn owl. This structure is formed by tightly packed feathers. The ear openings, located behind the preaural flaps, are connected to the surface of the ruff via two troughs running down the face to join below the beak. They collect high-frequency sound and funnel it into the ear canal. The preaural flap and ear opening is lower on the right side than on the left. Moreover, the right trough is tilted up, while the left trough is tilted down. These asymmetries lead to differently perceived intensities of sound by the left and right ears. For demonstration purposes, the owl's facial disc feathers, which normally cover the two troughs, were removed in this illustration. (After: Knudsen, E. I. (1981), and Knudsen, E. I. and Konishi, M. (1979).)

frequency sound. However, the ear openings are connected to the surface of the ruff via two troughs running through the ruff from the forehead to the lower jaw. The troughs serve a similar purpose to the external pinnae of the human ear: They collect high-frequency sounds and funnel them into the ear canals. Directional asymmetry of the two troughs is produced by the left trough being oriented downward, while the right trough is oriented upward.

Taken together, these asymmetries in the placement of the ears cause the left ear to be more sensitive to sounds coming from below, whereas the right ear is more sensitive to sounds from above.

The importance of the facial ruff is underlined by experiments in which the feathers are removed. The owl is then unable to locate sound in the elevation. Upon acoustic stimulation, it always faces horizontally—regardless of the true elevation of the source.

Head-orientation experiments employing the electromagnetic angle-detector system have also shown that the most important cues for locating sound in the azimuth are differences in the arrival times at the two ears, technically referred to as **interaural time differences** (Fig. 7.26). Such differences occur whenever the distance from the source of sound to the

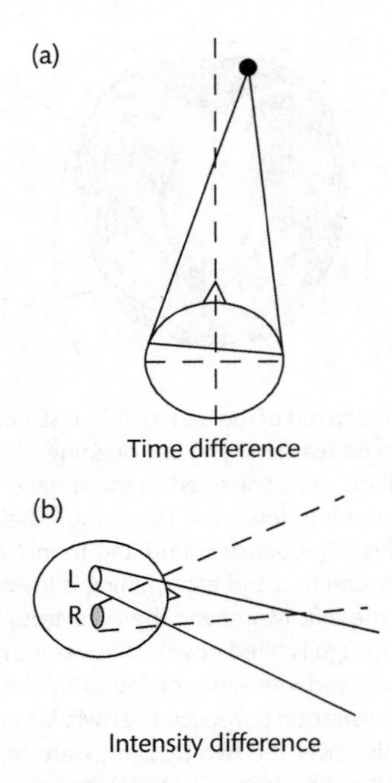

Figure 7.26 Physical parameters used by the barn owl for sound localization. (a) Top view of a schematically drawn head of an owl with a sound source. Sound originating from a source not directly in front of the owl results in a difference in the length of the sound path between the two ears, thus causing an interaural time difference. This difference increases with increasing incidence of the sound relative to the midsagittal plane of the head. The interaural time difference is maximal when the sound comes directly from one of the owl's sides. It is used to determine the azimuth component of the location of the sound source. (b) Side view of a schematically drawn head of an owl with left (L) and right (R) ears. In the barn owl, the left ear is located higher than the right ear, relative to the eye level. Moreover, the two ears are sensitive in different vertical directions, an effect enhanced by the structure of the facial ruff. This directional asymmetry leads to interaural intensity differences that enable the owl to localize the sound source in the elevation. (After: Konishi, M. (1990).)

two ears is different. Consequently, when the source is directly in front of the bird, there is no difference. When, on the other hand, the sound comes directly from one of the owl's sides, the difference is maximal. Within these two extremes, the interaural time difference varies with the incidence angle of the sound (measured relative to the midsagittal plane of the head) in a systematic way.

➤ Barn owls analyze interaural time differences and interaural intensity differences to locate sound in the azimuth and elevation, respectively.

Additional support for the hypothesis that interaural time and intensity differences are used to locate sound in the azimuth and elevation, respectively, comes from experiments in which miniature earphones were inserted into the owl's ear canal. Through this approach, acoustic stimuli can be generated that are (i) equal in intensity, but differ in the time of arrival at the two ears; (ii) differ in intensity, but are equal in arrival time; or (iii) different combinations of intensity and arrival time.

Such experiments evoke similar responses, as observed after the owl hears sound from outside sources (Fig. 7.27). If the intensity of the sound is held constant, but the sound is issued to one ear slightly before the other ear, the owl turns its head, mostly in the horizontal plane, in the direction of the leading ear. The longer the delay in delivering the sound to the second ear, the further the head turns.

Similarly, if the timing is held constant but the intensity is varied, the owl mostly moves its head in the vertical plane. Sound that differs in both intensity and arrival time between the two ears causes the owl to move the head in both the horizontal and vertical direction. In each case, the degree to which the owl turns its head corresponds to the degree of difference observed in these two parameters after generating sound at these imaginary locations. Thus, the owl does indeed appear to use interaural time and intensity differences to precisely locate sound in the horizontal and vertical planes.

The cochlear nuclei: Parallel processing of time and intensity information

As we have seen above, the owl determines the azimuth and elevation bicoordinate components of a sound source by analysis of interaural time and intensity differences. How is this achieved at the neuronal level?

The link between the ear and the brain is provided by the **auditory nerve**. Its axons originate from nerve cell bodies situated in the inner ear. Different sound frequencies are encoded by different fibers of the auditory nerve. By contrast, the codes for intensity and timing of the sound are not segregated within the auditory nerve. Rather, both sound parameters are carried by each fiber. Encoding of these parameters is achieved through variation of the rate and timing of

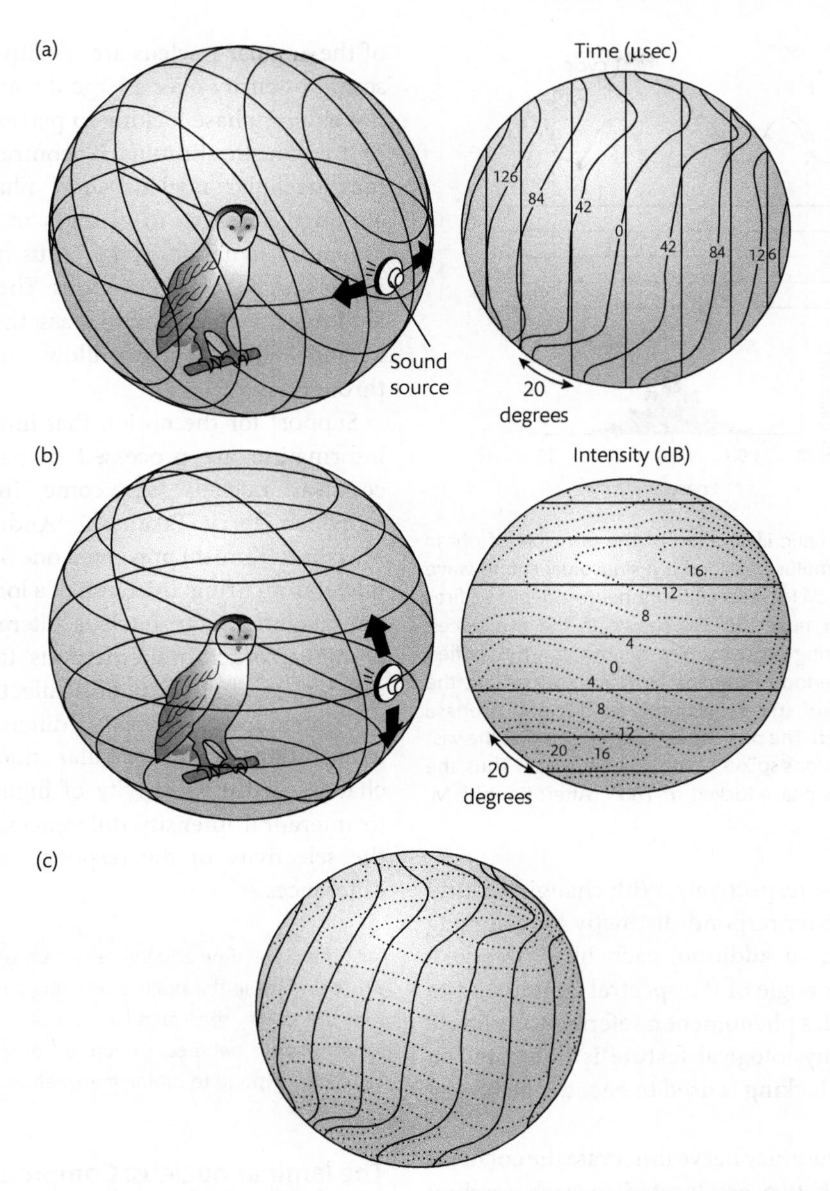

Figure 7.27 Barn owls use interaural time differences and interaural intensity differences to determine the azimuth and elevation components, respectively, of a source of sound within a bicoordinate system. For visualization, this bicoordinate system can be plotted along the surface of an imaginary globe encapsulating the owl's head. (a) An interaural time difference occurs when the sound source is not directly in front of the owl. This difference increases by 42μsec for every 20° that a sound source moves laterally. Thus, this parameter provides the horizontal (azimuth) component of the positional information of a sound source. (b) Interaural intensity differences determine the vertical (elevation) component of a source of sound. Sound originating from sources above eye level is perceived more intensely by the right ear. This difference increases with increasing vertical position of the sound source, as indicated by the respective decibel levels. On the other hand, sound originating from sources below eye level is perceived more intensely by the left ear. Again, the extent of this difference depends on the vertical position of the sound source, as indicated by the respective decibel levels. (c) Combination of the azimuth and elevation information defines each location in space. When the owl is stimulated by sound coming from such a location, it turns its head in the direction predicted by the model. (After: Konishi, M. (1993).)

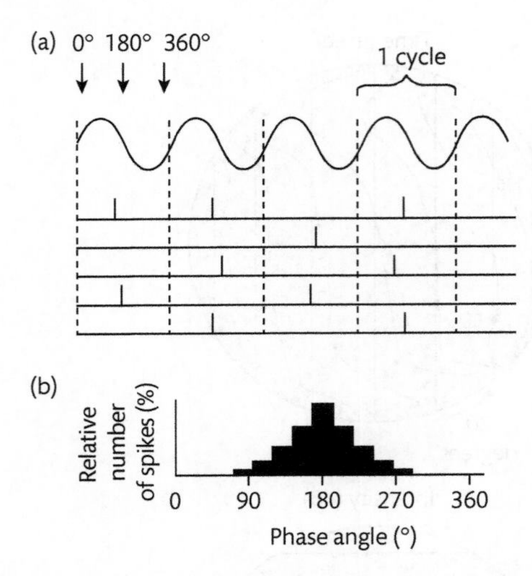

Figure 7.28 Schematic illustration of the principle of phase locking. (a) The stimulus provided is a sinusoidal sound wave running over four cycles. The auditory neuron depicted fires preferentially at or near 180°, as shown in the five traces. (b) The phase-locking property of a neuron can be verified by analysis of a period histogram. For such histograms, the relative frequency of spikes occurring at the various phase angles is determined. The plot shown here reveals that the vast majority of the neuron's spikes occur at or near 180°. Thus, the neuron analyzed is phase locked to 180°. (After: Konishi, M. (1992).)

action potentials, respectively. With changing sound intensity, the fibers respond distinctly by changing their spike rate. In addition, each fiber fires at a particular phase angle of the spectral component to which it is tuned, a phenomenon referred to as **phase locking**. This physiological feature is illustrated in Fig. 7.28. Phase locking is used to encode the timing of the signal.

Fibers of the auditory nerve innervate the **cochlear nuclei**, of which two are located in each cerebral hemisphere. The neurons of the two cochlear nuclei in each hemisphere differ in both morphology and physiology. Neurons of one subpopulation define the **magnocellular nucleus**, whereas neurons of the other subpopulation are located more laterally and form the **angular nucleus**. Each fiber of the auditory nerve divides into the collaterals. One of them innervates the magnocellular nucleus, while the other enters the angular nucleus (Fig. 7.29).

The branching of the auditory nerve fibers provides the structural basis to separate processing of phase and intensity information within the cochlear nucleus. A subpopulation of the neurons of the angular nucleus are sensitive to variations in sound intensity over a large dynamic range, but do not exhibit phase locking to particular phase angles of the acoustic stimulus. By contrast, neurons of the magnocellular nucleus show phase locking, but, although sensitive to changes in intensity, exhibit a smaller dynamic range. Thus, the two cochlear nuclei can be viewed as filters: The angular neurons let intensity information pass through, while the magnocellular neurons allow timing information through.

Support for the notion that intensity and timing information are processed in parallel within the cochlear nucleus has come from experiments in which Terry Takahashi, Andrew Moiseff, and Masakazu Konishi prevented one of the two cochlear nuclei from firing. Injection of a local anesthetic into the magnocellular nucleus altered the selectivity of higher-order brain neurons to interaural time differences, but did not affect their response to interaural intensity differences. Similarly, inactivation of the angular nucleus resulted in changes in the selectivity of higher-order neurons to interaural intensity differences, without altering the selectivity of the response to interaural time differences.

➤ Interaural time and intensity data are segregated in the first stations in the brain processing auditory information: Neurons of the angular nucleus process intensity information, whereas neurons of the magnocellular nucleus respond to timing information.

The laminar nucleus: Computation of interaural time differences

The next processing station after the magnocellular nucleus is the **laminar nucleus** in the brain stem. It receives input both from the magnocellular nucleus on the same (ipsilateral) side of the head and from its counterpart on the opposite (contralateral) side. Thus, it is in the laminar nucleus that information from both sides of the ear converges for the first time within the brain.

As a detailed anatomical and physiological study conducted by Catherine Carr and Masakazu Konishi has shown, the laminar nucleus plays a crucial role in **measuring and encoding interaural time differences**

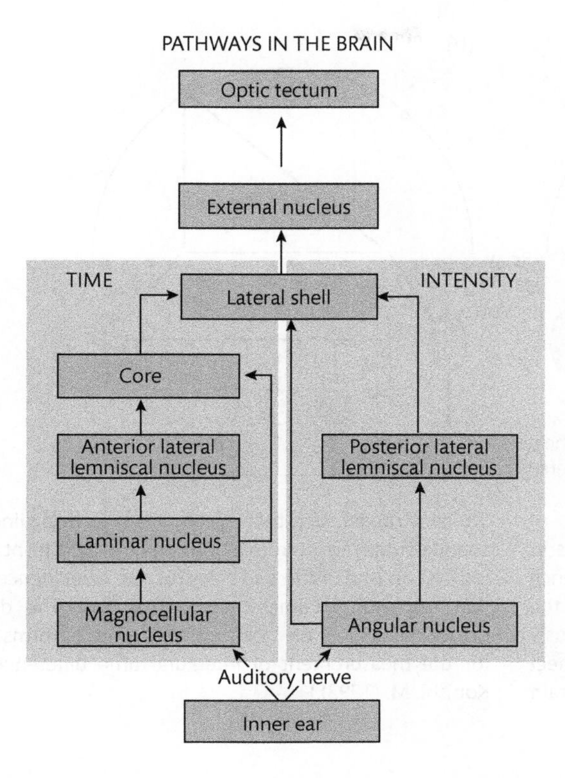

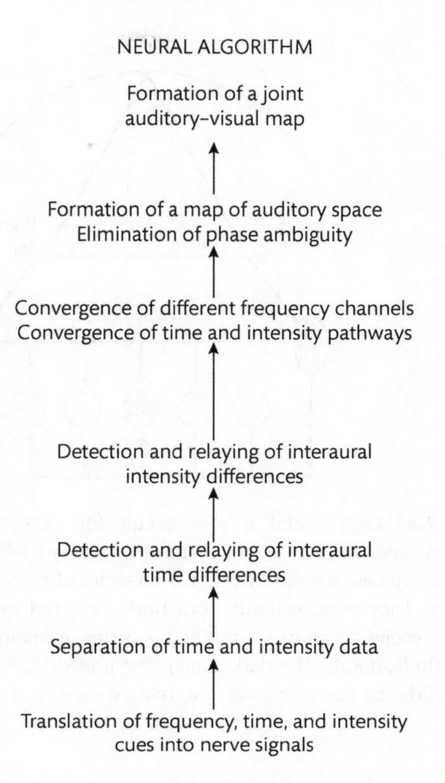

Figure 7.29 The neural circuitry involved in sound localization. To the right, the corresponding steps of the neural algorithm are summarized. The neural pathway starts in the inner ear, whose auditory hair cells send their axons, collectively called the auditory nerve, to two cochlear nuclei—the magnocellular nucleus and the angular nucleus. From this hierarchical level on, interaural time differences and interaural intensity differences are processed in parallel pathways. Time data are transmitted by the magnocellular nucleus, the laminar nucleus, the anterior lateral lemniscal nucleus, the core of the central nucleus of the inferior colliculus (here simply referred to as the 'core'), and the lateral shell of the central nucleus of the inferior colliculus (here simply referred to as the 'lateral shell'). Intensity data are conveyed by the angular nucleus to the posterior part of the dorsal lateral lemniscal nucleus, from where they reach the lateral shell. In the latter, time and intensity pathways converge. From the lateral shell, information is sent to the external nucleus of the inferior colliculus (here simply referred to as the 'external nucleus'). In the external nucleus, an auditory map is formed by a topographic arrangement of the so-called space-specific neurons. Projection of neurons of the external nucleus to the optic tectum leads to the formation of a combined auditory–visual map. (After: Konishi, M. (1992).)

through analysis of the corresponding interaural phase differences. How is this task achieved? As early as in 1948, the American psychobiologist Lloyd Alexander Jeffress (Fig. 7.30), in a theoretical study, proposed a neural cross-correlation network capable of extracting interaural time differences. Jeffress (1900–1986), who served on the faculty of the University of Texas at Austin throughout his professorial career, developed this model circuit while he was visiting professor at the California Institute of Technology in 1947–48. His proposal was published in a seminal paper, entitled 'A Place Theory of Sound Localization,' in the *Journal of Comparative and Physiological Psychology*.

Figure 7.30 Lloyd Alexander Jeffress. (Courtesy: Frank Armstrong, University of Texas at Austin, News and Information Service.)

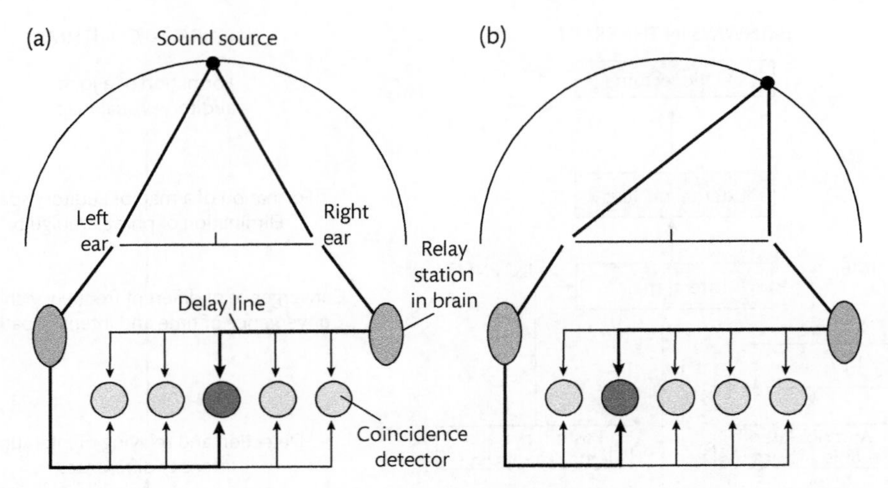

Figure 7.31 Generalized model circuit for detection of interaural time differences, as proposed by Lloyd Jeffress in 1948. A crucial element of this circuit is a series of coincidence detectors. They receive input from both ears, but only the one that receives impulses from the two sides simultaneously will fire (indicated by the dark circle). The fibers that connect the coincidence detectors with the relay station in the brain are constructed such that they function as delay lines. As the sound source moves from a position directly in front (a) to one located on one of the two sides (b), the coincidence detector that responds maximally changes as well. This differential responsiveness of the coincidence detectors forms the basis for the measurement of interaural time differences. (After: Konishi, M. (1993).)

The circuit, which became known as the **Jeffress Model**, is sketched in Fig. 7.31. Its two key elements are **delay lines** and **coincidence detectors**. Jeffress proposed that coincidence detectors receive input from both ears. However, the time of transmission of signals varies between the ears, due to the existence of the delay lines. These reflect the differences in arrival time at the two ears of an acoustic signal if the sound source is not placed directly alongside the midsagittal plane of the head. Thus, different coincidence detectors encode different interaural time differences. Coincidence detectors fire most strongly if the phase-locked impulses generated at lower brain levels reach the detector simultaneously. By contrast, asynchronous arrival of the impulses at the coincidence detector results in relatively weak firing of the detector. Thus, the relative firing rate of the coincidence detectors indicates the direction in azimuth where a given sound comes from.

Delay line In electric circuits, a device that introduces a specific delay time in transmission of a signal. Functionally similar structures exist in neural networks.

In the barn owl's brain, the axons of the magnocellular neurons serve as delay lines. Neurons within the laminar nucleus function as coincidence detectors. This basic neural circuit is shown in Fig. 7.32. When two impulses locked to a certain phase angle of the respective frequency component arrive at roughly the same time, the input-receiving neuron of the laminar nucleus fires at its maximal rate. With a decreasing degree of coincidence, its firing rate drops, until it finally ceases firing. As a result of this operation and of a systematic topographic arrangement of coincidence-detector neurons in the laminar nucleus, certain interaural time differences are encoded by certain laminar neurons defined by their location within the nucleus.

On the other hand, laminar neurons fire maximally not only when hit synchronously by impulses marking a particular interaural time difference; they also respond at maximum rate if this time difference is delayed or advanced by one full cycle (360°) or by integer multiples of this cycle of the sound wave, as shown in Fig. 7.33. This phenomenon is referred to as **phase ambiguity**. Obviously, phase ambiguity, as it persists in the laminar nucleus, imposes an enormous challenge upon the owl's ability to compute interaural

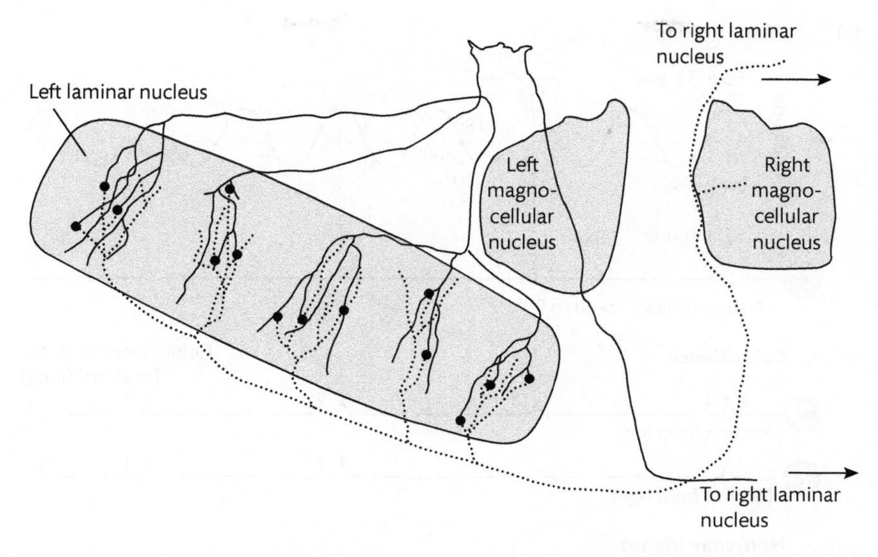

Figure 7.32 Schematic drawing of the dorsal brain stem area with the left laminar nucleus, as well as the left and right magnocellular nuclei. A reconstruction of an ipsilateral and contralateral magnocellular neuron is superimposed over this drawing. This axonal arbor was labeled using a neuronal tract-tracing technique (cf. Chapter 2). The fibers arising from the cell bodies in the magnocellular nucleus serve as delay lines. Neurons of the laminar nucleus (black dots) act as coincidence detectors. They fire at maximum rate when impulses carried by axons arising from the ipsilateral and contralateral magnocellular nucleus reach them simultaneously. For the sake of clarity, axons originating from the left and right magnocellular nuclei are distinguished by solid and dotted lines, respectively. (After: Konishi, M. (1993).)

time differences. However, as we will see in the next subsection, this problem is overcome at higher levels of sensory processing.

➤ Interaural time differences are computed in the laminar nucleus. This is achieved by neurons of this nucleus functioning as coincidence detectors and axons arising from the magnocellular neurons as delay lines.

Another complication arises from the fact that coincidence detectors, such as the neurons within the laminar nucleus of the barn owl, respond not only to **binaural stimuli**, but also to **monaural stimuli.** Discrimination between these two types of stimuli is observed only in higher-order neurons, the so-called space-specific neurons in the inferior colliculus (see 'External nucleus,' later in this section).

After processing in the laminar nucleus, timing information is relayed to higher brain centers on the opposite side of the head. These centers include the **anterior lateral lemniscal nucleus** and the **core of the central nucleus of the inferior colliculus** (see Fig. 7.29). The latter nucleus is here simply referred to as the 'core'.

The posterior lateral lemniscal nucleus: Computation of interaural intensity differences

Similarly, intensity information continues to be carried by a separate pathway after having been processed in the angular nucleus. Among these higher processing centers is the **posterior part of the dorsal lateral lemniscal nucleus.** This brain area receives excitatory input from the contralateral angular nucleus and inhibitory input from the contralateral posterior part of the dorsal lateral lemniscal nucleus. Thus, these two nuclei are reciprocally inhibitory.

➤ Interaural intensity differences are computed in the posterior part of the dorsal lateral lemniscal nucleus.

The posterior part of the dorsal lateral lemniscal nucleus is the brain region where intensity differences between the two ears are computed. The structural basis for this comparison of the intensity between the two ears is provided by the projection of each of the two dorsal lateral lemniscal nuclei to their contralateral counterparts. At the synaptic level, this

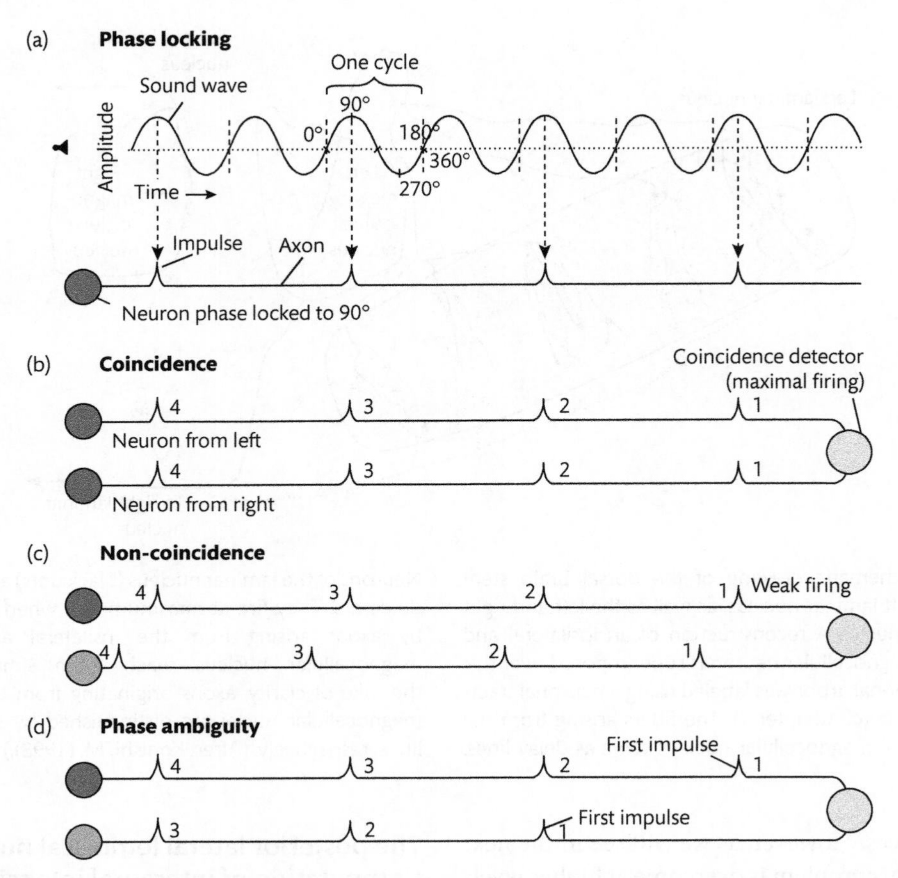

Figure 7.33 Illustration of the principles of phase locking, coincidence, non-coincidence, and phase ambiguity. (a) As a stimulus, a sinusoidal sound wave of a specific frequency is used. An auditory neuron tuned to this frequency produces action potentials at a certain phase angle (here 90°). In other words, the neuron is phase locked to 90°. However, instead of firing every time the wave reaches this phase angle, the neuron depicted here fires only every other time. (b) In both ears, auditory neurons tuned to the same frequency lock to the same phase angle. If these impulses reach the coincidence detector simultaneously, this neuron will fire at maximum rate. (c) If the impulses reach the coincidence detector asynchronously, the neuron will fire rather weakly. (d) The coincidence detector is prone to phase ambiguity. In the case illustrated, the first impulse transmitted by the upper delay line is advanced by the time of two full cycles of the sound wave relative to the first impulse carried by the lower delay line. Despite this ambiguity, the coincidence detector will fire strongly. For the sake of illustration, the impulses have been numbered. (After: Konishi, M. (1993).)

comparison is made by the two different types of input to neurons of the lemniscal nucleus: Inhibitory input received via the contralateral posterior part of the dorsal lateral lemniscal nucleus from the ipsilateral ear, and excitatory input from the contralateral ear. The difference between the strength of the inhibitory input and that of the excitatory input determines the rate at which neurons of the lemniscal nucleus fire.

However, as has been shown by Geoffrey Manley and Christine Köppl in collaboration with Masakazu Konishi, the threshold and strength of the inhibition varies systematically along the dorsoventral axis of the posterior part of the dorsal lateral lemniscal nucleus. This gradient results in neurons arranged in an orderly fashion within the nucleus depending on their selectivity for different interaural intensity differences. In the left nucleus, ventral neurons respond maximally when sound is louder in the left ear; conversely, neurons in the dorsal portion fire most strongly when sound is louder in the right ear. Similarly, in the right nucleus, ventrally located neurons respond maximally when sound is louder in the right ear; neurons in the dorsal part exhibit maximal responsiveness when sound is louder in the left ear.

The lateral shell: Convergence of timing and intensity information

Both the core of the central nucleus of the inferior colliculus and the posterior part of the dorsal lateral lemniscal nucleus project to the **lateral shell of the central nucleus of the inferior colliculus**, here simply referred to as the 'lateral shell' (see Fig. 7.29). This projection pattern suggests that **timing and intensity information converge** at this level of sensory processing. This hypothesis is strongly supported by recordings from neurons within this nucleus. Such experiments have demonstrated that most neurons respond to both interaural time differences and interaural intensity differences. However, localization of a sound source in space is not yet possible, as phase ambiguity still persists.

➤ After parallel processing over several brain levels, time and intensity signals converge in the lateral shell.

External nucleus: Formation of a map of auditory space

Phase ambiguity is overcome only at the next level of sensory processing, namely in the **external nucleus of the inferior colliculus** (see Fig. 7.29). This nucleus is here simply referred to as the 'external nucleus.' The so-called **space-specific neurons** in the external nucleus respond to acoustic stimuli only if the sound originates from a restricted area in space. Sound coming from a source outside this area either does not evoke a response at all or does so only at a low level of firing, by the same neuron. The area in space to which neurons within a certain brain area respond is called the **receptive field.** Neurons of the left external nucleus have their corresponding receptive fields primarily in the right auditory space, whereas neurons of the right external nucleus process information predominantly in the left auditory space, although some overlap in regions near the midsagittal plane of the head does occur.

Space-specific neurons Respond only to acoustic stimuli originating from a restricted area in space.

Another property of the external nucleus is that neighboring space-specific neurons have receptive fields representing neighboring regions in space. This leads to a systematic arrangement of these neurons such that the sound azimuth is arrayed mediolaterally and the sound elevation is mapped dorsoventrally. Overall, as a result of this arrangement, a **neural map of auditory space** is formed.

On each side of the brain, the neural map formed by neurons of the external nucleus represents an auditory region extending from 40° contralaterally to 15° ipsilaterally in the azimuth and from 20° above to 20° below eye level in the elevation. However, within this range, different space regions are represented to a different extent. For example, the region within 20° of the midpoint of the face in both the azimuth and elevation is proportionally heavily represented. On the other hand, the number of neurons devoted to processing of information from regions near the lateral, dorsal, and ventral edges of the auditory space is relatively low. This causes the spatial resolution to be much better if sound comes from an area in front of the owl than if the sound originates from other areas. Similar results have been obtained through behavioral experiments.

An essential prerequisite in eliciting a response from a specific space-specific neuron is that both the time difference *and* the intensity difference fall within the range to which the neuron is normally tuned. Stimulation by the correct time difference (or the intensity difference) alone is not sufficient for a response. These time and intensity difference combinations correspond to the horizontal and vertical location within the bicoordinate system of the neuron's receptive field. Stimulation of the owl through miniature earphones with correct combinations of the parameters evokes not only a response in the corresponding space-specific neurons, but also triggers a turning of the owl's head toward the correct location in auditory space.

➤ Through systematic arrangement of space-specific neurons, a neural map of auditory space is formed in the external nucleus of the inferior colliculus.

The final step in sensory processing: Formation of an auditory–visual map

Under most natural conditions, or if not kept in complete darkness, the barn owl uses both the auditory and the visual system to localize prey. How is information shared between these two systems coordinated?

As investigations by Eric Knudsen (then at Stanford University) have shown, the auditory map provided by neurons of the external nucleus of the inferior colliculus projects to the **optic tectum** (which is homologous to the superior colliculus of mammals). There, a joint **auditory–visual map** is formed. Each neuron of this map responds to both auditory and visual stimuli arising from the same point in space. In this map, the representation of the frontal region of space is greatly expanded. This corresponds to the behavioral observation that, in this region, the highest accuracy in localization of prey is achieved.

> ➤ In the optic tectum, a joint auditory-visual map is formed.

How is the joint auditory–visual map formed? Research by Eric Knudsen and co-workers have suggested that **the alignment of the two sensory maps is controlled during ontogeny by visual instruction of the auditory spatial tuning of neurons in the optic tectum**. The demonstration of such plasticity of the auditory orienting behavior has been achieved through elegant experiments in which young barn owls at various ages were exposed to visual fields displaced by prismatic spectacles. As expected, before the prisms are attached, the visual and auditory responses are very accurate and similar (Fig. 7.34(a)). After one day of experience with the visual field displaced, the auditory responses are still fairly accurate (Fig. 7.34(b)). However, after several weeks of continuous experience with the prisms, the auditory responses are significantly altered (Fig. 7.34(c)). Now, they match the optical displacement of the visual field imposed by the prisms. This learning enables the owls to bring the visual and auditory worlds back into mutual alignment. On the other hand, when the prisms are removed after several weeks of experience with chronic prismatic displacement of the visual field, the auditory and visual fields are again misaligned (Fig. 7.34(d)), and become realigned only gradually, over several weeks.

Where along the auditory localization pathway do the experience-dependent plastic changes take

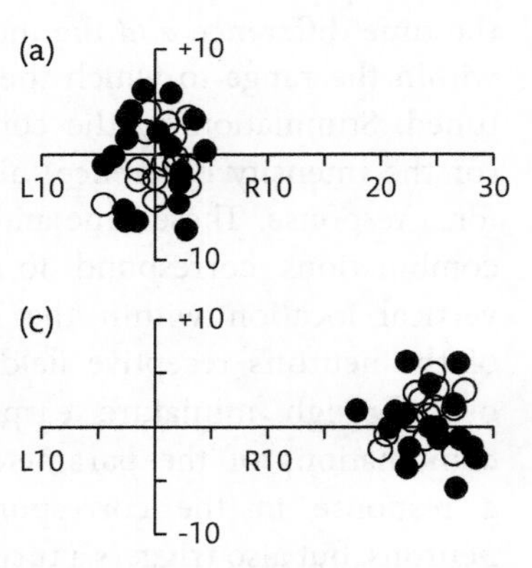

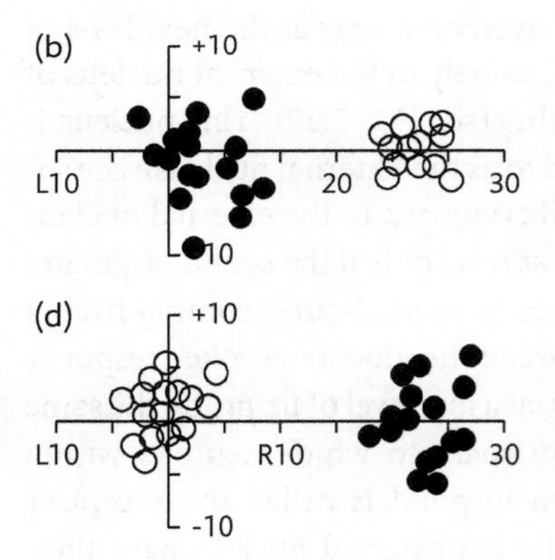

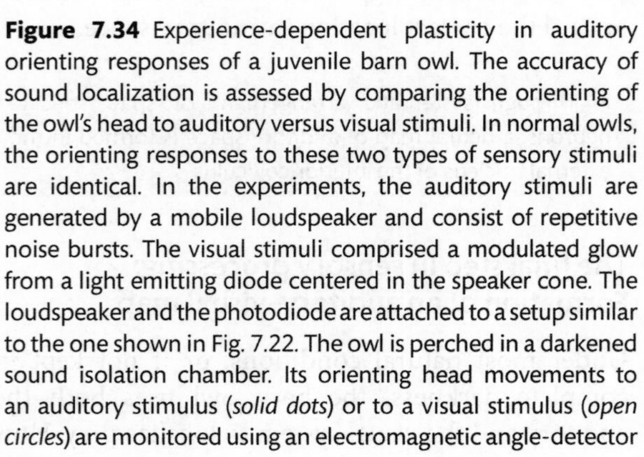

Figure 7.34 Experience-dependent plasticity in auditory orienting responses of a juvenile barn owl. The accuracy of sound localization is assessed by comparing the orienting of the owl's head to auditory versus visual stimuli. In normal owls, the orienting responses to these two types of sensory stimuli are identical. In the experiments, the auditory stimuli are generated by a mobile loudspeaker and consist of repetitive noise bursts. The visual stimuli comprised a modulated glow from a light emitting diode centered in the speaker cone. The loudspeaker and the photodiode are attached to a setup similar to the one shown in Fig. 7.22. The owl is perched in a darkened sound isolation chamber. Its orienting head movements to an auditory stimulus (*solid dots*) or to a visual stimulus (*open circles*) are monitored using an electromagnetic angle-detector system (for a detailed description, see text). All responses are plotted relative to the true location of the stimulus source. Sensory manipulation consists of chronic exposure of the owl to a visual field displaced by prismatic spectacles. (a) Before the prisms are attached. Auditory and visual responses are similar. (b) After 1 day of experience with the visual field displaced 23° to the right. The auditory responses are still fairly accurate. (c) After 42 days of continuous experience with the 23° right-displacing prism. The auditory orienting responses are altered so that the optimal displacement of the visual field is matched. (d) After removal of the prisms. The auditory and visual fields are again misaligned, this time with the auditory responses shifted to the right. (After: Knudsen, E. I., © 2002, Rights Managed by Nature Publishing Group.)

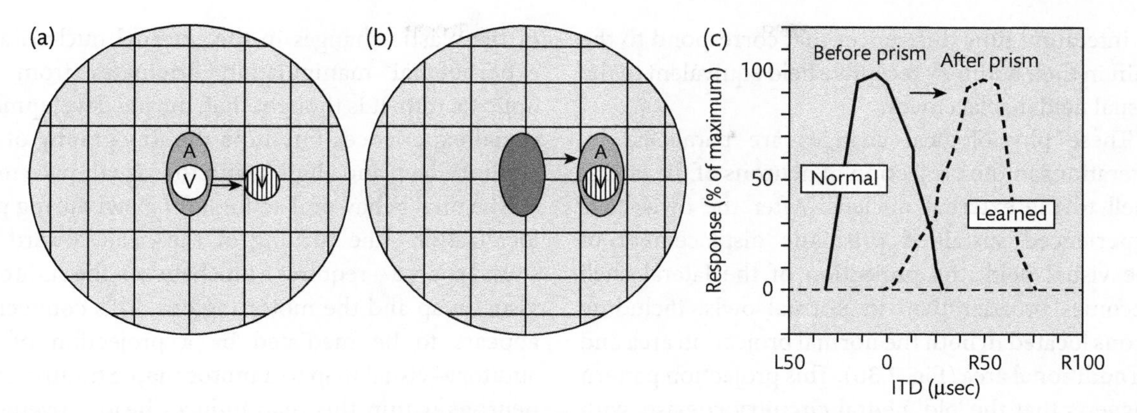

Figure 7.35 Changes in auditory tuning in the optic tectum after prismatic displacement of the visual field. (a) After attachment of prisms, the location of the visual receptive field (*V*) is displaced horizontally to the right. (b) Exposure of the owl to this changed visual receptive field for 8 weeks results in a shift of the auditory receptive field (*A*) so that the latter aligns with the displaced visual receptive field. (c) The tuning of tectal neurons for interaural time differences is changed through exposure to prism-displaced visual fields, as demonstrated by the responses of two units. Each of these units has a visual receptive field centered at 0° in the horizontal plane. Before prism experience, the unit is tuned to an interaural time difference of 0μsec (*solid line*). After attachment of left-shifting 23° prisms for more than 8 weeks, the interaural time difference tuning has shifted to 50-μsec right-ear leading (*dotted line*). *L*, left ear leading; *R*, right ear leading. (After: Knudsen, E. I., © 2002, Rights Managed by Nature Publishing Group.)

place? Research by the laboratory of Knudsen has shown that the site of this adaptive plasticity is the external nucleus. In owls that have experienced displacement of the visual field by prisms, the tuning to sound of neurons in this nucleus and the optic tectum is altered within a few weeks (Fig. 7.35). After experimental horizontal displacement of the visual field through prisms, these neurons soon respond

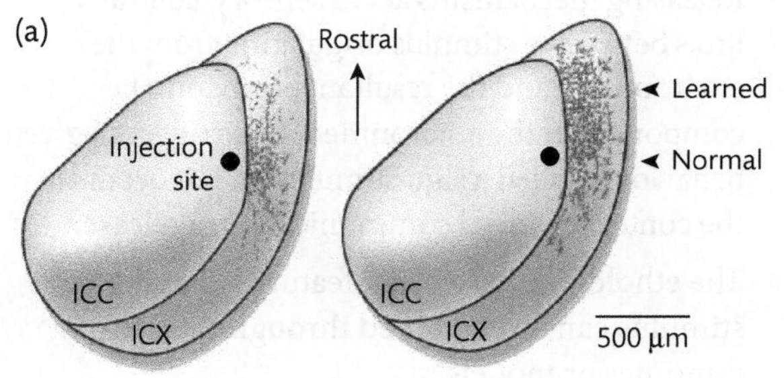

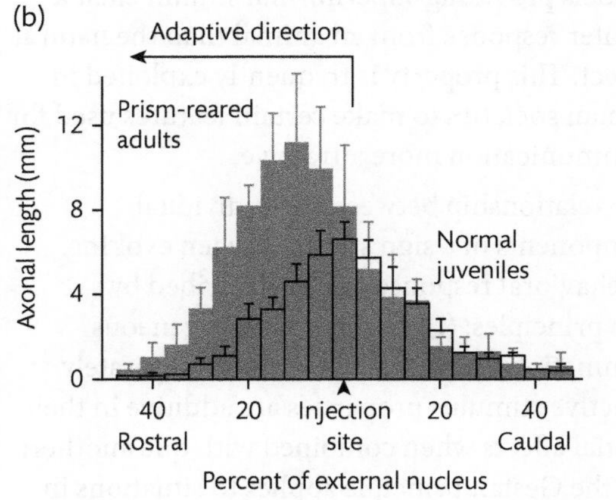

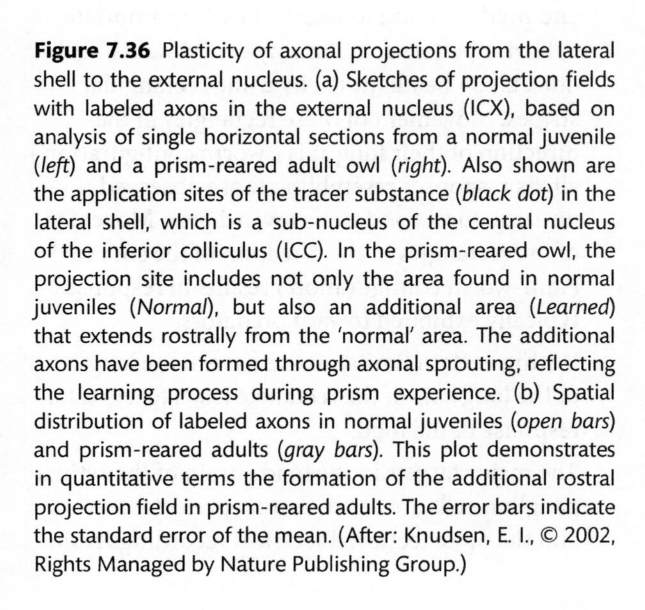

Figure 7.36 Plasticity of axonal projections from the lateral shell to the external nucleus. (a) Sketches of projection fields with labeled axons in the external nucleus (ICX), based on analysis of single horizontal sections from a normal juvenile (*left*) and a prism-reared adult owl (*right*). Also shown are the application sites of the tracer substance (*black dot*) in the lateral shell, which is a sub-nucleus of the central nucleus of the inferior colliculus (ICC). In the prism-reared owl, the projection site includes not only the area found in normal juveniles (*Normal*), but also an additional area (*Learned*) that extends rostrally from the 'normal' area. The additional axons have been formed through axonal sprouting, reflecting the learning process during prism experience. (b) Spatial distribution of labeled axons in normal juveniles (*open bars*) and prism-reared adults (*gray bars*). This plot demonstrates in quantitative terms the formation of the additional rostral projection field in prism-reared adults. The error bars indicate the standard error of the mean. (After: Knudsen, E. I., © 2002, Rights Managed by Nature Publishing Group.)

to interaural time differences that correspond to the shift in their auditory receptive field equivalent to the visual field displacement.

These physiological changes are paralleled by alterations in the projection of neurons of the lateral shell to the external nucleus. After the owls have experienced sustained prismatic displacement of the visual field, the projection of the lateral shell becomes broader than in normal owls, including axons located in both the normal projection area and an additional area (Fig. 7.36). This projection pattern suggests that the 'old' neural circuitry coexists with the 'new' circuitry formed through axonal sprouting and the generation of new synapses. The primary source of the instructive signals for the induction of the plastic changes in the external nucleus after experimental manipulation originates from the optic tectum. It is thought that, during development, visual experiences fine-tune the topography of the auditory map and align it with the visual space map.

The final behavioral action of the owl during prey localization—the turning of the head toward the sound source—requires a link between the auditory–visual map and the motor regions. This connection appears to be mediated by a projection of the auditory–visual map to a **motor map**. Stimulation of neurons within this map induces head movements in the direction of the corresponding receptive field represented by the auditory–visual map.

Summary

- The information flow arising from the environment of an animal is dramatically reduced as a result of sensory and central filters called releasing mechanisms. Thus, the Umwelt of an animal reflects only a tiny portion of the absolute physical environment.
- Releasing mechanisms act as sensory/central links between a stimulus originating from the environment and the resultant behavior. The component of the environment that triggers a given behavior is called a sign stimulus or, if it occurs in the context of social communication, a releaser.
- The ethologically relevant features of a sign stimulus can be examined through the use of dummies or models.
- Models providing supernormal stimuli elicit a greater response from an animal than the natural object. This property is frequently exploited in human societies to make certain features used for communication more attractive.
- The relationship between the individual components of a sign stimulus when evoking a behavioral response can be described by two principles: (i) the law of heterogeneous summation states that two or more separately effective stimulus properties are additive in their partial effects when combined with one another; (ii) the Gestalt principle applies to situations in which the combined stimulation regime is more effective than the sum of its parts.
- The effectiveness of a sign stimulus depends not only on the features of the stimulus, but also on the condition of the recipient.
- Besides having an immediate releasing effect, sign stimuli may also produce long-term changes in the receiver's motivation to generate the corresponding behavioral patterns.
- The first model system discussed in this chapter is based on the ability of toads to recognize prey and predators and to respond with appropriate behavior patterns. In dummy experiments, such objects can be identified by using rectangular stripes. Movement of these rectangles in the direction of their long axis ('worm configuration') elicits responses resembling those observed under natural conditions toward prey. Movement of the rectangles in the direction of the short axis ('anti-worm configuration') results in responses typically exhibited toward predators.
- Within certain limits, an increase in the length of the long axis of the rectangle leads to a greater response of the toad.
- The major targets in the toad's brain of the retinal ganglion cells are the thalamic–pretectal area and the optic tectum. Electrical recordings from

these brain regions, as well as the ganglion cells, have revealed a rather poor sensitivity of retinal ganglion cells to worm- and anti-worm-like stimulus configurations. By contrast, in both the thalamic–pretectal area and the tectum, there are populations of neurons that are activated by more complex stimulus configurations. One cell type called TP3, in the thalamic–pretectal area, is best activated by anti-worm-like stimuli. Conversely, a cell type termed T5(2) in the tectum is activated by a worm-like stimulus configuration.

- In agreement with the results of the recording experiments, prey-catching behavior is released by electrical stimulation of the optic tectum, whereas stimulation of the thalamic–pretectal region activates escape behavior.
- The configurational selectivity of T5(2) neurons in the tectum appears to depend crucially on inhibitory input received from the thalamic–pretectal area. Lesioning of this connection results in the toad being unable to discriminate between prey and predator objects.
- The second model system discussed in detail in this chapter centers around barn owls. As nocturnal hunters, they use the sense of hearing to localize prey, predominantly field mice. Upon hearing the noise generated by the prey animal, the owl turns its head in a rapid flick so that it directly faces the source of sound.
- The owl's head-turning response can be monitored with an electromagnetic angle-detector system consisting of (i) search coils mounted on top of the owl's head, and (ii) induction coils between which the bird is positioned.
- The flick of the head is initiated approximately 100msec after the onset of the sound. While moving the head, the owl does not use any feedback information. Thus, computational decisions regarding the head movement are made under open-loop conditions.
- Under optimal conditions, the barn owl can localize the source of sound within 1–2° in both the horizontal plane (azimuth) and the vertical plane (elevation).
- The owl's ability to precisely localize the source of sound is based on the analysis of interaural time differences and interaural intensity differences. Interaural time differences occur when the sound source is not directly in front of the owl; they define the location of the sound in the azimuth. Interaural intensity differences are the result of a directional asymmetry of the owl's ears and the facial ruff; they determine the elevation coordinate of a sound source.
- Sound intensity and timing are encoded in each fiber of the auditory nerve by variation of the rate of firing and the locking of the action potentials to a particular phase angle of the spectral component to which the respective fiber is tuned.
- Each fiber of the auditory nerve divides into two collaterals. One of these innervates the magnocellular nucleus, while the other enters the angular nucleus. Both nuclei are subdivisions of the cochlear nucleus. Whereas neurons of the angular nucleus are sensitive to intensity information, neurons of the magnocellular nucleus process timing data.
- The next processing station is the laminar nucleus. This receives input from both the ipsilateral and contralateral magnocellular nuclei. The main function of the laminar nucleus is to compute and encode interaural time differences. This is achieved by (i) the axons of the magnocellular nucleus serving as delay lines, and (ii) the neurons of the laminar nucleus functioning as coincidence detectors. However, the timing information extracted is not unambiguous due to the existence of phase ambiguity at this level of sensory processing.
- Interaural intensity differences are computed in the posterior lateral lemniscal nucleus, which receives input both from its contralateral counterpart and from the contralateral angular nucleus.
- Timing and intensity information converge in the lateral shell of the central nucleus where neurons respond to both interaural time differences and interaural intensity differences.
- While phase ambiguity still persists in the lateral shell, it is overcome at the next level of sensory processing, namely in the external nucleus. In this brain region, space-specific neurons respond only to acoustic stimuli originating from their receptive field. Overall, these neurons form a

neural map of the auditory space. Within this map, regions near the midpoint of the face are represented by significantly more neurons than are more lateral areas. This leads to high resolution if the sound comes from sources directly in front of the owl.

- The final step in sensory processing is the formation of an auditory–visual map achieved through projection of the neurons in the external nucleus to the optic tectum. The alignment of the two sensory maps is controlled during ontogeny by visual instruction of the auditory spatial tuning of neurons in the optic tectum.
- The auditory–visual map projects onto a motor map, stimulation of which induces head movements.

The bigger picture

The conceptual relationship between stimulus and behavior underwent significant changes over the centuries. For Descartes (see Chapter 3), stimulus and response were the only two factors defining behavior—the response being the necessary and automatic reaction of the 'animal machine' to the stimulus. This concept was radically modified by ethologists. They demonstrated that, out of the entire stimulus regime, just a minor portion is required to elicit a specific behavioral response. This observation gave birth to the 'sign stimulus' concept. Equally important, there is no automatism that a sign stimulus leads to a behavioral response. Depending on the animal's motivation and prior experience, the resultant behavior may vary tremendously. In the most extreme case, the animal does not respond at all, although at a different time the same stimulus evokes a vigorous response. On the neuroethological frontier, scientists succeeded in identifying neurons that respond selectively to specific sign stimuli, thus acting as the long-sought feature detectors. Significant advances in the characterization of such neurons include elucidation of how they interact with other brain structures, and how neural plasticity enables them to adapt to changes in the stimulation regime.

Recommended reading

Ewert, J.-P. (1974). The neural basis of visually guided behavior. *Scientific American* 230:34–42.

A good introduction to the research led by Jörg-Peter Ewert on the neural basis of object recognition in toads.

Ewert, J.-P. (1980). *Neuroethology: an Introduction to the Neurophysiological Fundamentals of Behavior*. Springer-Verlag, Berlin/Heidelberg/New York.

The first textbook of neuroethology. As many examples have been taken from Ewert's own research, this book is especially well suited for students who seek a comprehensive yet easy-to-read review of the neural mechanisms underlying object recognition in toads.

Ewert, J.-P. (1997). Neural correlates of key stimulus and releasing mechanism: a case study and two concepts. *Trends in Neurosciences* 20:332–339.

An update of Ewert's work and an attempt to interpret the results of his classic work from a more modern point of view. Although sometimes difficult to read, it is useful as a supplement to the above two references.

Knudsen, E. I. (1981). The hearing of the barn owl. *Scientific American* 245:82–91.

A good starting point for an introduction to the work on behavioral mechanisms governing sound localization in barn owls.

Knudsen, E. I. (2002). Instructed learning in the auditory localization pathway of the barn owl. *Nature* 417:322–328.

A stimulating summary of the author's research on barn owls that has shown how the association of auditory cues with the location of the physical space is shaped and modified through experience.

Konishi, M. (1992). The neural algorithm for sound localization in the owl. *The Harvey Lectures* 86:47–64.

A review of the neural mechanisms underlying sound localization in the owl. Suitable for more advanced readers.

Konishi, M. (1993). Listening with two ears. *Scientific American* 268:34–41.

The best overview available to introduce students to the work of Mark Konishi and associates.

Short-answer questions

7.1 Define, in one sentence, the ethological term 'sign stimulus.'

7.2 What is a 'supernormal stimulus?'

7.3 Describe two socially relevant features in humans that are often made supernormal.

7.4 In an experiment, the successive presentation of the individual components of a dummy elicits overall the same response from an animal as does the simultaneous presentation of these components using a single dummy. Based on this result, does the animal's response follow the law of heterogeneous summation or the Gestalt principle?

7.5 Give an example of the frequently made observation that the effect of a sign stimulus does not only depend on the features of the stimulus, but also on the condition of the recipient.

7.6 Briefly describe the experiment of Walter Heiligenberg and Ursula Kramer through which they demonstrated in the Burton's mouthbrooder (*Haplochromis burtoni*) that releasers not only have an immediate releasing effect upon a behavior but may also lead to long-term changes in the underlying motivation.

7.7 Define, in one sentence, the term 'feature detector'.

7.8 In the toad's sensory world, 'prey' can be described as objects characterized by the following two features:

a) ....

b) ....

7.9 What does the term 'retinotectal projection' describe?

7.10 'On the surface of the optic tectum, a retinotopic map is established.' Explain what the latter term in this sentence refers to.

7.11 In toads, neurons that qualify as feature detectors responding best to worm-like stimuli are located in the .... (specify brain region). These neurons belong to the subclass of .... (specify) neurons within this brain region.

7.12 In the toad, what type of behavioral pattern is evoked by electrical stimulation of the thalamic–pretectal region? What behaviors by electrical stimulation of the optic tectum?

7.13 Barn owls localize the source of sound in the azimuth by analyzing interaural .... differences, and in the elevation by analyzing interaural .... differences.

7.14 The two key elements of the Lloyd Jeffress model circuit for detection of interaural time differences are:

a) ....

b) ....

7.15 In the barn owl's brain, the computation of interaural time differences is achieved in the .... nucleus (specify brain region).

7.16 Complete the following sentence: In the .... nucleus (specify brain region) of the barn owl, ... neurons (specify neuronal type) respond only to acoustic stimuli originating from their receptive field. These neurons form a neural map of the .... space.

7.17 In which brain structure of the barn owl is a joint auditory–visual map formed?

7.18 Briefly describe the mechanism that controls the alignment of the auditory and visual maps during ontogeny. Is this alignment reversible?

7.19 **Challenge question** Supernormal stimuli have been suggested to play a role not only in intraspecific interactions of animals, but also as lures for attracting prey. The hydromedusa *Olindias formosus*, for example, has fluorescent patches on the tips of its tentacles, which attract prey fish. Speculate on why the fish might be attracted to such bright fluorescent color, without any obvious advantage.

7.20 **Challenge question** In the mouse retina, high-resolution ganglion cells have been identified that can detect small moving objects, but only if the background is featureless or stationary. It has been postulated that these cells serve as highly selective feature detectors for encoding the presence of aerial predators. Why is it important for the functioning of these 'alarm neurons' that the background is stationary? What natural scene would a global image flow indicate?

Essay questions

7.1 What is the Umwelt concept of Jakob von Uexküll? To illustrate your definition, use the prey-versus-predator recognition of toads as an example.

7.2 What signals do toads use to recognize prey and enemies? How is their response linked to the properties of cells in the retina, and in specific parts of the brain?

7.3 Many studies examining prey localization in barn owls have employed an electromagnetic angle-detector system. What can be measured using this system? How does it work? What are the advantages and disadvantages of such a system, compared with studying owls in free flight?

7.4 Barn owls are able to localize prey based solely on noise generated by the prey animal. How is this achieved at the behavioral level? How are the physical parameters of sound essential for localization of the sound source processed and integrated in the brain?

7.5 In a number of sensory systems, 'parallel processing' has been demonstrated. Explain this term. Illustrate your answer by using the processing of various sound parameters in barn owls as an example.

Advanced topic Adaptive plasticity in the central nervous auditory system

Background information

Many circuits in the central auditory system are not statically fixed, but can be shaped and modified by experience. This form of plasticity enables these circuits to adapt their functional properties to different conditions of the auditory world. The capacity for this plasticity is particularly pronounced during sensitive periods of development of young animals, when a wide range of acoustic stimuli can induce alterations in the fundamental architecture and synaptic connectivity of auditory circuits. However, some central auditory areas also retain their potential for plasticity during adulthood. In contrast to the earlier stages of development, the acoustic stimuli that are capable of inducing structural changes in the adults have to be behaviorally relevant.

Essay topic

In an extended essay, describe how different functional properties of the auditory system, such as frequency tuning and temporal sequence tuning, can be shaped by experience. Include in your review of the literature results of studies that have indicated that the training of musicians has a profound effect on the functional organization of the auditory cortex. Compare the adaptive auditory plasticity during sensitive periods in young animals with those in adults. What is known about the mechanisms that mediate auditory plasticity?

Starter references

Keuroghlian, A. S. and Knudsen, E. I. (2007). Adaptive auditory plasticity in developing and adult animals. *Progress in Neurobiology* **82**:109–121.

Pantev, C., Ross, B., Fujioka, T., Trainor, L. J., Schulte, M., and Schulz, M. (2003). Music and learning-induced cortical plasticity. *Annals of the New York Academy of Sciences* **999**:438–450.

To find answers to the short-answer questions and the essay questions, as well as interactive multiple choice questions and an accompanying Journal Club for this chapter, visit **www.oup.com/uk/zupanc3e.**

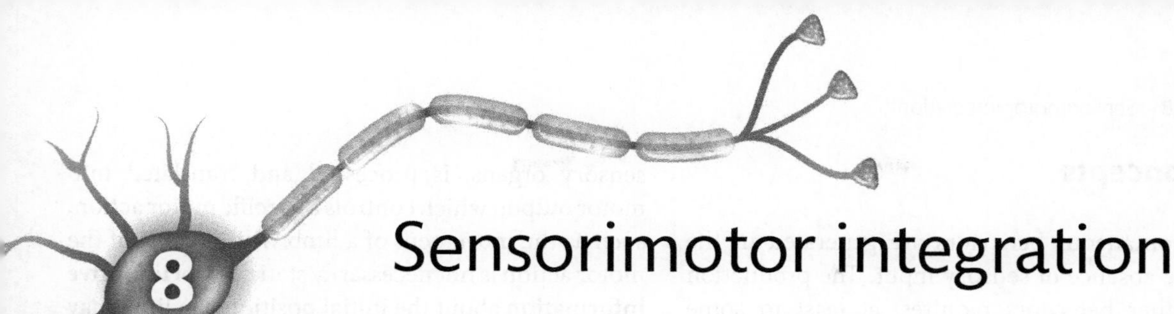

Sensorimotor integration

Introduction

Many behaviors are triggered by specific sensory stimuli. As we discussed in Chapter 7, the initial steps of this process include extraction of behaviorally relevant features from such stimuli at the levels of both the sensory receptors and the corresponding sensory areas in the brain. In toads, for example, worm-like prey objects are distinguished from predator-like objects by two important features: The area dimensions of the object, and the movement relative to these dimensions. This sensory information is transformed and integrated to generate the behavioral output.

The reason that a whole chapter is dedicated to the discussion of sensorimotor integration is *not* that such a process is not relevant for the generation of the behaviors discussed in the previous chapters of this book. The turning of the head toward prey triggered by worm-like stimuli in toads does, of course, require sensorimotor integration—the activation of the muscles that move the head toward an object is achieved by integration of the sensory information that indicate the presence and location of prey. Rather, the division of the text into separate chapters covering sensory processing of behaviorally relevant information, neural control of motor output, and sensorimotor integration, respectively, is motivated by the differences in research focus, as also reflected by the different model systems. Due to the complexity of the underlying processes, the entire neural pathway involved in generating sensory-guided behavior is known in sufficient detail for only a few model systems, and only in these model systems we can fully appreciate the process of sensorimotor integration. Other model systems focus on particular components of the pathway, for example sensory processing or motor output control. One of the model systems that is well characterized in terms of the entire sensorimotor pathway is the jamming avoidance response of the weakly electric fish *Eigenmannia*. The major findings of the research using this model are presented in this chapter.

Key concepts

Only spontaneous behaviors are generated in the complete absence of sensory input. The production of all other behaviors requires, at least to some degree, a certain form of sensory guidance. In the simplest case, the sensory-input-to-motor-output integration consists largely of a feed-forward process (Fig. 8.1(a)). Sensory input, possibly from multiple sensory organs, is processed and translated into motor output, which controls a specific motor action, such as the movement of a limb. This control of the motor action is not necessarily static. **Proprioceptive information** about the initial position of a limb may be used to compute the exact limb trajectory, when, for example, an animal catches an object or grooms a particular site of the body (for a definition of proprioception, see Chapter 6).

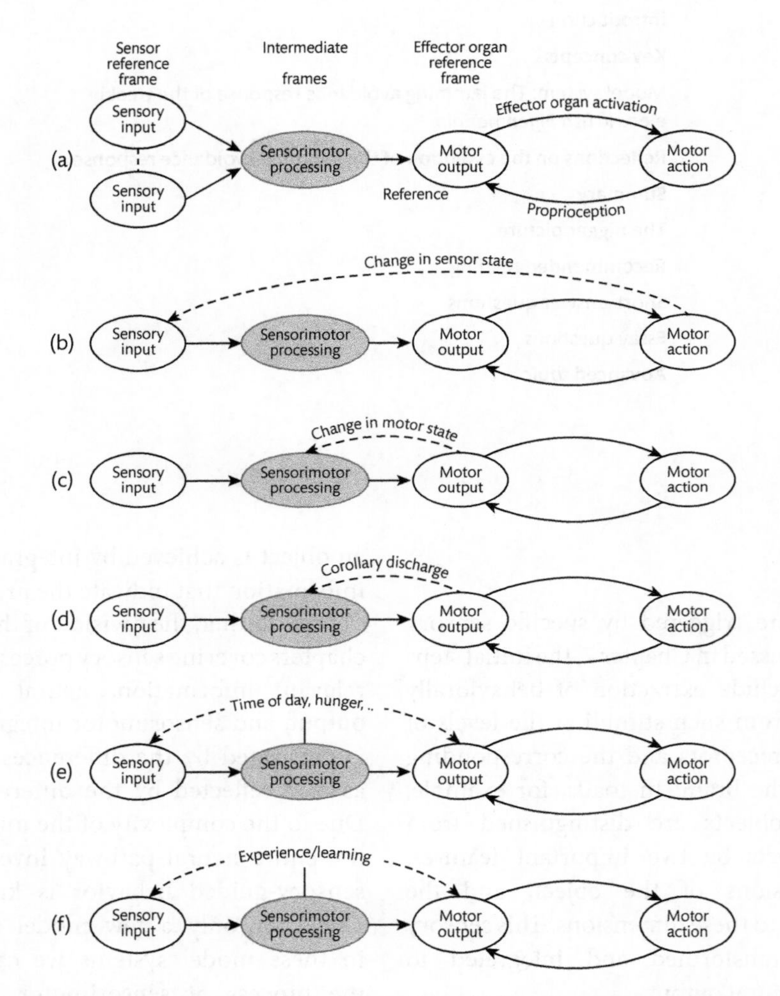

Figure 8.1 Modes of sensorimotor integration. (a) In the simplest form, transformation of sensory input into motor output is predominantly feedforward. The motor output may be adjusted according to proprioceptive information received from the effector organ, such as muscles. (b) During active sensation, signals are emitted to probe the environment, or sensors are moved in a search pattern. These specific motor actions may lead to changes in the motor state. (c) Sensorimotor integration may be fine-tuned according to differences in the motor state, as they are encountered, for example, when the animal exhibits locomotor activity compared to when it stands still. (d) If the animal's motor action affects the sensory input in a predictable way, then a corollary discharge may be produced to subtract the predictable component of the sensory input from the overall sensory input received. (e) Sensorimotor integration may be influenced by the animal's motivation or may show diurnal variability. (f) Similarly, the transformation of sensory input into motor output may depend on previous experience and learning. (After: Huston, S. J. and Jayaraman, V. (2011), © 2002, Rights Managed by Nature Publishing Group.)

In certain behavioral situations, animals actively change the sensor state to probe the environment by emitting a signal that can be detected by their own sensors, or by moving the sensors in a search pattern. An example is the emission of signals by bats, and the perception and analysis of the returning echo by sensory receptors and brain structures specialized in the processing of such sensory stimuli. Information obtained through this processing may then be used to modulate the motor output, as illustrated in Fig. 8.1(b). Greater horseshoe bats, for example, adjust the frequency of the emitted sound when relative movement occurs between them and an object. This strategy, called Doppler shift compensation, is used to keep the frequency of the returning echo within the range to which the sensory structures are optimally tuned (cf. Chapter 5).

Sensorimotor integration may also be modulated according to the motor stage (Fig. 8.1(c)). For example, the demands for sensory processing are much higher during movements of an animal than they are when the animal stands still. Thus, adaptations of the properties of sensory structures to these different behavioral states help the animal to save energy.

Similarly, modulation of sensorimotor integration is used to distinguish sensory input arising from the outside world from that generated by the animal's own actions (Fig. 8.1(d)). This is achieved by producing **corollary discharges**. As we will learn in Chapter 12, crickets use such a mechanism to selectively suppress sensory responses during their own chirping.

Corollary discharge A copy of a motor command that does not produce any motor action. Instead, it is routed to sensory structures where it influences sensory processing.

Another important factor that has a major impact on the various components of sensory processing, sensorimotor integration, and generation of motor output is the internal state of the animal. This state is determined by a multitude of factors, among which thirst, hunger, and time of the day are of particular importance (Fig. 8.1(e)). Modulation by circadian input is, for example, of high relevance during navigation based on sun-compass orientation (cf. Chapter 11).

Last but not least, the chain of processes involved in the perception of sensory stimuli and the generation of proper behavioral output may exhibit experience-dependent plasticity (Fig. 8.1(f)). An example of this property is when an animal displays a certain motor behavior in response to a sensory stimulus only if this stimulus was previously associated with reward.

Model system: The jamming avoidance response of the weakly electric fish *Eigenmannia*

Establishment of the jamming avoidance response as a neuroethological model system

One of the masterpieces of neuroethological research, which will be described here, is the behavioral and neurobiological analysis of the **jamming avoidance response** of the knifefish, *Eigenmannia* sp. This behavior was discovered by two Japanese scientists, Akira Watanabe and Kimihisa Takeda, both from the Tokyo Medical and Dental University. They described their findings in a paper published in the *Journal of Experimental Biology* in 1963.

Shortly after its discovery, Theodore ('Ted') Bullock (cf. Chapter 3) of the Scripps Institution of Oceanography of the University of California, San Diego, began to study this seemingly exotic behavioral pattern. He assumed the function of this response to be maintenance of a private frequency by each individual fish for detecting objects, such as obstacles in the closer vicinity. He therefore named it the jamming avoidance response.

Together with a German postdoctoral fellow, Henning Scheich, Ted Bullock carried out a thorough characterization of the jamming avoidance response and even made a computer model that predicted accurately the dynamics of the input–output relationship for a range of different stimuli. Their model included several key parameters of the stimulus, but lacked a measure of the assumed function, namely the accuracy of detecting objects. Such a test was devised by another German scientist, Walter Heiligenberg (see Box 8.1), who came to the Scripps Institution of Oceanography at the beginning of the 1970s. Over a period of 20 years, together with his group and collaborators, he studied the jamming avoidance response in great detail. They succeeded in revealing both the principal behavioral rules and the major neural components underlying this behavior. The results of this research provide an excellent example to illustrate **how sensory information and motor programs are integrated** to generate a biologically important behavior pattern in response to a stimulus.

Electric organs and electroreceptors

Eigenmannia sp. (in most studies, the species *Eigenmannia virescens* is used) is a teleost belonging to the order Gymnotiformes. Figure 8.2 shows a photograph of this fish. Members of the genus *Eigenmannia* live in freshwater habitats in South and Central America. Like all other members of its order, these fish are distinguished by their ability to produce **electric fields.** This is achieved by discharging an **electric organ** in the tail. The electric organ is composed of modified muscle cells called **electrocytes,** each of which produces a discharge of some tens of millivolts. As shown in Fig. 8.3, many of these cells are arranged in series. By this arrangement, and the synchronous depolarization of the electrocytes, the rather low voltages of the individual discharges add up to a sum potential of the order of 1V.

Box 8.1 Walter Heiligenberg

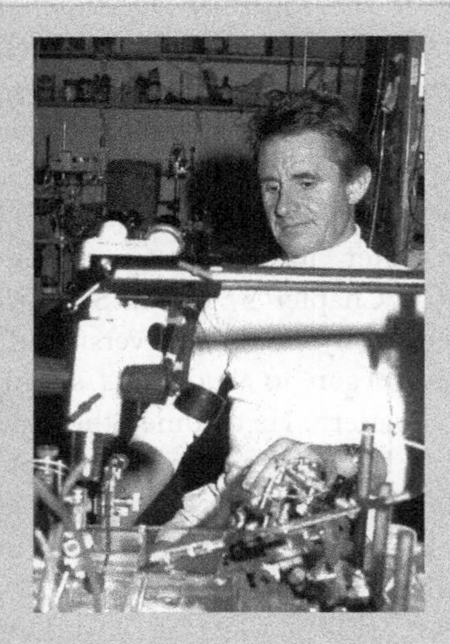

Walter Heiligenberg in his laboratory at the Scripps Institution of Oceanography. (Courtesy: G. K. H. Zupanc.)

One of the best and most complete case studies ever performed in neuroethology is the work of Walter Heiligenberg on the jamming avoidance response of weakly electric fish. Heiligenberg was born in Berlin (Germany) in 1938. At the age of only 15, he met Konrad Lorenz (see Box 3.1), who at the time was Head of the Max Planck research group in Buldern (Westphalia, Germany). Under the guidance of Lorenz, and while still at school, Heiligenberg performed his first behavioral observations on fish and birds. During that time, Lorenz, together with Erich von Holst (see Box 3.5), established the Max Planck Institute for Behavioral Physiology in Seewiesen. Heiligenberg followed Lorenz to do his Ph.D. work, an analysis of the influence of motivational factors on the occurrence of behavioral patterns in a cichlid fish. This study, completed in 1963 under the guidance of Konrad Lorenz and the well-known sensory physiologist Hansjochem Autrum of the University of Munich, already revealed features that became characteristic of Heiligenberg's work: Exploration of a biological phenomenon of general interest through the application of a strict analytical and quantitative methodology. In using this approach, Heiligenberg was years ahead of his time. The approach had its roots in Heiligenberg's keen interest in mathematics and physics, subjects, which in addition to zoology and botany, he also formally studied at the university.

During his postdoctoral years in the laboratory of Horst Mittelstaedt at the Max Planck Institute for Behavioral Physiology, Heiligenberg continued to analyze motivational processes. He was the first to succeed in a quantitative demonstration of the law of heterogeneous summation (see Chapter 7). In 1972, upon invitation by Theodore Bullock, he joined the Neurobiology Unit of the Scripps Institution of Oceanography of the University of California, San Diego. At the same institution, he established his own laboratory in 1973, and he remained there as a member of the faculty until his death in 1994. It was during this period of roughly 20 years that he performed his work on the jamming avoidance response.

Walter Heiligenberg was not only an extraordinarily gifted researcher of enormous devotion to, and enthusiasm for, science; he was also an outstanding craftsman who liked to design and assemble his own mechanical and electronic devices. After his first wife died of cancer in 1991, he married a musician who lived in Munich. Following their marriage, he traveled back and forth between San Diego and Munich. On one of these trips, on his way to San Diego, he arrived in Chicago earlier than expected. He decided to take a flight different from the one for which he was scheduled. A few minutes before landing at Pittsburgh, the plane went into a nose dive and crashed. All 132 people aboard the aircraft, including Walter Heiligenberg, perished.

Figure 8.2 The knifefish, *Eigenmannia virescens.* This weakly electric gymnotiform fish, as well as other species within its genus, have become one of the premier model systems in neuroethology. (Courtesy: G. K. H. Zupanc.)

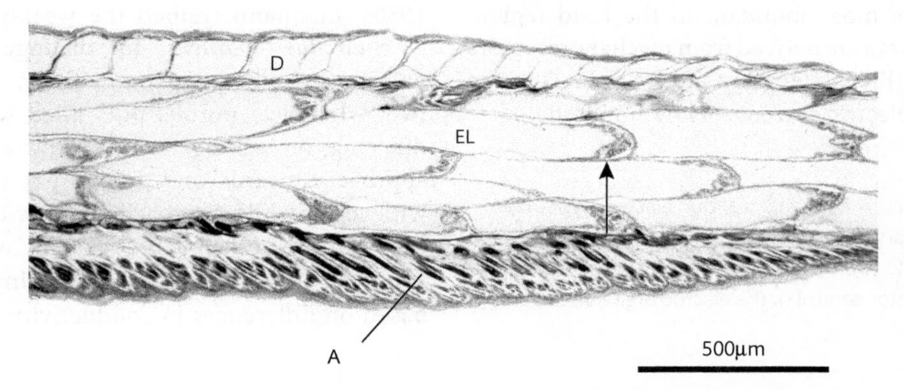

Figure 8.3 Parasagittal section through the electric organ of *Eigenmannia* near the caudal end of the anal fin (orientation: dorsal side, up; ventral side, down; rostral end, left; caudal end, right). The electric cells or electrocytes (EL) are distinguished by their long, slender shape. They are arranged in series along the longitudinal axis of the body. At their caudal end, where the membrane exhibits pronounced evaginations (arrow), the electrocytes are innervated by electromotoneurons. A, Muscle fibers of anal fin; D, dermis with scales. (Courtesy: G. K. H. Zupanc and F. Kirschbaum.)

The electric signals can be monitored by placing two stainless steel electrodes close to the head and tail of the fish, respectively. The electric organ discharges approximately 200–500 times/sec. The corresponding frequency range of 200–500 Hz is specific to the species. Although different individuals discharge at different frequencies within this range, the frequency of each individual fish is extremely regular. Moreover, and as demonstrated by Fig. 8.4, these so-called **electric organ discharges** are, in terms of their waveform, rather simple, resembling a train of sine waves.

The regularity of the electric organ discharges is determined by the action of an endogenous oscillator in the medulla oblongata, an area located in the brainstem of these fish. This oscillator, called the **pacemaker nucleus**, shows a similar high degree of regularity to the electric organ. It sends one volley of spikes down the spinal cord to trigger one discharge cycle of the electric organ discharges.

The fish are able to sense their own electric discharges, as well as electric signals of other fish or of abiotic sources, through various types of

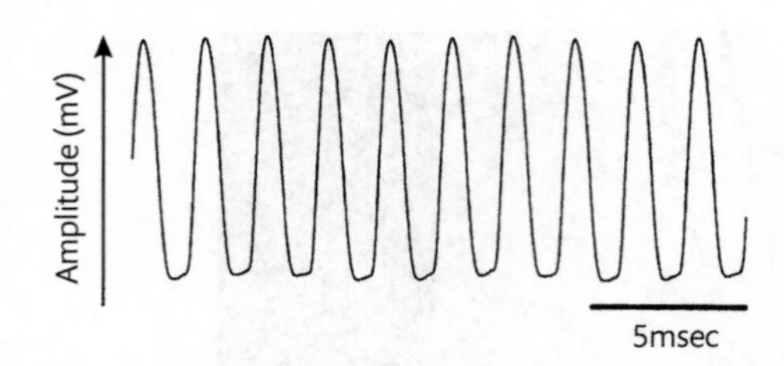

Figure 8.4 Electric organ discharge of *Eigenmannia*. The recording shows ten discharges produced within approximately 20msec, thus corresponding to a frequency of roughly 500Hz. Note the enormous degree of regularity in frequency. (Courtesy: G. K. H. Zupanc.)

electroreceptor. Two of the major receptor types—tuberous and ampullary—are shown in Fig. 8.5. This illustration is based on a photomicrograph taken from a histological section through the skin of *Eigenmannia*, and thus reveals the morphological structure of these receptors. Both types of electroreceptor are distributed all over the body, but are most abundant in the head region. Electroreceptors are derived from mechanoreceptors of the lateral line system, but are specialized in the detection of electric currents of low amplitude.

> ➤ The major components of the electric system are electroreceptors, the electric organ, and central structures devoted to the processing of electrosensory information and to the motor control of the electric organ discharge.

Electrolocation

As the fish is continuously 'surrounded' by its own electric field, electric signals of neighbors are perceived as perturbations in the feedback from its own discharges, rather than as discrete electric events. Similarly, an object in the fish's vicinity distorts the electric field and alters the current pattern perceived by the array of electroreceptors closest to the object. This is shown schematically in Fig. 8.6 by demonstrating the effect of a non-conducting object on the current pattern. The alteration in the current pattern on the fish's body surface represents the **electric image** of the object. The analysis of such images enables *Eigenmannia*, like other electric fish, to **electrolocate** objects in their own environment. Electrolocation can therefore be considered a form of 'seeing' with the body surface.

The electrolocation capability was first demonstrated by Hans Werner Lissmann (Box 8.2) of the University of Cambridge (U.K.) in the 1950s. Lissmann trained the weakly electric fish *Gymnarchus niloticus* to distinguish between objects of different conductivity, for example two cylindrical porous pots filled with solutions differing in salinity. In the course of training, approach towards one porous pot was rewarded with food, while approach towards the other was followed by punishment consisting of chasing the fish away. The fish's ability to discriminate the objects based on differences in conductivity was tested by

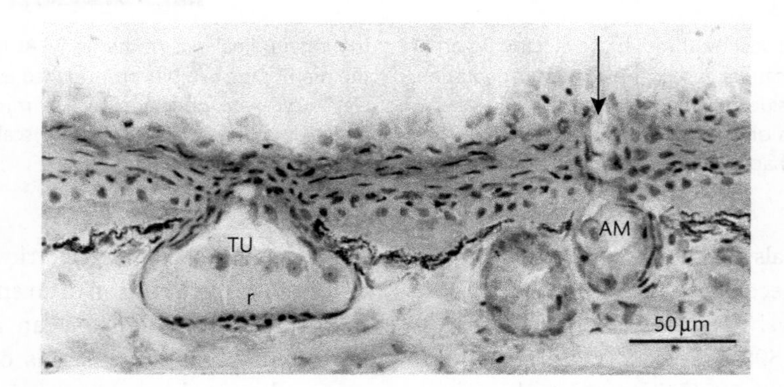

Figure 8.5 Two types of electroreceptor organ in the skin of *Eigenmannia*. In the left half of the photomicrograph, a tuberous (TU) electroreceptor organ, composed of individual receptor cells (r), is shown. This type of electroreceptor is tuned to AC signals with frequencies in the range of the fish's own electric organ discharge. In the right half of the photomicrograph, an ampullary electroreceptor organ (AM) is visible. This receptor organ type is characterized by a long, large-diameter canal (arrow). Ampullary receptors respond maximally to DC and low-frequency AC signals. (Courtesy: G. K. H. Zupanc and F. Kirschbaum.)

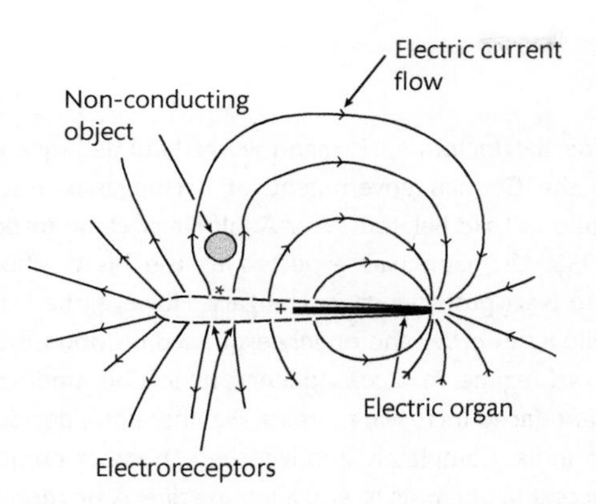

Figure 8.6 Schematic drawing of an electrolocating electric fish. The black bar indicates the location of the electric organ. Electroreceptors are found in pores of the body surface. Their density is highest in the rostral part of the body. While the interior of the body is of relatively low resistance, the resistance of the skin is high, forcing the current to flow through the pores occupied by the electroreceptors. An object with an impedance different from that of the surrounding water (in the drawing, a non-conducting object is shown) causes a distortion of the electric field, as represented by the lines of current flow. This leads to an alteration in the pattern of transepidermal voltage in the area of the skin nearest to the object (indicated by *). This alteration, which represents the electric image of the object, is monitored by the electroreceptors. (After: Heiligenberg, W. (1977).)

analyzing the differences in approach when the porous pots were presented without rewarding the fish for a 'correct' response or punishing it for an 'incorrect' response. The positive results of these and other experiments demonstrated that the fish are, indeed, capable of electrolocation, and that they can distinguish objects based solely on their electric properties—even when the differences in conductivity are minor. As key experiments in the discovery of a 'sixth sense' of electric fish, Lissmann's findings paved the path for numerous subsequent investigations on this animal group.

The jamming avoidance response

Problems in the ability to electrolocate arise if a neighboring fish with a discharge frequency similar to the fish's own electric organ discharge frequency comes close. Then, **the two electric fields interfere with each other**, resulting in phase and amplitude modulations of each of the two electric signals, as will be shown in detail below. It has been demonstrated that these modulations severely impair the fish's ability to electrolocate. This effect is most detrimental if the difference between the fish's frequency and the frequency of the neighbor is approximately 20Hz.

To avoid the detrimental interference of the two electric fields, the fish shifts its own frequency away from the neighbor's frequency such that it maximizes the frequency difference. This behavior is called the **jamming avoidance response**. It leads to the following behavioral pattern:

- If the neighbor's frequency is higher than the fish's own frequency, the fish lowers its discharge frequency.
- If the neighbor's frequency is lower than the fish's own frequency, the fish raises its own discharge frequency.

➤ The jamming avoidance response involves a shifting of the fish's own frequency away from the frequency of the interfering signal.

These observations demonstrate that the fish is able to determine the **sign of the frequency difference**. This difference, commonly termed 'Df,' is defined as the neighbor's frequency minus the fish's frequency.

Frequency difference (Df) In wave-type electric fish, frequency of the neighbor's signal minus frequency of the fish's signal.

Example: *(1) The frequency of the fish is 400Hz and that of its neighbor is 410Hz. Thus, the frequency difference is +10Hz, with the sign of Df being positive. This will cause the fish to lower its frequency. (2) The frequency of the fish is, again, 400Hz, but the frequency of the neighboring fish is now 397Hz. Thus, the frequency difference is −3Hz, with the sign of Df being negative. As a result, the fish will raise its frequency.*

How does the fish determine the sign of the frequency difference, and how does it implement these behavioral rules at the neural level?

Box 8.2 Hans Werner Lissmann

Hans Werner Lissmann. (Courtesy: Nicholas Sinclair.)

His life was marked by personal hardship and political oppression. To survive, he had to develop a strong and independent personality. It was the same kind of character traits–perseverance and independence in thinking–that enabled him to make a key discovery in zoology in the twentieth century: the phenomenon of 'electrolocation'–the ability of weakly electric fish to orient in their environment by detecting differences in the conductive properties of objects through analysis of distortions of their self-generated electric fields.

Hans Werner Lissmann was born in Nikolayev, a Black Sea port near Odessa, in 1909. His parents were Germans living in Imperial Russia. After the outbreak of the First World War, he and his family were interned as aliens in a village on the edge of the Urals. There, deprived from formal education, he was taught arithmetic and languages by his mother, and biology by two interned biologists. In 1919, the family tried to escape the Russian Revolution by heading on foot with a horse-drawn cart toward Germany. Three years later and after enormous suffering, the family arrived in Germany and finally settled in Hamburg. In 1928, Hans started to study zoology at the University of Hamburg. After the father's business had gone bankrupt, Hans earned money by working in the docks, sweeping roads, or helping with the maintenance of the Hamburg Aquarium. Despite the everyday struggle to have enough to eat, he obtained his Ph.D. in 1932, under the mentorship of Jakob von Uexküll (see Chapter 3), with an experimental study on the behavior of Siamese fighting fish (*Betta splendens*).

After his doctorate, Lissmann worked, funded by a grant from the German government, at a Hungarian research institute at Lake Balaton. After Adolf Hitler came to power in 1933, Lissmann was expected to use his position to spread Nazi propaganda in Hungary. However, he refused to follow this order and openly expressed his opposition to the Nazi regime. As a consequence, he lost his funding and job, and faced increasing threats. He, therefore, decided to go to India. Completely impoverished and after numerous unsuccessful attempts to find a job in science, he received a letter in Calcutta from James Gray who invited Lissmann to work with him on locomotion in the Department of Zoology at the University of Cambridge. Gray at that time played a leading role in England in managing the transition of zoology from its comparative and evolutionary positioning to a more functional orientation. Over the years following his arrival in Cambridge, Lissmann made seminal contributions to this development. His papers ranged from studies of the mechanisms of locomotion in leeches, amphibians, gastropod mollusks, dogfish, and snakes to locomotory reflexes in earthworms. Despite this scientific success, Hans Lissmann encountered continued hostility of some of his colleagues to foreigners, particularly Germans, after the outbreak of the Second World War. Like many Germans living in Britain at that time, he was interned and finally transferred to Canada. Only the support of his sympathizers secured his release and return to Cambridge in 1943.

After the war, Hans Lissmann's investigations were instrumental in establishing a new line of research in behavioral biology–exploration of the functions of the discharges generated by the electric organs of weakly electric fish. By conducting conditioning experiments in the laboratory, and complemented by field observations in both Africa and South America, he discovered that the fish can detect differences in the conductive properties of objects. Objects differing in conductivity cause perturbations of the fish's self-generated electric field, which the fish are able to analyze with their electroreceptive system. This ability to 'electrolocate' enables weakly electric fish to orient in turbid waters, which are common in their natural habitat.

Lissmann's achievements in experimental zoology were recognized by his election to the Royal Society in 1954. One year later, he was appointed Lecturer (comparable to Assistant Professor in the U.S. academic system) in Zoology at the University of Cambridge, followed by promotion to Reader (comparable to Associate Professor) and Director of the Sub-Department of Animal Behavior.

Hans Werner Lissmann died in Cambridge in 1995.

Determination of the sign of the frequency difference without internal reference

If a human engineer were confronted with the task of determining the sign of the frequency difference, the likely solution would be as follows. As the fish's own discharge frequency is determined by the frequency of the pacemaker nucleus, the fish could use the frequency of this oscillator as an internal reference. Thus, by comparing this frequency with the frequency of the interfering signal produced by the neighbor, the fish would be able to determine the sign of the frequency difference between the two electric organ discharges.

Very much to the disappointment of the biologist, however, this is not what *Eigenmannia* does. The following two experiments have ruled out this possibility and revealed some details of the behavioral mechanism underlying the jamming avoidance response.

In the first experiment, the fish's electric organ discharge was silenced by the application of **curare**. (This is possible because the electric organ of *Eigenmannia* is derived from muscle cells, and curare acts as a relaxant of skeletal muscles. On the other hand, curare does not affect neurons; thus, the frequency of the neuronal pacemaker nucleus continues to oscillate at a normal rate.) The electric organ discharge of the fish was then replaced by a mimic of similar amplitude and frequency. This was done by placing one electrode into the fish's mouth and another electrode at the tip of its tail (Fig. 8.7). Such an arrangement of electrodes results in an electric field geometry similar to that produced by the natural electric organ discharge. It is sufficient to use an **electric sine wave**, as it completely mimics the fish's discharges. Similarly, the electric field of a neighboring fish can be perfectly mimicked by an electric sine wave applied through a pair of external electrodes straddling the fish.

Curare Generic name for various types of unstandardized extracts derived mainly from the bark of the tropical plants *Strychnos* and *Chondrodendron*. It is prepared for use as an extremely potent arrow poison by Indians in South America. The physiologically active ingredient of curare is the alkaloid tubocurarine, which is employed as a relaxant of skeletal muscles during surgery to control convulsions. The muscle-relaxant effect is caused by interference of the alkaloid with the action of acetylcholine at the neuromuscular junction.

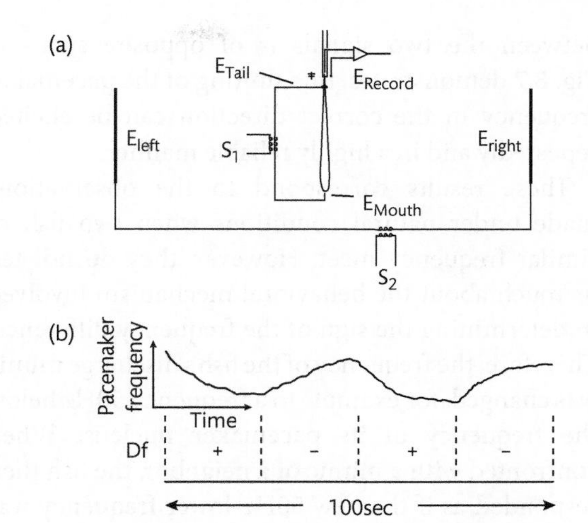

Figure 8.7 (a) Jamming avoidance response evoked in a curarized fish after its silenced electric organ discharge has been replaced by a sine wave substitute (S_1) applied through electrodes between the fish's mouth and tail (E_{Mouth} and E_{Tail}, respectively). The fish is stimulated with a second sinusoidal electric stimulus (S_2) simulating the electric field of a neighbor. This stimulus is delivered transversely through one electrode placed on the left side (E_{left}) and another electrode placed on the right side (E_{right}). During the experiment, the pacemaker frequency is monitored by recording the spinal volley traveling from the pacemaker nucleus to the electric organ through a suction electrode placed over the tip of the tail (E_{record}). (b) The sign of the frequency difference of the fish's discharge mimic and the neighbor's mimic (defined as Df) is positive if the frequency of the neighbor is higher than that of the fish. It is negative if the frequency of the fish's mimic is higher than that of the neighbor's mimic. In the experiment shown, the sign of Df is switched every 25sec. The results demonstrate that the fish executes a correct jamming avoidance response. It lowers its pacemaker frequency if the neighbor's frequency is higher (positive Df), and it raises its pacemaker frequency if the neighbor's frequency is lower (negative Df) than the frequency of the fish's mimic. A similar result can be obtained even if the frequency of the fish's substitute differs from the frequency of the pacemaker. This demonstrates that the fish uses the external electric field frequency rather than the 'true' internal frequency of the pacemaker nucleus as a reference. (After: Heiligenberg, W., Baker, C., and Matsubara, J. (1978).)

When the fish is exposed to this experimental condition, it will execute a correct jamming avoidance response. It will lower its pacemaker frequency if the mimic of the neighbor's signal has a frequency slightly higher than the mimic of its own discharges. On the other hand, it will raise its pacemaker frequency if the frequency difference between the two signals is of opposite sign. As Fig. 8.7 demonstrates, this shifting of the pacemaker frequency in the correct direction can be elicited repeatedly and in a highly reliable manner.

These results correspond to the observations made under natural conditions when two fish of similar frequency meet. However, they do not tell us much about the behavioral mechanism involved in determining the sign of the frequency difference. Therefore, the frequency of the fish's discharge mimic was changed, for example to a frequency 50Hz below the frequency of its pacemaker nucleus. When confronted with a mimic of a neighbor, the fish then responded as if the new 50Hz-lower frequency was its own frequency. It obviously used the electric field frequency as a reference, rather than the 'true' internal frequency of the pacemaker nucleus.

In the second experiment, the fish was placed into a two-compartment chamber. Around its pectoral region, an electronically tight seal was fitted so that practically no signal of its own electric organ discharge could be detected at its head surface. In this region, the density of the electroreceptors is higher than in any other part of the body surface.

When a jamming stimulus, that is, a stimulus within a frequency range of 20Hz above or below the fish's own discharge frequency, was presented to the head alone, the fish did not perform a jamming avoidance response. However, when the jamming signal was allowed to leak into the chamber containing the trunk, a proper jamming avoidance response was evoked. Similarly, a response was elicited by allowing the fish's electric organ discharge to leak into the chamber with the head.

Indirectly, this experiment supported the notion that the fish does not make internal reference to the frequency of the pacemaker. More importantly, this result demonstrated that the fish needs to be exposed to a mixture of its own signal and the interfering signal on some parts of its body surface in order to execute a jamming avoidance response.

Behavioral rules governing the jamming avoidance response

The first of the two experiments described in the previous subsection demonstrates that a potential interference of the electric fields of two fish with similar frequency does not require application of the natural discharge signals. Rather, it can be adequately simulated by replacing the fish's discharges by a sine wave applied through electrodes placed into the mouth and at the tail, and by applying the neighbor's mimic through a separate pair of electrodes. Due to this arrangement, the two electric fields have different geometries. This causes the electric current originating from the neighboring fish to affect the electroreceptors on different parts of the body to different degrees. As can be seen from Fig. 8.8, this is mainly due to differences in the angle by which

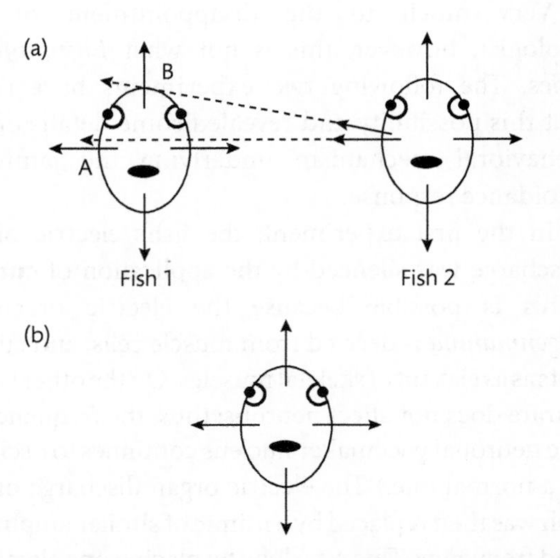

Figure 8.8 Separate (a) and identical (b) geometry conditions. Under separate conditions, the two electric fields are produced by two separate electric current sources, for example two separate technical dipoles or, as shown here, two individual electric fish. As the current of the fish's discharge originates from an internal source, it penetrates the body surface perpendicularly and at similar intensities at all points. This also holds for substitutes of the electric organ discharge supplied through electrodes placed in the fish's mouth and at its tail. On the other hand, electroreceptors only sense the current component oriented perpendicularly to the local body surface. As a consequence, electroreceptors located at points A and B of fish 1 are hit by electric current (indicated by dotted lines) from fish 2 at different angles. Point A is more strongly affected by the neighbor's current than point B. Therefore, the combined signal is more strongly modulated in A than in B. By contrast, under identical geometry conditions, the mimics of the two discharge stimuli are added electronically, and their sum is presented through a pair of electrodes placed in the fish's mouth and at the tail. As a consequence of this stimulation situation, electroreceptors at each part of the body are affected in a similar way. (After: Heiligenberg, W. (1991).)

the current of the neighbor hits the electroreceptors, and differences in the distance from the neighbor to the different parts of the fish. This ensuing difference in the current paths of the two fields also causes the mixing ratio of the two signals to vary across the fish's body surface.

The variation in the mixing ratio is important for the execution of the jamming avoidance response. This can be shown by a simple modification of the above experiment. The mimics of the two discharge stimuli are added electronically, and their sum is presented through the pair of electrodes in the fish's mouth and at the tail. The result is that a jamming avoidance response can no longer be elicited. Under this **identical geometry** condition, the two electric fields differ in frequency, but are spatially identical. This is in contrast to the natural situation, under which the two electric fields exhibit **separate geometry**, that is, they are *not* spatially identical. As a consequence of the identical geometry condition, the mixing ratio of the two currents no longer varies across the fish's body surface.

Various experiments suggest that *Eigenmannia* uses the following mechanism, solely based on afferent information, to determine the sign of the frequency difference. To illustrate this mechanism, it is sufficient to represent the fish's electric organ discharge by a sine wave stimulus. We will call this stimulus S_{Fish}, and the electric organ discharge of the neighbor, which is also represented by a sine wave stimulus, $S_{Neighbor}$.

As S_{Fish} originates from an internal current source—the electric organ in the tail—it will recruit all electroreceptors on the fish's body surface evenly and at nearly the same phase of the stimulus cycle. By contrast, the effect of $S_{Neighbor}$ is quite different. As this stimulus originates from an external source—the electric organ of the neighboring fish—it will affect only some areas on the body surface of the fish, namely those where the current flow has a component perpendicular to the skin surface (Fig. 8.8). Moreover, as the neighbor is at some distance, and the intensity of an electric field in water decreases dramatically with distance, the amplitude of $S_{Neighbor}$ is normally smaller than that of S_{Fish} at every point of the fish's surface.

Due to these features of the two stimulus signals, the mixing of S_{Fish} and $S_{Neighbor}$ causes **modulation of amplitude and phase** of the mixed signal in reference to the pure S_{Fish} signal. The degree of modulation depends on the amplitude ratio of the two interfering signals. If the ratio is not equal to 1, the phase shift will cycle periodically within the 'beat' cycle. The frequency of this beat cycle is equal to the frequency difference between the two signals.

> ➤ The mixing of the fish's electric organ discharge and the interfering signal produced by a neighboring fish causes modulations of both amplitude and phase of the mixed signal relative to the fish's signal.

Does the pattern of the amplitude and phase modulation contain information about the sign of the frequency difference between S_{Fish} and $S_{Neighbor}$? In other words, using this information, can we determine which of the two signals has the higher frequency?

The answer is 'yes.' We can demonstrate this by using an approach that requires a certain understanding of the physical principles underlying the interference of two waves. Before discussing the rather complex biological situation, we will first demonstrate the principle using a very simple theoretical example (Fig. 8.9).

Suppose two sine waves, S_1 and S_2, interfere with each other. The maximum amplitude of the first sine wave is termed A_{1max} and the maximum amplitude of the second sine wave A_{2max}. The two waves have the same frequency and are exactly in phase (i.e. when the first wave reaches its minimum or maximum, the second wave does so as well).

We can determine the amplitude of the mixed signal, S_{super}, by adding the instantaneous amplitudes of the two waves. Furthermore, we know that the phase difference between the two waves, at any given point, is zero, because they were chosen such that they are exactly in phase. When we now plot the phase difference on the x-axis and the amplitude values obtained at any time point within one complete cycle of the mixed signal on the y-axis, the resultant function represents a straight line parallel to the y-axis at the x value of zero. This line extends from $-(|A_{1max}| + |A_{2max}|)$ to $+(|A_{1max}| + |A_{2max}|)$. These two extreme values are reached at phase angles of 270° and 90°, respectively. An amplitude of zero results at phase angles of 0°, 180°, and 360° (the latter value is

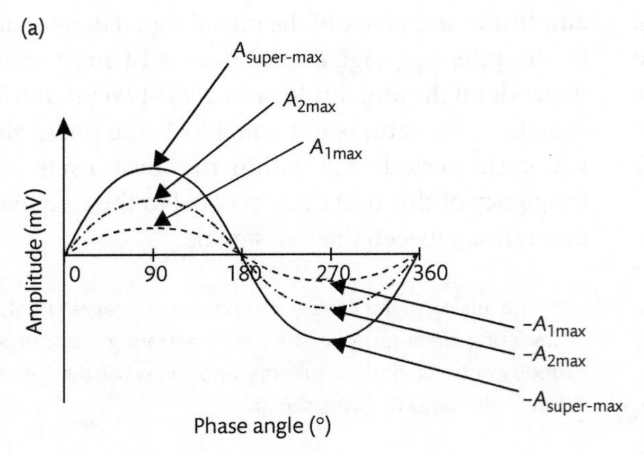

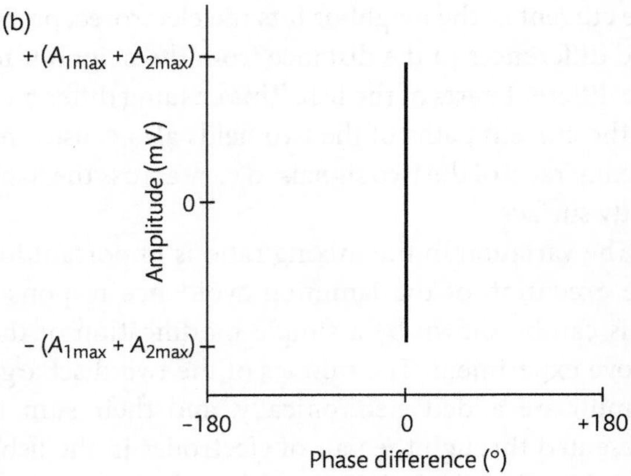

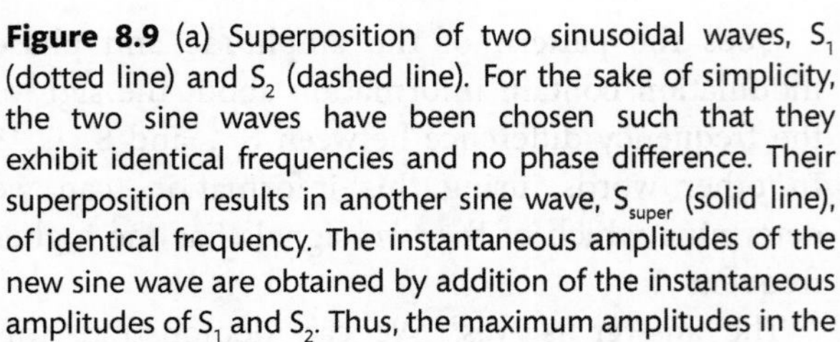

Figure 8.9 (a) Superposition of two sinusoidal waves, S_1 (dotted line) and S_2 (dashed line). For the sake of simplicity, the two sine waves have been chosen such that they exhibit identical frequencies and no phase difference. Their superposition results in another sine wave, S_{super} (solid line), of identical frequency. The instantaneous amplitudes of the new sine wave are obtained by addition of the instantaneous amplitudes of S_1 and S_2. Thus, the maximum amplitudes in the positive and negative directions, $A_{super\text{-}max}$ and $-A_{super\text{-}max}$, are equivalent to $A_{1max} + A_{2max}$ and $(-A_{1max}) + (-A_{2max})$. (b) A plot of the instantaneous amplitudes over a complete cycle of the mixed signal as a function of the phase difference results in a straight line parallel to the y-axis. The two end points of this line are represented by the sum of the maximum amplitudes of the two original waves in both negative and positive directions. (Courtesy: G. K. H. Zupanc.)

equivalent to 0° of the subsequent cycle). Over the time of one complete cycle of the mixed signal (i.e. between 0° and 360°), the resultant amplitude travels from zero to $(|A_{1max}| + |A_{2max}|)$, then back to zero, further to $-(|A_{1max}| + |A_{2max}|)$, and finally back to zero.

Now back to the more complex mixed signal occurring during the jamming avoidance response. Here, typically, not only is the amplitude of the mixed signal different from either of the two original signals, but the phase angle between the mixed signal and the fish's discharge is also unequal to zero. Similar to the very simple example discussed above, the amplitude and phase values of the mixed signal are recorded for successive S_{Fish} cycles. These values are then plotted in a two-dimensional, amplitude-versus-phase plane (Fig. 8.10). As a result, a circular trajectory is obtained, which repeats itself at a rate equal to the beat frequency of the mixed signal. The sense of rotation reflects the sign of Df: clockwise rotation indicates that the frequency of $S_{Neighbor}$ is lower than the frequency of S_{Fish}; counterclockwise rotation reflects an opposite sign of the frequency difference, that is, the frequency of S_{Fish} is lower than the frequency of $S_{Neighbor}$.

Example: *If S_{Fish} displays a frequency of 400Hz and $S_{Neighbor}$ a frequency of 402Hz, then the amplitude–phase plot defines a circular trajectory that rotates counterclockwise. This trajectory completes two full cycles per second, as it has a beat frequency of 2Hz.*

The amplitude-versus-phase plot shows that if *Eigenmannia* could determine the direction of rotation of the resulting graph, it could immediately gauge the direction in which it has to shift its frequency to perform a correct jamming avoidance response. It should raise its frequency for clockwise rotation and lower its frequency for counterclockwise rotation.

One crucial aspect of these considerations is the question of how the fish could extract the phase information out of the mixed signal. This would be relatively simple if the fish could sample a pure S_{Fish} signal (i.e. a signal without interference of a $S_{Neighbor}$ signal) somewhere from its body surface. It could then use this pure S_{Fish} signal as a reference to determine the relative phase shifts associated with the mixed signal over one beat cycle. It has been shown, however, that the fish is able to determine the relative phase, even if no pure representation of S_{Fish} is available anywhere on its body surface.

The solution for the fish is to sample amplitude and phase information from many points on its body

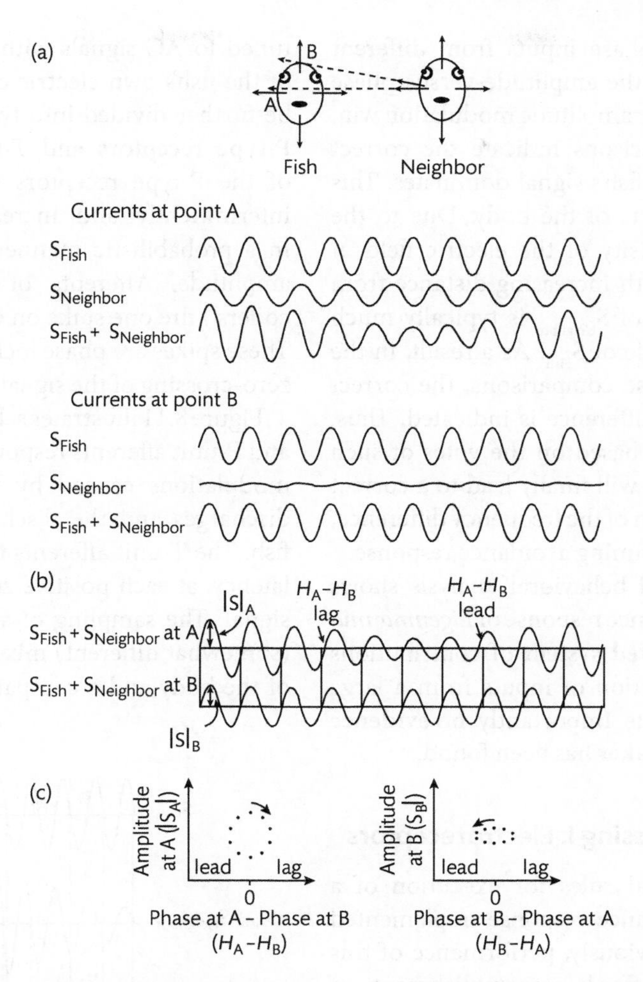

Figure 8.10 Schematic representation of the interference pattern resulting from the addition of two sine waves of slightly different frequencies. (a) The two sine waves, S_{Fish} and $S_{Neighbor}$, are used as mimics of a fish's electric organ discharge and the discharge of a neighboring fish, respectively. The amplitude of the sine wave S_{Fish} has been chosen such that it is dominant compared with the amplitude of $S_{Neighbor}$. This resembles the situation where the fish's own signal interferes with the neighbor's signal near the fish's body. The resulting signal, $S_{Fish} + S_{Neighbor}$, exhibits cyclic modulations in both amplitude and phase compared with the fish's signal. However, the combined signal, $S_{Fish} + S_{Neighbor}$, is more strongly modulated at point A than at point B, due to differences in the angle at which the neighbor's signal hits the fish's signal. The length of one modulation cycle of the mixed signal's envelope (e.g. the length between two minima of the amplitude) is referred to as the length of the beat cycle. (b) Comparison of the combined signals perceived at point A and point B. The diagrams show that the differences in the timing of their zero crossings ('phase', *H*) are modulated over the course of the beat cycle. Compared with the signal at point B, the signal at point A lags in phase while its amplitude falls, and it leads while its amplitude rises. This particular order in amplitude and phase modulation is caused by the fact that, in the example chosen, the neighbor's discharge frequency is lower than the fish's own frequency. If, on the other hand, the neighbor's frequency were higher, then a fall in the amplitude of the combined signal at point A would be paired with a phase lead, and a rise in amplitude would be paired with a phase lag. (c) The joint modulations in amplitude and phase can be recorded in a two-dimensional state plane, with the amplitude plotted on the y-axis and the differential phase (i.e. the phase in one point relative to the phase in the other point) plotted on the x-axis. This results in a circular trajectory, which repeats itself at a rate equal to the beat frequency of the mixed signal. If the more strongly modulated signal at point A is analyzed with reference to the less strongly modulated signal at point B, a clockwise rotation of the circular trajectory results. In contrast, if the signal at point B is analyzed with reference to point A, then an opposite sense of rotation is obtained. However, the latter graph is characterized by a smaller amplitude modulation. By application of the rule that the graph with the larger amplitude modulation wins in this pair-wise comparison, the conclusion drawn by the fish would be that the neighbor discharges its electric organ at a lower frequency than the fish itself does. (After: Heiligenberg, W. (1991).)

surface, compare the phase inputs from different pairs of points, and let the amplitude-versus-phase plot that shows the larger amplitude modulation win. These pair-wise comparisons indicate the correct rotation, as long as the fish's signal dominates. This is the case in most parts of the body. Due to the rapid decrease in intensity of the electric field of the neighboring fish with increasing distance from the fish, the amplitude of $S_{Neighbor}$ is typically much lower than the amplitude of S_{Fish}. As a result, in the majority of the pair-wise comparisons, the correct sign of the frequency difference is indicated. Thus, a 'democratic' decision based on the votes of such individual comparisons will finally lead to a correct determination of the sign of the frequency difference, and thus to a correct jamming avoidance response.

Overall, this detailed behavioral analysis shows that the jamming avoidance response of *Eigenmannia* is driven by a **distributed system** of contributions arising from the evaluation of inputs from a large number of pairs of points. Importantly, **no evidence of a central decision-maker** has been found.

Electrosensory processing I: Electroreceptors

How are the behavioral rules for execution of a correct jamming avoidance response implemented at the neural level? Obviously, performance of this behavioral response affords accomplishment of two major tasks: First, extraction of the sign of the frequency difference between the fish's own discharge frequency and the frequency of the neighboring fish by **electrosensory processing** of phase and amplitude information; and second, translation of the determination of the sign of the frequency difference into a **change in the motor output**, that is, of the pacemaker frequency. This will then lead to a correct shift in the frequency of the electric organ discharge.

In *Eigenmannia*, electrosensory information is perceived by two types of electroreceptor: ampullary receptors and tuberous receptors. The morphological differences between the two types have already been shown in Fig. 8.5. These receptors also differ in their physiological properties. The **ampullary receptors** are tuned to DC and low-frequency AC signals. Their possible role in the jamming avoidance response is, at present, unclear. The **tuberous receptors** are tuned to AC signals with frequencies in the range of the fish's own electric organ discharge. They can be further divided into two subtypes referred to as **P-type receptors** and **T-type receptors.** Afferents of the P-type receptors ('probability coders') fire intermittently and increase their rate of firing, in a probabilistic manner, with a rise in stimulus amplitude. Afferents of T-type receptors ('time coders') fire one spike on each cycle of the stimulus. These spikes are phase locked, with little jitter, to the zero-crossing of the signal.

Figure 8.11 illustrates schematically how the T-unit and P-unit afferents respond to phase and amplitude modulations caused by interference of the fish's discharges and the discharges of the neighboring fish. The T-unit afferents fire one spike, with a fixed latency, at each positive zero-crossing of the mixed signal. The sampling of such zero-crossings of the (somewhat different) mixed signal at different parts of the body and a comparison of the differences in

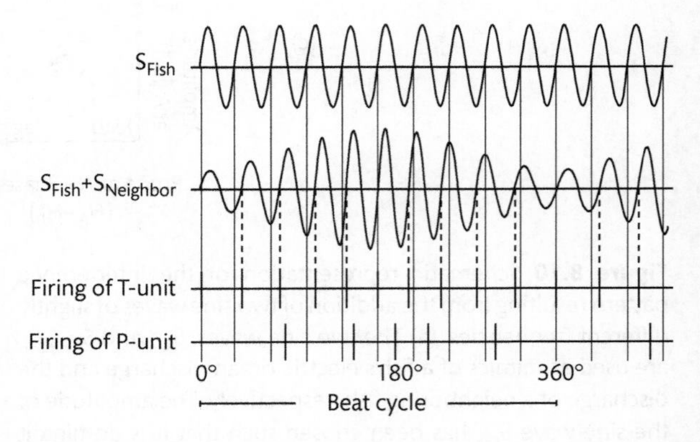

Figure 8.11 Schematic representation of the coding of phase and amplitude modulations by T-type and P-type tuberous electroreceptors. The top trace represents the fish's electric organ discharges and the second trace represents the mixed signal resulting from the addition of the fish's signal and the neighbor's signal. This interference pattern is characterized by modulations of both the instantaneous amplitude and the instantaneous phase. The latter type of modulation can best be appreciated by comparing the timing of the zero-crossings (indicated by broken lines) of the mixed signal compared with that of the fish's pure signal. The T-units encode the phase by firing a single action potential within each cycle of the signal and at a fixed latency with reference to the timing of the positive zero-crossings. The P-units encode the amplitude of the signal's envelope by modulating the probability of firing of action potentials with the instantaneous amplitude modulations. (After: Heiligenberg, W. (1991).)

firing of action potentials of the T-units reflect the **differential phase** between the respective mixed signals on two parts of the body surface. The firing rate of the P-unit afferents reflects the **amplitude** of the mixed signal. They generate more spikes at large amplitudes than they do at low amplitudes.

When the responses of the P-units and the T-units to the mixed signal are plotted over one period of the beat cycle in a similar way as the amplitude–phase values have been plotted in the two-dimensional plane (see Fig. 8.10), a similar circular trajectory is obtained. This shows that the pattern of amplitude and phase modulations is reflected by the joint activation of P-units and T-units.

> ➤ Tuberous electroreceptors can be divided into two types: P-type receptors encode the amplitude, whereas T-type receptors encode the phase of the perceived electric signal.

Electrosensory processing II: Electrosensory lateral line lobe

The primary afferents of the tuberous electroreceptors project to three maps of the **electrosensory lateral line lobe** in the hindbrain. According to their anatomical locations, these maps are referred to as lateral, centrolateral, and centromedial segments. The primary afferents of the ampullary electroreceptors project to a fourth map called the medial segment. Each of these **maps** is **somatotopically ordered**, that is, they preserve the spatial order of electroreceptors on the body surface. In these maps, the head region of the fish is by far over-represented, thus reflecting the high density of electroreceptors in this region compared with other areas of the body surface. Each primary afferent of the tuberous electroreceptors provides, via collaterals, input to each of the three maps. This suggests that electrosensory information received via tuberous electroreceptors is present in triplicate form within the electrosensory lateral line lobe.

Within each of the three maps, information encoded by T-type and P-type receptors is processed separately (Fig. 8.12). This mode of action is referred to as **parallel processing**. Input from the T-type receptors is received by **spherical cells**. One spherical cell is connected, via **electrotonic synapses**, to several T-units within its receptive field. To initiate the generation of spikes in spherical cells, several afferent action potentials have to arrive within a narrow time window. This ensures that individual afferent signals that arrive out of synchrony with the majority of the signals from a receptive field do not contribute to the production of action potentials. Consequently, spherical cells encode the phase of the stimulus with less jitter than do individual T-unit afferents.

> ➤ In the electrosensory lateral line lobe, amplitude and phase information are processed in parallel.

P-unit afferents provide input to two cell types of the electrosensory lateral line lobe. First, they form excitatory synapses on the basal dendrites of **basilar pyramidal cells**. Secondly, they inhibit **non-basilar pyramidal cells** through local interneurons. As a result of this connectivity, a rise in stimulus amplitude, causing an increase in P-unit firing, will excite basilar pyramidal cells directly and inhibit non-basilar pyramidal cells indirectly. By contrast, a fall in stimulus amplitude, which causes a decrease in P-unit firing, will release non-basilar pyramidal cells from inhibition and lead to an increase in their rate of firing. Thus, excitation of basilar pyramidal cells reflects a rise in the amplitude of the stimulus signal, whereas excitation of non-basilar pyramidal cells indicates a fall in stimulus amplitude.

Lesioning experiments have shown that the centromedial segment of the electrosensory lateral line lobe is both necessary and sufficient for the jamming avoidance response. The 'ampullary' medial segment also appears to have certain effects on this behavior. However, the exact role of ampullary information for the jamming avoidance response is currently unclear, especially as current theories can sufficiently explain this behavior based only on tuberous information.

Electrosensory processing III: Torus semicircularis

The next station in processing of electrosensory information is the **torus semicircularis** in the midbrain. The major connections between this

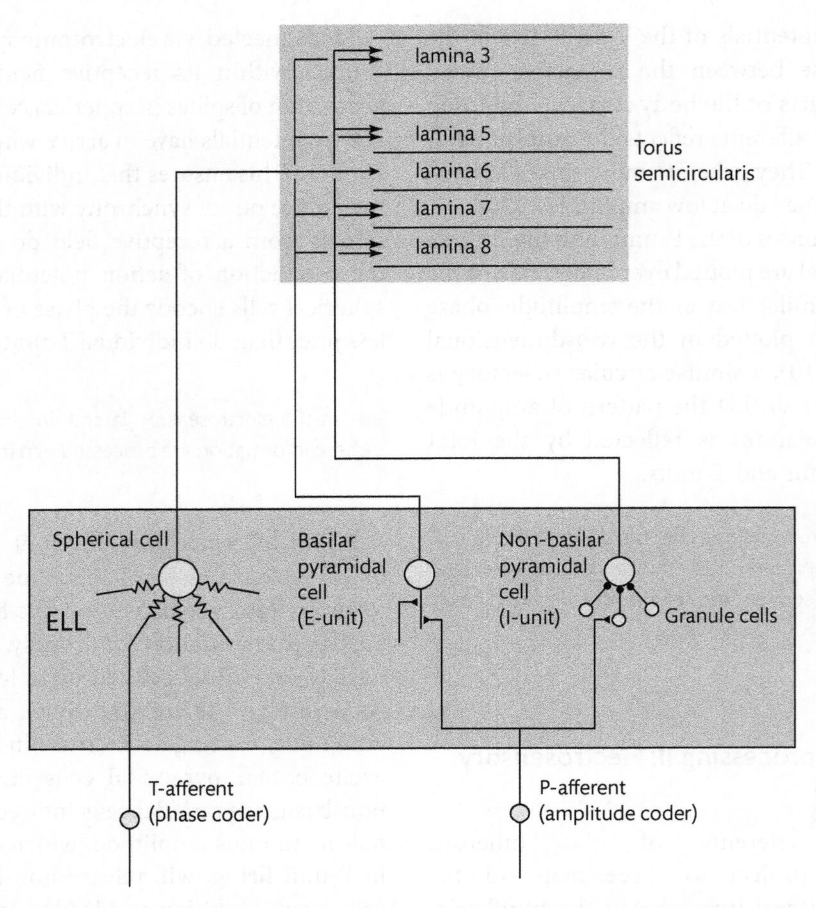

Figure 8.12 Encoding of phase and amplitude information in the electrosensory lateral line lobe (ELL). Information encoded by the T-type and P-type receptors is processed separately. Primary afferents from the T-type receptors (phase coders) form electrotonic synapses (indicated by zigzag lines) with spherical cells in the ELL. A single spherical cell may receive input from several T-afferents. Primary afferents from P-type receptors (amplitude coders) form excitatory synapses (indicated by triangles) on the basilar dendrites of basilar pyramidal cells (called E-units). Via collaterals, they also excite granule cells, which, in turn, inhibit (indicated by circles) non-basilar pyramidal cells (called I-units). As a result, E-units fire in response to a rise in stimulus amplitude, whereas I-units fire in response to a fall in stimulus amplitude. Spherical cells project exclusively to lamina 6 of the torus semicircularis. Pyramidal cells project to various laminae above and below lamina 6. (After: Heiligenberg, W. (1991).)

structure and the electrosensory lateral line lobe are shown in Fig. 8.12. As this wiring diagram indicates, the torus is divided into various **laminae**. The phase-coding spherical cells of the electrosensory lateral line lobe project exclusively to **lamina 6** of the torus. By contrast, both types of amplitude-coding pyramidal cells send axons to various laminae below and above this layer.

A network of neurons within lamina 6 computes phase differences between any two points on the body surface, and thus the differential phase. Neurons in the layers below and above lamina 6 show a variety of response characteristics: Some cells respond only to amplitude modulations, others only to phase information, and still others to both forms of modulation. The **convergence of amplitude and phase information** appears to be achieved by vertical connections between the different layers. While some cells in laminae 5 and 7 have dendritic fields limited to their respective layer, other neurons extend their dendrites into lamina 6. These cells respond to modulations in differential phase. Neurons within laminae 5 and 7 then project to deeper laminae, notably lamina 8c. Cells in these deeper layers of the torus are the first in the neuronal hierarchy that are able to recognize the sign of the frequency difference between the fish signal and the signal of the neighboring fish. However, the **sign selectivity** of these neurons depends on the orientation of the jamming signal, which is determined by the orientation of the neighbor relative to the fish. Thus, the sign selectivity information is still ambiguous.

> ➤ In the torus semicircularis, amplitude and phase information converge.

Electrosensory processing IV: Nucleus electrosensorius

Cells that encode the sign of the frequency difference between the fish's signal and the discharges of the neighbor unambiguously (i.e. independently of the orientation of the jamming stimulus) are found in the **nucleus electrosensorius.** This area in the diencephalon receives input from various layers of the torus semicircularis. However, the somatotopic order associated with the various toral layers is lost in the nucleus electrosensorius, suggesting pronounced spatial convergence of the projection from the torus to the nucleus electrosensorius. Moreover, the sign-selective neurons in the latter nuclear region are more sensitive to interfering signals than the sign-selective neurons of the torus. While the most sensitive neurons found in the torus can discriminate phase modulations as small as approximately 10μsec, the sensitivity of neurons in the nucleus electrosensorius is increased by roughly one order of magnitude. They can discriminate phase modulations as small as 1μsec. These, as well as other response properties, make the sign-selective neurons of the nucleus electrosensorius resemble the jamming avoidance response itself. This suggests that the nucleus electrosensorius is essential for execution of the jamming avoidance response.

> ➤ In contrast to the torus semicircularis, neurons of the nucleus electrosensorius encode the sign of the frequency difference between the fish's signal and the neighbor's discharges unambiguously.

Stimulation experiments combined with morphological analysis have revealed several distinct clusters of neurons within the nucleus electrosensorius. Two of these clusters are involved in the control of the jamming avoidance response (Fig. 8.13). A small dorsal area can be stimulated by iontophoretic application of L-glutamate to cause smooth rises in the frequency of the electric

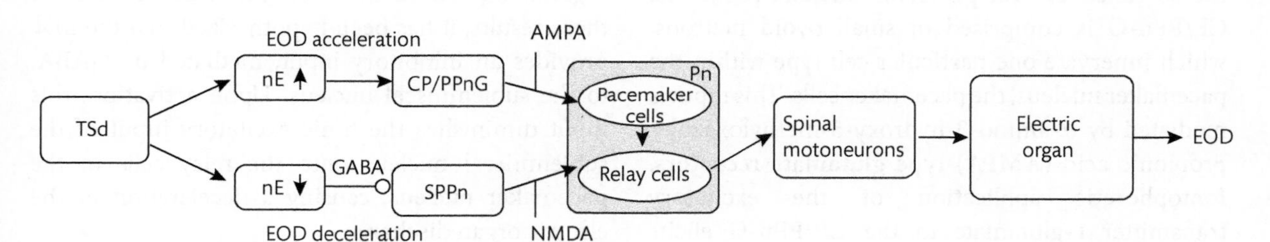

Figure 8.13 Flow diagram of the motor control of the jamming avoidance response in *Eigenmannia*. Arrowheads indicate excitatory synapses and open circles inhibitory connections. Various layers of the torus semicircularis (TSd) project to the nucleus electrosensorius (nE). Stimulation of one area, the nE↑, causes accelerations of the electric organ discharge, while stimulation of a different area, the nE↓, results in decelerations. The nE↑ innervates, via excitatory synapses (indicated by arrow), the central posterior/prepacemaker nucleus (CP/PPn) in the dorsal thalamus. The nE↓, on the other hand, provides, via inhibitory (GABAergic) synapses (indicated by a circle), input to the sublemniscal prepacemaker nucleus (SPPn). Final motor control is achieved in the pacemaker nucleus (Pn) of the medulla oblongata. Neurons of the CP/PPn innervate pacemaker cells in the pacemaker nucleus. This input is mediated by AMPA-type glutamate receptors. On the other hand, neurons of the SPPn innervate relay cells, which are also situated within the pacemaker nucleus. This input is mediated by NMDA receptors. As a final step, the relay cells project to spinal motoneurons that innervate the electric organ. Synchronous depolarization of the electrocytes comprising the electric organ generates the electric organ discharge (EOD). (After: Metzner, W. (1999).)

organ discharges. As the nucleus electrosensorius is commonly abbreviated 'nE,' this area is referred to as nE↑ (pronounced 'en-ee-up'). Conversely, L-glutamate stimulation of a more ventrally located area, called nE↓ (pronounced 'en-ee-down') causes a smooth fall in discharge frequency. The time course of both of these frequency alterations is similar to those observed during the jamming avoidance response. Bilateral lesions of these two areas eliminate frequency shifts from the jamming avoidance response, thus providing further support for the above hypothesis that the nE is a crucial link in the control of the jamming avoidance response.

Motor control

The above dissection of the individual steps involved in sensory processing of information relevant to the jamming avoidance response has shown that two separate pathways lead to the recognition of positive and negative frequency differences, respectively. Similarly, motor control of the resulting frequency changes is also mediated by two distinct pathways. Neurons of the nE↑ (excitation of which, finally, leads to rises in discharge frequency) project to a small region in the dorsal thalamus called **CP/PPn-G**. 'PPn' stands for '**prepacemaker nucleus**' and 'G' for 'gradual frequency rises.' The PPn-G is the dorsolateral portion of a larger cell assembly, the so-called **central posterior nucleus** (CP). The CP/PPn-G is comprised of small ovoid neurons, which innervate one particular cell type within the pacemaker nucleus, the pacemaker cells. This input is mediated by α-amino-3-hydroxy-5-methyloxazole-propionic acid **(AMPA)-type glutamate receptors.** Iontophoretic application of the excitatory transmitter L-glutamate to the CP/PPn-G elicits gradual accelerations of the electric organ discharges. Conversely, bilateral lesions of this cell group abolish frequency rises of the jamming avoidance response. Injection of 6-cyano-7-nitroquinoxaline-2,3-dione (CNQX), which selectively blocks AMPA-type receptors, into the pacemaker nucleus has an effect similar to lesions applied to the CP/PPn-G; it results in elimination of a jamming avoidance response to negative frequency differences.

Neurons of the nE↓ (excitation of which, finally, leads to decreases in discharge frequency) project to a different region, the **sublemniscal prepacemaker nucleus** in the mesencephalon (Fig. 8.13). Cells of this nuclear assembly, in turn, innervate the relay cells within the pacemaker nucleus. This input is mediated by *N*-methyl-D-aspartate **(NMDA)-type glutamate receptors**. Interestingly, bilateral lesions of the sublemniscal prepacemaker nucleus not only obliterate decelerations of the electric organ discharge during the jamming avoidance response, but also reduce the resting frequency of the discharges (i.e. the frequency when no jamming signal is presented). This indicates that neurons of the sublemniscal prepacemaker nucleus provide a **tonic input** to the pacemaker nucleus. Injection of 2-amino-5-phosphonovaleric acid (APV), an antagonist of NMDA-type glutamate receptors, into the pacemaker nucleus has shown results similar to lesions of the sublemniscal prepacemaker nucleus.

The hypothesis that neurons of the sublemniscal prepacemaker nucleus provide a tonic input to the pacemaker nucleus can, indeed, be confirmed by injection of the inhibitory transmitter γ-aminobutyric acid, commonly known as **GABA**, into the sublemniscal prepacemaker nucleus. Such experiments result in decreases in the discharge frequency, thus resembling the effect observed after L-glutamate stimulation of the nE↓. Furthermore, injection of the $GABA_A$-receptor antagonist bicuculline blocks any such effects of L-glutamate stimulation of the nE↓. Based on these results, it has been hypothesized that the nE↓ provides an inhibitory input, mediated by GABA, to the sublemniscal nucleus. Upon activation, this input diminishes the tonic excitatory input of the sublemniscal nucleus onto the relay cells in the pacemaker nucleus, causing a deceleration of the electric organ discharge.

The final motor control of the jamming avoidance response is mediated through input to the pacemaker nucleus, originating from the 'G' portion of the central posterior/prepacemaker nucleus and the sublemniscal prepacemaker nucleus.

While fibers of the CP/PPn-G appear to terminate mainly on the dendrites of the pacemaker cells, fibers of the sublemniscal prepacemaker nucleus synapse upon the somata of the relay cells (Fig. 8.13). Strong electrotonic coupling between the relay cells and the pacemaker cells ensures that the input arising from the sublemniscal nucleus affects not only the relay cells, but also the pacemaker cells.

In the final stage of motor control, the pacemaker cells drive the relay cells, which, in turn, transmit the command pulses generated by the pacemaker cells to spinal motoneurons of the electric organ. This is done in a one-to-one fashion, resulting in one discharge of the electric organ at each command pulse produced by the pacemaker nucleus.

> ➤ Motor control of the jamming avoidance response is mediated through input to the pacemaker nucleus in the medulla oblongata. This input originates from the 'G' portion of the central posterior/prepacemaker nucleus and the sublemniscal prepacemaker nucleus.

Reflections on the evolution of the jamming avoidance response

Both the behavioral and neural mechanisms governing the jamming avoidance response are far from perfect. As we have seen, an engineer would design a neural network for determination of the sign of the frequency difference between the fish's own discharge and that of a neighbor very differently by making use of the pacemaker frequency as an internal reference. However, evolution does not exhibit a goal-oriented, long-term planning approach to produce simple and elegant solutions. When, in the course of the phylogenetic development of *Eigenmannia*, the need for a jamming avoidance response arose, neural implementation of this behavior was only possible by modification of pre-existing structures. Obviously, none of these pre-existing structures allowed the fish simply to incorporate internal information about the discharge frequency.

On the other hand, other neural elements, used for the control of different behaviors, could readily be modified to subserve specific functions in the context of the jamming avoidance response. Such considerations are supported by comparative investigations. The genus *Sternopygus* within the order Gymnotiformes does not perform a jamming avoidance response. However, these fish already have brain neurons that encode amplitude and phase information. These neurons are used for electrosensory analysis of object movements in the context of electrolocation. It appears plausible that, through a few modifications in the associated neural network, these 'ancestral' neurons could easily be employed for the control of the jamming avoidance response. This is what might have happened in the course of the evolution of *Eigenmannia*.

> ➤ The imperfection of the jamming avoidance response can be explained by its likely evolutionary development.

Summary

- The generation of the vast majority of behaviors requires, at least to some degree, the transformation of the processed sensory input into proper motor action. This sensorimotor integration exhibits a remarkable degree of plasticity, and numerous factors may modulate the individual steps of this process.
- One of the premier model systems to study the process of sensorimotor integration is the jamming avoidance response of the weakly electric fish *Eigenmannia*. Fish of this genus continuously generate, by means of an electric organ, wave-type electric discharges. The discharge rate is determined by the frequency of the pacemaker nucleus, an endogenous oscillator in the medulla oblongata. The fish sense their own electric currents, as well as those of neighbors, through electroreceptors.
- Electrolocation involves the analysis of perturbations of the fish's own electric field by objects in the close vicinity.
- Object-induced changes in the electric current patterns of the fish's own electric field may be masked by modulations caused by a neighbor's electric organ discharges, especially if the frequency of the neighbor is close to the fish's

own frequency. The fish avoids the detrimental effects of such a signal interference, or 'jamming,' by shifting the frequency of its own electric organ discharge away from the frequency of the interfering signal. This behavior is called the 'jamming avoidance response.' Thus, if the neighbor's frequency is higher than the fish's own discharge frequency, the fish lowers its frequency. If the neighbor's frequency is lower than the fish's own discharge frequency, the fish raises its frequency.

- The fish does not determine the sign of the frequency difference between its own signal and that of the neighbor by making reference to the pacemaker frequency. Rather, it evaluates afferent information contained in the interference, or 'beat' pattern, which results from the mixing of its own discharge and that of the neighbor.
- Both amplitude and phase of the mixed signal are different from the fish's own signal. This difference is analyzed by comparing the differential phase in different parts of the body surface. Thus, the jamming avoidance response is driven by a distributed system of contributions resulting from evaluations of inputs from pairs of points. There is no evidence of a central controller or decision-maker.
- Extraction of differential phase and amplitude information, essential to determine the sign of the frequency difference, is mediated by two types of tuberous electroreceptors, T-type receptors ('time coders') and P-type receptors ('amplitude coders').
- Primary afferents of the T-type receptors terminate on spherical cells in the electrosensory lateral line lobe of the hindbrain. In the same brain region, primary afferents of P-type receptors form excitatory connections with basilar pyramidal cells and inhibitory connections, via local interneurons, with non-basilar pyramidal cells.
- Spherical cells project exclusively to lamina 6 of the torus semicircularis in the midbrain. Both types of pyramidal cell project to various laminae above and below lamina 6. Within this layer, the differential phase is computed. Convergence of the differential phase and amplitude information is achieved through vertical connections between the laminae. Although neurons within the torus exist that fire at a higher rate for one sign of the frequency difference than for the opposite sign, their response is still ambiguous, as it depends on the orientation of the jamming signal.
- Unambiguous sign-selective neurons occur in the next station of electrosensory processing, the nucleus electrosensorius in the diencephalon. Excitation of a subdivision of the nucleus electrosensorius, the nE↑, raises the discharge frequency, whereas excitation of a second subdivision, the nE↓, leads to frequency decreases.
- Frequency increases and frequency decreases are mediated by two separate motor pathways. Frequency increases are controlled by a subnucleus of the diencephalic central posterior/prepacemaker nucleus, the CP/PPn-G, whose neurons innervate the pacemaker neurons of the pacemaker nucleus. Frequency decreases are controlled by the mesencephalic sublemniscal prepacemaker nucleus, which in turn projects to the relay cells of the pacemaker nucleus.
- Command pulses generated by the pacemaker cells drive the discharges of the electric organ in a one-to-one fashion.

The bigger picture

As mentioned in the introduction to this chapter, the common division of neural pathways involved in generation of behavior into sensory and motor components, and the linkage of the two by the process of sensorimotor integration, is a somewhat artificial construct. It reflects, to a certain degree, the widely-observed division of research carried out by different investigators, each of them focusing on a single component. Another factor that has certainly contributed to this rather arbitrary fragmentation is the limitation in the availability of adequate tools for simultaneous exploration of the neural mechanisms that encode sensory information, and the mechanisms that translate this information into corresponding behavioral patterns. Due to this methodological restriction, individual aspects associated with this processes have, traditionally, been examined sequentially. However, the reconstruction of the overall process based on such fragmented information is inferior to the information that would be obtained through simultaneous analysis. It is important to keep this limitation in mind when reading the present chapter. Nevertheless, simultaneous examination of the multitude of processes required to generate a specific behavior upon sensory stimulation is within realistic reach. The enormous advances made in recent years in the area of global high-resolution mapping of neural activity, and the possibility of combining such physiological information with global structural information compiled through high-resolution brain mapping (see Chapter 2), will provide unprecedented opportunities to better understand how the brain as a whole integrates sensory and motor information to generate behavior.

Recommended reading

Heiligenberg, W. (1991). *Neural Nets in Electric Fish*. MIT Press, Cambridge, Massachusetts.

The classic synopsis of the work of Walter Heiligenberg and associates on the jamming avoidance response. Written for an informed audience, it is suitable only for advanced graduate students or instructors who would like to gain a deeper understanding of the subject.

Kawasaki, M. (2009). Evolution of time-coding systems in weakly electric fishes. *Zoological Science* **26**:587–599.

Time coding is a type of neural coding in which information about a stimulus is represented as the temporal pattern of spikes in neural pathways. This mechanism plays an important role for a variety of behavioral functions in weakly electric fish, including the jamming avoidance response. This review surveys electric behaviors that require time processing and the underlying neural mechanisms, and places these time-coding systems into an evolutionary perspective.

Metzner, W. (1999). Neural circuitry for communication and jamming avoidance in gymnotiform fish. *Journal of Experimental Biology* **202**:1365–1375.

A good summary of the research on the jamming avoidance response in weakly electric fish.

Short-answer questions

8.1 What are corollary discharges? Define this term in one sentence.

8.2 How can knifefish produce, with their electric organ, discharges of several volts, although each electrocyte generates only a few tens of volts?

8.3 Which brain structure in the knifefish *Eigenmannia* drives the electric organ discharge and determines its frequency?

8.4 'Electric fish "see" the environment with their entire body.' Explain this statement by making reference to the distribution of electroreceptors on the fish's body surface.

8.5 Complete the following sentence: The electrolocation capability enables weakly electric fish to distinguish between objects differing in .... [specify property].

8.6 What is the jamming avoidance response of weakly electric fish? Characterize the behavior performed in one sentence.

8.7 What is the presumed function of the jamming avoidance response of weakly electric fish?

8.8 What is parallel processing? As an example, use the electrosensory system of knifefish.

8.9 In what brain nucleus of the electrosensory system of the knifefish *Eigenmannia* do amplitude and phase information converge? Is the information about the sign of Df (frequency difference between the fish's own electric organ discharges and the discharges of a neighbor fish) computed by neurons of this nucleus unambiguous? What critical observation provides information about possible ambiguity?

8.10 Sketch in a simple diagram the neural pathways underlying frequency increases and frequency decreases as part of the jamming avoidance response in the weakly electric fish *Eigenmannia*. Start these pathways with the nucleus electrosensorius and end them with the pacemaker nucleus.

8.11 **Challenge question** It has been hypothesized that auditory feedback of one's own voice is critical for maintaining phonetic precision in humans. To address this aspect of sensorimotor integration in speech processing, is it desirable to study individuals with prelingual or postlingual deafness? Justify your answer.

Essay questions

8.1 Electric eels are approximately 2m-long gymnotiform fish that can generate discharges of several hundreds of volts. How is such an enormous voltage possible, as each electrocyte of the electric organ in the trunk of the fish is capable of producing a discharge of only some tens of millivolts?

8.2 How do weakly electric fish electrolocate? Suggest a behavioral experiment to verify the hypothesis that, on the basis of their electrolocation ability, the fish can analyze various features of objects, including size and direction of movement.

8.3 The weakly electric fish *Eigenmannia* sp. is able to extract time and amplitude information from an electric stimulus. How are these parameters encoded at the level of the electroreceptors? How does the activity of neurons of the electrosensory lateral line lobe reflect the processing of this information?

8.4 What is the jamming avoidance response of gymnotiform fish? Describe the computational rules and the neural correlates underlying this behavior.

8.5 During the jamming avoidance response, and some other behaviors, the weakly electric fish *Eigenmannia* sp. displays gradual changes in frequency of the electric organ discharge. Describe the motor control pathways that mediate the increases and decreases of the discharge frequency. What transmitters and receptors are involved in conveying information along these pathways?

Advanced topic Dynamical systems modeling of the jamming avoidance response

Background information

The jamming avoidance response is one of the best studied behaviors in neuroethology. It has been examined at all levels of organization, from the definition of the behavioral rules underlying its execution to the morphology and physiology of its neural correlates. Nevertheless, the dynamical systems modeling of the behavioral output in response to a jamming sensory stimulus has proven notoriously difficult. One reason for this difficulty lies in the intrinsically unstable nature of the jamming avoidance response—during the execution of this behavior, the fish's frequency is shifted away from the discharge frequency of the neighboring fish. This challenge has been addressed in the modeling study by Madhav and co-authors published in 2013.

Essay topic: Dynamical system modeling of the jamming avoidance response of the weakly electric fish *Eigenmannia*

After giving a brief overview of the jamming avoidance response and its underlying sensorimotor circuit, describe in your essay how the problem of the intrinsic instability of this behavior was overcome in the modeling study by Madhav and co-authors. Based on this strategy, how was the linear model obtained? What does this linear model achieve, and where are its limits? Using the linear model as a starting point, describe in both qualitative and mathematical terms the composition of the non-linear model. Discuss the 'snap-through' bifurcation predicted by the non-linear model. Then read the research article by Watanabe and Takeda, published in 1963, which is of particular historical interest as it is the original report of the discovery of the jamming avoidance response. Can you find in this paper any experimental evidence of a snap-through response of fish exposed to jamming stimuli?

Starter references

Madhav, M. S., Stamper, S. A., Fortune, E. S., and Cowan, N. J. (2013). Closed-loop stabilization of the jamming avoidance response reveals its locally unstable and globally nonlinear dynamics. *Journal of Experimental Biology* **216**:4272–4284.

Watanabe, A. and Takeda, K. (1963). The change of discharge frequency by A.C. stimulus in a weak electric fish. *Journal of Experimental Biology* **40**:57–66.

To find answers to the short-answer questions and the essay questions, as well as interactive multiple choice questions and an accompanying Journal Club for this chapter, visit **www.oup.com/uk/zupanc3e.**

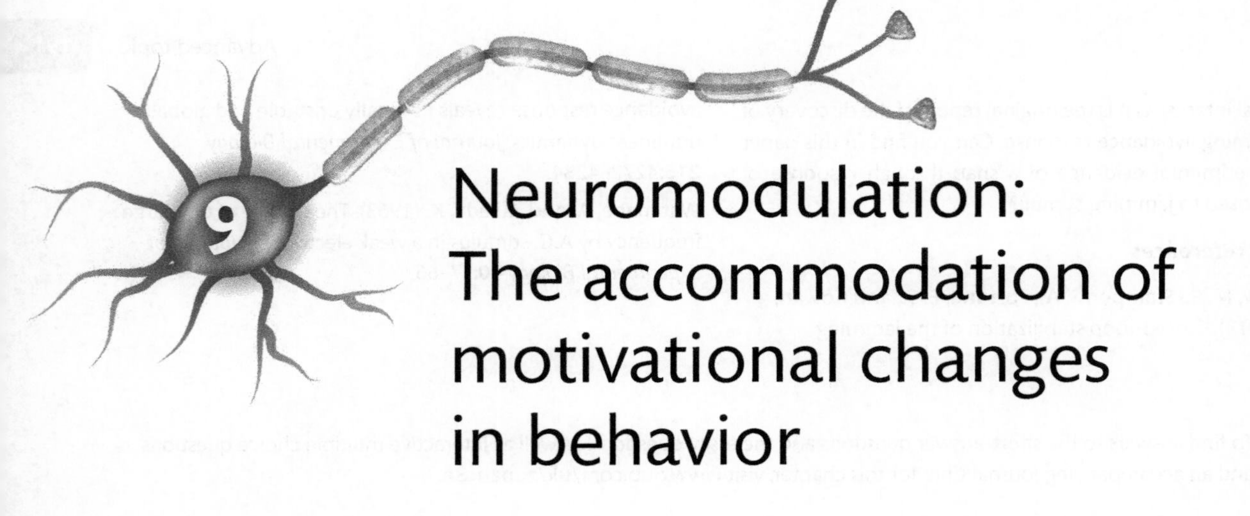

Neuromodulation: The accommodation of motivational changes in behavior

Introduction

As has been demonstrated in previous chapters, major factors that determine the behavior of an animal are stimuli originating from the environment. Different stimuli typically elicit different behaviors, as different intensities of a stimulus may elicit different frequencies, durations, or intensities of a behavioral pattern. In the case of an 'ideal' reflex (which probably does not exist in the real world, but is a helpful construct for theoretical considerations), these environmental stimuli are the only factors that determine the behavioral response. This situation is summarized schematically in Fig. 9.1(a). In this diagram, the animal is represented by a 'black box.' Input to the black box indicates the stimulus originating from the environment, whereas the output represents the animal's response. Very often, however, this input/output relationship is not fixed.

Example: *Seasonally breeding animals are characterized by dramatic changes in many behaviors, particularly those associated with courtship and aggression. During the breeding season, males may respond to a proper 'female' stimulus with courtship behavior, whereas outside the breeding season, the identical stimulus rarely, or almost never, elicits courtship.*

The variation in the animal's response, even under identical stimulation regimes, has led to the assumption that variable internal causal factors exist that mediate the relationship between external stimuli and resulting behaviors. The influence of such internal factors on the neural control of the respective behavioral pattern is referred to as **motivation**. Structures within the central nervous system from which a modulatory influence originates may be regarded as 'motivational' components of the neural network involved in the generation of the behavior. This situation is outlined in Fig. 9.1(b).

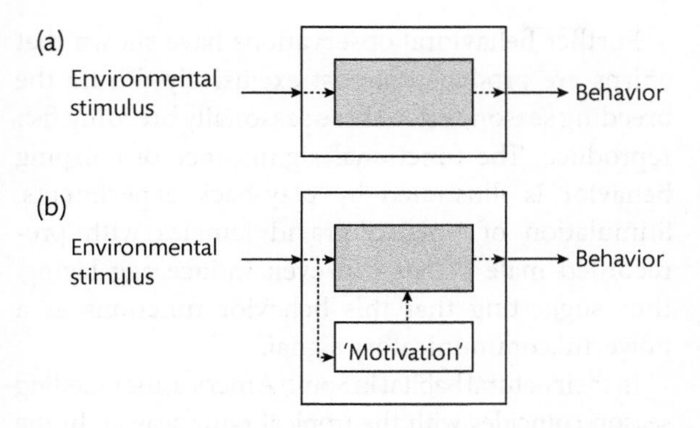

Figure 9.1 Input–output relationship in 'ideal' reflexes and motivational systems. (a) Block diagram of a hypothetical 'ideal' reflex. The behavioral response of an animal is determined solely by a given stimulus. This stimulus evokes, whenever applied, an identical response in the animal. (b) Block diagram of a motivational system. The behavioral response of an animal is determined not only by a stimulus, but also by the motivational state of the animal. Identical stimuli applied at different times can therefore evoke different responses. Note that the stimulus may also affect the motivation of the animal. (Courtesy: G. K. H. Zupanc.)

Motivation The physiological state of an animal that defines the frequency and intensity of occurrence of a behavior when elicited by a given endogenous or exogenous stimulus.

The proposal of the existence of motivational components does not make any predictions about where the modulatory influence exerts its effect within the neural chain connecting the sites of sensory perception with those that initiate motor action. In particular, the existence of a single central structure devoted to the exclusive control of the level of motivation (we could call such a structure a 'motivational center') is not required. In a given instance, the only effect of motivational factors could, for example, be to modulate the perception and processing of sensory stimuli. Central structures controlling the motor output may then modify their response solely on the basis of altered sensory information relayed to them, rather than due to a direct influence of motivational factors.

In the following sections, we will discuss neural mechanisms that mediate motivational changes in behavior. The underlying principles will be illustrated by referring to some well-studied model systems.

Key concepts

Motivational changes in behavior are possible only if the neural network underlying this behavior exhibits the potential for **neural plasticity**. This is not the case in the 'ideal' reflex. Therefore, and as mentioned above, the response in this type of behavior is determined solely by the stimulus originating from the environment. In contrast, plastic neural networks can produce different behavioral outputs, even under identical stimulation regimes.

➤ Motivational changes in behavior are possible only in networks exhibiting the potential for neural plasticity.

Research on a variety of different behavioral systems suggests that the neural mechanisms mediating motivational changes in behavior can be grouped into two major categories. These categories are subsumed under the terms **structural reorganization of neural networks** and **biochemical switching of neural networks.**

➤ Two major neural mechanisms that mediate motivational changes in behavior are structural reorganization of neural networks and biochemical switching of neural networks.

Structural reorganization

The principle of structural reorganization involves a variety of structural modifications in neural networks underlying behavior. They include:

- Generation of new neurons and glial cells.
- Elimination of older cells.
- Changes in the dendritic structure of neurons.
- Retraction and outgrowth of axons.

Until a few decades ago, it was believed that the adult brain of mammals, and particularly of primates, is incapable of producing new neurons. However, in recent years the potential for the generation of new neurons (**neurogenesis**) has been demonstrated in the brain of all vertebrate classes, including that of primates and even of humans. In 'lower' vertebrates, this capability of adult neurogenesis is especially pronounced, both in terms of the overall number of neurons produced and the number of brain regions exhibiting proliferative activity.

Although the functional significance of adult neurogenesis still requires detailed analysis, this phenomenon obviously provides an excellent means for the restructuring of neural networks, and thus for the production of dramatic changes in their properties, ultimately leading to alterations in the behavioral output. An example of the behavioral significance of this mechanism will be presented at the end of this book, when we discuss the involvement of adult neurogenesis in the improvement of learning and the formation of new memories in Chapter 13.

Dendritic plasticity: Seasonal changes in chirping behavior of weakly electric knifefish

Another important mechanism leading to dramatic changes in the properties of a neural network involves alterations in the dendritic structure of neurons. This mechanism and its relationship to seasonal behavioral changes has been examined in detail in the knifefish, *Eigenmannia* sp. Normally, this weakly electric fish produces, by means of a specialized electric organ, electric discharges distinguished by their enormous regularity in both frequency and amplitude. Short-term modulations consist of complex changes in frequency and amplitude, which may be followed by a complete cessation of the electric organ discharge, as shown in Fig. 9.2. When transformed into an acoustic signal, these modulations resemble the sound produced by crickets and hence are called **chirps**.

> **Chirps** Transient frequency and amplitude modulations of the electric organ discharge.

Further behavioral observations have shown that chirps are produced almost exclusively during the breeding season, when these seasonally breeding fish reproduce. The functional significance of chirping behavior is illustrated by play-back experiments. Stimulation of isolated gravid females with pre-recorded male chirps can even induce egg laying, thus suggesting that this behavior functions as a powerful communicatory signal.

In their natural habitat in South America, the breeding season coincides with the tropical rainy season. In the laboratory, maturation of gonads and development of the associated behaviors can be induced by simulating the rainy season. This is achieved by the daily addition of deionized water ('rain water') to the aquarium, which leads to a rise in the water level and to a decrease in the conductivity of the aquarium water. Furthermore, circulated water is periodically sprinkled onto the surface of the aquarium water, thus imitating rain drops. Taken together, these individual steps, applied over several weeks, lead to changes in the environment of the fish similar to those caused by the heavy rainfalls during the tropical rainy season. The ability to induce gonadal recrudescence by subjecting the fish to such rainy regimes has enabled researchers to study the neural basis of the seasonal changes in chirping behavior.

The electric organ discharge is controlled by an endogenous oscillator in the medulla oblongata, the so-called **pacemaker nucleus**. The pacemaker nucleus triggers each discharge of the electric organ by one spike. The regularity of the oscillations is an intrinsic property of the pacemaker nucleus, as this structure continues to fire with a frequency similar to that in the living animal after being isolated and kept in a Petri dish containing oxygenated artificial cerebrospinal fluid. Therefore, it cannot be the pacemaker nucleus itself that controls chirping behavior. Rather, input to this nucleus must be responsible for the control of this behavioral pattern.

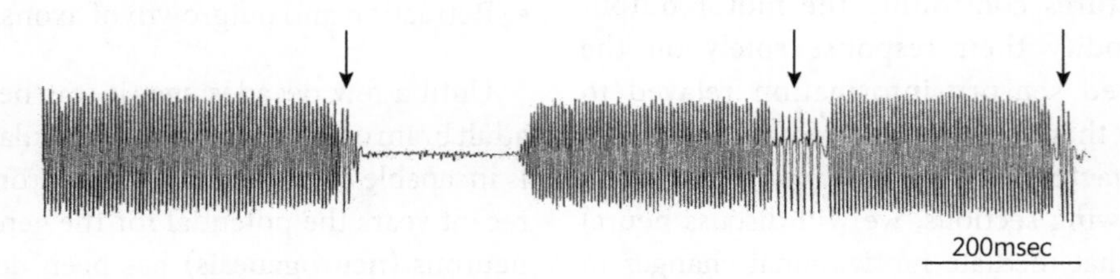

Figure 9.2 Electric organ discharge of the knifefish, *Eigenmannia* sp., during courtship. The oscilloscope trace reveals two types of discharge sequence: One with constant frequency and the other with brief periods of frequency modulations (arrows). The first modulatory event is followed by a complete cessation of the fish's discharge. (Courtesy: G. K. H. Zupanc.)

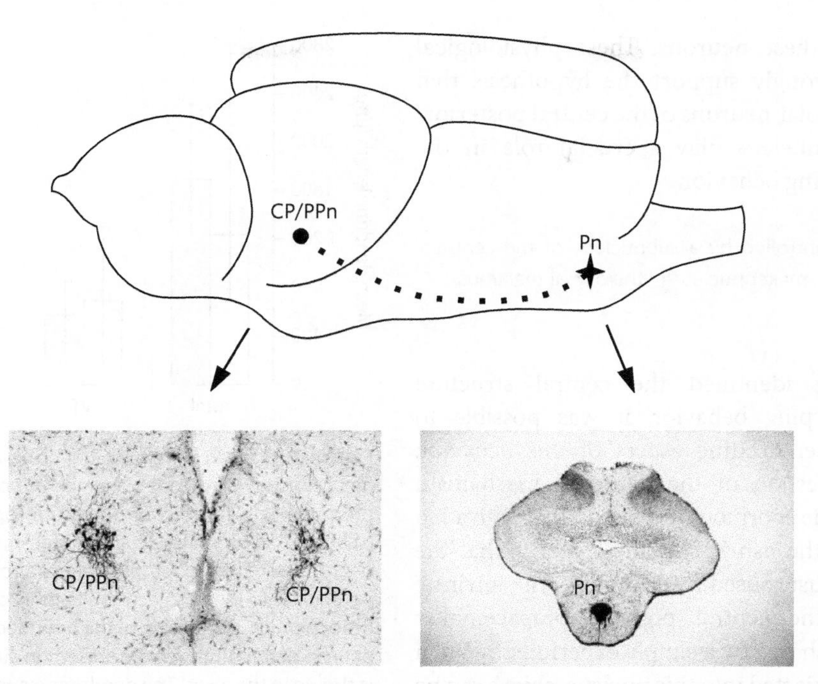

Figure 9.3 Identification of the structural correlate of input to the pacemaker nucleus by neuronal tract tracing. The drawing on top is a side view of the brain of *Eigenmannia* sp. The pacemaker nucleus (Pn) is located in the brainstem area and the central posterior/prepacemaker nucleus (CP/PPn) in the diencephalon. The two microphotographs below show cross-sections through the brain at the level of the pacemaker nucleus (right) and the central posterior/prepacemaker nucleus (left). In the experiment, the enzyme horseradish peroxidase was applied to the pacemaker nucleus in the anesthetized animal. After a survival period of three days, during which axons innervating the pacemaker nucleus took up the enzyme and transported it in the retrograde direction toward the cell bodies, the fish was killed and the brain cut into thin sections. After histochemical processing of these sections, the tracer became visible as a black precipitate. Screening of the brain sections revealed that the input to the pacemaker nucleus arises from a bilateral cluster of neurons in the diencephalon, called the central posterior/prepacemaker nucleus. On each side of the brain, a number of labeled cell bodies comprising this brain nucleus, together with part of their dendritic arbors, are visible. (Courtesy: G. K. H. Zupanc.)

Morphologically, input to a brain nucleus can be revealed by **neuronal tract tracing**, a technique described in detail in Chapter 2. Using this approach, the input to the pacemaker nucleus was identified by Walter Heiligenberg and his group at the Scripps Institution of Oceanography of the University of California, San Diego, at the beginning of the 1980s. Figure 9.3 summarizes the experimental procedure. First, Heiligenberg and his group injected horseradish peroxidase into the pacemaker nucleus of anesthetized fish. Next, following a post-injection survival time of several days (during which the enzyme is transported in the retrograde direction within the axon) and histochemical processing of the brain tissue, they screened sections of the brain for labeled cells. The neurons traced through this approach form a small bilateral cluster in the dorsal thalamus, which is part of the diencephalon. As this cellular assembly provides input to the pacemaker nucleus, it is referred to as the **prepacemaker nucleus.** Later studies showed that this nucleus is part of a larger complex called the central posterior nucleus. The region of the latter complex that provides input to the pacemaker nucleus is therefore now referred to as the **central posterior/prepacemaker nucleus.**

Detailed morphological and physiological analysis of the central posterior/prepacemaker nucleus has revealed that only approximately 100 neurons within this nucleus are involved in the control of chirping behavior. These neurons are rather large and give rise to three or four dendrites, which show a moderate degree of branching. Based on this morphological appearance, these neurons are categorized as **multipolar.**

Application of the excitatory transmitter L-glutamate to the subnucleus comprised of the large multipolar neurons elicits modulations of the electric organ discharge resembling the chirps in the intact animal. A similar effect is caused by electrical stimulation of these neurons. These physiological experiments strongly support the hypothesis that the large multipolar neurons of the central posterior/prepacemaker nucleus play a crucial role in the control of chirping behavior.

➤ Chirps are controlled by a subnucleus of the central posterior/prepacemaker nucleus in the dorsal thalamus.

After having identified the central structure controlling chirping behavior, it was possible to approach another exciting aspect of this behavior, namely, the question of the neuronal mechanism responsible for the enormous increase in the motivation to chirp when the fish is sexually mature. That the increase is not just caused by changes in the intrinsic properties of the central posterior/prepacemaker nucleus can be shown by a simple experiment. When L-glutamate is injected into this nucleus, chirps can be elicited in any post-juvenile fish—regardless of its state of maturity. So what causes the seasonally induced motivational changes in chirping behavior?

This question was tackled by Günther Zupanc in the laboratory of Walter Heiligenberg in the late 1980s. They first established the following two groups of fish: A first group that was kept under artificial rainy season conditions until the fish developed ripe gonads, and a second group that was kept under non-rainy ('dry') season conditions, so that the fish did not mature sexually. The large multipolar neurons of the central posterior/prepacemaker nucleus were then labeled by retrograde tracing from the pacemaker nucleus. Finally, individually labeled neurons were reconstructed to obtain a complete picture of their morphology.

Comparison of such reconstructed neurons in the two groups of fish revealed a close correlation between the dendritic morphology of the chirp-controlling neurons of the central posterior/prepacemaker nucleus and the state of sexual maturity of the fish. As Fig. 9.4 shows, neurons of sexually mature fish exhibit a well-developed dendritic arbor with dendrites extending in three directions—dorsolaterally, dorsomedially, and ventrally. In contrast, neurons of sexually immature fish almost completely lack dendrites in the dorsomedial field, while the dendrites in the dorsolateral and ventral territories of the dendritic arbor are developed similarly to those in mature fish.

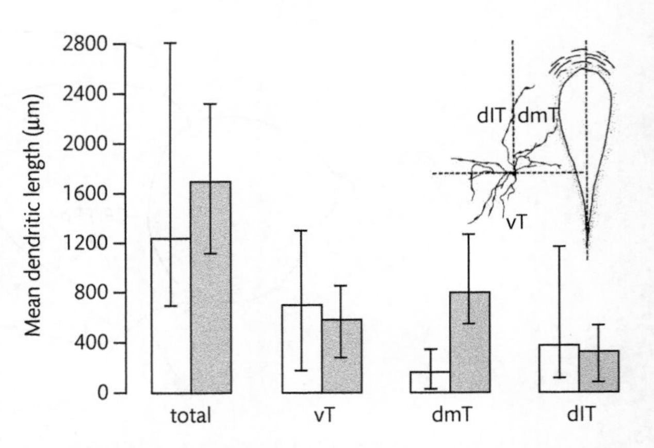

Figure 9.4 Mean length of dendritic arbors of individually reconstructed chirp-controlling neurons of the central posterior/prepacemaker nucleus of immature (open bars) and mature (shaded bars) females in the knifefish, *Eigenmannia* sp. The neurons were traced retrogradely by application of horseradish peroxidase to their projection site (the pacemaker nucleus) and histochemical processing of the brain sections. A reconstructed neuron of a mature female, based on such retrograde tracing, is shown in the inset. The dendrites extending from the labeled cell bodies define three different dendritic territories, called the ventral territory (vT), dorsomedial territory (dmT), and dorsolateral territory (dlT). The increase in overall length (in the diagram referred to as 'total') by approximately 40% during sexual maturation is mainly due to a significant proliferation in the dorsomedial territory. As shown in the inset, in mature fish some of the dendrites of the dorsomedial territory travel close to the wall of the third ventricle. The vertical lines on each bar indicate the range of individual data. (After: Zupanc, G. K. H. and Heiligenberg, W. (1989).)

Investigations at the electron microscopic level have shown that the dendrites of the large multipolar neurons of the central posterior/prepacemaker nucleus receive input mainly from excitatory synapses. By contrast, the cell body region is, to a large extent, contacted by inhibitory synapses. The excitatory input originates from neurons of another diencephalic region, the nucleus electrosensorius. Neurons within this sensory processing station respond specifically to various electric stimuli, including chirp-like modulations of the electric organ discharge.

➤ Seasonally induced changes in chirping behavior are accompanied by alterations in dendritic morphology of central posterior/prepacemaker nucleus neurons.

As dendrites of the dorsomedial territory comprise roughly half of the total dendritic arbor of neurons of the central posterior/prepacemaker nucleus, their retraction after the end of the breeding season leads

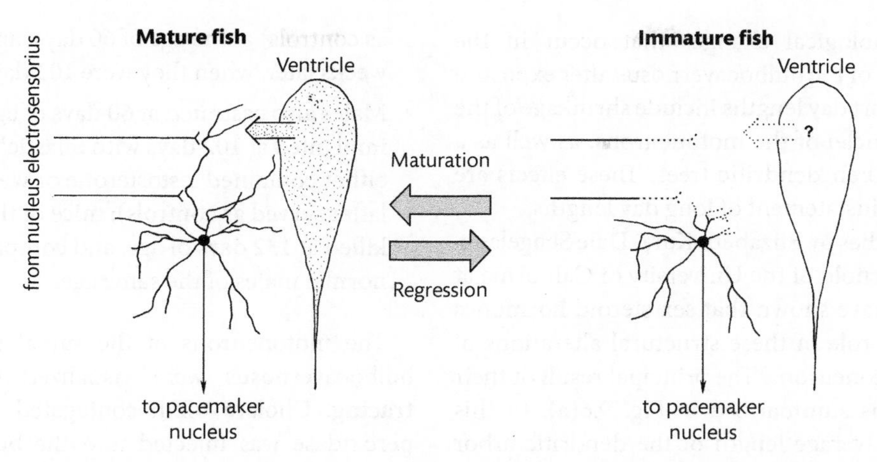

Figure 9.5 Model of the structural changes in the neural network underlying seasonal alterations in chirping behavior in the weakly electric knifefish, *Eigenmannia* sp. During the course of the rainy season, which triggers gonadal recrudescence, dendrites in the dorsomedial territory of the chirp-controlling neurons of the central posterior/prepacemaker nucleus grow out and make predominantly excitatory synaptic contact with axons originating from the nucleus electrosensorius. These structural changes may be induced by as yet unknown molecular factors diffusing from the ventricle into the brain parenchyma. After the onset of the dry season, the dendrites of the dorsomedial territory retract. As a result, the link between the sensory side (nucleus electrosensorius) and the motor side (chirp-controlling neurons of the central posterior/prepacemaker nucleus) is interrupted. This dramatically reduces the excitatory synaptic input received by the chirp-controlling neurons, thus leading to a cessation of chirping behavior. (Courtesy: G. K. H. Zupanc.)

to a dramatic reduction in the input received from the nucleus electrosensorius. On the other hand, sexual maturation during the breeding season re-establishes the interface between the sensory part and the motor part of this neural system. Figure 9.5 summarizes the structural changes taking place in the course of gonadal maturation and regression.

At the synaptic level, the interruption of the connection with axons of the nucleus electrosensorius, and thus the loss of predominantly excitatory synapses, leads to a shift of the balance between excitatory input and inhibitory input toward a higher degree of inhibition. The 'veto' exerted in the somatic regions of the large multipolar neurons of the central posterior/prepacemaker nucleus appears to cause the suppression of electrical activity in these chirp-controlling neurons outside the breeding season, thus resulting in the cessation of chirping behavior.

Seasonal variation in dendritic morphology of motoneurons in white-footed mice

Similar changes in dendritic morphology, as they occur in the central posterior/prepacemaker nucleus, have been observed in white-footed mice (*Peromyscus leucopus*), a species common in temperate habitats over wide areas of North America. This seasonal breeder exhibits marked fluctuations in reproductive behavior. Feral mice breed in the spring and summer, and are reproductively quiescent during the winter. As laboratory experiments have shown, gonadal recrudescence is stimulated by long day lengths, and gonadal involution takes place as the day length shortens. Nancy Forger and Marc Breedlove, then at the University of California, Berkeley, found that, in concert with these seasonal endocrine and behavioral changes, the motoneurons of a specific neuronal assembly, the **spinal nucleus of the bulbocavernosus**, also undergo pronounced alterations in morphology.

This nucleus is located in the lumbar region of the spinal cord. It consists of approximately 200 motoneurons in adult males, but of only 60 motoneurons in adult females. In males, the majority of these motoneurons innervate the muscles bulbocavernosus and levator ani. These muscles attach to the base of the penis, and are involved in producing an erection and ejaculation. Adult females lack the bulbocavernosus and levator ani. Instead, the motoneurons of their spinal nucleus of the bulbocavernosus innervate a sexually non-dimorphic muscle—the anal sphincter. This is also the case with the remaining spinal nucleus of the bulbocavernosus motoneurons in males.

The morphological changes that occur in the spinal nucleus of the bulbocavernosus after exposure of males to short day lengths include shrinkage of the somata and nuclei of the motoneurons, as well as a reduction in their dendritic trees. These effects are reversed by reinstatement of long day lengths.

Further studies by Elizabeth Kurz, Dale Sengelaub, and Arthur Arnold of the University of California at Los Angeles have shown that sex steroid hormones play a crucial role in these structural alterations of the spinal motoneurons. The principal result of their investigation is summarized in Fig. 9.6(a). In this diagram, the average length of the dendritic arbor per cell of the motoneurons has been plotted for the following experimental groups:

- Males at the age of 60 days.
- Males castrated (or sham-castrated, which served as controls) at the age of 60 days, and killed six weeks later, when they were 102 days old.
- Males also castrated at 60 days of age, but then implanted at 102 days with Silastic® tubes that either contained testosterone or were empty (the latter served as controls); mice of this group were killed at 132 days of age, and compared with normal males of the same age.

The motoneurons of the spinal nucleus of the bulbocavernosus were visualized by **retrograde tracing**. Cholera toxin conjugated to horseradish peroxidase was injected into the bulbocavernosus muscle (for a detailed discussion of neuronal tract-tracing techniques, see Chapter 2). Forty-eight hours later, the mice were killed. During this post-administration survival time, the tracer substance was taken up by the axonal terminals making

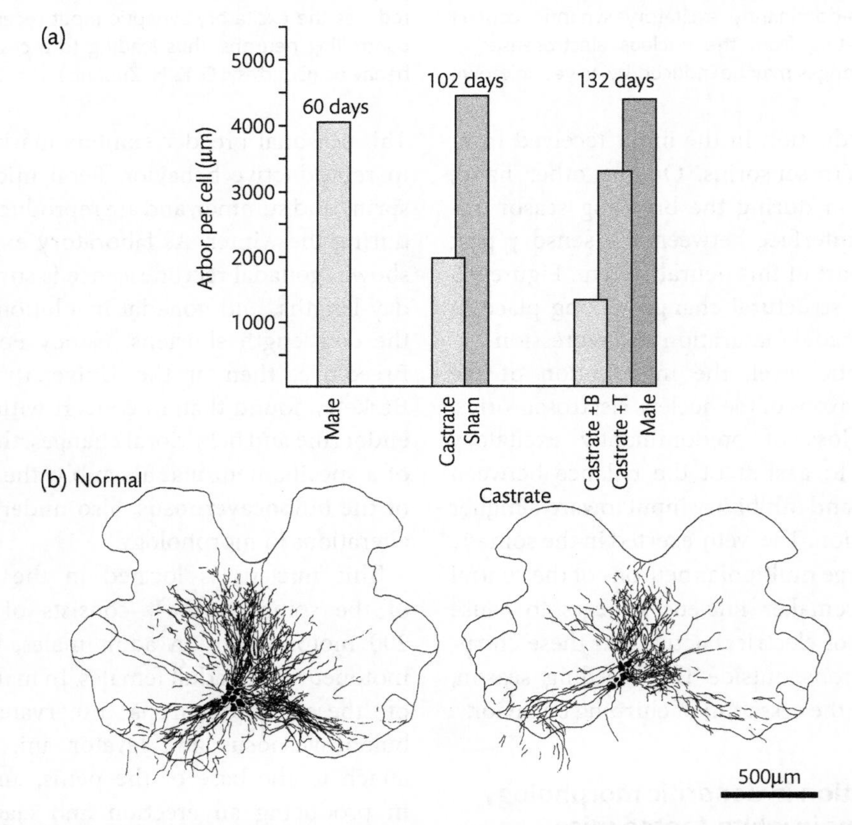

Figure 9.6 (a) Average length of dendritic arbor per motoneuron of the spinal nucleus of the bulbocavernosus in white-footed mice of different experimental groups: Normal males of 60 days of age, males castrated (or sham-castrated) at 60 days of age and killed at 102 days, and normal males at 132 days of age, as well as males castrated at 60 days of age, which received blank implants (Castrate +B) or testosterone implants (Castrate +T) at 102 days and were killed at 132 days. (b) Reconstructed dendritic arbors of the spinal nucleus of the bulbocavernosus motoneurons of a normal male (left) and a castrated male (right). The overall dendritic length is reduced in the castrated males. (After: Kurz, E. M., Sengelaub, D. R., and Arnold, A. P. (1986).)

synaptic contact with the muscle, transported in the retrograde direction towards the motoneuron cell body, from where it finally filled the dendrites. Sections through the spinal cord were processed histochemically so that the motoneurons labeled with tracer substance became visible.

It is evident from the diagram that the dendritic arbor was clearly reduced in castrated males killed six weeks after castration relative to that of sham-castrated males. The extent of this reduction is illustrated in Fig. 9.6(b), which shows reconstructed dendritic arbors of a normal and a castrated male.

The decrease in dendritic length could be reversed by treatment with testosterone, and after four weeks of testosterone treatment, the dendritic arbor of the cell was largely restored to normal levels. These changes in dendritic morphology are likely paralleled by a modulation of the number or the organization of synaptic inputs, and thus by profound alterations of the properties of the neural network involved in control of copulatory behavior.

A similar **endocrine-controlled mechanism** is thought to be at work in feral white-footed mice. According to this hypothesis, exposure to short day lengths causes regression of the testes, which reduces the titer of circulating androgens. This results in a number of morphological changes in androgen-sensitive structures, including changes in dendritic morphology and the size of the motoneurons of the spinal nucleus of the bulbocavernosus, as well as the size of the bulbocavernosus muscles. These changes then lead to the season-related changes in reproductive behavior.

Axonal/synaptic plasticity: A mechanism for accommodating variability in reproductive behavior during the estrous cycle of female cats

Structural reorganization of neural networks does not only occur in concert with seasonal (typically annual or biannual) changes in behavior, but also over shorter timescales. Such short-term changes in behavior and structure of the underlying network are well characterized in domestic cats (*Felis silvestris catus*), mainly due to the research of Veronique G. J. M. VanderHorst and Gert Holstege of the University of Groningen in the Netherlands. In female cats, the estrous cycle is highly variable, but most typically they exhibit multiple cycles during the year, each lasting for several days. Estrous cats display reproductive behavior, including **lordosis**, whereas such behavior is not seen in non-estrous cats. Lordosis behavior is induced by estrogen.

> **Lordosis behavior** Body posture of females in some mammals (including rodents and felines) during mating, triggered by mounting of the male. Characteristic features of this reflex-like behavior are a ventral arching of the vertebral column, lowering of the forelimbs, raising of the hips, and lateral or dorsal displacement of the tail so that the vagina is presented to the male.

Part of the lordosis behavior is controlled by a projection that originates in the **nucleus retroambiguus** in the caudal medulla of the brain and terminates in a group of lumbosacral motoneurons that innervate a distinct set of hindlimb, axial (= central-part), and pelvic floor muscles. These muscles are involved in lordosis and other reproductive behaviors. The nucleus retroambiguus–lumbosacral pathway is thought to constitute the final common pathway for controlling these behaviors in females.

Injection of the tracer wheat germ agglutinin–horseradish peroxidase into the nucleus retroambiguus (see Chapter 2 for details on neuronal-tract-tracing methods) revealed, in estrous females, a high density of fibers arising from these neurons and traveling to the motoneuron cell group in the lumbosacral region. By contrast, in non-estrous females the density of such fibers was markedly lower. Similarly, the number of synaptic terminals was almost nine times higher in estrous females, compared to non-estrous females. VanderHorst and Holstege have proposed that the increased number of fibers and synapses in estrous females is due to the outgrowth and/or sprouting of axons from nucleus retroambiguus cells, and to the formation of new synapses during estrus. This hypothesis receives strong support by the presence of **growth cones** in the target area of the nucleus retroambiguus projection in estrous females, whereas in non-estrous females such growth cones have not been found. Further experiments confirmed that the activation through structural reorganization of the pathway involved in lordosis behavior is, as expected, controlled by **estrogen**. Figure 9.7 illustrates the

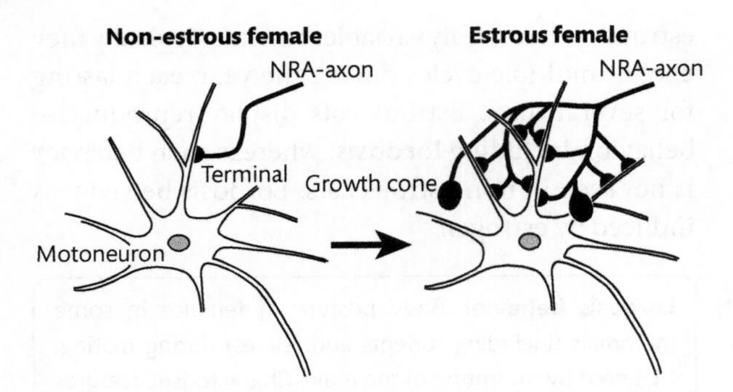

Figure 9.7 Model of the dynamics underlying the structural reorganization of the projection of nucleus retroambiguus (NRA) neurons to lumbosacral motoneurons. Estrogen induces outgrowth and sprouting of axons of the nucleus retroambiguus, and the formation of new synapses. The difference in density of axonal terminals shown in the diagram between non-estrous and estrous female cats reflects the difference found in histological material. (After: VanderHorst, V. G. J. M., and Holstege, G. (1997) © 1997 Society for Neuroscience.)

dynamics underlying the structural differences between estrous and non-estrous female cats.

Growth cone Highly motile structure at the tip of a growing axon. Growth cones play a critical role in the guidance of the extending axons.

➤ The dramatic differences in reproductive behavior between estrous and non-estrous female cats are thought to be accommodated by axonal/synaptic reorganization of the nucleus retroambiguus–lumbosacral tract, which constitutes the final common pathway for controlling several of these behavioral patterns.

Structural reorganization mediated by glial cells

Changes in neuronal excitation may be caused not only by changes in the number or the morphology of neurons, but also by alterations in the organization of the network provided by **glial cells.** An example of such a mechanism has been found in the **magnocellular hypothalamo-neurohypophysial system** of mammals, which forms part of the anterior hypothalamus. As electron micrographs have shown, the magnocellular neuronal somata are almost completely separated from their neighbors by fine processes of astrocytic glial cells, so that only approximately 1% of the total neural membrane is in direct apposition without intervening glial processes. In response to changes in the animal's physiological state, for example during periods of water deprivation or lactation, the astrocytes rapidly retract, thus resulting in a tremendous increase in the number of neuronal elements that are directly juxtaposed. This glial withdrawal leads to an enhancement of neuronal excitability, which is believed to be directly linked to the observed behavioral changes.

When to use 'structural reorganization'?

The strategy to produce motivational changes by modifying the structure of either neurons or glial cells, or both, and thus the properties of the associated neural network, appears to be used especially in the following cases:

- When changes in the motivation underlying a behavior are rather slow.
- When alterations at the behavioral level are dramatic.
- When a new motivational state becomes manifest for rather long periods of time.

All three features are characteristic of seasonally breeding animals. Therefore, it appears possible that structural reorganization of neural networks is a mechanism commonly employed in such animals to control seasonal changes in behavior, although this does not exclude other mechanisms.

Biochemical switching

Modulation of the stomatogastric ganglion

A second set of mechanisms employed by organisms to accommodate motivational changes in behavior can be subsumed under the term **biochemical switching**. This type has been particularly well studied in the **stomatogastric ganglion** of decapod crustaceans.

One advantage of the stomatogastric ganglion as a model system for studying the structure and function of neural circuits is its relatively low number of neurons—approximately 30. This feature has enabled researchers, even back in the early years of neuroethology, to achieve, through **simultaneous physiological recordings**, a comprehensive understanding of the physiological properties of this neural network. Further information on the contribution of the individual network components to the motor activity has been obtained by selectively eliminating single neurons within the ganglion through **photoinactivation** after intracellular injection of fluorescent dyes—a technique described in detail in Chapter 2. Equally important, the physiology of the stomatogastric ganglion can also be studied after its removal from the animal; the neural circuit generates then fictive motor patterns (for more information on fictive behaviors, see Chapter 6).

The stomatogastric ganglion is one of the four interconnected ganglia of the **stomatogastric nervous system**. The other three ganglia are a single **esophageal ganglion** and bilaterally paired **commissural ganglia**. Central pattern generators in this nervous system produce motor patterns that underlie the rhythmic movements of the esophagus, the cardiac sac, the gastric mill, and the pylorus. The two rhythms generated by the stomatogastric ganglion are the (fast) **pyloric rhythm** and the (slow) **gastric mill rhythm**.

Most of the neurons of the stomatogastric ganglion are motoneurons that make excitatory connections with the muscles of the stomach. There are no intervening premotor interneurons, a feature that distinguishes the stomatogastric ganglion from most other motor systems. This makes it possible to record the motor pattern directly from the neurons of the stomatogastric ganglion.

The muscles innervated by the stomatogastric ganglion neurons move different regions of the foregut:

- The **gastric mill** consists of three teeth. The rhythmic movements of its muscles determine the gastric mill rhythm.
- The **pylorus** abuts the gastric mill region. Its filtering and sorting movements are produced by a set of muscles that control both the pylorus and its valves. The corresponding rhythm is called the pyloric rhythm.

The commissural ganglia and the esophageal ganglion, as well as the brain, include neurons that project to the stomatogastric ganglion via the **stomatogastric nerve**. The fibers comprising this nerve contain a large number of modulators, although the exact nature of these substances may, to a certain extent, vary among species. Input to the neuropil of the stomatogastric ganglion originates also from the gastro-pyloric receptor neurons. The latter are sensory neurons that respond to the stretch of several of the stomach muscles. Since the stomatogastric ganglion is located anterior to the heart within an artery, additional modulatory substances reach the ganglion from neurosecretory structures, such as the pericardial organ. Figure 9.8 shows the location of the stomatogastric ganglion in relation to other body structures, and provides a partial list of **transmitters** and **neuromodulators** that have been identified in this ganglion.

> ➤ The stomatogastric ganglion of decapod crustaceans consists of individually identified neurons that control rhythmic movements of foregut structures.

In the early years of research on the stomatogastric ganglion, it was thought that the anatomical connections between the individual neurons define the function of the circuit of this ganglion. However, several decades of work by Eve Marder (see Box 9.1) and other investigators have shown that, although this information is necessary to understand the output pattern of the circuit, it is not sufficient. As these studies have demonstrated, the exact firing pattern of the neurons of the stomatogastric ganglion is defined by the actual modulatory environment. According to this model, the anatomical network of the stomatogastric ganglion provides only a physical backbone upon which the modulatory inputs can operate. This makes it possible that a single network can produce multiple variations in the behavioral output under different conditions. For this concept, Peter Getting and Michael Dekin of the University of Iowa coined the term **polymorphic network**. Although less well studied in vertebrates, overall there is compelling evidence that neuromodulation by a multitude of modulators is a universal mechanism to accommodate variable output of neural circuits.

Polymorphic networks Anatomically defined networks whose modulation results in multiple functional modes of operation.

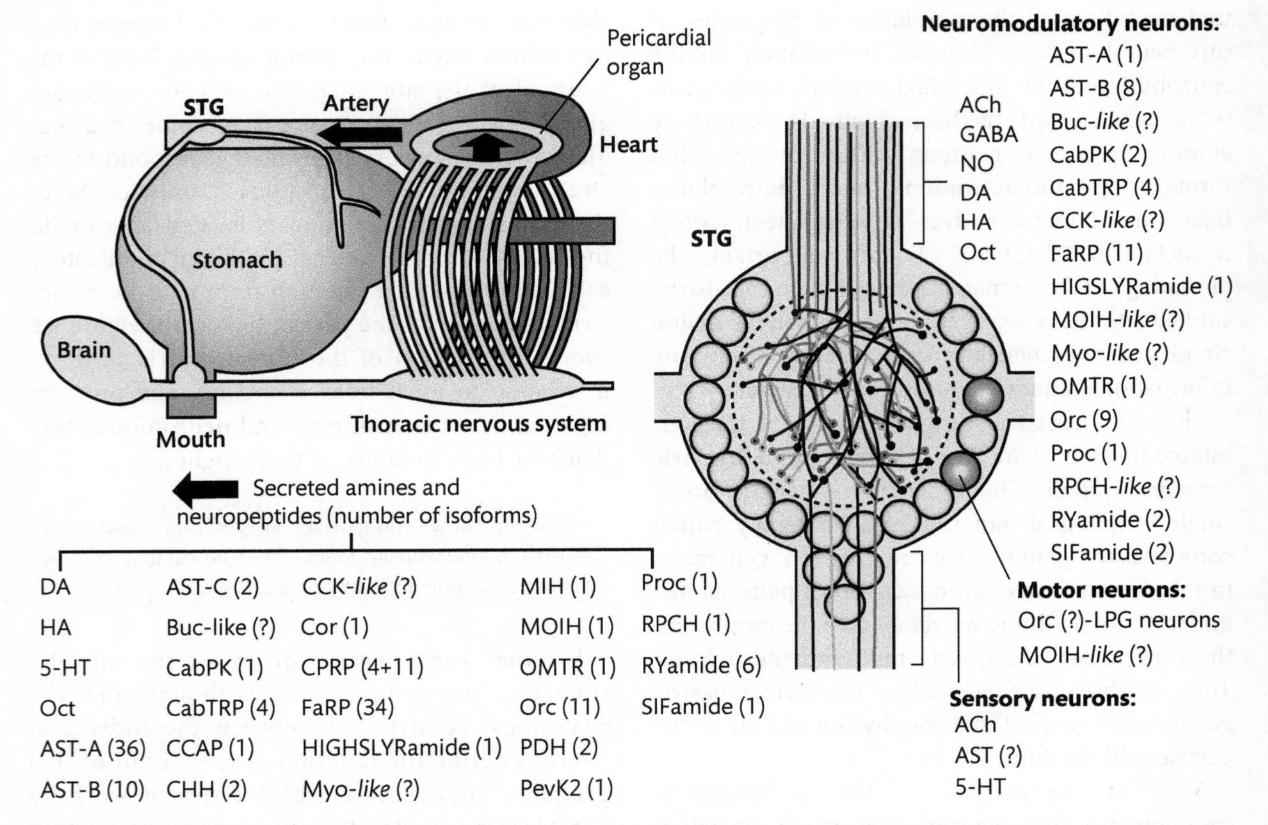

Figure 9.8 Location of stomatogastric ganglion and its neuromodulatory environment. The ganglion is located anteriorly to the heart within an artery. Through this artery, modulatory amines and peptides from neurosecretory structures, such as the pericardial organ, are supplied. A list of these secreted substances is shown on the bottom left. The names of some peptides denote their structure, using the one-letter amino acid code. Neuromodulatory neurons and sensory neurons deliver peptides and amines, through fibers, to the neuropil of the stomatogastric ganglion. A list of these substances is shown on the right. The number of isoforms of the neuropeptides is indicated in parentheses. (Courtesy of Dirk Bucher.)

An example of how neuromodulators can activate different motor rhythms in the stomatogastric ganglion is shown in Fig. 9.9. In this experiment, input from the anterior ganglia was removed by placing a Vaseline® well filled with sucrose on the stomatogastric nerve. This treatment blocks impulse traffic down the nerve. The major results are as follows:

- When applying saline as a control, only the so-called pyloric dilator neurons are rhythmically active.
- A bath application of the peptide TNRNFLRFamide produces a modest increase in the frequency and intensity of the motor pattern.
- Application of proctolin, crustacean cardioactive peptide, red pigment-concentrating hormone, and serotonin all strongly activate lateral pyloric neurons and increase the frequency of the pyloric dilator burst.

Modulation of crayfish aggressive behavior

Another example demonstrating the importance of neuromodulators in the setting of the motivational state of an animal also stems from work on crustaceans. Such investigations, mainly conducted by Robert Huber of the University of Graz, Austria, and Edward Kravitz of Harvard University in Boston, Massachusetts, make use of the fact that crayfish readily engage in aggressive encounters when placed together in an aquarium. The fights, which can easily be quantified, escalate in a probabilistic manner until one of the opponents retreats.

Under normal circumstances animals faced with much larger opponents quickly withdraw from the encounter. However, administration of **serotonin** into the hemolymph of freely moving individuals

Box 9.1 Eve Marder

Eve Marder in her laboratory at Brandeis University. (Courtesy Mike Lovett, Brandeis University.)

In the late 1970s, when Eve Marder started her faculty career, the common belief was that the connections in a neural circuit are hard-wired, and that its output is a single, predictable activity pattern. However, what she discovered as part of her research was an astonishing degree of plasticity defined by what has become known as 'neuromodulation'. Under the influence of neuromodulators—such as biogenic amines or neuropeptides—neural networks, even when their anatomical organization is fixed, are capable of generating multiple output patterns, thereby resulting in a range of behaviors. This discovery led to a paradigm shift in the neural sciences, with the concept of neuromodulation having become one of the fundamental principles underlying the function of neural networks.

Eve Marder was born in New York City in 1948. After graduating in biology from Brandeis University in Waltham, Massachusetts, in 1969, she went to the University of California at San Diego to do her Ph.D. thesis work in the laboratory of Allen I. Selverston. It was he who introduced Marder to the decapod stomatogastric ganglion as a model system to study the structure and function of simple neural networks. Following postdoctoral training at the University of Oregon and the École Normale Supériere in Paris, France, she joined the faculty of Brandeis University, where she progressed to the rank of full professor in 1990.

Together with a large number of Ph.D. students and postdoctoral fellows, she has continued research on neuromodulation in the stomatogastric ganglion throughout her career. Her achievements in this area have been widely recognized and include prestigious awards such as the W. F. Gerard Prize, the Karl Spencer Lashley Award, the Gruber Prize, and the Kavli Prize, as well as election to the American Academy of Arts and Sciences and the National Academy of Sciences of the U.S.A.

leads to an alteration of their behavioral strategy to retreat and act as subordinates. As a consequence, fights last considerably longer compared with control animals. Thus, serotonin injected into subordinate animals appears to change the aggressive motivation of the crayfish toward higher levels. At present, the central sites responsible for this effect are unknown.

> ➤ Administration of serotonin to the hemolymph of subordinate crayfish can change the aggressive motivation toward higher levels.

Modulation of the modulators

Although the examples given above suggest a rather unidirectional relationship between neuromodulators and behavior, the actual situation is certainly more complex. This has been indicated in a study conducted, again on crayfish, by Shih-Rung Yeh, Barbara Musolf, and Donald Edwards of the Georgia State University in Atlanta, Georgia. Their investigations have shown that the modulatory effect of serotonin on the **lateral giant interneuron**, a command neuron controlling escape, is itself modulated by the social status and the social history of the animal. In dominant crayfish, the response of the lateral giant interneurons triggered by serotonin is transiently increased, whereas in subordinates it is transiently inhibited. These slow, but reversible, modulatory alterations appear to result from changes in the population of serotonin receptors.

What makes modulators suitable for neuromodulation?

A number of properties make catecholamines and neuropeptides well suited to mediate motivational

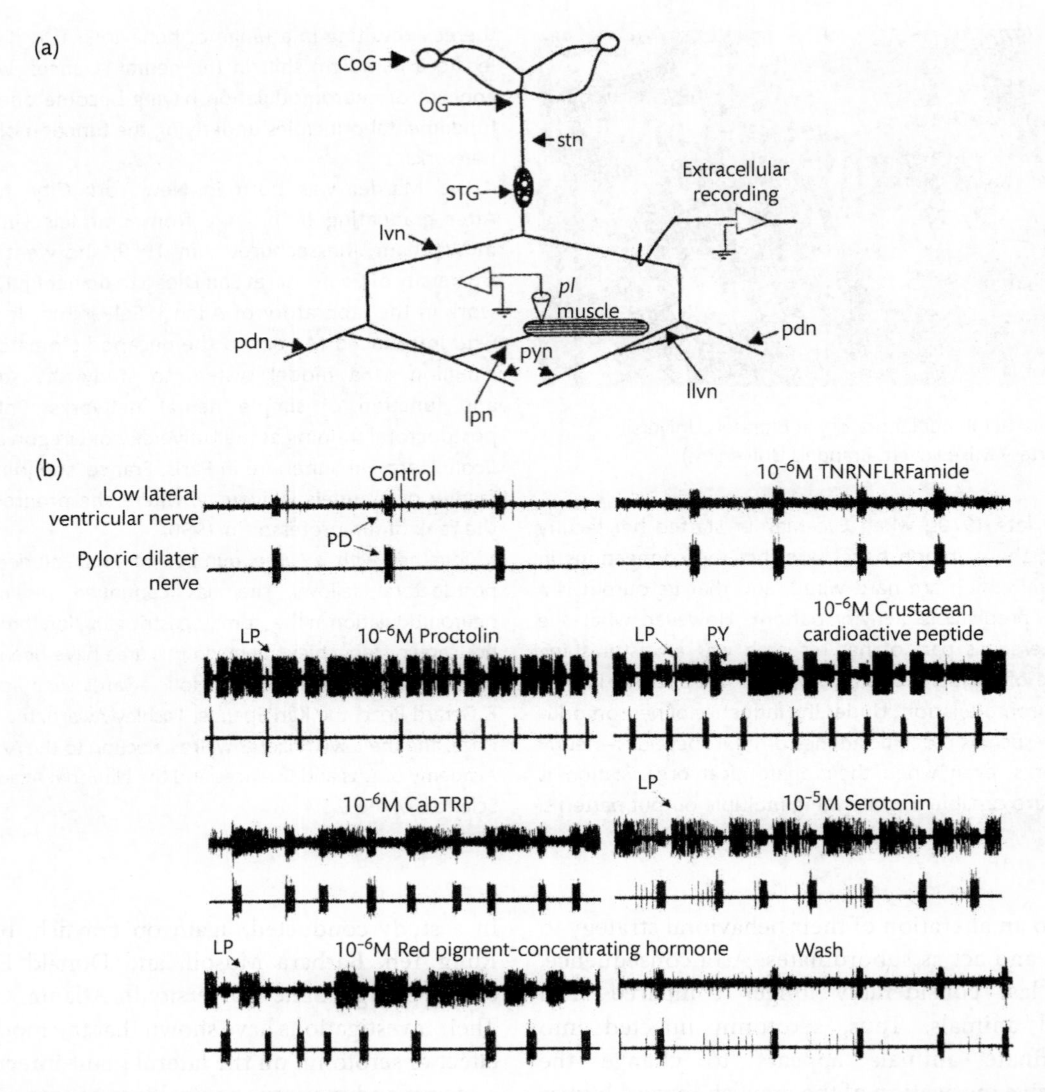

Figure 9.9 (a) Schematic organization of the stomatogastric nervous system of the American lobster (*Homarus americanus*). The paired commissural ganglia (CoG) and the single esophageal ganglion (OG) provide, via the stomatogastric nerve (stn), input to the stomatogastric ganglion (STG). Muscles of the foregut are innervated by neurons of the stomatogastric ganglion via several motor nerves, including the lateral ventricular nerve (lvn), the pyloric dilator nerve (pdn), the pyloric nerve (pyn), the lateral pyloric nerve (lpn), and the low lateral ventricular nerve (llvn). (b) The experiment shows the effects of various modulators on the motor patterns generated by the isolated stomatogastric ganglion. The ganglion is physiologically isolated by blocking the stomatogastric nerve. This results in rhythmic activity only of the pyloric dilator (PD) neuron of the stomatogastric nerve, as revealed by extracellular recordings from the low lateral ventricular nerve and the pyloric dilator nerve after bathing the ganglion in saline as a control. Bath application with the peptide TNRNFLRFamide leads to a modest increase in frequency and intensity of the bursts produced by the pyloric dilator neurons. In contrast, application of proctolin, crustacean cardioactive peptide, *Cancer borealis* tachykinin-related peptide (CabTRP), serotonin, and red pigment-concentrating hormone all strongly activate the lateral pyloric neurons, as seen on the trace of the recordings from the low lateral ventricular nerve. They also increase the frequency of the pyloric dilator (PD) burst, as seen on the trace of the recordings from the pyloric dilator nerve. Between each application, the preparation is washed with saline. The final application of saline recovers the original pattern of bursts generated by pyloric dilator neurons. (After: Marder, E. and Richards, K. S. (1999).)

influences within the central nervous system. Among them are, in particular, the following two features:

- Neuropeptides and catecholamines are frequently **released in a non-synaptic fashion**. This contrasts with 'classical' transmitters, such as acetylcholine, glutamate, or GABA, which are typically released at synaptic specializations. Such specialized synaptic zones are visible under the electron microscope as electron-dense (black) thickenings at the presynaptic membrane. These thickenings, commonly called active zones, represent specialized protein structures at which the synaptic vesicles dock to initiate fusion with the presynaptic membrane and release of the synaptic content.
- In the transmission process involving catecholamines and neuropeptides, there is often no focal relationship between the site of release and the location of the corresponding receptors. This again contrasts with the situation known from classical transmitters. At the neuromuscular junction, where acetylcholine is used as a transmitter, cholinergic receptors are located at high densities immediately opposed to the site of transmitter release on the postsynaptic membrane. The lack of such focal relationship between the site of release and the site of ligand–receptor interaction is referred to as **ligand–ligand receptor mismatch**.

Both properties—the non-synaptic release and the ligand–ligand receptor mismatch—lead to a 'diffuse' effect of catecholamines and neuropeptides. As a consequence, an endogenous ligand may interact with receptors at more than just one site in the central nervous system, and some of these sites may be fairly distant from the site of release. Not surprisingly, these sites may be involved in the control of more than just one behavior. Such a notion is in agreement with behavioral observations: Motivational factors that affect the occurrence of one type of behavior tend to influence the probability of other behavioral patterns as well.

When to use 'biochemical switching'?

The strategy to produce motivational changes through biochemical switching appears to be especially suitable in the following cases:

- When the changes in the motivation underlying a behavior are rather fast.
- When the alterations in the propensity to execute a behavior do not occur in an all-or-none function; rather, fine gradual differences within a broad range of possibilities are to be accommodated.

Such fast and gradual changes in motivation are typical, for example, of behaviors associated with feeding and ingestion. Indeed, biochemical switching has frequently been found in neural circuitries involved in the control of these behaviors.

Summary

- In an 'ideal' reflex, the relationship between a stimulus and the resulting behavior is fixed. Modulation of this input–output relationship is caused by internal, or motivational, factors.
- Motivational changes in behavior can occur only if the underlying neural network exhibits the potential for neural plasticity.
- Two major mechanisms that mediate motivational changes in behavior involve structural reorganization of the neural network underlying the control of the respective behavior, and chemical modulation of this network through neuromodulators, especially serotonin, catecholamines, and neuropeptides ('biochemical switching').
- Alterations in dendritic morphology of neurons, as one mechanism of structurally reorganizing neural networks, have been implied in playing a role in dramatic behavioral changes, such as those that occur in seasonally breeding animals.
- Structural reorganization may be mediated further by non-neuronal cells, for example,

by the degree to which glial cells intervene between neurons. This leads to alterations in the physiological properties of the neural network, and thus of the behavioral outcome.

- Biochemical switching as a mechanism to modulate behavior has, among other systems, been demonstrated in the stomatogastric ganglion of decapod crustaceans. The modulators, mainly neuropeptides, are contained in axons innervating this ganglion. These modulators operate on the 'polymorphic' network to produce different behavioral outputs.
- Modulators themselves may be modulated by the social status and social history of an animal.
- Among the cellular properties that make neuromodulators well suited to mediate motivational changes in behavior are the frequently observed release in a non-synaptic fashion and the lack of a focal relationship between the site of release and the site of ligand–receptor interactions ('ligand–ligand receptor mismatch').

The bigger picture

During its rather short history, neuroethology has made significant contributions to the evolution of the neural sciences. Several of these contributions are in the area of neural plasticity. For example, indication that the vertebrate central nervous system continues to generate neurons beyond embryonic stages of development had already been obtained previously, but it was the pioneering research of Fernando Nottebohm and his associates on songbirds in the 1980s and 1990s that finally led to the overturn of the 'no-new-neuron-in-old-brains' dogma—something that had dominated neuroscience for nearly one hundred years before (see Chapter 13).

Similarly, the studies of several neuroethologists, including Eve Marder, on the stomatogastric ganglion of decapod crustaceans (see this chapter) revealed not only an unexpectedly large number of neuromodulators in a neural structure, but also demonstrated how the action of such modulators can empower a hardwired neural circuit to produce multiple physiological (and behavioral) outputs. Drawing from the strength of neuroethology as an integrative discipline, such work has been seminal in relating structural and biochemical plasticity at the level of cells and networks to behavioral plasticity at the level of the organism.

Recommended reading

Harris-Warrick, R. M. and Marder, E. (1991). Modulation of neural networks for behavior. *Annual Review of Neurosciences* **14**:39–57.

An excellent review summarizing the concept of polymorphic networks. Most principles associated with this concept are illustrated using the stomatogastric ganglion of decapod crustaceans.

Marder, E. (2012). Neuromodulation of neuronal circuits: back to the future. *Neuron* **76**:1–11.

A stimulating review, written from a historical perspective by one of the pioneers in the field of neuromodulation.

Marder, E. and Calabrese, R. L. (1996). Principles of rhythmic motor pattern generation. *Physiological Reviews* **76**:687–717.

An exhaustive review of how rhythmic movements are controlled by central pattern-generating neural networks, and how the behavioral output is changed by modulatory input.

Zupanc, G. K. H. (1996). Peptidergic transmission: from morphological correlates to functional implications. *Micron* **27**:35–91.

A comprehensive review of morphological features characterizing synaptic and non-synaptic modes of neuropeptide transmission. An understanding of these features is essential to appreciate the biochemical switching mechanism.

Zupanc, G. K. H. and Maler, L. (1997). Neuronal control of behavioral plasticity: the prepacemaker nucleus of weakly electric gymnotiform fish. *Journal of Comparative Physiology A* **180**:99–111.

This review article provides an overview of chirping behavior and its neural substrate, the central posterior/prepacemaker nucleus, in weakly electric gymnotiform fish. Special emphasis is placed on the discussion of neural mechanisms accommodating motivational changes in this behavioral pattern.

Short-answer questions

9.1 Define the term 'motivation' in one sentence.

9.2 The two major brain mechanisms that mediate motivational changes in behavior involve the following:

i) ....

ii) ....

9.3 Summarize the essential features of the dendritic-plasticity model that provides a mechanistic explanation for the seasonal alterations in chirping behavior in the weakly electric knifefish, *Eigenmannia* sp.

9.4 In white-footed mice, motoneurons of the spinal nucleus of the bulbocavernosus undergo pronounced seasonal alterations in dendritic morphology. These changes are controlled by (indicate the correct statement):

i) testosterone;

ii) neuropeptide Y;

iii) L-glutamate.

9.5 What is the role of the anatomical network of the stomatogastric ganglion versus that of the modulatory environment in generating neural activity?

9.6 Define the term 'polymorphic network' in one sentence.

9.7 List two properties that make neuropeptides particularly suitable as neuromodulators.

9.8 **Challenge question** Seasonal differences in dendritic arborization have been demonstrated through neuronal tract tracing in several model systems. They are commonly interpreted as seasonally induced outgrowth and retraction of these dendrites. Provide an alternate explanation by taking into consideration the intracellular transport of the tracer substance(s) used. What additional experiments or observations could potentially provide evidence in support of one of these two hypotheses?

Essay questions

9.1 What is motivation? How can animals, at the neuronal level, accommodate short-term and long-term motivational changes in behavior?

9.2 White-footed mice are seasonal breeders that reproduce in spring and summer, and are sexually inactive during the winter. Describe the neural mechanism that is thought to mediate the seasonal variation in copulatory behavior in male mice. What experimental evidence is available to support the existence of such a mechanism?

9.3 Adult neurogenesis has been demonstrated in the central posterior/prepacemaker nucleus of weakly electric knifefish. Design experiments to verify the hypothesis that this phenomenon is causally linked to changes in chirping behavior.

9.4 Explain the concept of polymorphic networks, using the stomatogastric ganglion of lobsters and crabs as a model system. What is the advantage of such a network, instead of separate subpopulations of neurons dedicated to the control of distinct behaviors?

9.5 Why are neuropeptides much better suited to mediate motivational changes in behavior than 'classical' transmitter substances?

Advanced topic Neuromodulation of the respiratory rhythm

Background information

In mammals, the pre-Bötzinger complex in the brainstem is critical in generating respiratory activity. It is composed of at least two different types of pacemaker neurons, which, together with extensive synaptic connections, form the respiratory central pattern generator. Brain slices containing the pre-Bötzinger complex generate three types of physiological activity patterns that are thought to drive the production of three distinct forms of respiratory motor patterns—normal breathing, sighs, and gasps. Immunohistochemical studies have revealed a multitude of modulatory inputs received by the pre-Bötzinger complex. These neuromodulators regulate frequency, regularity, and amplitude of respiratory activity, as well as the transition between the different types of respiratory rhythms.

Essay topic

In an extended essay, provide an overview of the neuroanatomical structure of the pre-Bötzinger complex, including the different types of neuromodulators found in this brain area. Describe how these neuromodulators regulate the different parameters of a given respiratory activity, and how they differentially affect the activity of the two types of pacemaker neurons. Discuss, from a general perspective, the ability of neuromodulators to reconfigure the network activities of central pattern generators, comparing neuromodulation in the pre-Bötzinger complex and the stomatogastric ganglion.

Starter references

Doi, A. and Ramirez, J.-M. (2008). Neuromodulation and the orchestration of the respiratory rhythm. *Respiratory Physiology & Neurobiology* **164:**96–104.

Tryba, A. K., Peña, F., and Ramirez, J.-M. (2006). Gasping activity *in vitro*: a rhythm dependent on 5-HT_{2A} receptors. *Journal of Neuroscience* **26:**2623–2634.

Tryba, A. K., Peña, F., Lieske, S. P., Viemari, J.-C., Thoby-Brisson, M., and Ramirez, J.-M. (2008). Differential modulation of neural network and pacemaker activity underlying eupnea and sigh-breathing activities. *Journal of Neurophysiology* **99:**2114–2125.

Viemari, J.-C. and Ramirez, J.-M. (2006). Norepinephrine differentially modulates different types of respiratory pacemaker and nonpacemaker neurons. *Journal of Neurophysiology* **95:**2070–2082.

To find answers to the short-answer questions and the essay questions, as well as interactive multiple choice questions and an accompanying Journal Club for this chapter, visit **www.oup.com/uk/zupanc3e.**

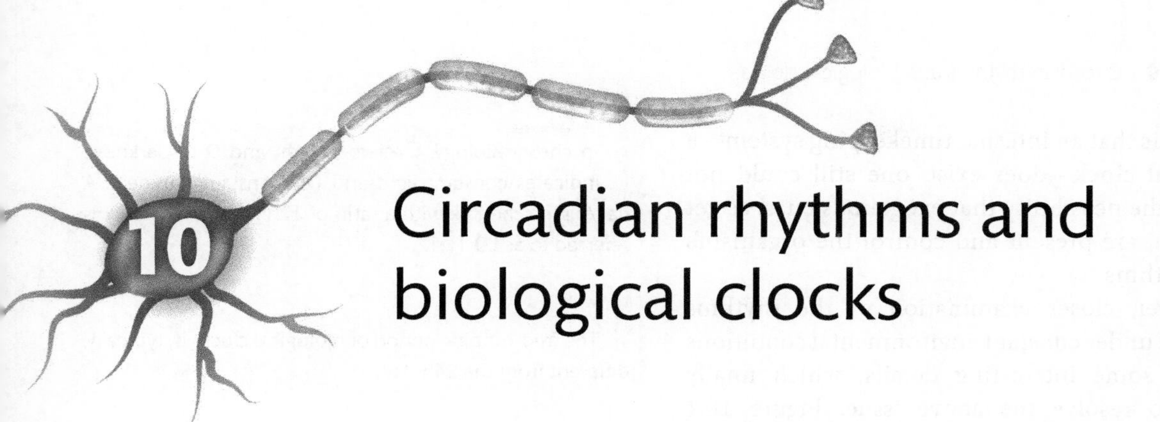

Circadian rhythms and biological clocks

Introduction

Many spontaneous behaviors and physiological activities show daily rhythmic oscillations. For example, locomotor activity in nocturnal animals such as hamsters or mice occurs predominantly during the night, whereas in diurnal animals such activity is largely restricted to the daytime. In the natural environment, these cycles of activity repeat themselves every 24h. How are such rhythms generated? Are they merely a response to the presence or absence of light? Or are they controlled by an internal timekeeping system?

In this chapter, we will address these questions. We will first take a look at some of the major behavioral characteristics of rhythmic patterns occurring daily, and find out how the observations made in the course of these studies have led to the concept of biological clocks. Then, we will analyze the molecular structure of these clocks, and discuss how the individual clock components integrate to produce an oscillatory output. Finally, we will summarize what is known about the localization of biological clocks at a systems level.

Key concepts

Until the 1950s, it was unclear whether the daily rhythms exhibited by virtually all organisms, including humans, were controlled by exogenous cues arising from the physical environment or whether an endogenous timekeeping system does exist. This issue remained unresolved, even when an increasing number of experiments demonstrated that such rhythms can persist in the absence of any obvious exogenous time signals, for example when an animal is kept in constant darkness. Although the persistence of rhythmicity under such conditions appeared to favor the hypothesis that an internal timekeeping system—a **biological clock**—does exist, one still could not rule out the possibility that exogenous cues, as yet unknown, are present and control the organism's daily rhythms.

However, closer examination of the rhythms exhibited under constant environmental conditions revealed some interesting details, which finally helped to resolve the above issue. Figure 10.1 shows data from one of the first studies in the then new discipline known as **chronobiology**. This investigation was published by Jürgen Aschoff (see Box 10.1) in 1952. Initially, Aschoff kept mice under natural day/night conditions and continuously monitored their locomotor activity over several days. This activity peaked in the early night hours and repeated itself every 24h. When Aschoff kept the mice in constant darkness (a condition commonly referred to as 'DD' in chronobiology), he still observed a rhythmic cycling of locomotor activity. However, the period between maxima was now somewhat shorter than the period exhibited under natural day/night conditions. Evidently, the 'biological period,' as Aschoff named it, was different from the 'physical period' of the 24h day. This 'biological period' is now commonly called the **free-running period**.

➤ In chronobiology, L refers to light and D to darkness. LL indicates constant light and DD constant darkness. A day/night schedule with a ratio of 12h light to 12h dark is referred to as LD 12:12.

➤ The free-running period of biological clocks is, typically, different from the 24h day.

The deviation of the free-running period from the 24h cycle defines the **circadian rhythm** (Latin *circa* = about, *dies* = day). As numerous studies on humans and animals have shown, the rhythms of many behaviors and physiological activities revealed under constant environmental conditions are close to, but not exactly, 24h. Most commonly, the values of these free-running periods are in the range of 23–25h. In most humans, the free-running circadian rhythms exhibit periods of approximately 25h, but in a minority of people the free-running period is shorter than 24h. The phenomenon of circadian rhythms provides a strong argument in favor of the existence of biological clocks. If an organism's rhythms were exclusively exogenously driven, then one would expect a free-running period of exactly 24h. However, this is almost never the case.

Chronobiology The scientific discipline that examines biological rhythms.

Circadian rhythm A rhythm of a biological phenomenon, driven by an endogenous time-keeping system, with a period close, but not equal, to 24 hours.

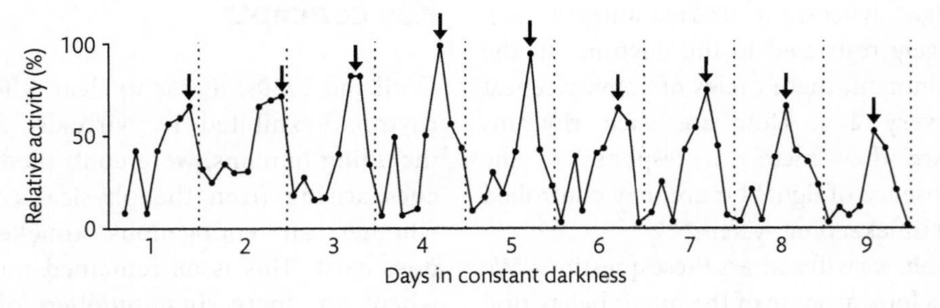

Figure 10.1 Circadian rhythm of locomotor activity in mouse, as demonstrated in a classic experiment by Jürgen Aschoff. Before the experiment, the mouse was exposed to natural day/night condition. The mouse was then kept in constant darkness for nine days and locomotor activity was monitored. Whereas under normal day/night conditions locomotor activity peaked in the early night hours, in constant darkness the maximum peak of activity gradually shifted to the earlier hours of the day. Thus, the free-running period in this individual was shorter than the 24-hour cycle of the day—the locomotor activity showed a circadian rhythm. (After: Aschoff, J. (1952).)

Box 10.1 Jürgen Aschoff

Jürgen Aschoff. (Courtesy: Max Planck Society.)

Jürgen Aschoff, one of the founders of chronobiology (together with Erwin Bünning and Colin Pittendrigh (see Box 10.2)), was born in Freiburg, Germany, in 1913. His father was the renowned pathologist Ludwig Aschoff, after whom a node of modified cardiac muscle tissue ('Aschoff-Tawara node') is named. Following medical studies at the University of Bonn, Jürgen Aschoff attended the University of Göttingen, where he worked as an assistant of Friedrich Hermann Rein on the physiology of thermoregulation. During that time, he studied the rhythmic variations of body temperature over the 24h day. This work triggered his lifelong interest in the mechanisms underlying such biological rhythmicity.

After the Second World War, Aschoff lectured at the University of Würzburg. In 1952, he joined his mentor Rein for a second time, after the latter had been appointed director of the Department of Physiology at the newly established Max Planck Institute for Medicine in Heidelberg. In 1958, Aschoff accepted an invitation by the Max Planck Society to become one of the directors of the newly formed Max Planck Institute for Behavioral Physiology, and to establish his own department dedicated to the study of biological timing in both humans and animals. He was particularly influenced by Gustav Kramer and Erich von Holst (see Chapter 3) at the institute. Kramer had discovered that birds use a circadian rhythm to compensate for the sun's apparent movement across the sky, a mechanism that enables them to determine compass direction (see Chapter 11). Von Holst, on the other hand, taught Aschoff how to analyze the properties of coupled oscillators—an approach that became central to Aschoff's own research.

One of the topics Aschoff studied time and again was how endogenous circadian rhythms are synchronized with the 24h environment by their response to external synchronizing factors, the most common of which is light. He coined the German word *Zeitgeber* for such factors, a term that is also used in English literature. He showed that the free-running period of an organism kept under constant light conditions is a function of the intensity of illumination. This effect is known as 'Aschoff's rule.'

Jürgen Aschoff applied to humans the experimental approaches used to study biological rhythms in animals. With the financial support of NATO, he built an underground bunker at Erling-Andechs. This bunker enabled him and other investigators to study rhythmic patterns in human subjects isolated from time cues. These investigations demonstrated that human physiology and behavior are controlled by similar endogenous circadian oscillators, as is the case in animals. The results obtained through these studies have far-reaching implications on a number of issues relevant to human welfare, including occupational stress caused by shift work, accidents related to circadian rhythms, optimal timing of drug administration, and treatment of sleep disorders.

After his retirement in 1983, Aschoff returned to Freiburg to live in the mansion of his parents and continued to work in the area of chronobiology. He died in 1998 at the age of 85.

Temperature compensation of free-running periods

The rate of most biochemical reactions depends strongly on temperature. Typically and within a certain temperature range, the reaction rate doubles or triples when the temperature increases by 10°C. If biological clocks were affected by changes in ambient temperature in a similar way, this would pose a serious problem regarding their reliability as timekeepers. This problem would be particularly severe in poikilotherms, as their natural environment is often characterized by dramatic variations in temperature over the day.

➤ A distinctive feature of biological clocks is their ability to compensate for variability caused by ambient temperature.

The problem of temperature dependence was already recognized in the early days of chronobiology. Then, the demonstration of the ability of organisms to compensate for such environmentally induced variability was viewed as critical in finding supportive evidence for the existence of biological clocks. In 1954, Colin S. Pittendrigh (see Box 10.2) published the results of an investigation in which he had addressed this problem using **eclosion** behavior in *Drosophila*. Among a population, eclosion—the emergence of the adult fly from the pupal case—occurs in bursts in the early morning hours, but is less frequent in the late morning and afternoon. This timing is an important adaptation to the environment that the young adults encounter after emergence. When they are still immature, they lose water at much higher rates, and thus the hot and dry times of the day impose an enormous risk on their chances of surviving the first few hours of their adult life. As Pittendrigh showed, the rhythmic occurrence of bursts of eclosion persists in constant darkness, with a free-running period of about 24h, and a length largely independent of variations in ambient temperature. Similar results indicating effective **temperature compensation** of free-running rhythms have been obtained in a number of other organisms and for various rhythmic phenomena.

Eclosion Emergence of adult insects from their pupal cases or of larvae from the eggs.

Entrainment of biological clocks by environmental cues

In addition to their ability to maintain a free-running rhythm under constant environmental conditions and to compensate for changes in ambient temperature, a third characteristic feature of biological clocks is that they can become aligned to external time cues arising from the organism's environment. This process is called **entrainment**. The entraining cues are often referred to as **zeitgebers**, a German term coined by Jürgen Aschoff that literally means 'time givers.' The best-examined zeitgeber is the 24h cycle of light and dark defined by the sun. However, this does not imply that other rhythmic exogenous stimuli are less important, or not important at all. Other types of zeitgebers that have been identified in various animals include food availability, temperature steps (like transitions from cold to warm or warm to cold, which have similar effects as lights-on and lights-off, respectively), and social cues.

➤ The circadian rhythm of a biological clock can be synchronized to the 24h cycle of the environment by the process of entrainment.

The result of the photic entrainment process is that the period of the biological rhythms becomes equal to the period of the light–dark cycle. Furthermore, at the end of this process, a stable relationship will be established between the phase of the light–dark cycle and that of the biological rhythm. Photic entrainment has been studied in the laboratory through numerous experiments. These investigations have shown that the biological clock responds differently to light stimuli in different phases of its cycle. This phase-dependent response can be described quantitatively in the form of a **phase–response curve**. In such a plot, the phase shifts of the circadian rhythm are shown as a function of the circadian time when the stimulus is delivered.

Figure 10.2 illustrates the principle of how a phase–response curve may be obtained. In the example, which, although hypothetical, is based on empirical data collected in various species, the rest–activity cycle of a nocturnal animal is analyzed in constant darkness. The black horizontal bars in Fig. 10.2(a) represent times of wheel running. This activity can easily be monitored using a running wheel equipped with a rotary sensor that records the number of wheel rotations.

In the first experiment, summarized in panel 1, the constant darkness condition started at day –4. The **actogram** shows that on each of the following days, the onset of activity is delayed by 1h, compared with the 24h clock time, thus reflecting a free-running period of 25h. On day 0, a 1h light pulse of defined intensity is presented in the middle of the subjective day. Evidently, this light pulse does not have any effect on the phase shift of 1h on each of the following days.

Actogram Plot of the recorded activity of an individual during a given time period.

Box 10.2 Colin Stephenson Pittendrigh

Colin Stephenson Pittendrigh. (Courtesy: Chuck Painter/ Stanford News Service).

Together with Jürgen Aschoff (see Box 10.1) and the botanist Erwin Bünning, Colin Stephenson Pittendrigh is regarded as one of the founders of chronobiology. Pittendrigh (pronounced PIT-ten-dree) was born in Whitley Bay in the north of England in 1918. In 1940, he graduated with a science degree from the University of Durham, and in 1942 he received an advanced degree from the Imperial College in Trinidad.

During the Second World War, he was assigned to wartime service as a biologist, working on a project sponsored by the Rockefeller Foundation and the government of Trinidad. This project aimed at controlling mosquitoes (which are vectors for malaria) near military bases in Trinidad. As part of his work, he studied different anopheline mosquito species that breed in water pooled in epiphytic bromeliads in the forest canopy. He observed daily rhythms in the activity patterns of these mosquitoes and noted that the times of peak activity differed among species and canopy levels. This work laid the foundation for his lifelong interest in biological rhythms.

After the Second World War, Pittendrigh moved to Columbia University in New York to do his Ph.D. research work in the laboratory of the evolutionary geneticist Theodosius Dobzhansky. He analyzed the daily activity rhythms of two *Drosophila* species in the forests of the Northern California Sierras. With this study, he established *Drosophila* as a major model system in chronobiology.

In 1947, a year before he formally received his Ph.D. from Columbia, he accepted an appointment as Assistant Professor at Princeton University. There, in collaboration with Jürgen Aschoff in Germany, he introduced the oscillator concept to chronobiology. Pittendrigh showed that circadian rhythms are based on endogenous oscillations, thus refuting the then widely held view that rhythmic behaviors are exclusively controlled by exogenous cues. He characterized these oscillations and demonstrated that they are temperature-compensated–an important property without which a reliable measurement of time would not be possible. His laboratory analyzed, in great detail, entrainment of circadian rhythms by the diurnal cycle of light and darkness, and constructed the first phase-response curves. Equally important as the careful and meticulous design and execution of his experiments was the development of the theory of biological clocks, including modeling of the endogenous oscillator.

In 1968, Pittendrigh joined the faculty of Stanford University, where he was one of the founders of the Human Biology program and Director of the Hopkins Marine Station in Pacific Grove. Other prominent activities outside the lecture hall and the research laboratory included service as Dean of the Graduate School at Princeton, Chair of a National Academy committee on the exploration of Mars, science advisor to NASA, and President of the American Society of Naturalists. Together with the renowned paleontologist George Gaylord Simpson he wrote the introductory textbook *Life*. Among the numerous honors he received were election to the National Academy of Sciences and to the American Academy of Arts and Sciences.

After his retirement, in 1984, he divided his time between Sonoita, Arizona, and Bozeman, Montana. He died of cancer in Bozeman in 1996, at the age of 77.

However, when the 1h light pulse is given late in the subjective day (panel 2) or early in the subjective night (panel 3), the animal's activity over the following days commences later than expected from the free-running rhythm. Now, phase delays of 1h (panel 2) or even 3h (panel 3) are caused. By contrast, when the same type of light pulses are delivered late during the subjective night (panel 4) or early in the subjective morning (panel 5), phase advances are produced.

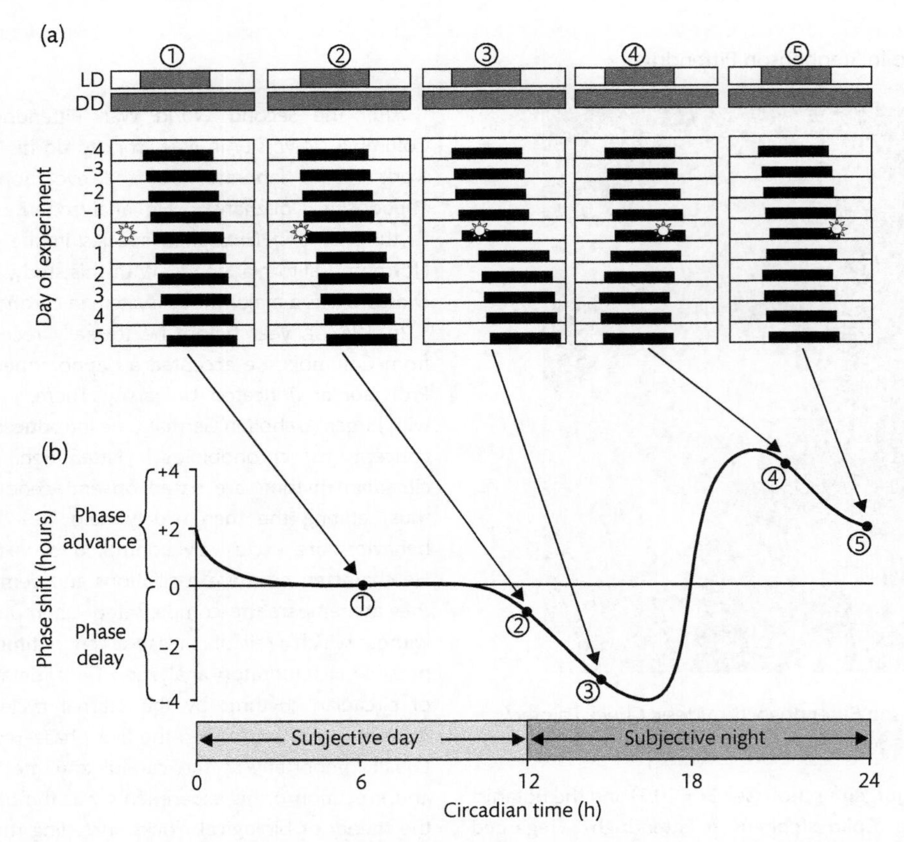

Figure 10.2 Photic entrainment of a nocturnal animal and derivation of a phase–response curve. (a) Five experiments (panels 1–5) performed with one individual animal are shown. Before the start of each experiment, the animal is kept in LD 12:12. On day −4, constant darkness (DD) starts, revealing a free-running period of 25 hours. On day 0, a one-hour light pulse (indicated by the sun symbol) is given at different phases of the subjective day or night, in each of the five experiments. The LD 12:12 and DD light schedules are shown above the actograms (white = light; gray = dark). Activity is represented by black bars in the five actograms. In the experiments, the light pulse is given at the subjective midday (1), late in the subjective day (2), early in the subjective night (3), late in the subjective night (4), and early in the subjective day (5). The light pulse at the subjective midday has no effect. Light pulses late in the subjective day and early in the subjective night produce phase delays of the activity rhythm, whereas light pulses late in the subjective night and early in the subjective day cause phase advances. (b) Phase–response curve, derived from the data in (a). (After: Moore-Ede, M. C., Sulzman, F. M., and Fuller, C. A. (1982), and Dunlap, J. C., Loros, J. J., and DeCoursey, P. J. (eds) (2004).)

Based on these data, a phase–response curve can be constructed (Fig. 10.2b). This plot shows that biological clocks of nocturnal animals are most responsive to solar entrainment stimuli during the subjective night. In contrast, light stimuli defined during the subjective day (when the animal is normally exposed to light) have little or no effect on resetting the biological clock. It is thought that the discrete light pulses used for stimulation in laboratory experiments mimic dawn and dusk transitions, which appear to play a major role as entraining cues in nature.

In addition to the phase-shifting property of light pulses, light may also affect the length of the free-running period as follows:

- The free-running period of nocturnal animals slows down in constant light.
- The free-running period of day-active animals slows down in constant darkness.

This effect, first described in a paper published by Jürgen Aschoff in 1952, has become known as **Aschoff's rule.** In an extended form, the rule states that, under constant light conditions, an increase in light intensity tends to cause a shortening of the free-running period of day-active animals, while lengthening the period of nocturnal animals.

➤ Aschoff's rule describes the effect of the intensity of constant light on the length of the free-running period.

Molecular nature of the clock

Despite the vast amount of behavioral data collected on biological rhythms in the 1950s and 1960s, and although a number of theoretical models of how biological clocks might work were proposed during that time, the molecular mechanisms controlling circadian rhythms remained elusive until the beginning of the 1970s. This situation changed when Seymour Benzer (Box 10.3) and one of his graduate students, Ronald Konopka, reported the identification of a **clock gene** in the fruit fly (*Drosophila melanogaster*) in 1971.

In their study, Konopka and Benzer made use of the eclosion behavior of *Drosophila*, which had earlier been characterized in great detail, particularly through the work of Colin Pittendrigh (see 'Temperature compensation of free-running periods,' above). Peak numbers of young adults emerge from their pupal cases in the first few hours after dawn. These bursts of eclosion activity within a population occur every morning for several days, until all adults have hatched.

A similar rhythmicity in the timing of eclosion, as observed in the natural habitat, can be evoked in the laboratory by entraining the pupae to a day/night cycle consisting of 12h of light, followed by 12h of darkness. This rhythm persists even when the pupae are transferred to constant darkness (Fig. 10.3a), indicating that this behavior is controlled by an endogenous clock.

Konopka and Benzer hypothesized that mutations in a gene forming part of the molecular clock could lead to alterations in the timing of eclosion. Initial observations had shown that, after **mutagenesis** of flies, the abnormal behavior was particularly pronounced in males, indicating that a potential clock gene involved in control of eclosion behavior might be located on a sex chromosome. To isolate such clock mutants, Konopka and Benzer mated mutagenized males individually to virgin females of a specific strain carrying attached X-chromosomes (see Box 10.4). This mating protocol produced a stock of F_1 males bearing identical X chromosomes, in addition to F_1 attached-X females free of any mutation generated on the X chromosome of their fathers. These females could therefore be used as internal controls; they showed the same eclosion behavior as normal males or females.

Mutagenesis The process by which a mutation occurs in nature, or is induced in the laboratory using mutagens (e.g. X-rays, UV light, specific chemicals).

The progeny of each mutagenized father were raised in separate bottles in LD 12:12 and examined to identify males that emerged abnormally from their pupal cases. This was the case in a few bottles. These candidate mutants were analyzed in more detail by mating these F_1 males to attached-X females and examining the eclosion behavior of the resultant pupae. These pupae were first raised in LD 12:12. Then, at the end of the light period, they were transferred to constant darkness, and the eclosion rhythm of the respective population was determined under free-running conditions, by recording the number of young adults emerging from the pupal cases every hour, for several days.

By using this approach, out of about 2000 F_1 males, three rhythm mutants were obtained. One mutant lacked an apparent eclosion rhythm (Fig. 10.3b). In another, the period of the rhythm was shortened from 24h, as displayed by the attached-X (control) females, to about 19h (Fig. 10.3c). In the third mutant, the period was markedly longer than in normal flies, approximately 28h (Fig. 10.3d). The aberrant features of each mutant strain could be hereditarily transmitted over many generations.

Subsequent experiments have shown that, despite the differences in the phenotype, the same gene on the X chromosome is affected in all three mutants. Konopka and Benzer called this gene *period* or *per*. Mutations of this gene alter not only the timing of eclosion, as shown by the mutant strain as a whole, but also the rhythmicity of behavioral patterns of

Box 10.3 Seymour Benzer

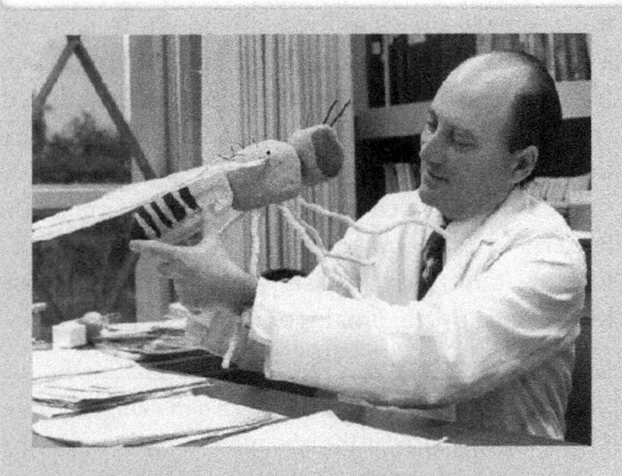

Seymour Benzer in his office at the California Institute of Technology in 1974, with a large model of *Drosophila*. (From: Harris, W. A. (2008).)

Seymour Benzer was one of the most versatile and ingenious scientists of his time. When he died in 2007, at the age of 86, he had pioneered research in fields as diverse as semiconductor physics, molecular biology, behavioral genetics, and neurogenetics. Yet, he has remained largely unknown outside his fields of study.

Benzer was born in 1921 in South Bronx, New York, to Polish immigrants. He grew up in Brooklyn and became the first of his family to attend college. After graduating from Brooklyn College with a major in Physics, he pursued graduate studies at Purdue University in Lafayette, Indiana, and joined the group of Karl Lark-Horovitz. The team worked on a secret war project to make germanium semiconductors usable for radar. Based on Benzer's patents, Bell Laboratories developed the first transistor.

Upon receiving his Ph.D. in Physics in 1947, Benzer joined the faculty of Purdue as an assistant professor in physics. However, inspired by reading Erwin Schrödinger's book *What is Life?*, he decided to switch to biology and study the physical nature of the gene. Although he had just started his faculty position, he immediately took leave-of-absence, and extended these sabbaticals time and again, overall totaling four years. Benzer's training during these sabbaticals with scientific giants, such as André Lwoff, François Jacob, and Jacques Monod at the Pasteur Institute in Paris, and Max Delbrück at the California Institute of Technology (Caltech) in Pasadena, California, proved to have prepared him very well to make the transition from physics to molecular biology. Back at Purdue, he used mutations in the gene *rapid lysis* (r) in the bacteriophage T4 to perform the first fine mapping of a gene. Using this approach, Benzer showed that a gene is not an indivisible unit, but one that is composed of linearly arranged building blocks—nucleotides—each of which can be subject to alteration. This was a pioneering achievement that gave the gene a physical meaning.

Despite his success in molecular genetics, Benzer decided to change research direction again. Fascinated to see his two daughters developing very different personalities, he wanted to explore how genes influence behavior. He took a sabbatical from Purdue and went to Caltech to learn neurobiology with Roger Sperry. At Caltech, he chose the fruit fly *Drosophila* as a suitable model system for his future studies. In 1967, he joined the faculty of Caltech. Over the following four decades, he and his group isolated a large number of behavioral mutants, including mutants with defects in phototaxis, circadian rhythms (see this chapter), learning, memory, courtship behavior, and other types of behavior. With this ground-breaking work, Benzer founded a new discipline—behavioral genetics.

After his first wife Dorothy passed away, Seymour Benzer married Carol Miller, a neuropathologist at the University of California, Los Angeles. Inspired by her work, he turned his attention, once more, to new problems, this time ones of immediate medical relevance. He and his associates discovered mutations that cause brain degeneration, or extend the lifespan of its carrier. This opened the possibility of studying neurodegenerative diseases, or longevity, using *Drosophila* as a model system.

Benzer received numerous prestigious awards for his research. Arguably, his greatest merit was to have opened so many new fields, not only for himself, but also for many other scientists. When, in an interview, Benzer was asked what made him move from one field to another, he replied: '*When a subject develops very thoroughly, there's too much you have to know. It gets sort of overwhelming . . . The big attraction is starting something new and being very stupid about it. Ask stupid questions, and you often get amazing answers.*'

the individual adult fly, such as locomotor activity and courtship singing. This indicates that different behaviors are not necessarily controlled by separate clock systems, and that a single system can determine the temporal patterning of various behavioral activities.

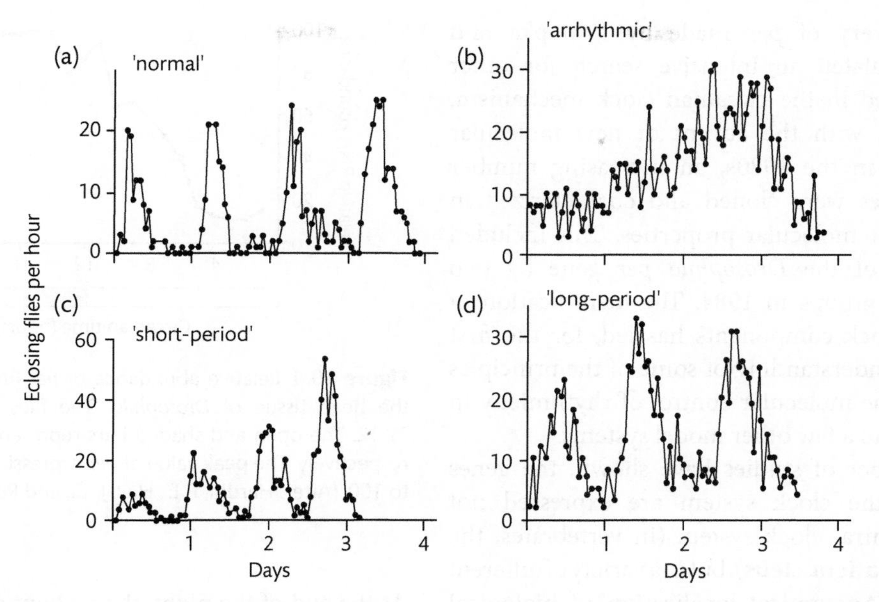

Figure 10.3 Eclosion rhythms for populations of rhythmically normal and mutant fruit flies (*Drosophila melanogaster*), as demonstrated by the classic experiment of Ronald Konopka and Seymour Benzer. Each of the different populations was previously exposed to an LD 12:12 cycle, but kept in constant darkness for the time of the recordings. (a) In attached-X females (see Box 10.4), which served as internal controls, adults emerged from pupae rhythmically. This pattern is indistinguishable from that of normal flies, which exhibit a 24-hour period, with peak frequencies of eclosion in the morning hours of the subjective day period. In the males examined, mutations at the locus of the clock gene *period* lead to several types of alteration in the timing of eclosion. (b) One mutant lacked a clear rhythmic pattern, and was hence called 'arrhythmic.' (c) Another mutant ('short-period' mutant) had a shortened eclosion period of about 19 hours. (d) A third mutant ('long-period' mutant) exhibited a long eclosion period of about 28 hours. (After: Konopka, R. J. and Benzer, S. (1971).)

Box 10.4 Attached-X chromosomes in *Drosophila*

Sperm from normal male

Eggs from attached-X female	X	Y
X^X	X^X, X	X^X, Y
Y	X, Y	Y, Y

Inheritance pattern of normal males (XY) mating with attached-X females (X^XY). (After: Keenan, K. (1983).)

On 12 February 1921, Lilian Vaughan Morgan, the wife of Thomas Hunt Morgan, discovered an unusual female fruit fly. Subsequent analysis showed that in such flies the two X chromosomes are physically linked in the centromeric region (to which the spindle fibers attach during cell division). These attached-X chromosomes (symbolized as 'X^X') are passed on to the next generation as a single unit. Attached-X females usually also carry a Y chromosome, which they inherit from their fathers. Thus, these females have the sex chromosome complement X^XY. Mating of such females with normal males (XY) results in four genotypes, X^XX, X^XY, XY, and YY (see figure). The X^XX and YY progeny are usually not viable. The X^XY flies are phenotypically normal females, whereas the XY flies are normal males. Due to its remarkable genotypic characteristics, the attached-X chromosome strain provides a powerful tool in genetics. Sons resulting from mating of an X^XY female with a XY male can inherit their X chromosome only from their fathers. This X chromosome is never next to another X chromosome (as could be the case after mating of an XX female with an XY male), and therefore the X chromosome never crosses over. As a consequence, by mating of a male to an attached-X female, certain combinations of characteristics, defined by genes on the X chromosome, can be maintained through generations.

The discovery of *per* made by Konopka and Benzer stimulated an intensive search for other genes involved in the circadian clock mechanism. Furthermore, with the advent of new molecular technologies in the 1980s, an increasing number of these genes were cloned and characterized in terms of their molecular properties. This included the cloning of the *Drosophila per* gene by two independent groups in 1984. The identification of molecular clock components has led, for the first time, to an understanding of some of the principles underlying the molecular control of rhythmicity in *Drosophila* and a few other model systems.

As a number of studies have shown, the genes comprising the clock system are expressed not only in a central clock system (in vertebrates, the **suprachiasmatic nucleus**), but in a variety of different tissues (see 'Anatomical localization of biological clocks,' below). Each cell within these tissues appears to contain all of the components necessary for functioning of the clock. Besides *per*, a second clock gene, *timeless* or *tim*, discovered by the laboratory of Michael W. Young of Rockefeller University in 1994, has proven to be an essential component of the *Drosophila* clock system. **Null mutations** in either of the two genes render the flies arrhythmic.

Suprachiasmatic nucleus A paired neuronal structure in the brain, at the base of the hypothalamus, just above the optic chiasm. It receives photic input from the retina via the optic nerve and plays an important role in the regulation of the body's circadian rhythms.

Null mutation Any mutation of a gene that results in a lack of function.

When fruit flies are kept under an LD 12:12 cycle, levels of RNA of both *per* and *tim* are low early in the morning, but increase over the day and reach peak values early in the night. Quantitative analysis has revealed a five- to tenfold difference in relative abundance of *per* RNA extracted from the head between trough and peak levels (Fig. 10.4). A similar cycling of *per* RNA levels was observed in flies kept under constant dark conditions.

The levels of the proteins encoded by *per* and *tim*, PER and TIM, lag behind the RNA oscillations by approximately 6h, reaching peak levels late in the night.

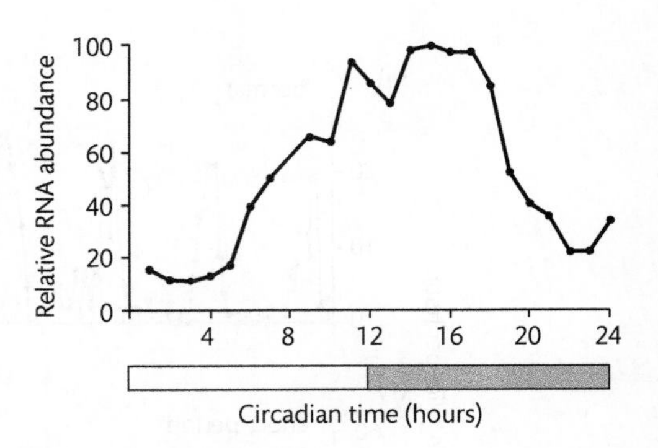

Figure 10.4 Relative abundance of *per* RNA extracted from the head tissue of *Drosophila*. The flies were kept in LD 12:12. The open and shaded bars represent light on and off, respectively. The peak value of *per* expression was normalized to 100. (After: Hardin, P. E., Hall, J. C., and Rosbash, M. (1990).)

At the end of the night, the amount of both proteins has increased. PER is phosphorylated in the cytoplasm by the action of a casein kinase, which is encoded by a gene called *doubletime* (*dbt*). The phosphorylation renders PER unstable until it binds to TIM so that the two proteins form a heterodimer, PER/TIM. The formation of this heterodimer leads to stabilization of PER and prompts PER/TIM to translocate from the cytoplasm into the nucleus later in the night (Fig. 10.5).

Upon entering the nucleus, PER/TIM sequesters the transcription factor dCLK (d for *Drosophila*, CLK for CLOCK). The latter protein forms a heterodimer with CYCLE (CYC). This complex, in turn, binds to a specific nucleotide sequence (CACGTG), called the E-box, within the promoter regions of *per*, *tim*, and some other *clock-controlled genes* (*ccgs*), thereby activating the transcription of these different genes. The binding of PER/TIM to dCLK negatively regulates the transcription of these genes, including *per* and *tim*, by inhibiting the DNA binding of the dCLK/CYC complex. This repression lasts until PER and TIM are degraded, thereby initiating a new round of transcription of *per*, *tim*, and *ccgs*.

The negative feedback loop defined by the oscillating protein products of *per* and *tim* is central to the understanding of the molecular mechanism determining circadian rhythmicity in *Drosophila*. The regulation of dCLK is less well understood. It is thought to occur through a second negative loop in which dCLK represses the transcription of its own gene.

Taken together, the rates of transcription of the clock genes *per* and *tim*, the synthesis of the

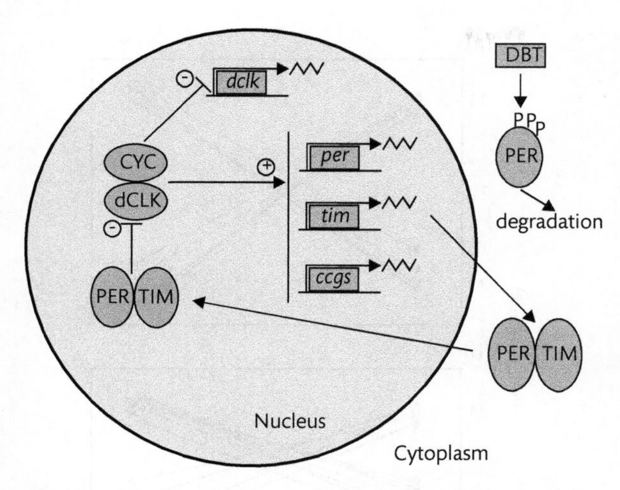

Figure 10.5 Simplified model of the molecular clock mechanism in individual cells of *Drosophila*. Following transcription of the genes *period* (*per*) and *timeless* (*tim*), the protein products of these two genes, PER and TIM, accumulate in the cytoplasm. PER is phosphorylated by Doubletime (DBT). PER and TIM form a heterodimer (PER/TIM), which protects PER from degradation, and translocate into the nucleus. In the nucleus, the transcription of *per*, *tim*, and other *clock-controlled genes* (*ccgs*) is activated by binding of a heterodimeric complex formed by the proteins CYCLE (CYC) and *Drosophila* CLOCK (dCLK) to the promoter regions of these genes. The activity of CYC/dCLK is inhibited by sequestration through PER/TIM. The transcriptions of the *Drosophila clock* gene (*dclk*) appear to be repressed by its own protein product. For further explanation, see text. (After: Meyer-Bernstein, E. L. and Sehgal, A. (2001).)

corresponding proteins PER and TIM, and finally the degradation of these two proteins last about 24h, thus leading to daily oscillations. The time information from this oscillator is thought to drive circadian output by regulating target-clock-controlled genes. The expression pattern of the clock genes and the clock-controlled genes, in turn, regulates the activity of many other genes. Gene-microarray studies have revealed that an astonishing number—at least 10%—of the total transcripts undergo circadian rhythm-like oscillations.

Anatomical localization of biological clocks

A vast amount of behavioral data suggests that a biological clock must exist within organisms. Molecular biological studies have revealed details, at the subcellular level, of the structure and function of the molecular machinery of oscillators producing circadian rhythms. Where, one must ask, is the biological clock located within the body? And is there just one clock, or do several clocks exist in different parts of the body? If several clocks do exist, does one of them function as a 'master clock,' or do the different clocks act as autonomous oscillators?

Since the early 1970s, a large number of studies have pointed to the suprachiasmatic nucleus as the site of the biological clock in mammals, and probably also in other vertebrates. Evidence in favor of the role of the suprachiasmatic nucleus as the master clock includes the following:

- Explants of tissue containing the suprachiasmatic nucleus continue to exhibit free-running rhythms of electrical activity for some time.
- Ablation of the suprachiasmatic nucleus by chemical or electrolytic techniques results in a loss of overt behavioral rhythmicity of the animal.
- As predicted by behavioral studies, the suprachiasmatic nucleus receives direct and indirect input from the eye, mediating the entrainment of circadian rhythms to photic stimuli.
- Isolation of the suprachiasmatic nucleus within the brain by surgically severing all input to, and output from, this structure renders the animal arrhythmic.
- Grafting of fetal brain tissue containing the suprachiasmatic nucleus into host animals, made arrhythmic by lesioning the suprachiasmatic nucleus, restores rhythmicity.

➤ In mammals, the suprachiasmatic nucleus acts as a master clock.

A particularly elegant modification of the latter experiment, performed by the group of Michael Menaker of the University of Virginia, Charlottesville, was published in 1990. For the transplantation, Menaker and his associates used grafts of the suprachiasmatic nucleus from hamsters with a specific mutation called *tau* (see 'Advanced topic,' below). The behavioral effect of this mutation is that the period of the circadian rhythm is reduced from 24h in wild-type animals to about 22h in heterozygotes and to about 20h in homozygotes. Grafting of suprachiasmatic nucleus tissue, obtained from such mutant donors, into arrhythmic wild-type hamsters whose own nucleus has been ablated, restores rhythms. The rhythms of the host animals always correspond to the rhythms of the donors with a free-running period of 20 or 22h. Similarly, if a mutant hamster with an ablated suprachiasmatic nucleus receives a graft from a wild-type hamster, a 24h rhythm is restored (Fig. 10.6).

By the mid-1990s, it was well established that, in mammals, the suprachiasmatic nucleus harbors the biological clock. It therefore came as quite a surprise when, in 1996, the group of Michael Menaker announced the discovery of a clock in a second localization—the **retina**. Michael Menaker and his associate Gianluca Tosini demonstrated the existence of such a clock by culturing retinas and measuring the amount of **melatonin** released by retinal cells into the culture medium. Over several days, such recordings showed clear circadian rhythms that persisted in constant darkness (Fig. 10.7). Mutations in the *tau* gene affected the release of melatonin from retinal cells in a similar way to how they changed rhythmic activities controlled by the suprachiasmatic nucleus. Furthermore, as Tosini and Menaker demonstrated, the retinal clock is temperature-compensated, and can be entrained by light. Thus, it fulfills all of the major criteria defining a circadian oscillator.

Melatonin A hormone secreted by the pineal gland in the brain and by some other organs, including the retina; its circulating levels exhibit daily cycles.

➤ Like the suprachiasmatic nucleus, the retina functions as a biological clock.

Perhaps one of the most exciting recent developments in the attempt to localize biological clocks has become possible through the application of **transgenic technology**. During the mid-1990s, a group of scientists from Brandeis University, Waltham, Massachusetts, the University of Virginia, Charlottesville, and Promega Corporation, Madison, Wisconsin, succeeded in coupling in *Drosophila* the promoter for the *per* gene to *luc*, the gene that encodes the bioluminescent protein **luciferase**. The result was a *per–luc* construct that controls *per*-driven bioluminescent oscillations in the transgenic organism. The bioluminescent signal can be monitored with photomultipliers and recorded as an actogram.

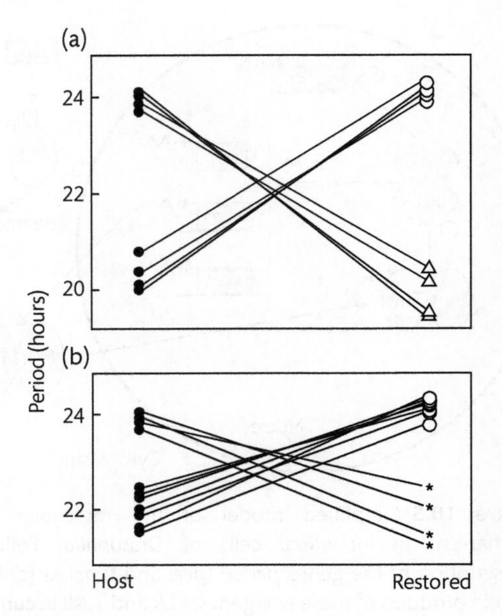

Figure 10.6 Transplantation of tissue containing the suprachiasmatic nucleus between wild-type hamsters and hamsters of the mutant strain *tau*. After the free-running period of activity in running wheels had been determined under constant light conditions (the closed circles on the left indicate the mean period of each hamster), the rhythm was eliminated by ablation of the suprachiasmatic nucleus. The rhythm was then restored by grafting into these host animals suprachiasmatic nucleus tissue from one of the following three donor types: wild-type animals (open circles), heterozygous mutants (asterisks), or homozygous mutants (open triangles). The restored periods of running activity after transplantation are shown on the right. (a) Reciprocal transplants between wild-type animals and homozygous mutants. (b) Reciprocal transplants between wild-type animals and heterozygous mutants. (After: Ralph, M. R., Foster, R. G., Davis, F. C., and Menaker, M. (1990).)

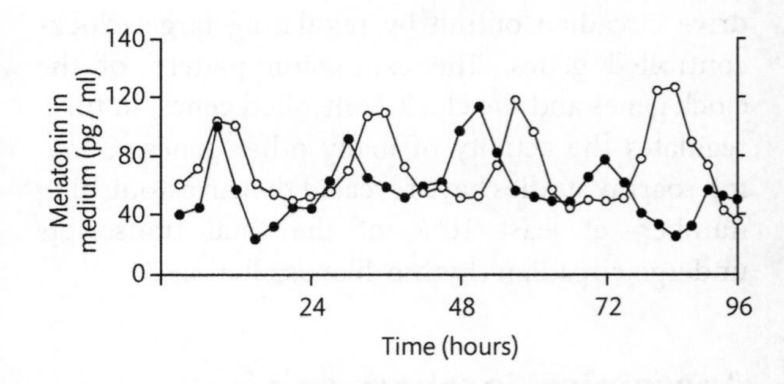

Figure 10.7 Free-running rhythms of melatonin release from retinal explants of wild-type hamsters (open circles) and *tau* homozygous mutants (closed circles). The retinas were cultured over the entire time of recording in constant darkness. The data are means of results from four retinas each. The actogram demonstrates that the rhythm of melatonin release persists under constant environmental conditions, and that the *tau* mutation leads to a shortening of the free-running period of this physiological activity. (After: Tosini, G. and Menaker, M. (1996).)

Luciferases Enzymes that catalyze the oxidation of luciferins to produce light in a bioluminescent reaction. The best-known luciferase is that produced by the North American firefly, which emits green light. Luciferase (*luc*) genes can be inserted into the DNA of other organisms so that the light produced acts as a 'reporter' for the activity of regulatory elements that control the expression of *luc*.

This approach revealed a surprisingly wide distribution of oscillators within the body of *Drosophila* (Fig. 10.8). Previous experimental evidence had pointed to a small cluster of lateral neurons between the lateral protocerebrum and the medulla of the optic lobes as the site of the biological clock. Now, using the *per–luc* reporter system, *per-*

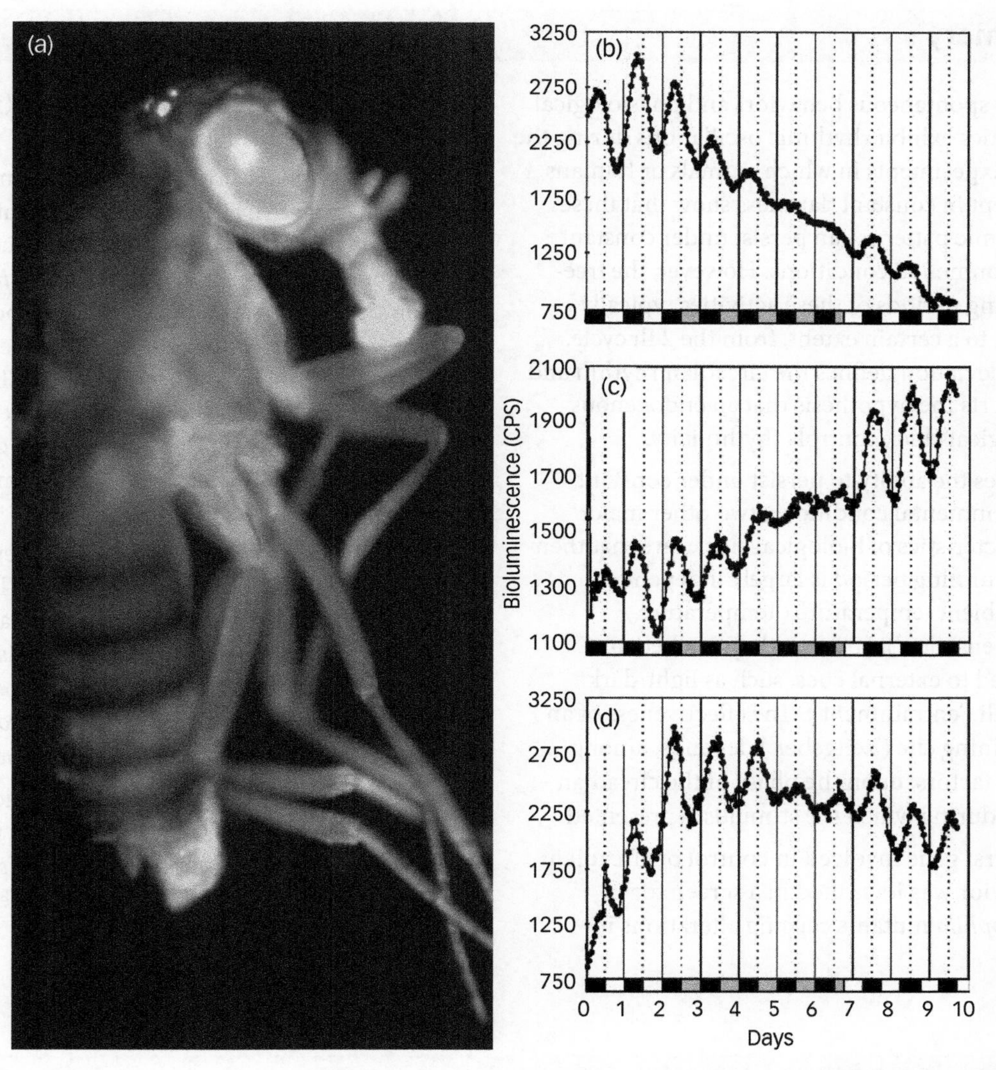

Figure 10.8 Generation of rhythmic bioluminescence by cultured body segments of *per–luc Drosophila.* (a) Bioluminescence is visible throughout the transgenic fly. (b–d) Bioluminescence actograms obtained from the heads (b), thoraxes (c), and abdomens (d). Each of these body segments was individually cultured and kept under LD 12:12, followed by DD, which was again followed by LD 12:12, as indicated at the bottom of each plot (filled bars = darkness; open bars = light; gray bars = subjective light condition). The *per*-driven bioluminescence expression levels were monitored using an automated system. The recordings show that each of the body segments produced rhythms during the initial LD 12:12 phase, and that these rhythms were maintained, to a large extent, during the following phase of complete darkness, with the amplitude of the oscillations gradually decreasing. However, the cultured body segments were able to re-entrain to the new LD 12:12 cycle employed at the end of the experiment. CPS, counts per second. (After: Plautz, J. D., Kaneko, M., Hall, J. C., and Kay, S. A. (1997).)

driven bioluminescent oscillations were found not only in the head (as expected), but also in the thorax and abdomen, and even in the proboscis and antennae. These oscillations free-ran in constant darkness and were independently entrainable by light stimuli. Therefore, it appears that the clock of *Drosophila* can operate in a cell-autonomous fashion, without the head functioning as the master oscillator.

How then are these independent oscillators synchronized? As each of the oscillators is photosensitive, and photic stimuli usually affect all parts of the fly simultaneously, it is thought that light serves as the master coordinator of this distributed clock system.

Summary

- Many spontaneous behaviors and physiological activities exhibit rhythmic oscillations during the day. Experiments in which animals or humans are kept in constant darkness show that these rhythmic patterns can persist under constant environmental conditions. However, the free-running periods of these activities typically differ, to a certain extent, from the 24h cycle. This deviation defines the circadian rhythm and supports the hypothesis that an endogenous biological clock controls rhythmicity.
- Besides the ability to persist under constant environmental conditions, two other major characteristics of biological clocks are that their free-running period is largely independent of ambient temperature ('temperature compensation'), and that they can become aligned to external cues, such as light/dark stimuli ('entrainment'). The effectiveness of an entraining cue ('zeitgeber') depends, among other factors, upon the phase of the circadian cycle during which the stimulus is presented.
- The first gene involved in control of circadian behavior was identified in a screen for *Drosophila* mutants causing alterations in rhythmicity. This gene, termed *period* (*per*), is X-chromosome linked.
- In the *Drosophila* clock system, *per* is an integral part of a negative feedback loop, generating oscillations in both the *per* gene product and a second protein encoded by the gene *timeless* (*tim*). These two proteins form a heterodimer that translocates into the nucleus and sequesters a complex formed by the proteins CYCLE and *Drosophila* CLOCK. The latter complex activates the transcription of *per*, *tim*, and other clock-controlled genes by binding to the promoter regions of these genes.
- Coupling of the promoter for the *per* gene to the *luc* gene, which encodes luciferase, has led to the generation of transgenic flies that display *per*-driven bioluminescent oscillations. This approach has revealed a wide distribution of autonomous biological clocks within the body of *Drosophila*, which appear to be coordinated by photic stimuli.
- In mammals, the principal site of the biological clock is the suprachiasmatic nucleus in the hypothalamus of the brain. Additional structures that generate circadian rhythms include the retina of the eye.

The bigger picture

Virtually all physiological and behavioral actions are governed by circadian rhythms. Although not explicitly mentioned in each instance, many behaviors covered in this book are also influenced by circadian rhythms. One well-studied example—sun-compass orientation—will be discussed in detail in the next chapter. We will show that this behavior depends critically on biological clock(s) to fulfill its function—enabling animals to navigate during homing and migration.

The omnipresence of circadian rhythms correlates with the expression of clock genes and clock-controlled genes in a variety of cells and tissues, and with the existence of circadian rhythm-like oscillations associated with numerous transcripts and molecular signaling pathways defined by the activity of these genes. Such findings have led to the notion that circadian rhythms observed at the organismal level are not dictated by a single body clock, as assumed in the early years of chronobiology. Instead, current models explain circadian rhythms as the output of a complex network of oscillators widely distributed in the body. As a consequence, the suprachiasmatic nucleus—once thought to harbor *the* biological clock—has lost its exclusive role as the sole driver of circadian rhythms, although this brain nucleus continues to assume a prominent position within the network of interacting oscillators.

Recommended reading

Dunlap, J. C., Loros, J. J., and DeCoursey, P. J. (eds) (2004). *Chronobiology: Biological Timekeeping*. Sinauer Associates, Sunderland, Massachusetts.

A comprehensive overview of biological timing from unicellular organisms to humans. This book provides both a review of major concepts and a more detailed treatise of specific topics in chronobiology.

Short-answer questions

10.1 Outline an experiment through which one could determine the free-running period of a hamster.

10.2 Define, in one sentence, the term 'circadian rhythm.'

10.3 The rate of many biological processes is temperature dependent. By contrast, free-running periods are temperature compensated. Why is such compensation important for the functioning of biological clocks?

10.4 Complete the following sentence: Biological clocks can become aligned to external cues, such as light/dark stimuli; this process is referred to as .... (one word).

10.5 Indicate which of the following statements is correct: Mutations in the *period* (*per*) gene of *Drosophila* may result in the following phenotypic changes, compared to the behavior of wild-type flies:

a) lack of an apparent eclosion rhythm;

b) shortening of the eclosion rhythm;

c) lengthening of the eclosion rhythm;

d) all of the above.

10.6 Explain how the negative feedback loop defined by the protein products of the genes *per* and *tim* can produce circadian oscillations in single cells of *Drosophila*.

10.7 Describe how transplantation experiments have advanced the notion that the suprachiasmatic nucleus functions as a master clock in mammals.

10.8 Define the anatomical evidence that has provided the structural foundation for the notion that the circadian rhythm generated by the suprachiasmatic nucleus is entrained to cues from the animal's environment.

10.9 Which discovery led to the idea that in mammals biological clocks exist in parts of the body outside the suprachiasmatic nucleus?

10.10 Transgenic technology has made possible the demonstration of a wide distribution of biological-clock-like oscillations in the body of *Drosophila*. Describe the construct used as part of this experimental approach.

10.11 **Challenge question** Several studies have demonstrated that transplants of the suprachiasmatic nucleus restore circadian activity rhythms to animals whose own suprachiasmatic nuclei have been ablated. Yet, the nature of the coupling signal from the grafted suprachiasmatic nucleus to the brain of the host animal is not known. Suggest two modes of how this signal could be transmitted. By what experiments could these alternate hypotheses be tested?

10.12 **Challenge question** Most people experience more severe jet-lag symptoms when traveling east than they do when traveling west. Considering the most commonly found free-running period in humans, how do you explain this difference?

10.13 Challenge question Melatonin is secreted for about 12h at night and can be considered a darkness signal for the biological-clock system. Realignment of the circadian clock can be promoted by administration of exogenous melatonin, and its effect is stronger when it does not overlap with endogenous secretion. At what time at the destination do you expect an east-west traveller to minimize jet lag by taking a short-acting dose (0.5mg) of melatonin? How does this treatment affect the biological clock?

Essay questions

10.1 What lines of behavioral, anatomical, physiological, and molecular evidence are in support of the notion that rhythmic physiological activities and behavioral patterns in animals and humans are controlled by internal biological clocks? What role does the external environment of an organism play in this process?

10.2 What are the molecular components of the biological clock in *Drosophila*? How do they integrate to produce an oscillatory output? Describe one approach to test the function of the individual components of the clock.

10.3 What experimental evidence supports the notion that the suprachiasmatic nucleus functions as a biological clock in mammals?

10.4 The activity of the suprachiasmatic nucleus in mammals is modulated by photic input from the eye. What approaches are available to verify this input? How could other types of input be identified?

10.5 Design an experiment to examine whether a certain type of tissue in mammals functions as a biological clock. What animal species would you select as a model system? Why?

Advanced topic Molecular biology of the mammalian clock system

Background information

The success in isolation of mutants affecting the circadian system in *Drosophila* and the molecular identification of the corresponding genes has stimulated attempts to dissect the molecular components of the clock system in mammals. In 1988, Martin R. Ralph and Michael Menaker, then at the University of Oregon, Eugene, reported the discovery of a mutation that shortens the period of the circadian locomotor system of the golden hamster. The locus at which the mutation occurred was called *tau*. Other mammalian circadian mutants were isolated subsequently, including *clock* by the laboratory of Joseph S. Takahashi of Northwestern University, Evanston, Illinois, in 1994. Three years later, the same group cloned *clock* as the first gene of the mammalian circadian system.

Essay topic

In an extended essay, describe the major findings made in the molecular characterization of the mammalian circadian system. What approaches were used to identify the individual components and to analyze their properties? Compare both the molecular structure and the function of these components in mammals with those of the *Drosophila* clock system.

Starter references

Antoch, M. P., Song, E.-J., Chang, A.-M., Vitaterna, M. H., Zhao, Y., Wilsbacher, L. D., Sangoram, A. M., King, D. P., Pinto, L. H., and Takahashi, J. S. (1997). Functional identification of the mouse circadian *clock* gene by transgenic BAC rescue. *Cell* **89**:655–667.

King, D. P., Zhao, Y., Sangoram, A. M., Wilsbacher, L. D., Tanaka, M., Antoch, M. P., Steeves, T. D. L., Vitaterna, M. H., Kornhauser, J. M., Lowrey, P. L., Turek, F. W., and Takahashi, J. S. (1997). Positional cloning of the mouse circadian clock gene. *Cell* **89**:641–653.

Ralph, M. R., and Menaker, M. (1988). A mutation of the circadian system in golden hamster. *Science* **241**:1225–1227.

Vitaterna, M. H., King, D. P., Chang, A.-M., Kornhauser, J. M., Lowrey, P. L., McDonald, J. D., Dove, W. F., Pinto, L. H., Turek, F. W., and Takahashi, J. S. (1994). Mutagenesis and mapping of a mouse gene, *clock*, essential for circadian behavior. *Science* **264**:719–725.

To find answers to the short-answer questions and the essay questions, as well as interactive multiple choice questions and an accompanying Journal Club for this chapter, visit **www.oup.com/uk/zupanc3e.**

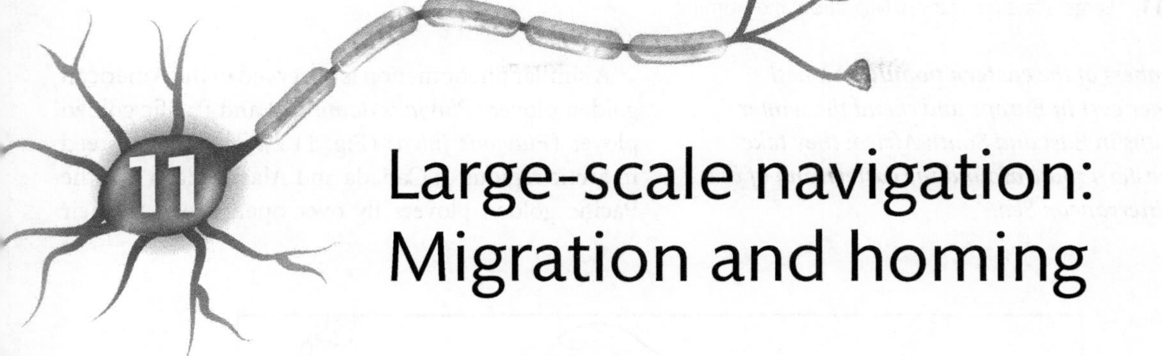

11 Large-scale navigation: Migration and homing

Introduction

The migration and homing of animals, some of which travel thousands of kilometers in the course of their journeys, have fascinated man for centuries. The best-known examples are the annual journeys of many migratory birds. They travel in the fall from the place of their birth to wintering grounds in warmer regions. In the spring, they make a similar migratory journey, but this time in the reverse direction.

Focusing on these long-distance migrations, in the first part of this chapter we will summarize some of the key concepts of these phenomena, such as the modes of migration and the specificity of homing. Then, we will take a closer look at the general methods used to study animal migration. The last, and largest, portion of this chapter will be devoted to a detailed discussion of the behavioral, sensory, and neural mechanisms involved in orientation during migration and homing, using representatives of two particularly well-studied animal groups—birds and fish.

Key concepts

During large-scale migrations, the entire species, or at least subpopulations of the respective species, use **specific routes** from their point of origin to their final target site.

> **Examples:** *Based on their distribution and migratory behavior, European white storks* **(Ciconia ciconia)** *can be divided into the following two populations (Fig. 11.1):*
>
> 1. *Members of the* ***western population*** *migrate to their wintering grounds in western parts of Africa through Spain and across the Straits of Gibraltar.*

2. *Members of the* ***eastern population*** *nest farther east in Europe and spend the winter months in East and South Africa; they take an eastern route around the eastern end of the Mediterranean Sea.*

A similar phenomenon is observed in the American golden plover (*Pluvialis dominica*) and Pacific golden plover (*Pluvialis fulva*) (Fig. 11.2). The adults breed in arctic regions of Canada and Alaska. Each fall, the Pacific golden plovers fly over open ocean to their

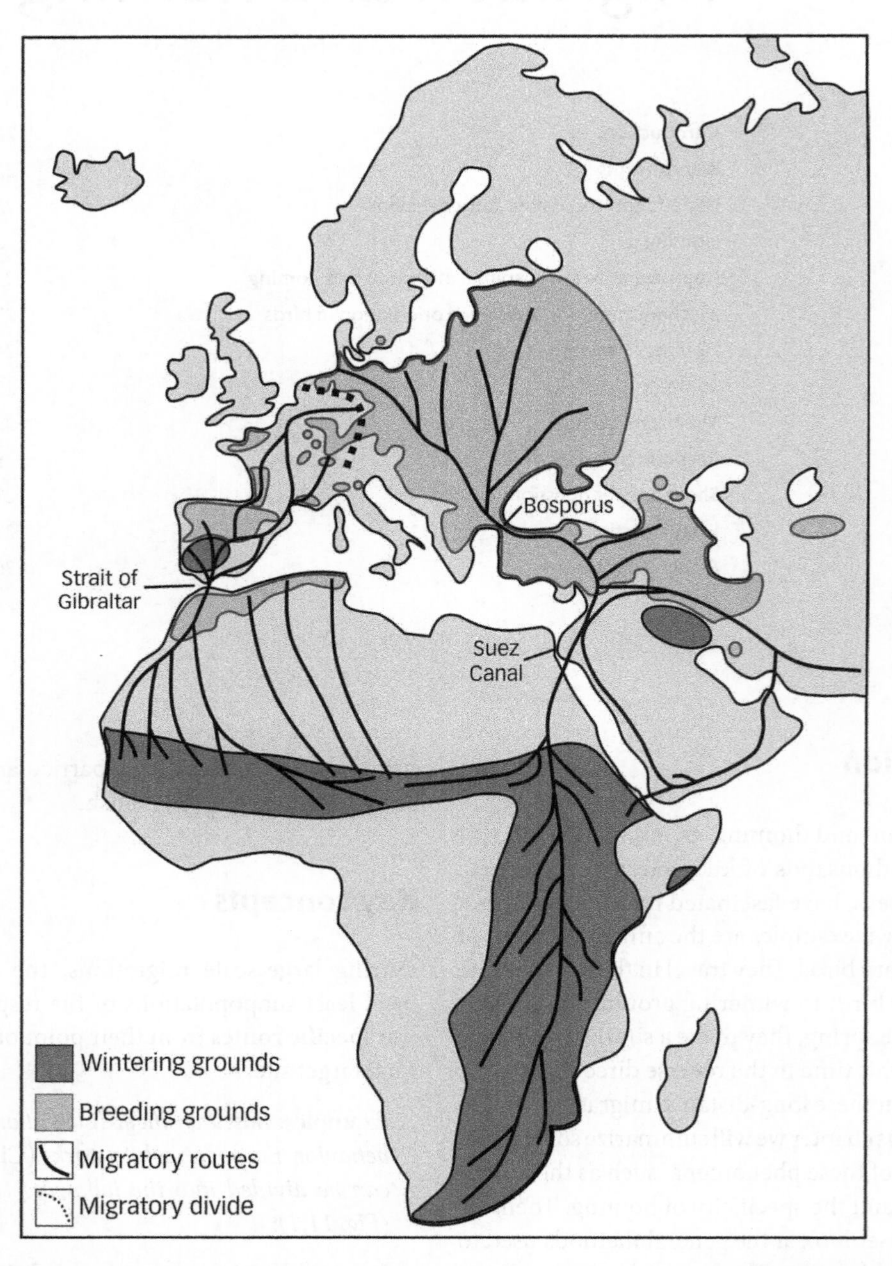

Figure 11.1 Differential distribution and migratory routes of the two major European populations of the white stork (*Ciconia ciconia*). While birds of the western population take a southwesterly route to their wintering grounds in the western parts of Africa, members of the eastern population fly around the eastern end of the Mediterranean to spend the winter in the eastern parts of Africa. (After: Michael-Ott-Institut im NABU, http://bergenhusen.nabu.de/zug/zug.jpg.)

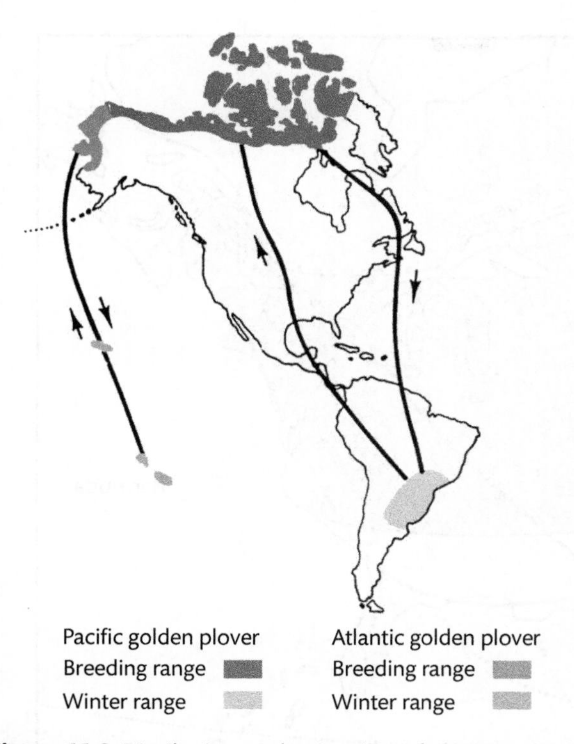

Figure 11.2 Distribution and migration of the American golden plover (*Pluvialis dominica*) and Pacific Golden Plover (*Pluvialis fulva*). Pacific golden plovers breed in Alaska and fly over open ocean to their winter range in Hawaii, the Marquesas Islands, and the Lower Archipelago. By contrast, American golden plovers breed more easterly and spend the winter in South America. Members of the latter species use different routes for the southbound and northbound migration. (After: Keeton, W. T. (1980).)

winter range in Hawaii, the Marquesas Islands, and the Low Archipelago. In the spring, the birds take the same route back. At the same time, the American golden plovers cross, in the course of their southbound journey, the Atlantic from Labrador and Nova Scotia to the Lesser Antilles and northeastern South America to continue to their winter range in temperate latitudes in northern Argentina, Uruguay, and Southern Brazil. In the spring, they take a different route, flying northwest through Central America and north over the Great Plains back to the Arctic. Interestingly, in contrast to the adults, almost all of the young American golden plovers also take the inland route via Central America during their first fall journey.

In a very few animal species, the **annual migratory cycle is completed by several generations**, rather than by one individual. Such animals represent rare cases of a biological rhythm that is longer than the life of the organism exhibiting this rhythm. Due to this complexity, details of such a cycle have been revealed in only a very few instances. One of the better-understood examples is the monarch butterfly (*Danaus plexippus*), a North American butterfly species of the family Nymphalidae. One large population of the monarch butterfly breeds east of the Rocky Mountains. In the fall, the butterflies migrate through Central Texas into Mexico where they follow the Sierra Madre Oriental to finally overwinter in high-altitude forests of the Transverse Neovolcanic Belt in Central Mexico (Fig. 11.3). During this journey, they migrate up to 3600km over a period of about 75 days. This corresponds to an average per-day traveling distance of approximately 50km.

In early spring, a rapid northward journey takes place so that by early April both sexes, now approximately six months old, reach the Gulf Coast states of the U.S.A. The females lay their eggs on the resurgent spring milkweeds (genus *Asclepias*) and die. The offspring of this first spring generation continue the migration northeastward to the Great Lakes region and southern Canada, where they arrive by early June. Along their way, they lay eggs, which give rise to the second spring generation of adults. This new generation migrates due east, again laying eggs along their way, thus resulting in a third generation. While one or two more short-lived breeding generations are produced, the butterflies spread eastward across the Appalachian Mountains. From late August to early September, the fall migration is again under way.

➤ An unusual example: In monarch butterflies, the annual migratory cycle is completed not by one, but several generations.

Genetic control of migratory behavior

It has been suspected for a long time that the **urge to migrate** and the **sense of migratory direction** are under **direct genetic control**. Such an assumption is not trivial, as, theoretically, the first departure of obligate migrants could be triggered by experienced conspecifics. However, in more recent years, convincing evidence in support of the inheritance

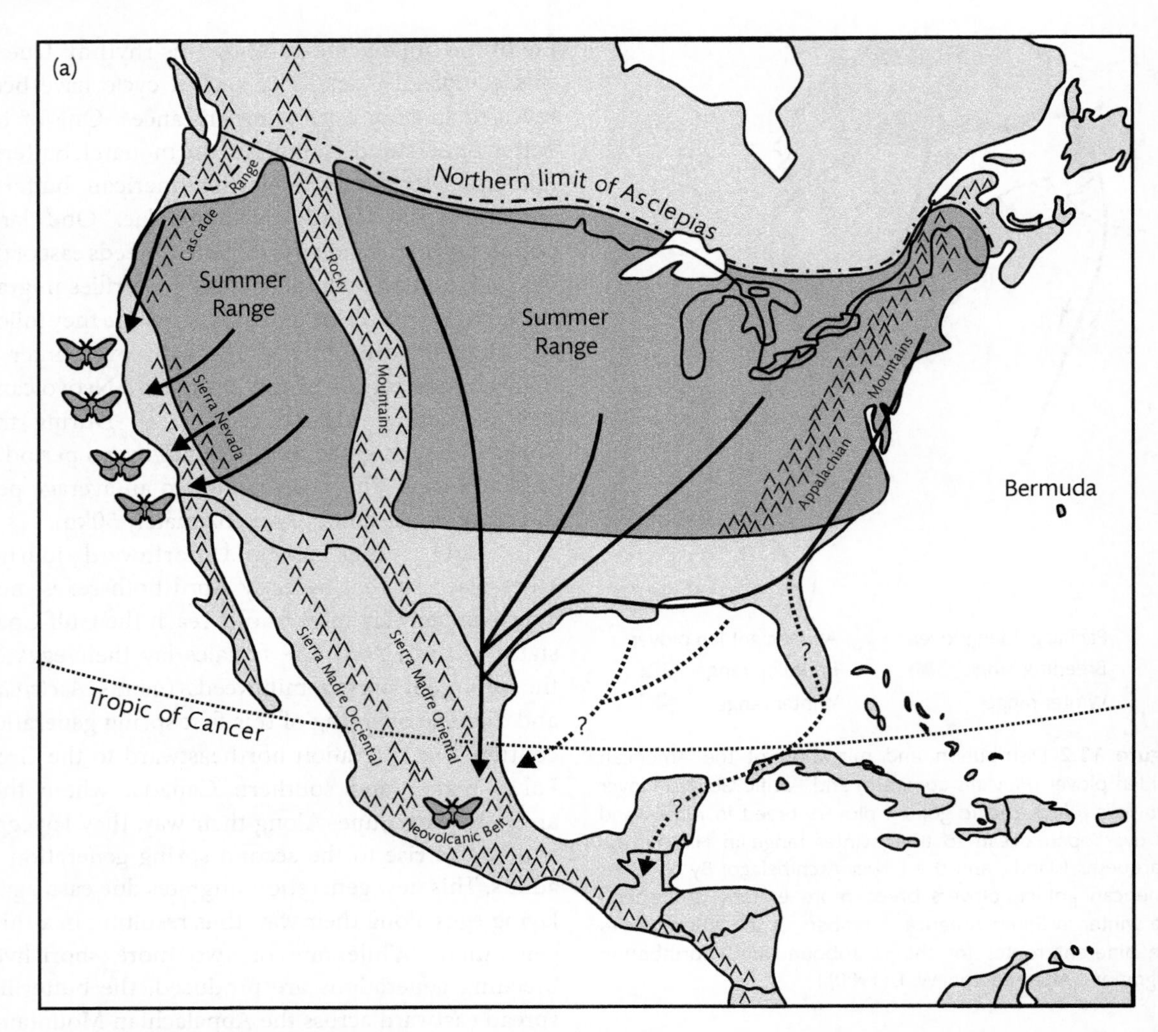

Figure 11.3 Migratory routes of the eastern population of the monarch butterfly. This population breeds, over several generations, east of the Rocky Mountains. (a) In the fall, the butterflies migrate southwards to overwinter in Central Mexico. (b) In the spring, they re-enter the U.S.A. along the Gulf Coast states, where the females lay eggs and die. The adults of the resulting first spring generation migrate further and produce the next generation. One large subpopulation of this second generation, mainly located in the Midwest, continues the migration eastwards over the Appalachians, where two or more summer generations follow. Also indicated is the northern limit of the milkweed plants (genus *Asclepias*). (After: Brower, L. P. (1996).)

hypothesis has been obtained. This is mainly due to the pioneering work of Peter Berthold and Andreas Helbig of the Ornithological Station in Radolfzell, Germany.

These investigations have taken advantage of the existence of various populations of blackcaps (*Sylvia atricapilla*), a European warbler. These populations differ in terms of their migratory behavior. For example, in Central Europe, blackcaps are fully migratory. By contrast, on the Cape Verde Islands (islands in the Atlantic, off Africa), a resident population exists. When these two populations were crossbred, 40% of the F_1 hybrids were migratory. This result demonstrates that the urge to migrate can be bred into a non-migratory bird population. However, the fact that not all hybrids become migratory indicates that it is not a single genetic locus that determines the urge to migrate. Rather, a multi-locus system has been proposed.

➤ Crossbreeding experiments using birds of different populations provide information about genetic factors controlling migratory behavior.

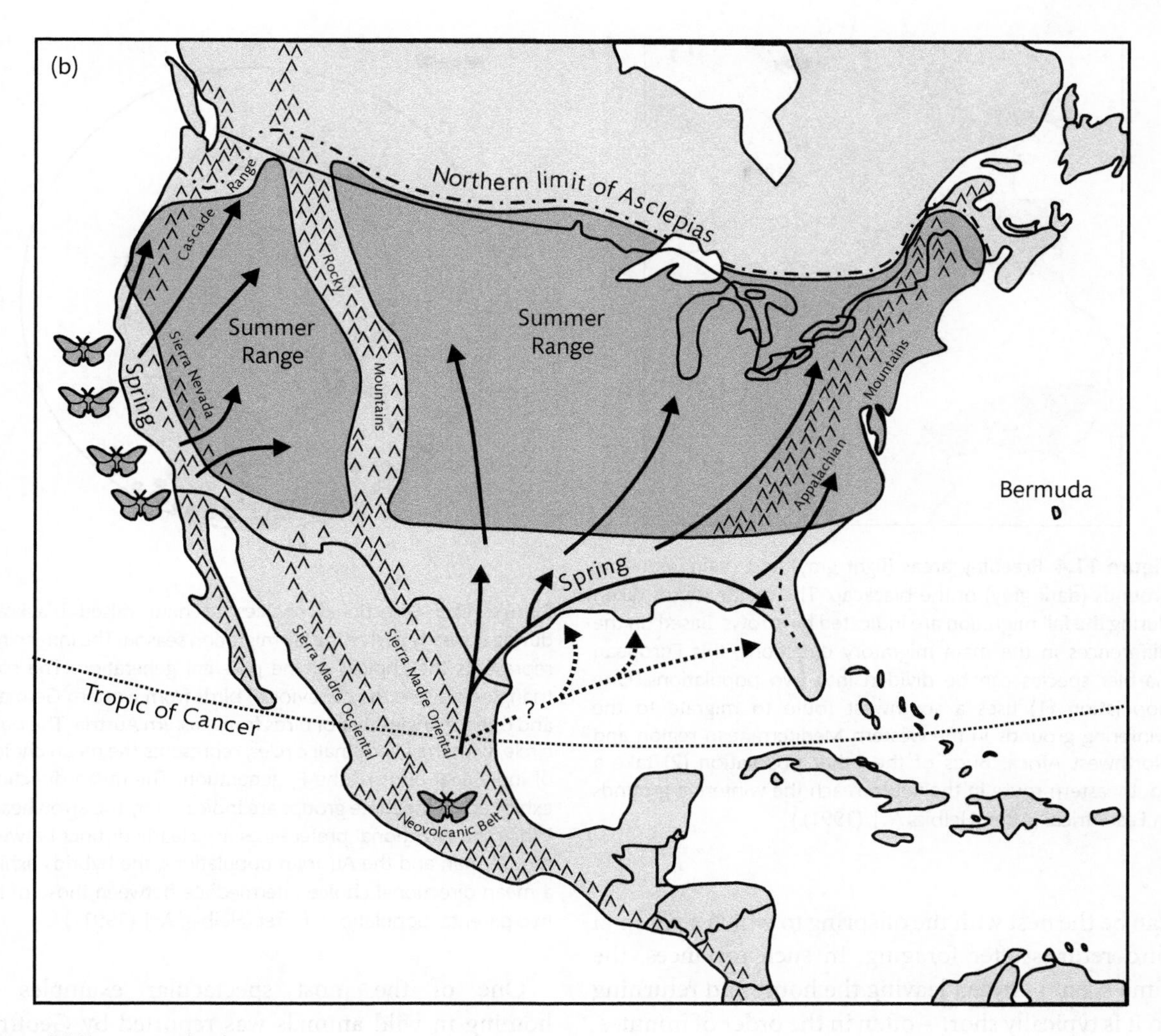

Figure 11.3 ... *contined*

In a second experiment, blackcaps from two Central European populations that differ considerably in the route they take during the fall migration were crossbred:

- One population breeding in Germany uses mainly a western route to the wintering areas in western Mediterranean regions and Northwest Africa (Fig. 11.4).
- A second population of blackcaps have their breeding areas in Austria. They take an eastern route to their wintering range in East Africa (Fig. 11.4).

Crossbreeding produced phenotypically intermediate offspring. These birds were oriented toward mean directions intermediate between those of the two parental populations (Fig. 11.5).

Based on these results, it is thought that migratory birds inherit information on the direction to travel and how long to fly in this direction. They then complete their first migration using a 'clock-and-compass' mechanism (see 'Mechanisms of long-distance orientation in birds' below).

Homing

While **migration** merely describes the directed locomotory activity of an animal during long-distance journeys, **homing** refers to the ability of an animal to specifically return to its 'home.' The home

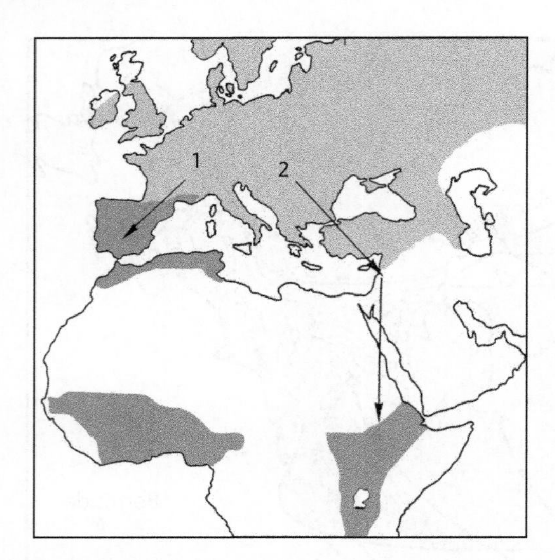

Figure 11.4 Breeding areas (light gray) and main wintering grounds (dark gray) of the blackcap. The major routes taken during the fall migration are indicated by arrows. Based on the differences in the main migratory directions, this European warbler species can be divided into two populations. One population (1) uses a southwest route to migrate to the wintering grounds in the western Mediterranean region and Northwest Africa. Birds of the other population (2) take a southeastern route in the fall to reach the wintering grounds in East Africa. (After: Helbig, A. J. (1991).)

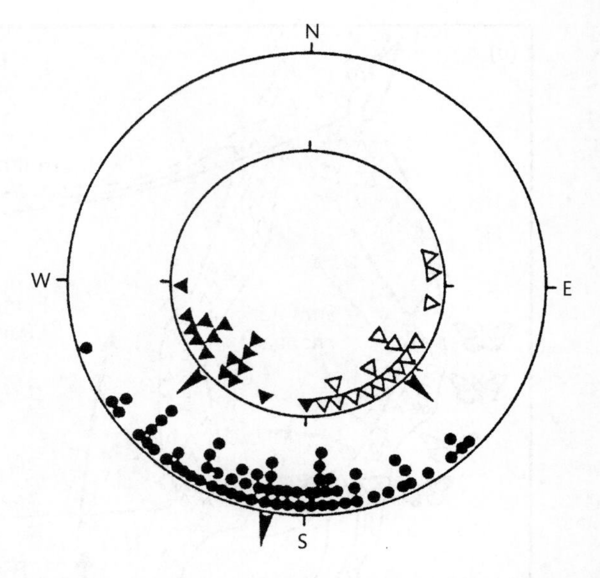

Figure 11.5 Directional choices of hand-raised blackcaps during the early part of the fall migration season. The inner circle represents the choices of the parental generation. The solid triangles indicate the behavior of birds from western Germany and open triangles that of birds from eastern Austria. The outer circle, with the filled small circles, represents the mean choices of individual birds of the F_1 generation. The mean directions exhibited by the three groups are indicated by the arrowheads. While the directional preferences are clearly distinct between the German and the Austrian populations, the hybrids exhibit a mean directional choice intermediate between those of the two parental populations. (After: Helbig, A. J. (1991).)

can be the nest with the offspring to which a parental bird returns after foraging. In such instances, the time spent between leaving the home and returning to it is typically short—often in the order of minutes. In other cases, a considerable amount of time, sometimes years, may elapse until the animal returns to its home. Sea turtles, for example, lay eggs in underground nests on the beach. Upon emerging from these nests, the hatchlings quickly move to the sea, until they reach the ocean. After migration across thousands of kilometers in the open sea, they return as adults to their natal beach to nest.

The **precision of homing** has been well studied in birds. Some avian species are even able to return to their nest after being transported in a closed box over considerable distances and released at a location unfamiliar to them. This may resemble the situation when a wild bird, after being displaced by strong winds from its nesting area, has to find its way back home. In the experiment, marking, often with a colored band, makes it possible to identify the bird unambiguously upon arrival and note the time of its return.

One of the most spectacular examples of homing in wild animals was reported by Geoffrey Matthews of the University of Cambridge, U.K., in 1953. Matthews performed a large number of displacement experiments on Manx shearwaters (*Puffinus puffinus*), ocean birds feeding on small fish and crustaceans, which they catch with their bills in the water. They can rest on the surface of the ocean, but most of the time they are seen flying near the water. Manx shearwaters come ashore only during the time of reproduction, when they breed in colonies in burrows on islands and the tops of cliffs. Between intervals of incubation or care of the young, they may make foraging excursions of a hundred or more kilometers in length. Furthermore, incubation is shared between the parents, and frequently one bird takes a spell of several days without feeding. This allows the investigator to remove one of the parents from the nest to conduct experiments without endangering the breeding success of the birds.

To test the homing ability of Manx shearwaters, Matthews released individuals at various distances from the nest. The most extreme experiment was conducted on a bird that was transported by airplane from its burrow off the coast of Wales to Boston, Massachusetts. Just 12.5 days later, this bird reappeared at its nest. During that time, it must have flown more than 4500km across the Atlantic, with an average speed of 360km per day!

➤ Manx shearwater, one of the 'champions' among birds: Homing over thousands of kilometers has been documented.

While homing experiments on wild birds are rather cumbersome to perform, the **homing of domestic pigeons** is more readily accessible. Experiments can be carried out throughout the year, rather than being restricted to the breeding season. Moreover, pigeons are highly tolerant to experimental manipulation. Therefore, numerous laboratory experiments have been performed, including many on their sensory capabilities. Some selected examples will be discussed below.

The enormous homing ability of pigeons is likely related to the breeding behavior of their ancestors. Domestic pigeons are thought to originate from wild Mediterranean rock doves (*Columba livia*), which nest on cliffs and forage for food on nearby fields. For centuries, these birds have been bred and selected for fast and reliable homing, especially in their function as carrier pigeons transporting messages in small capsules attached to their legs. They are raised in small sheds called **lofts**, which remain their permanent homes throughout life. Starting a few months after birth, the pigeons are systematically trained by transporting them to increasingly distant points from the home loft where they are released to fly back home. Upon completion of this training, the most capable pigeons are able to home from more than 2000km in two or three days.

Approaches to study animal migration and homing

To study migratory behavior in the field, research was initially restricted to the collection of **distribution data** about breeding, stopover, and winter ranges of the various species. Although these observations were, in the history of migration research, of enormous importance in establishing that many animals are migrating, numerous questions remained to be answered. For example, this approach could not answer the question of whether or not populations of a given species from different areas migrate to different target regions.

Therefore, the introduction of **tags** attached to the animal was a milestone in the historical development of migration research. For birds, cylindrical plastic or metal bands, loosely fitted around one leg, are employed. Each tag has inscribed a unique identification number and the name of an addressee who can be informed if the animal is found. Butterflies are typically tagged by gluing round paper discs to one wing. For fish, metal or plastic tags applied to jaws, opercula, or fins are often used.

Since the introduction of this method, millions of animals have been tagged. An enormous amount of information has been accumulated through the banding of birds. The corresponding data show that the **degree of recovery** is relatively high in large birds, such as ducks and geese, where it sometimes reaches 20% or higher. This is due to the fact that these birds are readily visible and many of them are heavily hunted. In small songbirds, on the other hand, the percentage of recovery is very low, with values typically far below 1%. This low recovery rate requires large-scale banding programs to yield meaningful data on the migrational pattern of a given species. Despite this limitation, the specificity of the tagging method has provided extremely valuable information on the migration and homing of animals.

➤ Most information available on animal migration and homing is based on tagging experiments.

Two further technologies successfully applied to navigation research are **radar** and **microtransmitters**. The radar echoes produced by aggregations of animals can be used either to screen for migratory activity within a certain area (**surveillance radar**) or to detect and follow a single target (**tracking radar**). As the resolving power of radar is relatively low, it typically allows the researcher only to follow aggregations of animals, but not single individuals. By contrast, radio-tracking through microtransmitters attached to animals provides the possibility of tracking individuals. Signals emitted by the microtransmitter are received by antennae that are stationary, carried on a vehicle or boat, or attached to an aircraft. More recently, this method has been further enhanced by **satellite-based radiotelemetry**. The major limitations of this extremely powerful method are caused by the costs associated with the satellite time, the lifetime of the batteries required to operate the microtransmitter, and the weight of the transmitter/battery, which currently restricts its use to animals with a relatively large body mass.

Mechanisms of long-distance orientation in birds

Among migratory animals, more studies on mechanisms involved in orientation have been conducted in birds than in any other animal group. These investigations have resulted in an impressive amount of behavioral data; however, due to the complexity of the mechanisms involved, interpretation of the data has often proved difficult. In the last couple of years, these behavioral data have increasingly been supplemented by the results of sensory, physiological, and neurobiological experiments.

Sun compass

Experimental evidence that the **sun** may provide an important visual cue for diurnal migrants was first reported by Gustav Kramer from the Max Planck Institute for Marine Biology in Wilhelmshaven, Germany. In 1950, he published, partially with his associate Ursula von St. Paul, a series of papers on this topic. Earlier observations had shown that caged migratory birds exhibit a pronounced **migratory restlessness** at the time when they normally start their journeys. To investigate this phenomenon further under controlled conditions, Kramer and von St. Paul placed the birds in circular cages that were designed in such a way that they displayed a perfect symmetry. Any visual landmark was excluded so that only the sky and the sun were visible. Thus, the birds had only celestial cues, but no landmarks, available for orientation.

When starlings (*Sturnus vulgaris*), diurnal migrants, were used in the experiments, the birds attempted to fly predominantly in the direction of their normal migratory path, as long as they could see the sun. If the sun was completely obscured by clouds, the headings of the birds became rather random. On the other hand, if the sunlight was deflected by a mirror at a certain angle, the birds altered the direction of the attempted flight accordingly (Fig. 11.6).

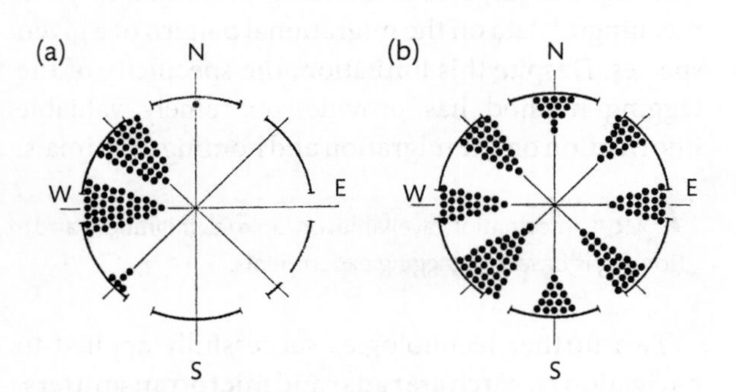

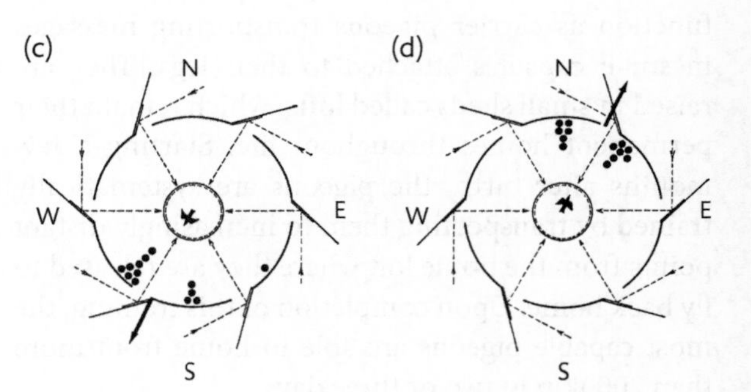

Figure 11.6 Spontaneous directional headings of a starling during the time of migratory restlessness, as examined in the circular cage used by Gustav Kramer. Each dot represents the mean direction of the bird's body exhibited during 15sec of migratory restlessness. Broken arrows show the incidence of light from the sky through the six windows. Thick arrows indicate the mean direction of activity. (a) When the sun is directly visible through one of the six windows, the bird heads in a northwest direction. (b) Under heavily overcast sky, the directional activity of the starling is rather random. In a modified setup, each window has an opaque screen, and a mirror deflects the light so that it appears to enter the cage at an angle approximately 90° counterclockwise (c) and clockwise (d) from its normal direction. Now the bird attempts to fly in a southwest or northeast direction, respectively. (After: Kramer, G. (1950).)

How do the starlings determine the direction they head to during the migration restlessness? To answer this question, Kramer and von St. Paul conducted a second set of experiments. To avoid the restrictions imposed upon the experimenter by the rather short seasonal period of migratory restlessness, a **food conditioning approach**, which could be employed at any time of the year, was used.

The birds were placed in a circular cage with an array of food chambers arranged at equal spacing around the periphery. The starlings were then trained to look for food in a certain chamber lying in a particular compass direction. Each food chamber was covered with a slotted rubber membrane. It was therefore impossible for the birds to know whether the feeder actually contained food, before they had thrust their bills through the slot and picked up any grain provided. Despite these difficulties, the birds maintained the training direction—even if the cage was rotated or the experiments were conducted at different times of the day.

Kramer and von St. Paul concluded from these results that the birds must use the sun for orientation and possess an **internal clock** to compensate for the apparent movement of the sun across the sky. In other words, the birds can extract compass information from the position of the sun, independent of whether the bird sees the sun low in the east in the morning or low in the west in the evening. This mechanism is referred to as a **time-compensated sun compass.**

> ➤ A time-compensated sun compass allows an animal to extract compass information from the position of the sun, independent of the time of day.

The time-compensation mechanism of the sun compass by an internal clock was confirmed in subsequent experiments carried out by Klaus Hoffmann at the Max Planck Institute for Behavioral Physiology, first in Wilhelmshaven and then in Seewiesen, Germany. He trained starlings in a circular cage to obtain food from a feeder in a certain direction using the sun as a cue, similar to the experiments by Kramer and von St. Paul. Then, he reset or shifted the bird's internal clock experimentally. This can be done by confining the bird to a light-proof room and exposing it to artificial photoperiods. This resulted in the bird misreading the sun compass in a predictable manner: A clockwise shift of the internal clock by xh leads to a counterclockwise misinterpretation of the sun compass by $(360/24)x$ degrees, and vice versa.

Example: *A starling is trained to look for food in a feeder located in a northwesterly direction. The bird is then exposed for four to six days to an artificial day beginning and ending 6h later than the natural day. When it is returned to the experimental cage and exposed to natural sunlight, it preferentially orients toward the northeast instead of the northwest (Fig. 11.7). The shift of its internal clock by 6h in a counterclockwise direction has resulted in a 90° clockwise misreading of the sun compass.*

Similar experiments in which the internal clock was shifted behind or ahead of the local photoperiod were conducted by Klaus Schmidt-Koenig from the University of Tübingen, Germany, on homing pigeons. After the clock shift, the birds were taken to a place far from the home loft and released singly. The direction in which the pigeon vanished from sight was recorded and the data were entered in a compass diagram. While control pigeons (i.e. pigeons that were not subjected to a clock shift) vanished roughly in the direction pointing to the home loft, time-shifted pigeons vanished with an error angle reflecting the direction and degree of the time shift as described above.

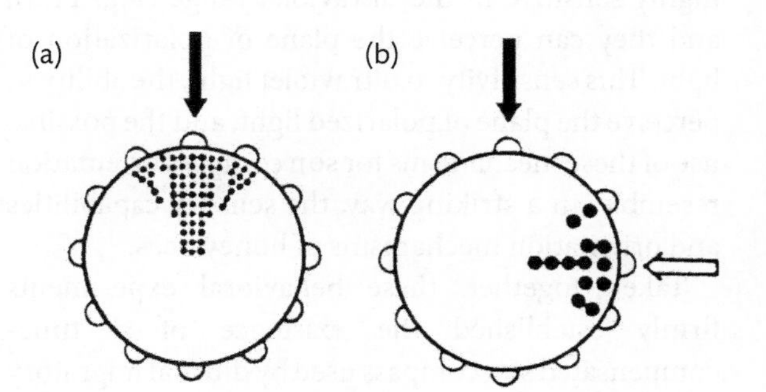

Figure 11.7 Change of orientation of a food-conditioned starling after experimental shifting of its internal clock. (a) Test during the training phase under natural-day conditions. (b) Test after exposing the starling for several days to an artificial day, which is 6h behind local time. Each dot represents one critical choice without food reward. The black arrows indicate the original training direction and the open arrow indicates the direction expected if the clock is successfully shifted. (After: Hoffmann, K. (1954) Versuche zu der im Richtungsfinden der Vögel enthaltenen Zeiteinschätzung. *Zeitschrift für Tierpsychologie* 11:453–475. Reprinted by permission of Blackwell Verlag GmbH.)

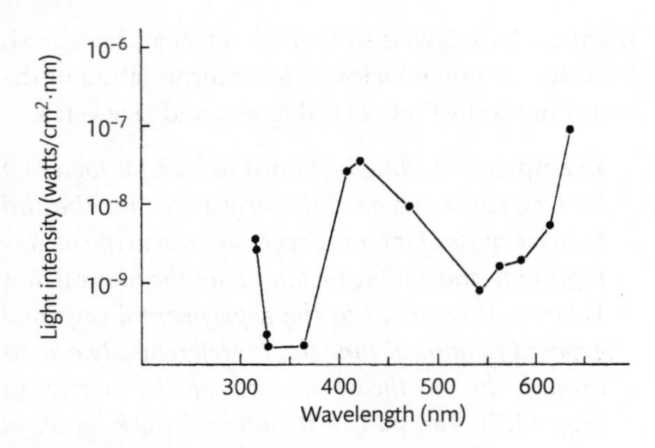

Figure 11.8 Behavioral sensitivity of a pigeon to light of different wavelengths. The pigeon shows high sensitivity to light in the ultraviolet range (i.e. at wavelengths between 325 and 360nm), as indicated by the low intensity of light necessary to elicit a behavioral response. (After: Kreithen, M. L. (1979).)

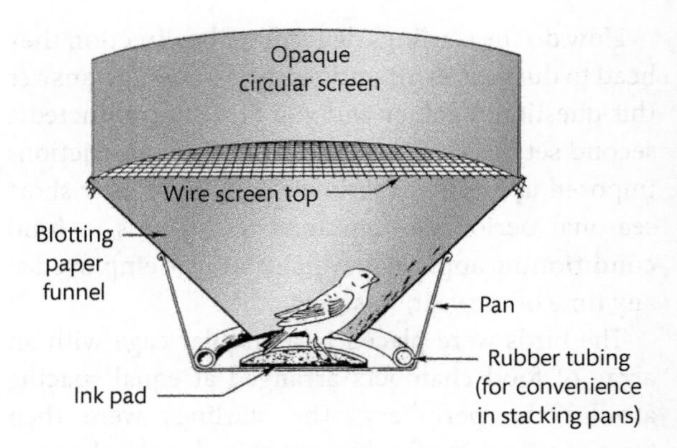

Figure 11.9 Schematic drawing of the 'funnel' experimental setup designed by Stephen and John Emlen. Single birds are kept in a closed, funnel-like cage. The bottom of this cage is covered with an ink pad. When the bird attempts to leave the cage, it produces footprints on the blotting paper lining the sides. (After: Emlen, S. T. and Emlen, J. T. (1966).)

It is likely that birds are able to determine the position of the sun and make use of this information for sun compass orientation, even if the sun is obscured by clouds. However, this is possible only if at least some patches of blue sky are still visible. It is believed that, under such circumstances, the birds use the polarization pattern of the blue sky for determining the position of the sun. This polarization vision is strongest in the ultraviolet range of light, where polarization is especially strong. Indeed, as revealed by conditioning experiments, pigeons are highly sensitive in the ultraviolet range (Fig. 11.8), and they can perceive the plane of polarization of light. This sensitivity to ultraviolet light, the ability to perceive the plane of polarized light, and the possible use of these mechanisms for sun compass orientation resemble, in a striking way, the sensory capabilities and orientation mechanisms of honey bees.

Taken together, these behavioral experiments firmly established the existence of a time-compensated sun compass used by diurnal migratory birds and homing pigeons for orientation.

Star compass

The experiments conducted by Gustav Kramer and confirmed by many others have demonstrated that birds can use the position of the sun for compass orientation. This can explain the directed restlessness of diurnal migrants. But how do **nocturnal migrants** orient?

This question is especially intriguing, as nocturnal migratory birds also exhibit a directed, rather than a random, migratory restlessness. To identify the cues directing this activity, Franz Sauer from the University of Freiburg, Germany, performed experiments on warblers (genus *Sylvia*)—nocturnal migratory birds—in circular cages with a glass top, similar to the setup used by Kramer. When these birds were kept in the cage indoors, their activity appeared disoriented. However, when they were kept in the same cage outdoors so that they could see the starry sky, their activity became oriented during the time when they normally started their seasonal migration. In the fall, they showed preference for southern directions, while in the spring they oriented predominantly northward. This indicated that cues provided by the night sky may be used for orientation.

To analyze this phenomenon quantitatively, Stephen Emlen, then a graduate student at the University of Michigan, Ann Arbor, and Stephen's father, John Emlen, from the University of Wisconsin, Madison, developed the **funnel technique** (Fig. 11.9). The bird is kept in a circular cage with an ink pad covering the bottom. Blotting paper lines the sides in a funnel-like fashion. When the bird attempts to leave the cage, ink footprints are produced on the blotting paper. These footprint data can be transcribed into vector diagrams by a variety of methods, for example by measuring the density of

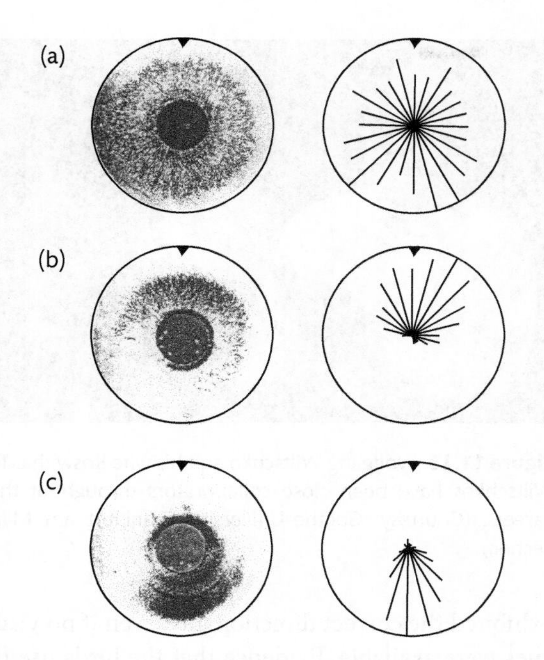

Figure 11.10 Footprint data produced by indigo buntings (*Passerina cyanea*) in the 'funnel' setup introduced by the Emlens. Left: Footprint raw data. Right: Corresponding transcription into vector diagrams. (a) Rather random orientation. (b) Orientation directed predominantly north-northeast. (c) Orientation directed predominantly south-southwest. (After: Emlen, S. T. and Emlen, J. T. (1966).)

the blackening using computer-aided image analysis (Fig. 11.10).

Using this method with indigo buntings (*Passerina cyanea*), Stephen Emlen demonstrated that the orientation of the birds was the same under the natural night sky and the stationary sky of a planetarium. When the planetarium sky was horizontally rotated by 180° in the spring, the indigo buntings exhibited their migratory restlessness predominantly to the south, rather than to the north (Fig. 11.11). This demonstrated the existence of a **star compass**. Further experiments have shown that time compensation is **not** used in order to take into account the movement of the stars during the night. It is thought that the birds learn, individually, the different star constellations near the pole star so that they can draw direction from them. It has been shown that indigo buntings do this by watching that part of the nocturnal sky that rotates the least. In the northern hemisphere, this is the area around **Polaris.** How exactly the birds obtain compass information from the constellation of the stars around Polaris

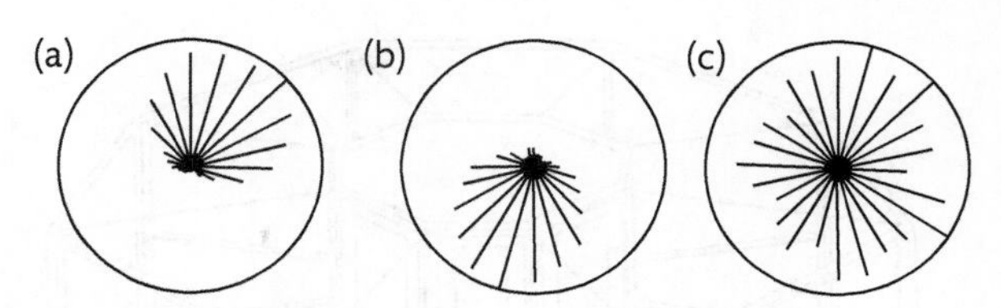

Figure 11.11 Preferential orientation of indigo buntings (*Passerina cyanea*) in planetarium experiments, as revealed by the funnel technique. The original footprint data have been transcribed into vector diagrams. (a) Orientation of indigo buntings in the spring under a spring planetarium sky. (b) Orientation in the spring under a spring planetarium sky horizontally rotated by 180°. The rotation of the starry sky results in the migratory restlessness exhibited in a direction roughly opposite to that before rotation. (c) Control experiment with the stars shut off. (After: Emlen, S. T. (1967a).)

is unknown. In principle, several star patterns are available to achieve this.

> ➤ Planetarium experiments have provided compelling evidence of the use of star-compass information by nocturnal migratory birds.

Emlen also demonstrated that indigo buntings learn to read the star compass in a **sensitive period** between the time when the birds leave the nest and the beginning of the fall migration. If they have not seen the nocturnal sky during the time before their first migration, they are unable to use the star compass—even if they are exposed to starry skies thereafter.

Magnetic compasses and maps: Behavioral evidence

Although the demonstrations of the existence of a sun and star compass can be considered major discoveries, these mechanisms cannot, by far, explain all observations. For example, these two mechanisms can be used to determine the direction of orientation (**compass function**), but there is no evidence that they enable the animal to determine its position relative to its goal. The latter task can be achieved only by mechanisms incorporating a **map function.** Theoretical consideration predicted that this requires a system in which two physical parameters are combined to a bicoordinate navigation system. Another observation that cannot sufficiently be

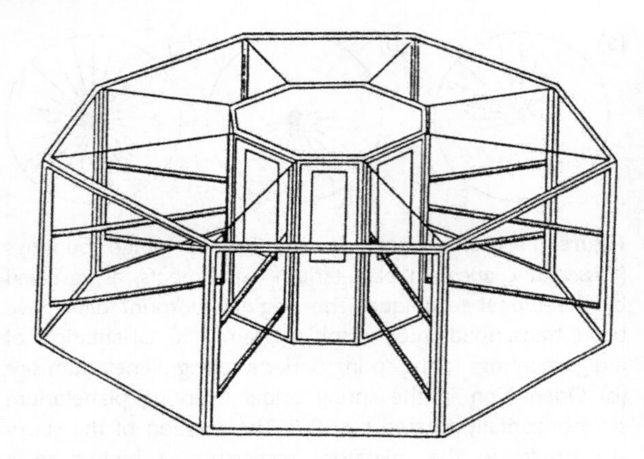

Figure 11.12 Schematic drawing of the octagonal cage used for testing migratory restlessness in European robins. The birds can move freely around the central structure, the inner part of which they can enter through holes to obtain food. Their activity is recorded electromechanically when they hop in the peripheral part of the cage onto one of the radial perches. (After: Merkel, F. W. and Fromme, H. G. (1958) Untersuchungen über das Orientierungsvermögen nächtlich ziehender Rotkehlchen, *Erithacus rubecula. Naturwissenschaften* 45:499–500. Reprinted by permission of Springer-Verlag. © 1958 Springer-Verlag.)

Figure 11.13 Wolfgang Wiltschko and his wife Roswitha. The Wiltschkos have been close collaborators throughout their careers. (Courtesy: Goethe-Universität Frankfurt am Main/ Lecher.)

explained by the existence of a sun and star compass is the fact that migratory birds frequently maintain the correct orientation, even if the sky is completely obscured by clouds. Such observations, as well as the proposal of a bicoordinate map mechanism, sparked early speculations that birds may use **geomagnetic cues** to fulfill these tasks.

The first indication that sensory and central structures for the detection of geomagnetic cues may indeed exist came from experiments on European robins (*Erithacus rubecula*), a passerine species that migrates at night. Captive individuals of this species become restless at the times of the year when their free-living conspecifics migrate, and prefer to stay at the side of the cage that points toward their migratory direction. To record this activity, the birds were kept in an octagonal cage with radial perches (Fig. 11.12). In this cage, the individuals could move freely around the central structure, which provided food and shelter in its inner part. Each hop is recorded by one of the radial perches through an electromechanical system. The 'hop scores' of one night are then processed vectorially to determine the mean direction of the restlessness shown by the bird.

In the first set of experiments it was confirmed that the migratory restlessness of European robins exhibited the correct directionality, even if no visual cues were available. Evidence that the birds use the Earth's magnetic field for orientation was obtained by Wolfgang Wiltschko (born 1938; Fig. 11.13) from the University of Frankfurt, Germany, in the late 1960s. When the magnetic north was artificially rotated by Helmholtz coils arranged around the octagonal test cage, the birds altered their directional preference accordingly (Fig. 11.14).

> ➤ Geomagnetic cues could, in theory, be used by magnetoreceptive animals to establish a bicoordinate map.

Analysis showed that the avian magnetic compass is notably different from the technical magnetic compass used by humans. To appreciate this difference, it is necessary to have a closer look at the **Earth's magnetic field**. At first approximation, the Earth's geomagnetic field can be described as a **dipole field** whose poles lie near the geographical poles. The field lines leave the ground at the southern magnetic pole and curve around the Earth to re-enter its surface at the northern pole (Fig. 11.15). In other words, the magnetic field vector points perpendicularly away from the ground at the southern pole, perpendicularly into the ground at the northern pole, and is parallel to the Earth's surface at the magnetic equator. At locations other than the poles and the equator, the angle between the magnetic vector and the horizon varies between these two extremes. This angle is called the **inclination** or **dip**.

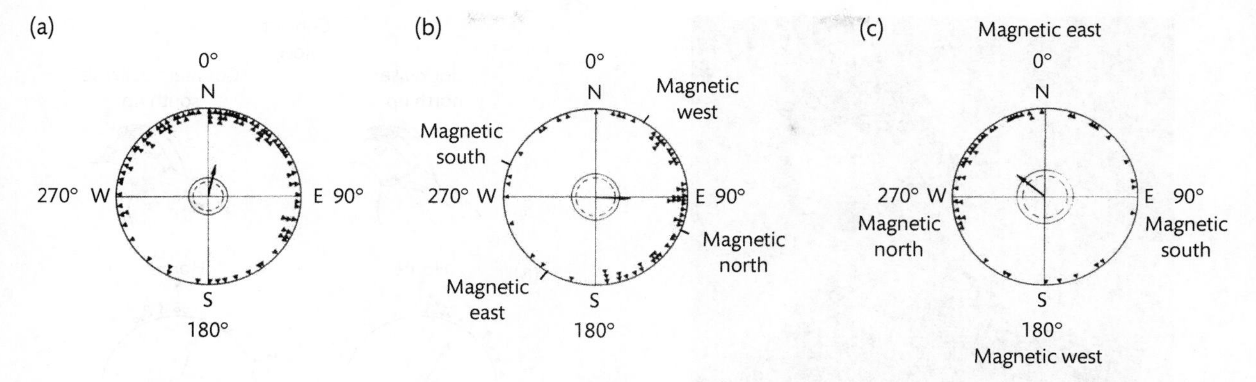

Figure 11.14 Orientation of European robins in various magnetic fields during the migratory spring season. Each triangle indicates the mean bearings of a single experimental night. The arrows represent the mean vector of all individual responses within one test group. When the normal Earth magnetic field with the magnetic north at 0° (a) is altered so that magnetic north is located at 120° (b) or 270° (c), the birds change the predominant direction of their activity accordingly. (After: Wiltschko, W. and Wiltschko, R. (1996).)

Several experiments have shown that, in contrast to the technical compass of humans, which points to magnetic 'north' and 'south' (and is therefore called a **polarity compass**), the magnetic compass of birds is an **inclination compass**. It defines 'poleward' as the direction along the Earth's surface where the angle between the magnetic field vector and the gravity vector becomes minimal. By contrast, 'equatorward' is the direction along which the angle formed between the magnetic field vector and the gravity vector becomes maximal. However, the inclination compass ignores the polarity of the magnetic field. This property has an important advantage over a polarity compass. As the polarity of the Earth's magnetic field has reversed repeatedly in the past, a polarity compass might have led to devastating misguidance of the birds over evolutionary times.

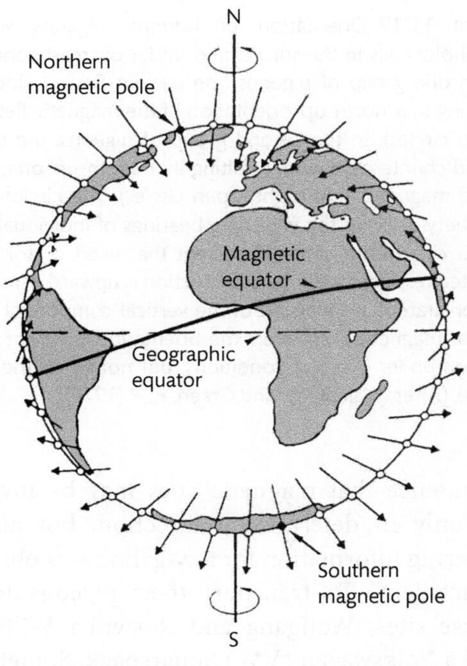

Figure 11.15 Schematic view of the geomagnetic field of the Earth. In the southern hemisphere, the magnetic field lines leave the Earth to re-enter it in the northern hemisphere. The arrows are a vector representation of the magnetic field lines. Their lengths indicate the intensities, and their orientation indicates the direction of the field lines. (After: Wiltschko, W. and Wiltschko, R. (1996).)

> In contrast to the human technical compass, the magnetic compass of birds is an inclination compass.

The finding of Wolfgang Wiltschko that the restlessness behavior of migratory birds can be altered by artificial magnetic fields inspired others to conduct similar experiments in homing pigeons. After failed attempts by several scientists, William Keeton (1933–1980; Fig. 11.16) from Cornell University in Ithaca, New York, reported in 1971 that bar magnets glued to the backs of pigeons resulted in disorientation, as long as either inexperienced birds were used, or experimental birds were released under total overcast conditions, that is, when the sun compass was not available. These findings also explained why so many previous studies had failed to produce positive results. As long as the sun is visible,

Figure 11.16 William T. Keeton with a Cornell pigeon. (Photographer unknown. Courtesy: Charles Walcott/Cornell University.)

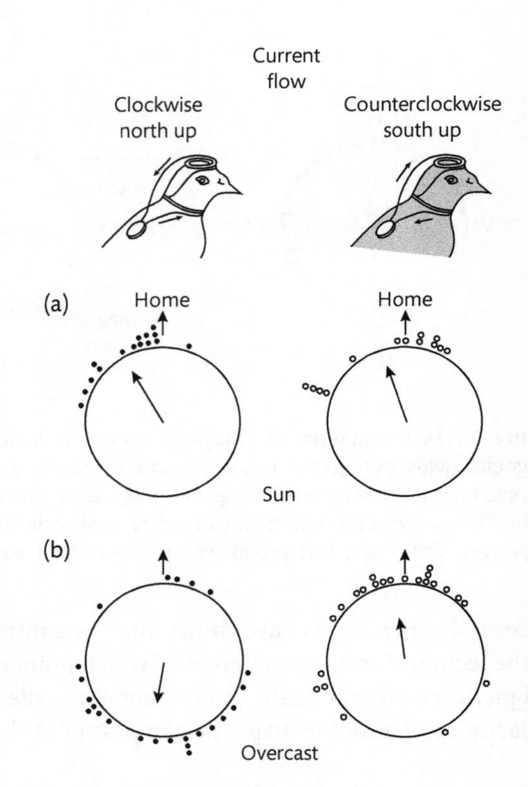

Figure 11.17 Orientation of homing pigeons wearing Helmholtz coils in the sun (a) and under overcast conditions (b). In one group of pigeons, the current flowed clockwise, resulting in a 'north-up' orientation of the magnetic field (left, closed circles). In the second group of pigeons, the current flowed counterclockwise, resulting in a 'south-up' orientation of the magnetic field (right, open circles). The circles in the periphery indicate the vanishing bearings of individual birds. The arrows in the center represent the mean vector of the oriented response. Homeward direction is upward. The results demonstrate that inversion of the vertical component of the natural magnetic field alters the orientation behavior of the pigeons under overcast conditions, but not when the sun is visible. (After: Walcott, C. and Green, R. P. (1974).)

the birds appear to prefer using the sun compass. Only when the sun compass cannot be used at all do pigeons rely entirely on magnetic cues.

The results of Keeton were further substantiated by Charles Walcott, then at the State University of New York at Stony Brook, now at Cornell University. Walcott attached battery-operated coils around the head of pigeons. Current flowing through the coils changed the magnetic field around the bird's head. Similar to Keeton's finding, the coils had little effect on the behavior of the pigeons when the sun was visible. Under overcast conditions, the effect depended upon the direction of the current. When the magnetic north of the induced magnetic field pointed upward ('north-up arrangement'), the test birds showed a tendency to fly in the opposite direction to their home loft (Fig. 11.17). Under this experimental condition, the resultant magnetic field shows roughly an inversion of the vertical component of the natural field. Correspondingly, if the resultant magnetic field pointed downward ('south-up arrangement'), the pigeons tested flew in the correct direction toward their home loft.

Evidence that magnetic cues may be involved not only in determining direction, but also in gathering information for navigation was obtained by accident. To transport their pigeons to the release sites, Wolfgang and Roswitha Wiltschko used a Volkswagen (VW) Squareback. Sometimes, they observed that the orientation of the pigeons was disturbed. This was regularly the case when the pigeons' crate was placed on top of the engine. As it turned out, it was the attached generator that produced a magnetic field and thus caused this 'VW effect.'

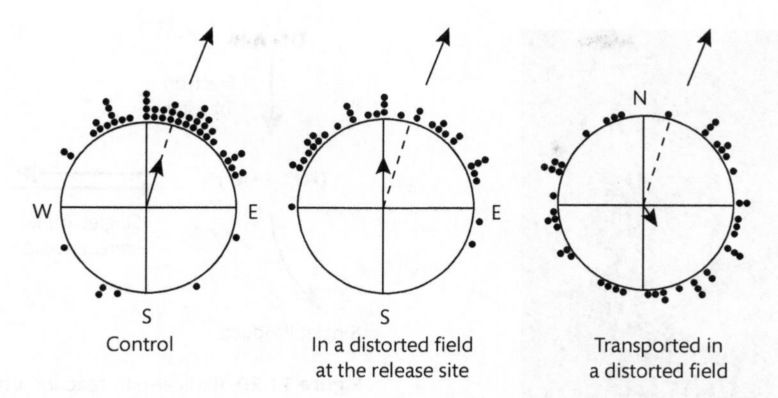

Figure 11.18 Effect upon orientation of young, inexperienced pigeons following exposure to distorted magnetic fields. The closed circles in the periphery indicate the vanishing bearings of individual birds. The arrows in the center represent the mean vector of the oriented response. Open arrows point in the direction of the home loft. While pigeons that are subjected to a distorted field only at the release site show a similar response to pigeons not exposed to any artificial magnetic field, the orientation of pigeons transported in a distorted field is severely impeded. (After: Wiltschko, R. and Wiltschko, W. (1978).)

This finding was confirmed in subsequent, systematically conducted experiments. When young, inexperienced pigeons were transported to the release site in a distorted magnetic field, they were disoriented. If, however, a second group of inexperienced pigeons was exposed to a distorted magnetic field for an equal amount of time at the release site, their homing behavior did not appear to be affected (Fig. 11.18). These findings show that it is not the exposure to a distorted magnetic field itself that causes disorientation, but **transport in the distorted field**. It appears as though the pigeons store information during the course of their outward journey. In the simplest case, they may integrate the different directions recorded by their magnetic compass over time during the outbound journey and define the reversed route as the correct way back home.

It is interesting that older and more experienced pigeons are barely affected by magnetic distortion during their outward journey. Apparently, young inexperienced pigeons and older experienced pigeons use different navigational strategies.

Magnetoreception: Sensory systems and neural correlates

The large body of behavioral evidence has led scientists to propose the existence of specific **magnetoreceptors** in migratory birds. The main questions arising from this proposal are as follows:

- What is the structure of the proposed magnetoreceptors?
- Where are they located?
- How do they function?
- How do they interact with other sensory and central systems?

Two models of the putative magnetic sense organ have been suggested. One is thought to be closely associated with photoreceptors in the retina and to involve a radical-pair process sensitive to magnetic fields. The other has been hypothesized to transduce magnetic information into neural activity using magnetite-based sensors. In the following two sections, evidence will be presented that supports, or is in conflict, with these two models.

Photoreceptor-based radical-pair mechanism

The photoreceptor-based magnetoreception model was proposed by Klaus Schulten and his group. Schulten (1947–2016; Fig. 11.19), a computational biophysicist at the Beckman Institute for Advanced Science and Technology at the University of Illinois at Urbana-Champaign, suggested in a series of theoretical papers published between 1978 and 2000 that a **radical-pair reaction** can serve as a chemical sensor for magnetic compass orientation. This mechanism is based on the influence of external magnetic fields on the **spin orientation** of an electron

Figure 11.19 Klaus Schulten. (Photograph by Thompson-McClellan.)

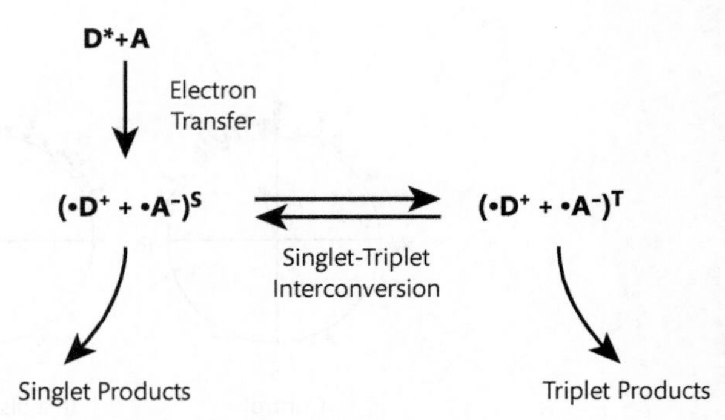

Figure 11.20 Radical-pair reaction under the influence of an external magnetic field. After light absorption (*) by a donor (D), a radical pair is generated by electron transfer from this donor molecule to an acceptor molecule (A). Usually, the photoexcited donor and the acceptor are in singlet states before electron transfer, and the resultant radical pair is also in a singlet (S) state. However, this overall singlet state can be converted into a triplet (T) state under the influence of an external magnetic field. The ratio of singlet vs. triplet states depends on the alignment of the radical pair with respect to the direction of the external magnetic field. The singlet and triplet pairs will react differently in the subsequent step, yielding distinct singlet and triplet products. (From: Ritz, T., Adem, S., and Schulten, K. (2000) © 2000 The Biophysical Society. Published by Elsevier Inc. All rights reserved.)

during its transfer from an excited donor molecule to an acceptor molecule (Fig. 11.20). The spin is caused by the electron rotating about its own axis, which results in a weak magnetic field. Electron-transfer processes leave both the donor molecule and the acceptor molecule with an unpaired electron. The pair of electrons can have either opposite spins (↑↓; called a singlet) or parallel spins (↑↑; called a triplet).

When, after a brief period, the transferred electron returns to the donor molecule, the new relationship of the two spins may, under the influence of external magnetic fields, differ from that before the transfer. This may be the case even if these fields are as weak as those of the Earth's magnetic field. However, a reversal of the original spin relationship occurs only if a number of requirements are met. This is frequently the case in electron-transfer processes induced by photoexcitation, that is, by the absorption of light. Alterations in spin orientation affect the chemical properties of the donor and the acceptor molecules. Thus, a singlet forms a different chemical product than a triplet does.

This quantum phenomenon has an important consequence: An external magnetic field can affect a chemical reaction. Schulten realized that such a mechanism could be used as a magnetic compass sensor by magnetoreceptive animals.

According to the radical-pair reaction model, a magnetoreceptor based on this mechanism has to meet two essential criteria: First, it should contain a donor/acceptor pair of molecules capable of a radical-pair reaction that can be influenced by a magnetic field of similar strength as the Earth's magnetic field. Second, the magnetoreceptor should be linked to a light-absorbing photoreceptor that initiates the radical-pair process upon excitation.

As a candidate for a radical-pair-based magnetoreceptor, Schulten and colleagues suggested **cryptochromes**. Absorption of light by these flavoproteins results in generation of radical pairs. The second criterion mentioned above is also met by this class of molecule: One form of cryptochrome, cryptochrome 1a, has been found in the ultraviolet/violet (UV/V) cones in the retina of several bird species. Cones of this particular receptor type are evenly distributed across the retina and exhibit sensitivity to all wavelengths of light, including those in the short-wavelength range.

In the UV/V cones, cryptochrome 1a is associated with the disk membranes in the outer segment, probably in an oriented manner. Such an arrangement would ensure that the reactions of neighboring cryptochrome 1a molecules to a magnetic field of a particular orientation do not cancel each other but add up to produce a joint and robust response. Moreover, the distribution across the retina might ensure that all spatial directions are represented, leading to a differential activation pattern in different parts of the retina in response to magnetic field stimulation.

The UV/V cones contain, in addition to cryptochrome 1a, the short-wave sensitive 1 (SWS1) opsin, which is sensitive to UV or violet light. Thus, these cones are involved in magnetoreception and perception of UV/violet light, although these two mechanisms appear to be largely independent of each other.

As predicted by the photoreceptor-based magnetoreception model, behavioral experiments have demonstrated that the functioning of the avian magnetic compass requires light, specifically from the short-wavelength portion of the spectrum. When birds were tested while the test cages were illuminated by artificial light of such wavelengths, they showed orientation in their migratory direction. However, the same birds became disoriented when they were exposed to long-wavelength (yellow or red) light. Notably, this photoreceptor-based magnetoreception mechanism operates under rather low light levels, corresponding to natural light levels as they occur approximately 45min before sunrise or after sunset, respectively.

The photoreceptor-based magnetoreception model can, furthermore, be tested by applying radio frequency fields in the megahertz (MHz) range while monitoring the orientation of the birds to magnetic fields. Radiation in this frequency range is known to interfere with the singlet–triplet interconversion. Depending on the alignment of the oscillating radio frequency field in respect to the vector of the local geomagnetic field, in experiments this manipulation either disrupted the conditioned directional responses of the birds to the magnetic fields, or left their orienting behavior unchanged (Fig.11.21), thereby providing support for the involvement of a radical-pair mechanism in magnetoreception.

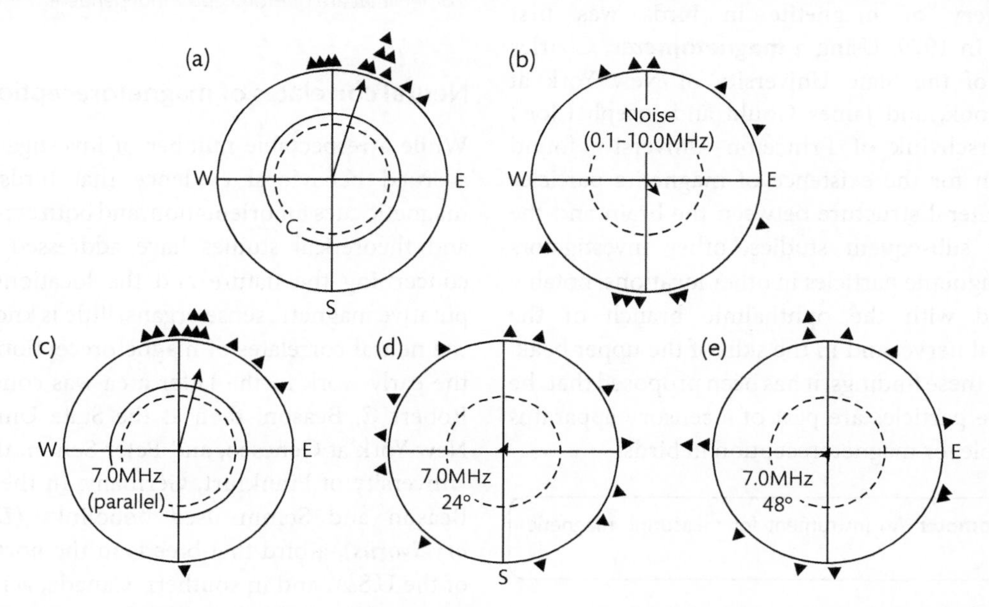

Figure 11.21 Effect of oscillating magnetic fields on the migratory orientation of European robins. (a) Control condition, with the Earth's geomagnetic field only. (b) Addition of broadband (0.1–10MHz) field to the Earth's geomagnetic field. (c–e) Addition of single-frequency (7MHz) fields oriented parallel (c), at a 24° angle (d), or at a 48° angle (e) to the Earth's geomagnetic field. Under the control condition, the birds exhibit northerly orientation, as is characteristic during spring migration. The arrowheads indicate the mean heading of each of the 12 individuals tested. The arrow represents the overall mean vector. When the birds were stimulated with a single-frequency field parallel to the geomagnetic field, the orientation of the birds was indistinguishable from the orientation under the control condition. However, in the presence of a broadband noise field, or of oscillating single-frequency fields at 24° or 48° relative to the geomagnetic field, the birds were disoriented. (From: Ritz, T., Thalau, P., Phillips, J. B., Wiltschko, R., and Wiltschko, W. (2004) © 2004, Rights Managed by Nature Publishing Group.)

Magnetite-based magnetoreception

The second mechanism proposed to mediate magnetoreception in migratory birds and homing in pigeons is based on **magnetite** (Fe_3O_4). The search for this magnetic mineral in tissues of animals was sparked after Richard Blakemore, then a graduate student at the University of Massachusetts at Amherst and working at the Woods Hole Oceanographic Institution, reported in 1975 that magnetotactic bacteria contain magnetite. In these bacteria, magnetic crystals of magnetite act like permanently magnetized bar magnets and are surrounded by a phospholipid bilayer to form a unique nano-sized organelle called a **magnetosome**, which is arranged in a linear chain. The magnetosomes mediate passive alignment of magnetotactic bacteria along the Earth's magnetic field lines. It is thought that this form of **magnetotaxis** (cf. Chapter 4) aids the bacteria in aquatic habitats in swimming towards zones for optimal growth and survival.

Discovery of magnetite in birds was first reported in 1979. Using a **magnetometer**, Charles Walcott of the State University of New York at Stony Brook, and James Gould and Joseph ('Joe') Lynn Kirschvink of Princeton University found indication for the existence of magnetite particles in a unilateral structure between the brain and the skull. In subsequent studies, other investigators found magnetite particles in other locations, notably associated with the ophthalmic branch of the trigeminal nerve, and in the skin of the upper beak. Based on these findings, it has been proposed that the magnetite particles are part of a sensory apparatus responsible for magnetoreception in birds.

Magnetometer An instrument for measuring magnetic fields.

However, the notion of magnetite particles involved in magnetoreception in birds has been challenged in a study published in 2012 by a multinational group of scientists from the Institute of Molecular Pathology in Vienna (Austria), University College London (U.K.), the Université de Strasbourg (France), and the University of Western Australia (Australia). The authors of this paper mapped the distribution of iron-rich cells, and determined their cellular identity, in the beak of pigeons. Although granules composed of **ferritin** (an iron-storing protein) were, indeed, found in such cells, several lines of evidence suggested that these cells are **macrophages**, rather than neurons. Iron is known to accumulate within macrophages during the catabolism of hemoglobin. It is, then, stored as ferritin. Presumably, the macrophages are recruited to the beak regions in response to tissue damage or host invasion.

Although this study failed to find any evidence to support a magnetite-based magnetoreception system in the beak of pigeons, it does not necessarily preclude the existence of such a sensory system. The authors speculated that magnetoreceptor cells associated with magnetite may reside in the olfactory epithelium, a sensory structure implicated in magnetoreception in the rainbow trout (see 'Homing in salmon', below).

➤ Magnetoreceptors in birds appear to fall into two categories: Ones associated with photoreceptors and others that are photoreceptor independent.

Neural correlates of magnetoreception

While a respectable number of investigations have accrued behavioral evidence that birds can use magnetic cues for orientation, and both experimental and theoretical studies have addressed questions concerning the nature and the location(s) of the putative magnetic sense organs, little is known about the neural correlates of magnetoreception. Some of the early work in the latter area was conducted by Robert C. Beason, then at the State University of New York at Geneseo, and Peter Semm, then at the University of Frankfurt, Germany. In their studies, Beason and Semm used bobolinks (*Dolichonyx oryzivorus*), a bird that breeds in the northern part of the U.S.A. and in southern Canada, across North America. In the fall, bobolinks migrate, primarily during the night, to their winter ranges in southern Brazil and northern Argentina. Due to the enormous distance between the summer and winter ranges, bobolinks have the longest migratory pathway of any New World landbird.

Experiments have demonstrated that, for orientation during their journey, bobolinks use both visual cues originating from the stellar constellation and magnetic cues derived from the Earth's magnetic field. Electrophysiological recordings from the **ophthalmic nerve** and the **trigeminal ganglion** have revealed the presence of units that are highly sensitive to small changes in the magnetic field. The most sensitive units respond to changes as small as 200nT (**nanotesla**), which corresponds to less than 0.5% of the Earth's total field. Most commonly, these units increase the rate of firing as a logarithmic function of the increased intensity in the ambient magnetic field (Fig. 11.22).

Tesla (T) Unit of magnetic field, also known as magnetic flux density. It is named after the Serbian-American inventor and engineer Nikola Tesla. Commonly, the geomagnetic field on the surface of the Earth is measured in nanotesla ($1nT = 10^{-9}T$).

Ophthalmic nerve The smallest of the three divisions of the trigeminal nerve. It supplies afferent branches, among other structures, to parts of the eye and the nasal cavity.

Additional support for the involvement of the ophthalmic nerve in magnetoreception has been obtained through behavioral experiments. Bobolinks were tested during the fall migratory season, when they exhibited a directional preference towards southeast. Magnetization of the birds with a brief magnetic pulse of high intensity and different polarity than the Earth's magnetic field resulted in a specific change in the direction of orientation. This effect of magnetization could be abolished by blocking the ophthalmic nerve with a local anesthetic.

In line with the notion that the ophthalmic branch of the trigeminal nerve is involved in magnetoreception are the results of a behavioral molecular mapping in European robins (*Erithacus rubecula*), carried out by a group of scientists under the leadership of Henrik Mouritsen of the University of Oldenburg, Germany. Immunohistochemical staining of **ZENK**, a neuronal activity-dependent marker protein, demonstrated, in response to a changing magnetic field, strong activation of neurons near the principal and spinal tract nuclei of the trigeminal brainstem complex—two brain regions known to receive primary input from the trigeminal nerve.

ZENK An acronym of the initial letters of Zif268, Egr-1, NGFI-A, and Krox-24. ZENK is the avian homolog of these four immediate-early gene products of the zinc finger family. ZENK expression is widely used as a marker of neural activity induced by relevant behavioral stimuli.

More direct support for the hypothesis that there is a neural substrate underlying magnetoreception in the central nervous system of birds has been provided by electrophysiological experiments by Le-Qing Wu and J. David Dickman at the Baylor College of Medicine in Houston, Texas. Single-cell recordings identified neurons in the brainstem of pigeons that encode direction, intensity, and

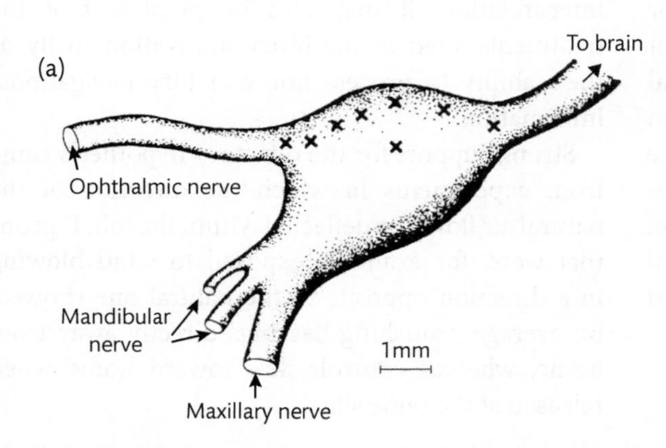

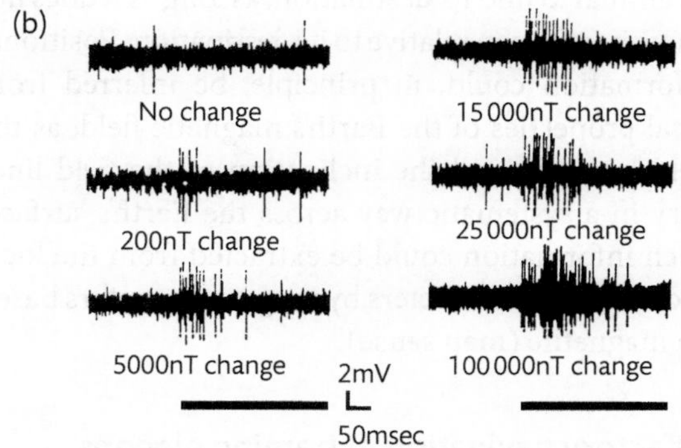

Figure 11.22 Electrophysiological recordings from the trigeminal ganglion of the bobolink during alterations in the ambient magnetic field. (a) Schematic structure of the trigeminal ganglion with the ophthalmic branch. The crosses indicate the locations from which recordings were taken. (b) Responses from one unit in the ganglion during changes in vertical magnetic-field intensity. The duration of the stimulus is indicated by the horizontal bars. The upper trace in the left column shows the spontaneous activity of the unit. The other traces show the responses of the same unit to changes in the magnetic field as indicated. (After: Semm, P. and Beason, R. C. (1990).)

polarity of magnetic fields. Since the brainstem area in which the neurons responsive to magnetic fields is innervated by afferents from the **lagena** receptors of the inner ear, one would expect that these neurons receive magnetic input from the inner ear. Indeed, indication that the lagena receptor organ may function as a magnetic sense apparatus was obtained in a previous study by the laboratory of David Dickman. However, other sources that supply magnetic information to the brainstem cannot be ruled out.

Lagena The third otolith organ, besides the utriculus and the sacculus, of the inner ear (cf. Chapter 4, 'Model system: Geotaxis in vertebrates'). It is present in elasmobranchs, teleosts, amphibians, reptiles, and birds. In mammals, it is found only in monotremes. The lagena has been implicated in auditory and vestibular functions, but may also be involved in magnetoreception.

Considering the overall behavioral, sensory physiological, and neurobiological evidence, it is possible that different sets of magnetoreceptors exist in the same animal, and that the information generated by these different magnetic sense organs is processed by different central pathways. The different magnetic sensory systems may subserve different functions. It has been hypothesized that magnetoreceptors associated with the visual system could provide directional information (**compass sense**). This information is, however, insufficient for an animal to find its destination, as long as it does not know its position relative to its destination. Positional information could, in principle, be inferred from local properties of the Earth's magnetic field, as the field intensity and the inclination of the field lines vary in a systematic way across the Earth's surface. Such information could be extracted from the local geomagnetic parameters by magnetoreceptors based on magnetite (**map sense**).

Olfactory navigation in homing pigeons

In 1971, the Italian scientist Floriano Papi and his group at the University of Pisa first proposed what has become known as the **olfactory hypothesis**. The essential idea of this hypothesis is best summarized in Papi's own words:

Pigeons acquire their homing ability at the loft by learning to recognize the odor of the loft area, as well as foreign odors carried by the winds. Associating these different 'foreign' odors with the direction from which they come, pigeons gain information about odors prevailing in the surrounding areas and build up their 'olfactory map.' This olfactory experience at the loft enables pigeons to use odors perceived during the outward journey and over the release area to establish the home direction. The sun or the magnetic compass will be employed to select the bearings deduced thereby.

Initial experimental evidence for this proposal came from experiments in which pigeons were subjected to olfactory deprivation. This was achieved by various methods, including sectioning of the olfactory nerve, plugging of the nostrils with cotton, insertion of tubes into the nasal passages, and application of local anesthetics to the nasal epithelium. These treatments resulted in a poor homing performance, as well as in rather random vanishing bearings. However, such negative effects were observed only when the pigeons were released at unfamiliar sites, but not at sites where they had learned certain features through prior experience.

While Papi and his group interpreted the behavior of anosmic pigeons released at unfamiliar sites as evidence for the hypothesis that olfactory cues are necessary for navigation, others (including William Keeton) raised doubts about the **specificity** of the approaches employed. According to Keeton's interpretation, it may also be possible that the treatments affected the birds' motivation to fly or their ability to process non-olfactory navigational information.

Strong support for the olfactory hypothesis came from experiments in which the direction of the natural air flow was deflected within the loft. Pigeons that were, for example, exposed to wind blowing in a direction opposite to the natural one showed, on average, vanishing bearings directly away from home, whereas controls flew toward home when released at the same site.

➤ Despite many experiments, the olfactory hypothesis of pigeon homing remains controversial.

Despite positive results, such as those produced by the wind-reversal experiments, the olfactory hypothesis has remained controversial. This is due to several problems that, regardless of considerable effort to resolve, have remained elusive. For example, it is unknown how gradients of odors, which are predicted by the olfactory hypothesis to exist, are maintained over hundreds of kilometers in the atmosphere. Furthermore, nothing is known about the exact chemical nature of the proposed olfactory cues in the atmosphere. Moreover, there are some indications that the answer to the question of whether or not a pigeon uses olfactory cues for navigation depends crucially on the environment to which the bird is exposed during the first few months of its life. Wolfgang Wiltschko and his group, for example, obtained evidence that pigeons reared in a loft sheltered from winds were largely unaffected by anosmia and were homeward-oriented, even when released at unfamiliar sites. Thus, birds raised under such conditions may use cues different from olfactory ones for navigation. If this is the case, it will add considerable complexity to the phenomenon of homing in pigeons.

Homing in salmon

Salmon are famous for their ability to return to their natal stream for reproduction after lengthy foraging migrations in the sea. A large number of extensive studies have suggested that the precision with which salmon find their way home is achieved by using different navigational mechanisms during their journey in the ocean and in the in-river migration. These mechanisms will be discussed in detail below.

Life cycle of salmon

One of the best-studied salmonid species is the coho salmon (*Oncorhynchus kisutch*). Its life cycle is summarized in Fig. 11.23. Coho salmon spawn in late October to December in fast-flowing streams along the Pacific coast of North America. The female digs nests (so-called **redds**) by lying on her side and rapidly beating with her tail until a nest pocket results. In each of the several nests, the female lays 500–1000 eggs, which are fertilized by the male. Using her mouth and tail, the female covers the eggs with gravel from the bed of the river. In this phase, which lasts a few days, the female guards the nests.

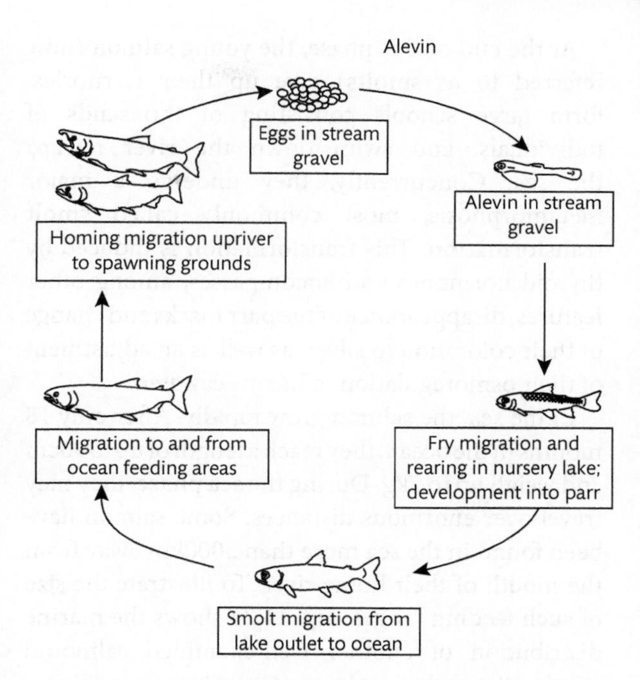

Figure 11.23 Schematic illustration of the life cycle of coho salmon (*Oncorhynchus kisutch*), indicating the main developmental stages. (Courtesy: G. K. H. Zupanc. Based on Childerhose, R. J. and Trim, M. (1979). *Pacific Salmon and Steelhead Trout.* Douglas & McIntyre, Vancouver; and Quinn, T. P. and Dittman, A. H. (1990). Pacific salmon migrations and homing: Mechanisms and adaptive significance. *Trends in Ecology and Evolution* 5:174–177. Reprinted by permission of Elsevier. © 1990 Elsevier.)

Both partners, by then exhausted, drift down the river and die.

During the time that follows, the gravel provides protection for the eggs. One to two months after egg deposition, the larvae hatch. The hatchlings are called **alevin** or **sac fry**. They remain buried in the gravel for another four months. During that time, they are nourished by the yolk sac.

After the yolk sac is absorbed, the alevin emerge from the gravel. However, only approximately 1% of coho salmon survive past the larval stage. The survivors, called **fry**, establish territories and feed on plankton and small insects. After another few months, now at the end of their first year of life, the fry develop camouflaged vertical stripes on the flanks of their bodies. They are now called **parr**. The parr remain in the natal stream until they are 18 months old. During that time, they reach a total length of approximately 10cm.

At the end of this phase, the young salmon (now referred to as **smolts**) give up their territories, form large schools consisting of thousands of individuals, and swim down the river toward the sea. Concurrently, they undergo a major metamorphosis, most commonly called **smolt transformation**. This transformation is induced by thyroid hormones and encompasses, among other features, disappearance of the parr marks and change of their coloration to silver, as well as an adjustment of their osmoregulation to life in seawater.

In the sea, the salmon grow rapidly. After only 18 months in the ocean, they reach a length of up to 90cm and weigh up to 7kg. During the **sea phase**, they may travel over enormous distances. Some salmon have been found in the sea more than 5000km away from the mouth of their home river. To illustrate the size of such feeding ranges, Fig. 11.24 shows the marine distribution of another well-examined salmonid species, the sockeye salmon (*Oncorhynchus nerka*).

The spawning migration begins in midsummer of the third year and involves an extensive journey from the open water into shore areas near the home-river system. Upon arrival near the coast, and for the next month or so, both re-adaptation of the salmon's osmoregulation to freshwater and full sexual maturation take place. While in early June the gonadal mass is still very small, typically comprising only 1–2% of the total body weight, by the end of August the gonads have grown tremendously, so that they contribute roughly 50% to the fish's total body weight. At that time, the fish also stop feeding and mobilize fat reserves.

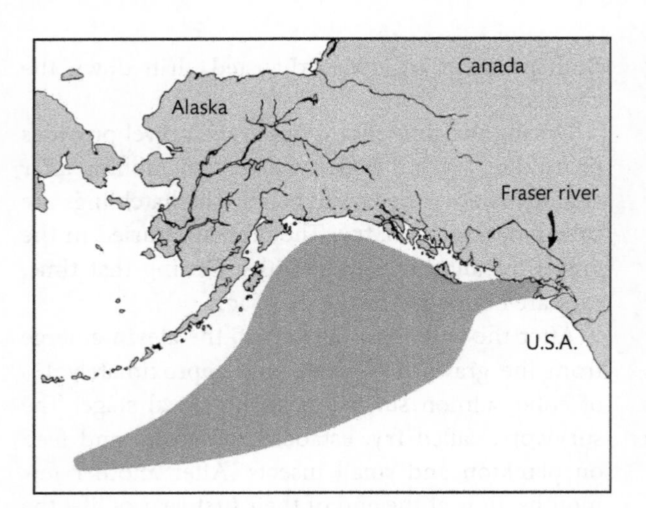

Figure 11.24 Marine distribution of a sockeye salmon (*Oncorhynchus nerka*) population originating from the Fraser River in British Columbia, Canada. The distribution has been inferred from results of tagging experiments and analysis of scale markings, which differ among fish from different watersheds. (After: Quinn, T. P. and Dittman, A. H. (1990).)

As the fish migrate upstream, they develop secondary sexual characteristics, such as brightly colored sides and hooked upper jaws. Moreover, their gastrointestinal tract is now completely absorbed and the entire body cavity, except for the heart and kidneys, is filled with ripe eggs or testes. After spawning, the adults die and leave it up to the offspring to repeat the cycle.

The above life cycle is characteristic of coho salmon. In other salmonid fishes, the pattern of migration and reproduction may vary considerably. In some salmonids, for example the Atlantic salmon (*Salmo salar*), the males often spawn before going to sea, but then, after a phase of significant growth, return to the home river to reproduce once more. In masu salmon (*Oncorhynchus masou*), males remain in freshwater, where they spawn, throughout their life. Only the females show the characteristic **anadromous** pattern, that is, the migration from seawater to freshwater for the purpose of reproduction. Yet other species spend most of the time of their sea phase in coastal waters near the mouth of their natal river, rather than migrating thousands of kilometers to oceanic feeding grounds. Moreover, salmon can even complete their life cycle without ever going to sea. This was artificially forced in a population of coho salmon by introducing them into the Great Lakes in the late 1960s. These salmon have accepted the freshwater lakes as a substitute for a marine environment during their adult life. However, similar to their Pacific conspecifics, they return to their natal stream to spawn.

➤ Salmon are anadromous migrants: Born in freshwater, they grow to adulthood in the sea and return to freshwater to reproduce.

Precision of homing

The precision of the salmon's homing ability has been studied by capturing the fish in their natural tributary before the seaward migration. The salmon are then marked with fin chips, external tags, or magnetic wire implants, and released back into the river.

The results of a large number of such studies are in remarkable agreement. They indicate that, because of high mortality in the ocean, only about 0.5–5% of the original downstream migrants survive to spawn. Of these survivors, roughly 95% return to their natal streams. The remaining 5% or so stray and spawn in non-natal rivers. It is thought that straying ensures colonization of new habitats. Straying can also occur as an avoidance response to degradation in water quality, as was observed when volcanic ash contaminated streams following the eruption of Mount St. Helens, in the state of Washington, in 1980.

Remarkably, **meta-analysis** of the data of several independent studies has indicated that the rate of successful homing increases (while the rate of straying decreases) with population abundance. It is thought that anadromous salmon use **collective navigation** during their homeward migration. This hypothesis is also supported by simulations based on a simplistic theoretical model for collective binary decisions—similar as the fish face at each river confluence when they have to find the correct branch for their ascending journey. These simulations have shown that a collective strategy improves the capability of the salmon to return to their natal sites. It is unknown how the collective behavior is coordinated among the individuals.

Meta-analysis Quantitative statistical analysis of several separate, but conceptually similar, studies. The outcome may be a more precise estimate of an effect examined in these previous studies, or the assessment of an effect not done in any of the previous investigations.

Transplantation experiments

To examine whether the salmon's homing ability is based on inherited or learned components, a series of transplantation experiments have been performed. In these studies, the fish were transplanted from their natal tributary to a different river. The results of these experiments can be summarized as follows:

- Salmon transplanted before undergoing smolt transformation return to the river of release.
- Salmon transplanted after smolt transformation return to their natal tributary.
- The apparent learning of certain features of the home river takes place rapidly and irreversibly, and thus resembles the process of **filial imprinting** known particularly in birds (see following subsections).

The olfactory imprinting hypothesis

What are the cues of the home river that salmon become imprinted to? A breakthrough in answering this question was made in 1951 when Arthur Davis Hasler (see Box 11.1), a pioneer in the research on homing of salmon, and his student Warren Wisby, of the University of Wisconsin in Madison, presented their **olfactory imprinting hypothesis**. The core element of this hypothesis is the proposal that, in a sensitive period during the smolt phase, salmon become imprinted to the odor of their home river and during adulthood use this characteristic odor to find their way back home. Today, it is thought that during the homeward migration the detection of the imprinted odor is combined with a positive **rheotactic response** (cf. Chapter 4), i.e. an upstream swimming of the fish.

Olfactory imprinting hypothesis The notion that salmon become imprinted to the odor of their home river during the smolt phase; after reaching sexual maturity, they use this information to find their way back home.

In formulating this hypothesis, Hasler was inspired by the work of two Austrian scientists working in Germany, Karl von Frisch and Konrad Lorenz. Von Frisch had discovered that the skin of schooling minnows, if injured by a predator, releases tiny amounts of a specific chemical substance called ***Schreckstoff*** (alarm substance). This substance is contained in alarm cells in the skin of the fish (Fig. 11.25). Release of *Schreckstoff*, which occurs only if the skin is broken, causes the other members of the school to disperse and hide. As this substance is present in the water after the injury only in minute amounts, von Frisch's finding suggested that fish must be able to smell with high sensitivity.

This enormous sensitivity was subsequently confirmed by the German zoologist Harald Teichmann. He showed that European eels (*Anguilla*

Box 11.1 Arthur Davis Hasler

Arthur Davis Hasler (right) and his student Peter Hirsch (left) performing a physiological experiment on a salmon. (Courtesy: F. Albert.)

Arthur Davis Hasler, although best known for his research on the homing of salmon, made major scientific contributions in a variety of biological disciplines. Hasler was born in Lehi, Utah, in 1908. After completing his undergraduate degree from Brigham Young University, he attended graduate school at the University of Wisconsin, Madison. Although determined to become a limnologist, he decided first to obtain training in a more rigorous subject, physiology, as limnology at that time was largely confined to descriptive studies. In his thesis, he investigated the physiology of digestive enzymes of copepods and cladocerans. In 1937, he received his Ph.D., and in the same year he joined the faculty of his *alma mater*, as an instructor. By 1948, he had reached the rank of full professor. During his tenure, he established the Laboratory of Limnology in Madison, of which he was the director until his retirement in 1978.

Based on his own roots—his grandfather had immigrated to the U.S.A. from Switzerland—Hasler maintained close ties to Europe, especially to scientists in Germany. Among them, he was mostly inspired by the work of Karl von Frisch, whom he met for the first time in 1945, immediately after the end of the Second World War, in his capacity as an officer of the United States Air Force Strategic Bombing Survey, in Germany. These interactions led to the establishment of the study of sensory physiology of fish as one major area of his own research. The second area where he gained international reputation was limnology. He was instrumental in the shift from the classical, descriptive approach toward a new experimental orientation, in which the rigor of the experiments in the laboratory was applied to research in the field. However, most of his effort was devoted to elucidation of the mechanisms underlying homing in salmon (see text). Besides his academic achievements—more than 100 doctoral and master's degree students graduated from his school—Hasler also became widely known for his initiatives in conservation ecology and international peace programs.

Hasler died in 2001 at the age of 93.

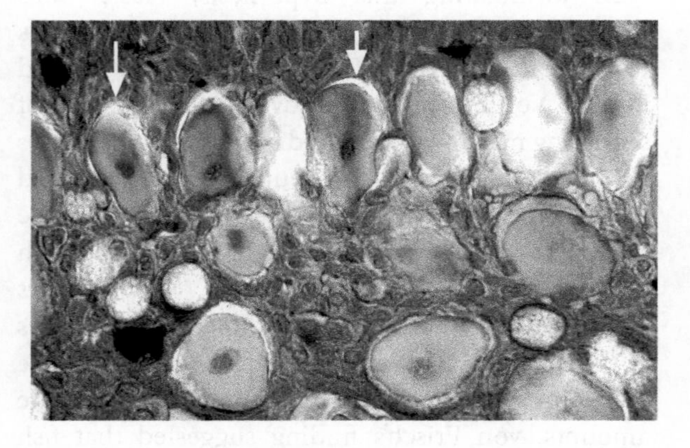

Figure 11.25 Two club-like cells containing *Schreckstoff* in the skin of tench (*Tinca tinca*), indicated by arrows. This alarm substance is released only if the skin is broken after injury, for example, caused by the bite of a predator. (Courtesy: G. K. H. Zupanc and H. Altner.)

anguilla) can detect β-phenylethyl alcohol in concentrations as low as 1 part alcohol in 3×10^{18} parts of water! This corresponds to only three molecules of β-phenylethyl alcohol in the nostril of an eel. This enormous sensitivity is partly achieved by much folding of the olfactory epithelium, which leads to the accommodation of a large number of olfactory cells in the nasal cavity of the fish (Fig. 11.26).

The second prominent scientist who had a profound impact on the work of Arthur Hasler and his group was Konrad Lorenz. By investigating geese, he found a process of rapid and irreversible learning during a critical period of development. Lorenz called this learning process **imprinting**. In geese, the critical period occurs shortly after birth. The gosling forms a permanent attachment to the first

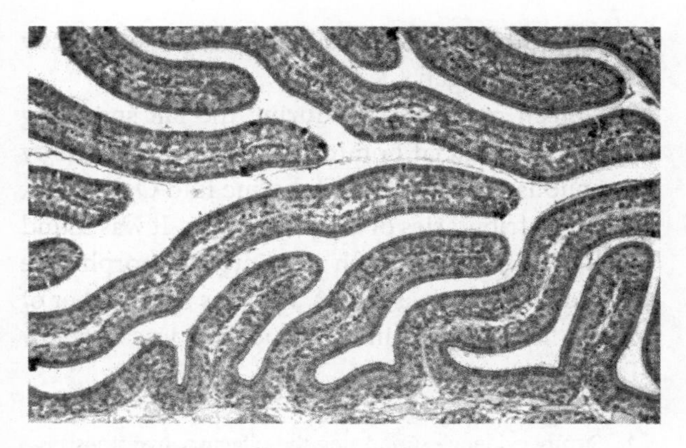

Figure 11.26 Section through the nasal cavity of a blackmouth catshark (*Galeus melastomus*). Accommodation of the numerous olfactory cells is achieved by an enormous folding of the olfactory epithelium. (Courtesy: G. K. H. Zupanc and H. Altner.)

moving object seen—normally its mother, but in an experimental situation, it can be almost any object, including a human observer.

Hasler's merit was to have the idea that imprinting, usually involving visual or acoustic sensory channels, could also take place by making use of olfactory cues. There are three prerequisites for this phenomenon:

1 Every stream has a characteristic and persistent odor that the fish can perceive.
2 The salmon are able to discriminate between the odors of different streams.
3 The salmon remember the typical odor of the home stream when they return to freshwater after the sea phase.

Although Arthur Hasler formulated the olfactory imprinting hypothesis more than half a century ago, it is still unclear what chemical substances in the water cause the predicted characteristic odor of each natal stream. Hans Nordeng (1918–2010) of the University of Oslo, Norway, suggested in the 1970s that during the downstream migration smolts release a **population-specific pheromone**, which guides the upstream swimming adult fish to their natal sites. However, this hypothesis has received limited support. The majority of investigators believe that the unique odor of the natal stream is **produced by the local environment**. There is some indication that dissolved free amino acids, presumably produced by microbial communities, play a role in in this process.

Laboratory experiments

As a first step to verify the olfactory imprinting hypothesis, and by using a reward/punishment paradigm, Hasler and Wisby trained bluntnose minnows (*Pimephales notatus*) and coho salmon to discriminate between waters collected from two streams. These **conditioning experiments** showed that the fish could distinguish between different water samples, even if the water was collected in different seasons. The latter is an important point, as it suggests that the factor detected by the fish is long-lasting and remains present throughout the year. However, when the nasal sacs were cauterized, the fish were no longer able to discriminate between different water samples. Also, when the organic fraction of the water was removed, the fish failed to discriminate different waters.

Electrophysiological studies have, in general, confirmed the conclusion that salmon can distinguish their home water from other waters. In these experiments, migrating adult salmon are captured after arriving in their natal tributary. **Electroencephalograms** are taken by inserting a recording electrode into the olfactory bulb. The nasal sacs are then flushed with water from the stream in which the fish are captured and with water from other streams. Typically, high-intensity electroencephalograms are elicited by presentation of home-stream water, but not by water from other streams.

Imprinting to artificial substances

Strong evidence that supports the hypothesis formulated by Hasler and Wisby comes from experiments with artificially imprinted salmon. In the first set of such experiments, Hasler's group used morpholine (Fig. 11.27). This heterocyclic amine is not found in natural waters and acts neither as a deterrent nor as an attractant. Furthermore, it is cheap, highly stable in the natural environment, extremely soluble in water, and can be detected by coho salmon even at very low concentrations.

For the experiments, salmon from the population artificially introduced to Lake Michigan were used. Smolts of identical genetic stock, hatched and raised under uniform conditions, were divided into two groups. One group was exposed for 30 days to water

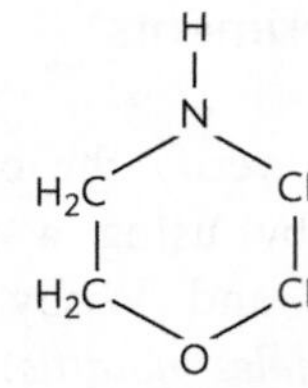

Morpholine

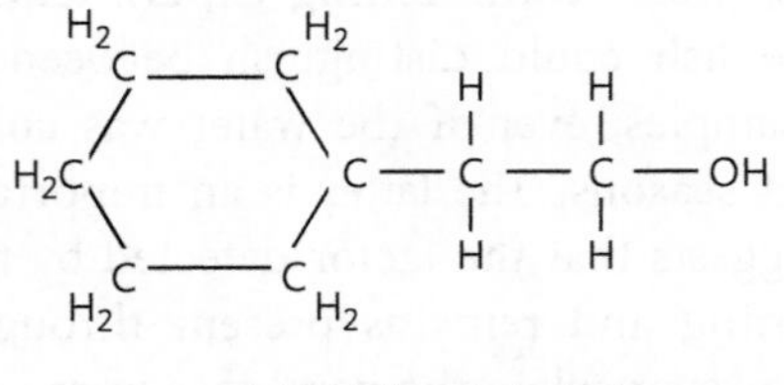

β-Phenylethyl alcohol

Figure 11.27 Chemical formulae of morpholine and ß-phenylethyl alcohol, two chemical substances widely used for artificial olfactory imprinting of salmon.

piped from Lake Michigan containing morpholine at a very low but perceptible concentration. The second group was also kept in Lake Michigan water for the same length of time, but no morpholine was added to this water. This latter group served as a control. Neither of the two groups had been exposed to water from any tributary of Lake Michigan during their early life history. The fish from each group were marked with different fin clips and released directly into Lake Michigan.

Eighteen months later, during the fall spawning season, an artificial homing stream was created by continuous addition of morpholine into Oak Creek, one of the tributaries of Lake Michigan. It was found that a large number of fish imprinted to morpholine entered Oak Creek as adults, whereas the number of control fish captured in the same river was very low (Table 11.1).

➤ Strong evidence supporting the olfactory hypothesis is provided by experiments in which salmon are imprinted to artificial substances.

In an additional control experiment, smolts were exposed to morpholine, and an equal number were left unexposed. In contrast to the previous series of experiments, no morpholine was added to Oak Creek or any other tributary of Lake Michigan during the fall of the following year. This time, only relatively few salmon were captured in Oak Creek, and the number was very similar in the two groups (Table 11.1). This underlines that it is morpholine, and not any other feature of Oak Creek, that attracts the imprinted salmon during adulthood.

Table 11.1 Census record of coho salmon at Oak Creek

Experiment (census year)	Treatment and number of salmon released	Morpholine added to Oak Creek	Number of salmon recovered	Percentage of fish stock
1 (1972)	Imprinted: 8000	Yes	218	2.73
	Not imprinted: 8000		28	0.35
2 (1973)	Imprinted: 5000	Yes	437	8.74
	Not imprinted: 5000		49	0.98
3 (1973)	Imprinted: 5000	Yes	439	8.78
	Not imprinted: 5000		55	1.10
4 (1973)	Imprinted: 8000	Yes	647	8.09
	Not imprinted: 10 000		65	0.65
5 (1974)	Imprinted: 5000	No	51	1.02
	Not imprinted: 5000		55	1.10

After: Hasler, A. D. and Scholz, A. T. (1983).

In a second, more refined artificial imprinting experiment, coho smolts were divided into three equally large groups. The first group, held at a hatchery, was exposed to morpholine, the second to β-phenylethyl alcohol, and the third was left unexposed. Fish of all three groups were released into Lake Michigan, midway between the mouths of two test streams, the Little Manitowoc River and Two Rivers, located 9.4km apart.

During the following year, morpholine was metered into one of the two test streams and β-phenylethyl alcohol into the other. Both streams and an additional 17 other locations were surveyed for marked fish. This experiment was conducted twice, using a total of 45 000 fish. The results are compelling: Of the morpholine-exposed fish recovered, 95% were captured in the morpholine-scented stream; of the β-phenylethyl alcohol-exposed fish recovered, 92.5% were captured in the β-phenylethyl alcohol-scented stream. By contrast, the number of untreated (control) fish was much lower than the number of correctly homing salmon in each of the two test streams.

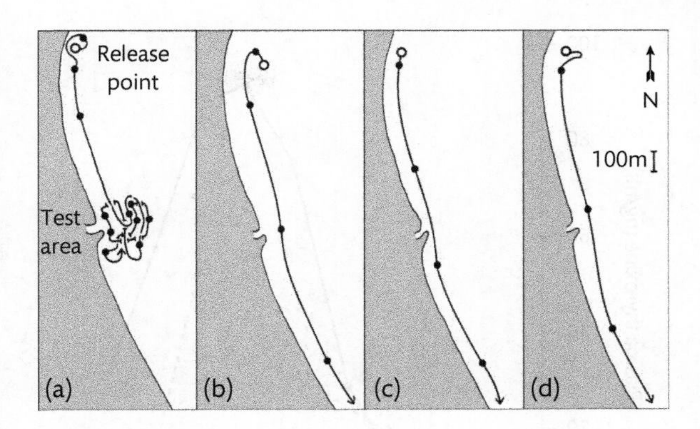

Figure 11.28 Tracking the movements of coho salmon in a test area when encountering the scent of morpholine previously used for imprinting. The movements of the salmon were tracked by ultrasonic transmitters inserted into their stomach. Recordings were taken at 15min intervals. (a) Response of an imprinted salmon when morpholine was present in the test area. (b) Response of an imprinted salmon when morpholine was absent. (c) Response of an imprinted salmon when a chemical other than morpholine was present. (d) Response of a non-imprinted salmon when morpholine was present. (After: Hasler, A. D. and Scholz, A. T. (1983).)

Ultrasonic tracking

To monitor the behavior of morpholine-imprinted salmon when they encounter the morpholine scent during adulthood, the following experiment was conducted. Adult homing salmon that had been exposed to morpholine as smolts were captured in the course of the experiments described above. Before releasing them, an **ultrasonic transmitter** was inserted down their esophagus into their stomach. Each individual was then tracked with a receiver on a boat that followed the signal emitted by the fish. Morpholine was introduced into an area located several hundred meters south of the release point. This area was located at the mouth of a small stream flowing into Lake Michigan. Through the water flow of the river, an 'odor barrier,' extending from the mouth of the river to about 100m offshore, was created. Thus, fish following the shoreline in a southerly direction had to swim through this area.

When morpholine was present in the test area, morpholine-imprinted fish stopped their migration and remained in the area for roughly the time it took for the water currents to dissipate the chemical (Fig. 11.28a). When morpholine was not introduced into the test area, morpholine-treated fish swam through without stopping (Fig. 11.28b). A similar behavior was observed when a chemical other than morpholine was added to the water (Fig. 11.28c), or when non-imprinted salmon were tested in the morpholine-scented area (Fig. 11.28d).

Hormonal regulation of olfactory imprinting

Which neuronal and endocrine factors mediate olfactory imprinting in salmon? Measurements of **thyroid hormones** in the blood serum by radioimmunoassay have shown that triiodothyronine and thyroxine increase roughly five- to tenfold at the beginning of the smolt transformation, compared with the level during the pre-smolt and post-smolt stages (Fig. 11.29). This, as well as other observations, suggests that the level of thyroid hormones may play a role in the process of olfactory imprinting.

This hypothesis has been confirmed by Allan Scholz, a student and co-worker of Arthur Hasler. Scholz artificially elevated the levels of triiodothyronine and thyroxine of pre-smolt coho salmon by injection of thyroid-stimulating hormone (TSH). The fish were then injected with gonadotropic

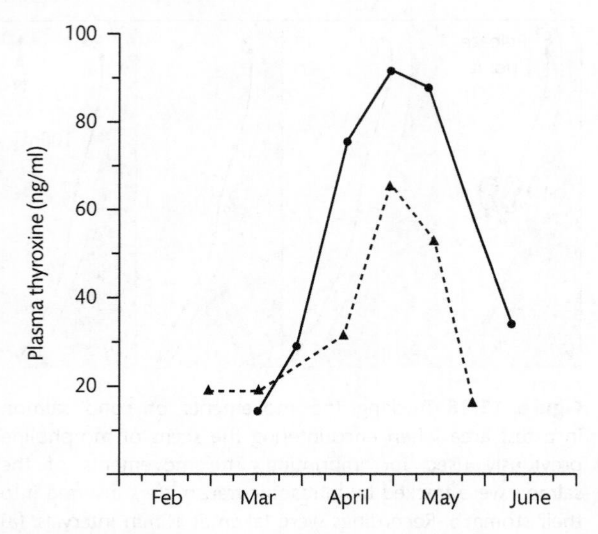

Figure 11.29 Plasma concentrations of thyroxine of coho salmon from two hatcheries (indicated by circles/solid lines and triangles/dashed lines, respectively) during smolt transformation. Each symbol represents the mean of ten samples. The thyroxine surge occurs just prior to smolt transformation. (After: Dickhoff, W. W., Folmar, L. C., and Gorbman, A. (1978).)

hormone for 12 weeks to bring them into a migratory disposition and to mimic the physiological state of naturally spawning salmon. The behavioral tests were conducted in the Ahnapee River at Forestville, Wisconsin. The fish were released 150m before the junction of two tributaries (Fig. 11.30). The water of the two tributaries arises from the same reservoir further upstream. The two tributaries are thus very similar in terms of their natural olfactory composition. This arrangement allowed Scholz to introduce morpholine or β-phenylethyl alcohol into either arm of the river, while keeping all other factors constant.

The outcome of the experiments demonstrated that pre-smolts treated with TSH and simultaneously exposed to an artificial odor were able to select the tributary with the correct odor when in migratory condition. In other words, they had become imprinted. Salmon injected with saline instead of TSH, or uninjected fish, swam downstream instead of upstream and did not select either of the two tributaries. This downstream behavior is typical of fish that do not become imprinted, or of imprinted fish if the correct odor is not present in either tributary. Therefore, TSH injections appear to mimic

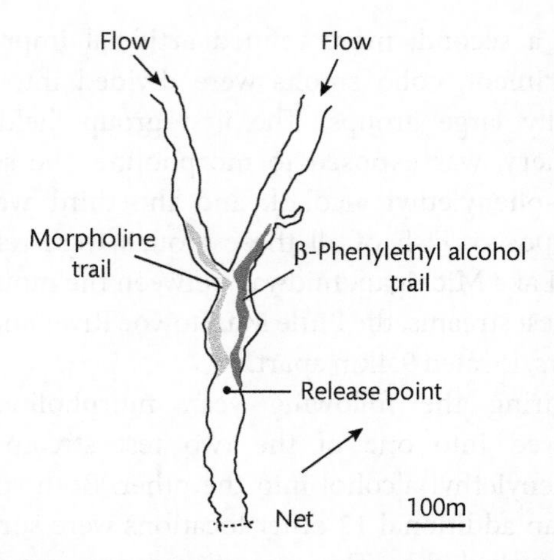

Figure 11.30 Study site on the Ahnapee River at Forestville, Wisconsin, used to test the response of salmon to imprinting substances. The experimental fish were released 150m below the junction of the two tributaries. Addition of morpholine and β-phenylethyl alcohol into either tributary produced odor trails traveling along opposite shores for some distance below the confluence. (After: Hasler, A. D. and Scholz, A. T. (1983).)

the events that, under natural conditions, activate olfactory imprinting in smolts.

The model

In wild salmon, a complex interaction between thyroid hormones, migratory activity, and imprinting appears to take place. It has been shown that changes in thyroid hormone levels not only occur during certain stages of development, but are also caused by a number of environmental factors. For example, exposure to new environments, changes in water temperature or water flow rate, photoperiod, and lunar phase contribute to increases in thyroid hormone levels. On the other hand, artificially elevated thyroxine levels induce migration in salmon. It has therefore been hypothesized that developmentally and environmentally induced increases in thyroid hormone levels cause the salmon to migrate during smolt transformation. The migratory activity, in turn, exposes the fish to new environments and this leads to further rises in thyroid hormone levels. The resulting thyroid hormone surges increase the tendency to learn odors through the process of imprinting (Fig. 11.31).

The appealing feature of this model is that migration itself is involved in the control of

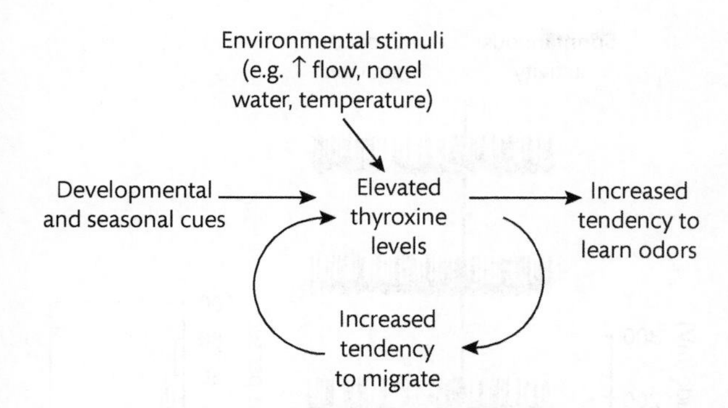

Figure 11.31 Hypothetical relationship between developmental factors, environmental stimuli, migratory behavior, thyroid hormones, and olfactory imprinting. Both developmental and environmental cues stimulate the production of thyroid hormones. This causes an increase in migratory activity, which leads to a further elevation of thyroid hormone levels. The resulting thyroid hormone surges exert, in turn, a positive effect on the tendency to learn odors. (After: Dittman, A. W. and Quinn, T. P. (1996). Homing in Pacific salmon: Mechanisms and ecological basis. *Journal of Experimental Biology* **199**:83–91. Reprinted by permission of the Company of Biologists, Ltd.)

imprinting. This could enable the salmon to learn a series of olfactory intermediary points, rather than just a single odor. This is known as the **sequential imprinting hypothesis**. As adults, they could trace this odor sequence to find the correct natal stream.

Sequential imprinting hypothesis The notion that salmon learn a series of olfactory intermediary points in the course of their downstream migration.

Open-sea navigation: Sun-compass orientation

While the olfactory imprinting hypothesis, in combination with positive rheotaxis behavior, explains how adult salmon find their way from coastal areas to their natal stream, it cannot account for how they orient in the open sea. Hasler proposed **sun-compass orientation** as one possible mechanism. Together with Wolfgang Braemer and Horst Schwassmann, two German behavioral physiologists who had joined his laboratory, he demonstrated the use of the sun for orientation in several freshwater fish. However, direct evidence for such a mechanism mediating open-sea migration in a salmonid species is still lacking.

➤ During the open-sea phase, salmon may use a sun compass and magnetic information for navigation.

Open-sea navigation: Magnetoreception

A second mechanism possibly used by salmon for long-distance navigation is based on a **magnetic sense.** A seminal discovery supporting this hypothesis was reported by the laboratory of Michael Walker of the University of Auckland, New Zealand, in 1997. For their study, Walker and associates examined rainbow trout (*Oncorhynchus mykiss*), a salmonid species. The fish were trained to discriminate between the presence and absence of a magnetic anomaly induced by a Helmholtz coil and superimposed on the background of the Earth's magnetic field. After a short training period, the fish responded consistently differently to the reinforced stimulus (which could be either the presence or absence of the magnetic anomaly) compared with the non-reinforced stimulus (Fig. 11.32). This demonstrated that rainbow trout can discriminate different magnetic fields.

As possible structural correlates of this magnetic sense, cells located near the basal lamina of the

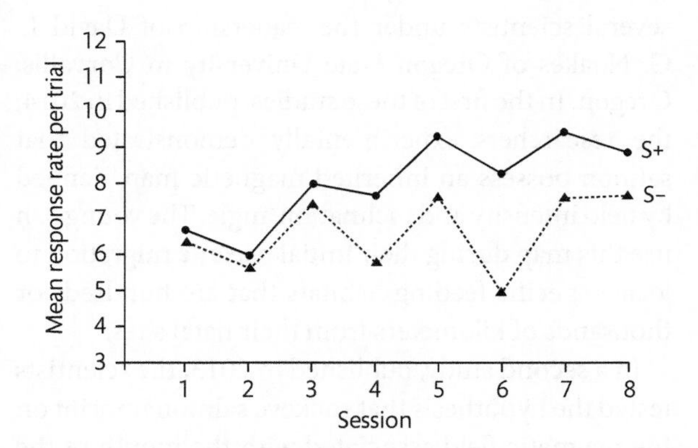

Figure 11.32 Results of magnetic discrimination experiments performed on rainbow trout (*Oncorhynchus mykiss*). Employing a reinforced stimulus (S+) versus non-reinforced stimulus (S–) paradigm, the fish were trained to distinguish between the presence and absence of a magnetic anomaly superimposed on the Earth's magnetic field. The response measured was the rate at which the fish struck a target in anticipation of a food reward, or of lack of reinforcement, at the end of each trial. Five S+ and S– trials were given in a balanced quasi-random order in each training session. The data shown are the mean response rate per trial exhibited by these fish. (After: Walker, M. M., Diebel, C. E., Haugh, C. V., Pankhurst, P. M., Montgomery, J. C., and Green, C. R. (1997).)

olfactory epithelium have been identified. These cells contain iron-rich crystals of approximately 50nm length, resembling the chains of magnetite present in magnetotactic bacteria. These presumptive **magnetoreceptive cells** are innervated by fine processes of the superficial ophthalmic ramus (also referred to as the ramus ophthalmicus superficialis, or **ros**) of the trigeminal nerve. As the trigeminal nerve is the fifth cranial nerve and is indicated by the roman number 'V,' this ramus is called **ros V**. The cell bodies of this nerve, together with other somata, make up the anterior ganglion in the brain.

Stimulation of the fish with various magnetic fields (differing in direction, intensity, or both) and electrophysiological recordings from ros V have revealed units that respond to changes in the magnetic field presented. For example, addition of a magnetic field intensity of 50μT (microtesla) to the background field of 25μT, without changing the field direction, elicits excitatory responses immediately following the onset of the stimulus from certain units (Fig. 11.33).

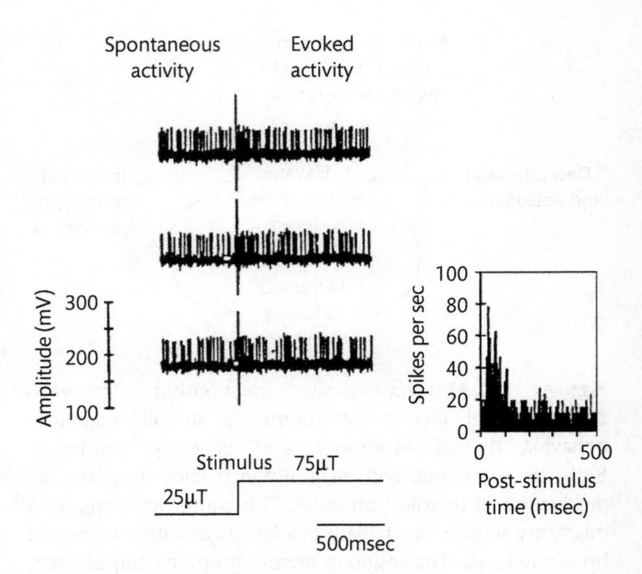

Figure 11.33 Electrophysiological responses to a step-like increase in magnetic field intensity from 25 to 75μT, as indicated at the bottom on the left side. The three traces show the peri-stimulus activity of a single unit in ros V of the rainbow trout. Before the onset of the stimulus, the unit is spontaneously active in the background magnetic field. After the onset of the stimulus, the traces show the activity of the unit for 1sec. After the stimulus step, the firing rate increases for the first 100msec. This is also evident from the post-stimulus time histogram of the responses exhibited by the same unit to 128 presentations of the stimulus (right). (After: Walker, M. M., Diebel, C. E., Haugh, C. V., Pankhurst, P. M., Montgomery, J. C., and Green, C. R. (1997).)

Evidence that salmonids indeed use magnetic cues during their migratory phases in the open ocean has been obtained through two studies conducted by several scientists under the leadership of David L. G. Noakes of Oregon State University in Corvallis, Oregon. In the first of these studies, published in 2014, the researchers experimentally demonstrated that salmon possess an inherited magnetic map, defined by field intensity and inclination angle. The young fish use this map during their initial oceanic migration to locate specific feeding habitats that are hundreds or thousands of kilometers from their natal sites.

In a second study, published in 2013, the scientists tested the hypothesis that sockeye salmon imprint on the magnetic field associated with the mouth of the Fraser River when they enter the sea. The correlative data suggest that the specific route—either a northern or a southern passage around Vancouver Island—taken by the salmon when they return to freshwater after the open-ocean phase can be predicted by geomagnetic field drifts as they occur over the years. Besides its significance for the understanding of the navigation mechanism of migrating salmon, the proposed geomagnetic model might also be useful for fisheries to forecast the movements of salmon.

Summary

- Many migratory animals travel, in annual cycles, over extremely long distances between their breeding grounds and the non-breeding ranges. During these large-scale migrations, the entire species, or subpopulations of it, use specific routes.
- In some species, the annual migratory cycle is completed not by one individual, but by several generations instead.
- Crossbreeding experiments using populations of birds differing in their migratory behavior have shown that the urge to migrate, as well as

the migratory direction, is under direct genetic control.

- Some animals exhibit a remarkable homing ability, which enables them to return home after foraging excursions or (natural or experimental) displacement.
- Migratory behavior and homing can be studied in the field using a variety of methods, including analysis of distribution data and tagging, as well as by use of radar and microtransmitters.
- In birds, several mechanisms have been shown, and others are likely, to be involved in navigation. They include the use of a time-compensated sun compass in diurnal migrants and homing pigeons; and use of a star compass in nocturnal migrants. Olfactory cues have been hypothesized to play a role in the homing of pigeons reared under certain conditions.
- Some birds, including homing pigeons, are also able to detect cues arising from Earth's geomagnetic field, presumably through specific magnetoreceptors. One of these proposed receptor types is associated with photoreceptors and based on a radical-pair mechanism, while the other is photoreceptor independent and thought to be based on magnetite. Electrophysiological recordings have identified neurons that respond, with high sensitivity, to changes in the ambient magnetic field. Magnetite-based magnetoreceptors could provide directional information (compass function), whereas photoreceptor-associated magnetoreceptors could mediate analysis of local information in relation to the bird's destination (map function).
- Equally well examined as the homing of pigeons is the ability of salmon to return as adults to their natal stream after extended periods in the sea. One of the best-examined species is the coho salmon. Its life cycle consists of the following stages. The eggs are laid in upstream areas of rivers. The hatching alevins, as well as the later emerging parrs, remain in the natal stream until the fish are approximately 18 months old. They then transform into smolts that swim down the river toward the sea. In the sea, the salmon grow until, after another 18 months, when they reach adulthood, they migrate back to their natal stream to reproduce.
- Tagging experiments have shown that approximately 95% of the salmon that survive into adulthood return to their natal stream. The remaining 5% stray and spawn in non-natal rivers. The rate of successful homing increases with population abundance.
- Experiments pioneered by the laboratory of Arthur Davis Hasler have suggested that the homing ability of salmon is based on imprinting to olfactory cues of the natal stream during smolt transformation. Strong evidence in favor of this hypothesis has been provided by imprinting of salmon to artificial chemicals, such as morpholine and β-phenylethyl alcohol.
- Smolt transformation, and thus olfactory imprinting, is under the control of thyroid hormones. As these hormones also induce migration, smolting salmon learn, within a rather short sensitive period, a series of olfactory intermediary points. As adults, they trace this odor sequence to find the correct natal stream. At the neuronal level, the process of imprinting may be mediated by thyroid-hormone controlled postnatal neurogenesis in the olfactory epithelium.
- Correlative and experimental evidence suggests that salmon use magnetic cues for navigation during the open sea migrations. Possible structural correlates of this putative magnetic sense are magnetite-containing cells in the olfactory epithelium. These cells are linked to the brain via processes of the superficial ophthalmic ramus of the trigeminal nerve. Electrophysiological recordings from this nerve have revealed responses to changes in the magnetic field.

The bigger picture

Migration and homing of some animals have been known to humans for a long time, yet the scientific study of these behaviors and the underlying sensory and neural mechanisms has remained challenging. Fair progress has been made at the behavioral level, thanks in part to the technological advances made in the tracking of animals, and in part to the possibility of examining some key components of the behavior under the controlled conditions of the laboratory. Less is known about the sense organs involved in the navigational tasks accomplished by migrating and homing animals. This difficulty is largely caused by the multitude of sensory modalities that contribute to long-distance orientation, and by the involvement of a novel and still poorly characterized modality—magnetoreception—in this behavior. The significant lack of information on the sensory side has also hampered progress at the neurobiological frontier, although the demonstration of neurons in the brain that respond selectively to magnetic cues is encouraging. Nevertheless, a comprehensive understanding of the phenomenon of animal migration and homing will require not only identification and characterization of the missing link in the sensory processing chain, but in the long run also elucidation of the neural mechanisms that integrate the sensory input with other critical components of the orientation system, such as spatial learning and biological clock functions.

Recommended reading

Hasler, A. D. and Scholz, A. T. (1983). *Olfactory Imprinting and Homing in Salmon: Investigations into the Mechanism of the Imprinting Process*. Springer-Verlag, Berlin/Heidelberg/New York/Tokyo.

A review of the classic research conducted by the laboratory of Arthur Hasler.

Quinn, T. P. (2005). *The Behavior and Ecology of Pacific Salmon and Trout*. University of Washington Press, Seattle.

The most authoritative review currently available on the behavior and ecology of pacific salmonids, written by one of the leading researchers in the field. The main focus of this book is on salmon migration and homing, but the author also covers many other aspects of the biology of these fishes in its nearly 400 pages.

Walker, M. M., Diebel, C. E., Haugh, C. V., Pankhurst, P. M., Montgomery, J. C., and Green, C. R. (1997). Structure and function of the vertebrate magnetic sense. *Nature* 390:371–376.

This seminal research article provides an excellent example of how behavioral, anatomical, and physiological experiments can be combined to gain an integrative understanding of an important biological phenomenon.

Wiltschko, R. and Wiltschko, W. (2003). Avian navigation: From historical to modern concepts. *Animal Behaviour* 65:257–272.

In this essay, the authors reflect on how historical models, foremost the 'map-and-compass' model of Gustav Kramer proposed in the 1950s, have influenced the modern concepts of avian navigation.

Wiltschko, R. and Wiltschko, W. (2013). The magnetite-based receptors in the beak of birds and their role in avian navigation. *Journal of Comparative Physiology A* 199:89–98.

A summary of the research on avian magnetoreception systems based on magnetite. A good introduction to the topic, although the authors leave it to the reader to review findings that are not in line with the magnetite-based receptor mechanism located in the beak of birds.

Wiltschko, R. and Wiltschko, W. (2014). Sensing magnetic directions in birds: Radical pair processes involving cryptochrome. *Biosensors* 4:221–242.

A review of the radical-pair model of magnetoreception from a biological point of view, written by two of the pioneers in this field of avian navigation.

Short-answer questions

11.1 List three methods that can be used to infer (directly or indirectly) migratory behavior of an animal species.

11.2 Indicate the correct statement. The monarch butterfly is unusual in that:

a) only the females but not the males migrate;
b) only the males but not the females migrate;
c) the migratory cycle is completed by several generations.

11.3 Experiments in blackcaps have demonstrated that migratory birds

a) inherit from their parents information on the direction to travel;
b) acquire information about the travel direction through learning from older birds.

11.4 Which of the two statements is true? Indicate the correct statement. Homing pigeons have been shown to use the following mechanism for finding the way back to the home loft:

a) only time-compensated sun compass information;
b) only magnetoreception-based compass/map information;
c) both time-compensated sun compass and magnetoreception-based compass/map information.

11.5 A diurnal migratory bird preferentially orients toward the northeast during the migratory season. The bird is then transferred to a light-proof chamber and exposed for 5 days to an artificial day beginning and ending 6 hours earlier than the natural day. When the bird is exposed to natural daylight after this experimental manipulation, in which direction is it expected to preferentially orient under blue-sky conditions? What inferences can be made from this experiment about the mechanism underlying long-distance orientation?

11.6 Sketch the life cycle of salmon. Which developmental stage plays a critical role in forming memories of the natal stream?

11.7 Complete the following sentence. The homing of salmon is thought to be based on imprinting to .... cues (specify sensory modality associated with these cues) of the natal stream during the freshwater phase and .... orientation (indicate specific mechanism) and .... (specify sensory modality) during the open-sea phase.

11.8 Why is it important in homing experiments in which salmon are imprinted on an artificial substance to use a chemical that does not exist in natural waters?

11.9 **Challenge question** Tagging studies have shown that numerous shark species are highly migratory. It is widely thought that the sharks use the Earth's magnetic field for long-distance navigation, but most of the evidence is circumstantial. Suggest a behavioral (laboratory or field) experiment to demonstrate that sharks are capable of sensing magnetic cues.

11.10 **Challenge question** The prehistoric shark genus *Bandringa* is commonly thought to consist of two species, the marine *B. rayi* and the freshwater *B. herdinae*. Reexamination of the fossil material has, however, led to the suggestion that *B. rayi* and *B. herdinae* represent a single species that exhibited seasonal migratory behavior. What evidence would you expect to find to support this hypothesis?

Essay questions

11.1 What sensory cues are involved in the homing of pigeons and in the annual migrations of migratory birds? What role does an internal clock play in this ability?

11.2 Experiments aimed at verifying the involvement of magnetoreception in the navigation of migratory birds and homing pigeons have sometimes been difficult to reproduce. Discuss possible reasons for this difficulty.

11.3 Sketch the life cycle of a salmon. How do the adult fish find their way back to their place of birth? What sensory modalities are involved in homing, and what environmental cues are used for this capability?

11.4 Great scientific discoveries rarely occur in isolation. Illustrate this statement by describing how the ethological work of Konrad Lorenz and the behavioral physiological work of Karl von Frisch influenced the development of the olfactory imprinting hypothesis formulated by the school of Arthur Davis Hasler.

11.5 A magnetic sense likely to exist in many, if not all, salmonids has been demonstrated in rainbow trout. Although it has frequently been suggested that such a sensory capability may be involved in long-distance navigation of salmon during the open-sea stage, experimental evidence supporting this hypothesis is sparse. Propose experiments through which one could verify the use of magnetoreception by salmon for open-sea navigation.

Advanced topic Technological advances have revolutionized the tracking of animals

Background information

Like many other areas of research, the study of animal migration and navigation is highly dependent on advances in other disciplines of science and engineering. Two lines of technological innovation that have had an enormous impact include the miniaturization of transmitter devices and the possibility of satellite tracking. For example, the development of radio transmitters with mass of the circuitry of less than 100mg has enabled investigators to study much smaller animals than was previously possible. Such animals include arthropods, amphibians, and reptiles. Similarly, through satellite telemetry, global monitoring not only of location and speed of movement of an animal, but also of additional physiological and environmental parameters, has become feasible.

Essay topic

In an extended essay, describe how new technological devices help biologists to trace the movements of animals. Illustrate the power of technological development in this area by using examples from different animal taxa.

Starter references

Cohn, J. P. (1999). Tracking wildlife: High-tech devices help biologists trace the movements of animals through sky and sea. *BioScience* **49**:12–17.

Fuller, M. R., Seegar, W. S., and Schueck, L. S. (1998). Routes and travel rates of migrating peregrine falcons *Falco peregrinus* and Swainson's hawks *Buteo swainsoni* in the Western Hemisphere. *Journal of Avian Biology* **29**:433–440.

Naef-Daenzer, B., Früh, D., Stalder, M., Wetli, P., and Weise, E. (2005). Miniaturization (0.2g) and evaluation of attachment techniques of telemetry transmitters. *Journal of Experimental Biology* **208**:4063–4068.

Schaffer, S. A., Tremblay, Y., Weimerskirch, H., Scott, D., Thomson, D. R., Sagar, P. M., Moller, H., Taylor, G. A., Foley, D. G., Block, B. A., and Costa, D. P. (2006). Migratory shearwaters integrate oceanic resources across the Pacific Ocean in an endless summer. *Proceedings of the National Academy of Sciences U.S.A.* **103**:12799–12802.

Wikelski, M., Moskowitz, D., Adelman, J. S., Cochran, J., Wilcove, D. S., and May, M. L. (2006). Simple rules guide dragonfly migration. *Biology Letters* **2**:325–329.

To find answers to the short-answer questions and the essay questions, as well as interactive multiple choice questions and an accompanying Journal Club for this chapter, visit **www.oup.com/uk/zupanc3e**.

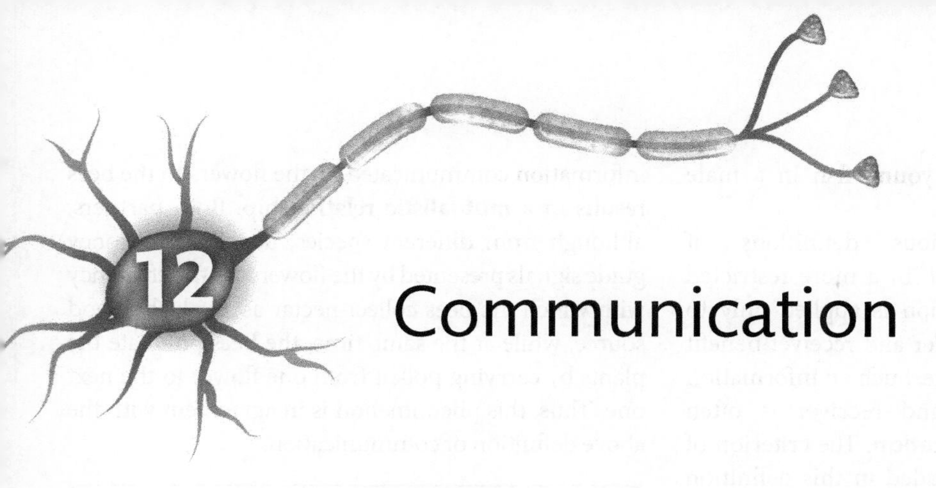

Communication

Introduction

In their book *Principles of Animal Communication*, Jack W. Bradbury and Sandra L. Vehrencamp refer to communication as 'the glue that holds animal societies together'. This picturesque, though adequate, description underscores the importance of communication. It is, therefore, not surprising that legions of ethologists and neuroethologists have put the behavioral analysis of communication systems and the exploration of the underlying neural mechanisms at the center of their research.

In this chapter, we will first review the biological definition of communication and some of the sensory modalities involved in the detection of communication signals. Then, we will focus on two intensively studied model systems: the neuroethology of cricket song and the development of bird song.

Key concepts

What is communication?

In order to communicate, animals need to **exchange information.** The vehicle used to transfer information is called the **signal.** Signals are produced by a **sender** and are transmitted to a **receiver.** Upon detection, the signal invokes a response in the receiver. In many instances, this is an **immediate behavioral response.** However, as it appears to be true especially in cases where a stereotyped social signal is repeatedly broadcast, a **tonic and motivational effect** may also be exerted upon the receiver. An intriguing example of such an effect was revealed through the experiment performed by Walter Heiligenberg and Ursula Kramer described in Chapter 7. Through repeated presentation of a dummy incorporating signals typical of a territorial male conspecific, Heiligenberg and Kramer were able to induce long-term changes in the motivation to attack young fish in a male cichlid fish.

In the literature, various definitions of communication can be found. In a more restricted sense, the term communication is applied only to situations in which both sender and receiver benefit from the information exchange. Such an information exchange between sender and receiver is often referred to as **true communication**. The criterion of mutual benefit has been included in this definition of communication to exclude the quite frequent case where the stimulus produced by an animal is used by a receiver to the detriment of the sender. We discussed one such scenario in Chapter 7: The noise that is inadvertently generated by mice when they move or feed is utilized by owls to localize and eventually capture them. In this case, the provision of the acoustic information benefits the receiver, but certainly not the sender! To distinguish between situations in which true communication takes place, and situations in which an animal inadvertently generates a signal without benefiting from the transfer of the associated information, the term **cue**, instead of signal, is used.

True communication Transfer of information from a sender to a receiver benefiting both partners.

The distinction between cue and signal is particularly important when we consider the evolution of communication systems. Like signals, cues can provide information about the sender. However, in contrast to signals, cues are often by-products of a certain activity or the animal's morphology. Unlike signals, they are *not* shaped by natural selection to convey the information associated with this behavioral activity or morphological pattern so that the behavior of the receiver is influenced, and ultimately the fitness of both sender and receiver is affected in a positive way. On the other hand, cues can evolve into signals by specialization of certain aspects to transmit the information from the sender to the receiver more effectively.

Inclusion of the mutual benefit of the sender and the receiver as a criterion in the definition of communication does not necessarily restrict the information transfer to members of the same species. Foraging bees, for example, are guided to the center of flowers, where the nectar is, by the **honey guides**—the radiating lines on the petals of many flowers (see Chapter 3). This piece of information communicated by the flowers to the bees results in a **mutualistic relationship**: Both partners, although from different species, benefit—the honey guide signals presented by the flowers aid the efficiency with which the bees collect nectar as a valuable food source, while at the same time, the bees pollinate the plants by carrying pollen from one flower to the next one. Thus, this phenomenon is in agreement with the above definition of communication.

Mutualism Two organisms of different species interacting, each benefiting from this relationship.

In any communication system, the transmitted signal has to be designed in such a way that it can be detected and 'understood' by the receiver. Using the conveyed information, the receiver makes a decision about how to respond. As mentioned above, the response of the receiver affects the fitness of both the sender and the receiver in a beneficial way. This is not necessarily the case in every instance, but it happens on average over many interactions between the two individuals. The requirement that the receiver understands the signal produced by the sender implies that there must be a certain degree of agreement between the sender and the receiver.

Sensory modalities involved in communication

Based on the sensory modalities involved in their detection, different types of communication signals are distinguished. Many animals use **visual signals** for communication. For example, the banded jewelfish (*Hemichromis fasciatus*) indicates its readiness to flee or attack through different spotting patterns in its coloration (Fig. 12.1). This is done by moving pigments within melanophores in the skin. Such signals allow an intruder to assess the likelihood of the owner of a nesting site defending its site. This assessment then forms the basis for the intruder to decide whether or not it is economical to attack the owner and try to take over his territory.

Morphological structures involved in the provision of communicatory information are often especially conspicuous and exaggerated far beyond their need for normal function. In male fiddler crabs, one claw is enormously enlarged and frequently brightly colored (Fig. 12.2). To amplify the signal even more,

Figure 12.1 Colorations of the banded jewelfish (*Hemichromis fasciatus*) as indicators of aggressive motivation. The spotting pattern in the top drawing is typical of a neutral fish without a territory. The following three drawings represent fish of increasing aggressiveness. The coloration on the bottom is characteristic of a highly aggressive fish that is very likely to defend its territory against an intruder. (After: Wickler, W. (1978).)

this claw is waved rhythmically during courtship and aggressive encounters with other males.

Acoustic signals have the advantage that they allow the sender not only to quickly start communicating but also to quickly stop sound production. The latter property is particularly important in case a predator is present—in such a situation it could be deadly to continue displaying a conspicuous courtship signal!

Another advantage of acoustic signals is that, in order to convey different messages in different behavioral contexts, they can be varied relatively

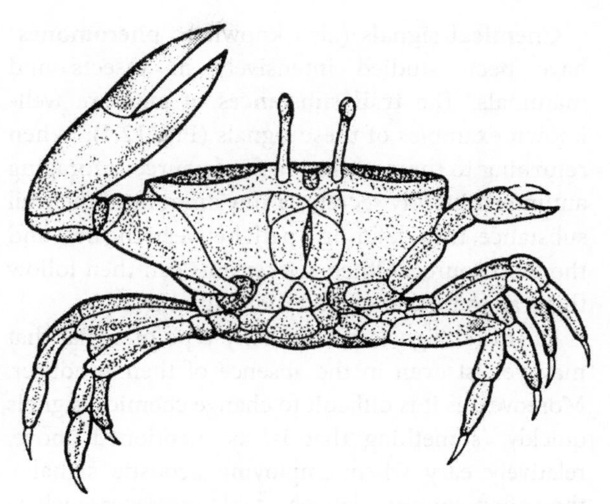

Figure 12.2 Male fiddler crab with one claw enormously enlarged. This claw is used as a signal in courtship and aggressive encounters. (After: Manning, A. and Stamp Dawkins, M. (1992).)

easily. Common ways of modulating acoustic signals involve changes in frequency, amplitude, and temporal structure of the sound produced.

Acoustic signaling has been intensively studied in mammals, birds, amphibians, fish, and insects. Figure 12.3 shows an Australian red-eyed tree frog (*Litoria chloris*) with a fully inflated **vocal sac**. This outpocketing of the buccal cavity serves as a resonator to amplify calls produced by the larynx and the vocal cords.

Acoustic communication in crickets and songbirds will be discussed in detail in the two model systems sections of this chapter.

Figure 12.3 Red-eyed tree frog with vocal sac inflated. (Courtesy: Froggydarb (CC BY-SA 3.0).)

Chemical signals (also known as **pheromones**) have been studied intensively in insects and mammals. The **trail substances** of ants are well-known examples of these signals (Fig. 12.4). When returning to the nest from a food source, a foraging ant intermittently secretes a tiny amount of the trail substance, thus defining a path between the nest and the food source. Other worker ants can then follow this trail to further exploit the food source.

Chemical signals are the only type of signal that may persist even in the absence of their producer. Moreover, as it is difficult to change chemical signals quickly (something that is, as mentioned above, relatively easy when employing acoustic signals), they often convey relatively stable messages, such as the sexual condition of a female or the ownership of a territory. Also, many chemical signals (referred to as **primer pheromones**, in contrast to **releaser pheromones**) produce long-term alterations in the physiological condition of the receiver. They do not necessarily generate any immediate behavioral change. Rather, because they affect the receiver's motivation, the effect of primer pheromones becomes visible as measurable behavior only later, when encountering communication signals. It has been demonstrated, for example, that the estrous cycle of female mice can be synchronized by the odor of a male mouse—even in the absence of this mouse. Similarly, the odor of a strange male mouse can lead to the termination of the pregnancy of a newly impregnated female mouse.

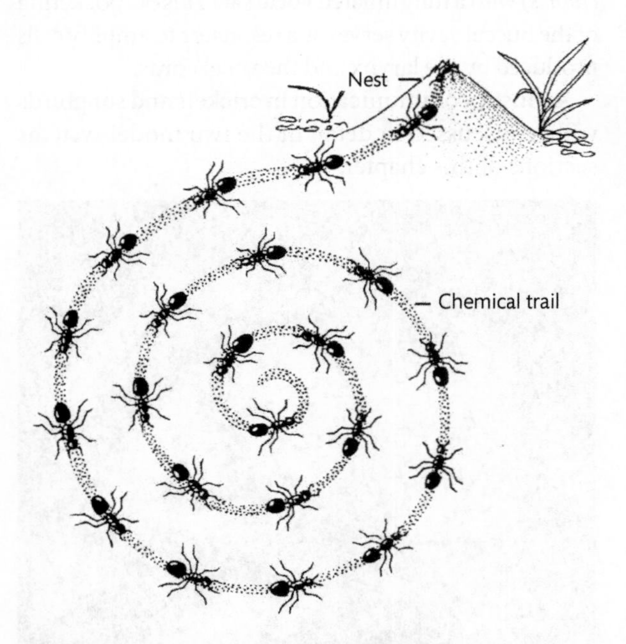

Figure 12.4 Demonstration of the efficacy of a trail substance as a chemical signal. This pheromone was laid down by an experimenter in a spiral-like fashion. Ants encountering this signal follow the spiral path. (After: Keeton, W. T. (1980).)

Tactile signals are commonly transmitted either through a substrate or by touching. Males of web-building spiders generate, as part of their courtship behavior, distinctive substrate-borne vibrations when entering the female's web. These mechanical vibrations are referred to as shudders and are generated by quickly rocking in the web several times. Experiments, in which Anne E. Wignall and Marie E. Herberstein of Macquarie University in Syndney, Australia, studied orb-web spiders (*Argiope keyserlingi*), have shown that these courtship signals delay female predatory behavior, thus reducing the risk of the males facing pre-copulatory cannibalism.

An intriguing example of tactile signals transmitted by touching can be observed in Titi monkeys (genus *Callicebus*). These monogamous South American monkeys live in a group typically consisting of one pair of adults and their offspring. The adult partners coordinate many of their activities and remain within close proximity. The tactile behaviors exchanged by individuals include huddling, grooming, holding hands, and **tail-twining**. The latter behavior is displayed whenever two to four individuals sit side by side. Sometimes, the tails are looped around one another making only one turn. At other occasions, they make several turns, as depicted in Fig. 12.5. It is thought that this tactile signal helps to strengthen friendly social relations.

Communication based on **electric signals** is found in two orders of freshwater fish, the South American Gymnotiformes and the African Mormyriformes. During social interactions, specific aspects of these signals may be modulated. Further information, including figures, on this so-called **electrocommunication** can be found in Chapters 1 and 8.

Finally, signals displayed in communication may also be multimodal. An instructive example is the vocal sac of frogs. Its extension enables males to attract females through vocalization. However, while it is likely that the vocal sac initially evolved under selection for acoustic function, this structure may have been co-opted as a visual cue. Particularly when inflated, the vocal sac differs in color from the surrounding area of the body. Experiments have shown that, in at least some species,

Figure 12.5 Tactile communication in Bolivian Gray Titi Monkeys (*Callicebus donacophilus*). (a) Cuddling family, with three individuals displaying tail twining. (b) The three intertwined tails shown at higher magnification. Courtesy: Anita Yantz.

this cue is used in visual communication, suggesting its evolutionary development into a visual signal.

Model system: The neuroethology of cricket song

The behavior

An important feature of communication is that the information transmitted from a sender to a receiver is specific to a given behavioral situation. This is also true for acoustic communication in crickets. Figure 12.6 shows three different songs of male crickets produced in different behavioral situations. **Calling songs** are generated to attract sexually receptive females. **Courtship songs** entice the female to mate. **Aggressive songs** form part of the aggressive behavior during encounters with other males.

Based on the calling songs of cricket males, and on the taxis response of females, we will, in this section, take a closer look at the auditory communication of crickets—one of the first neuroethological model systems in which a detailed and comprehensive analysis of the neurobiological basis of communication was conducted. We will start with a biophysical analysis of the calling songs and will examine how these signals are produced by the male, and how they are perceived and processed by the female. Finally, we will discuss mechanisms ensuring unambiguous transmission of information in situations when the communication system is challenged by environmentally induced variability.

Biophysics of cricket songs

Like any sound, the songs of crickets can be analyzed by oscilloscopes. The result of such analysis can be displayed in form of an **oscillogram**. Figure 12.7 shows the oscillogram of a **calling song** of a male cricket. On a compressed time scale (Fig. 12.7a), it becomes evident that this type of song consists of brief bursts of pulses, followed by intervals of silence. The individual pulses are often referred to as **syllables**. One burst of syllables is called a **chirp**. On an expanded timescale (Fig. 12.7b), which allows the researcher to resolve the finer details of the sound, the oscillogram reveals that each syllable is comprised of an uninterrupted train of sound waves. These waves vary in amplitude over the duration of the syllable, but are rather constant in frequency. The latter parameter defines the **carrier frequency** of the song and is determined as the reciprocal of the duration of one complete cycle of the wave in seconds. The unit of the carrier frequency is known as Hertz (Hz), where 1Hz denotes one cycle per second.

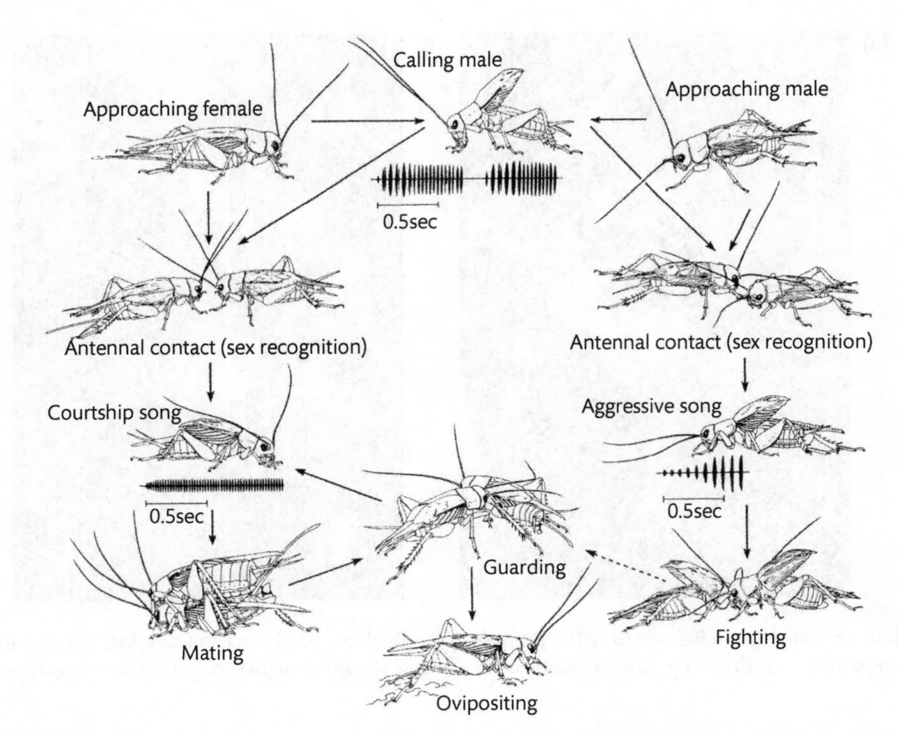

Figure 12.6 Communication in the black field cricket, *Teleogryllus commodus.* Males attract receptive females by producing a calling song. Upon arrival of the female, the two mating partners touch antennae, a behavior involved in sex recognition. The male then generates a courtship song to entice the female to mate. During copulation, the female mounts the male, and the male transfers a small bag of sperm (called the spermatophore) to the end of the female's abdomen. Subsequently, the male exhibits mate guarding, a behavior thought to ensure that the female does not remove and eat the externally attached spermatophore before the sperm are passed into the internal sperm receptacle. The female permits her eggs to be fertilized and, using her long ovipositor, places fertilized eggs onto a proper substrate. Other males that may also have been attracted by the calling song of the singing male are inspected by tactile contact via the antennae. Typically, the production of aggressive song and fighting between the two males follows. (After: Loher, W. and Dambach, M. (1989).)

Oscillogram Trace on the panel display of an oscilloscope of instantaneous voltage of an electric signal as a function of time. Since a microphone converts the sound pressure of an acoustic signal into voltage, the oscillograms of such a signal represents a plot of instantaneous sound pressure over time. The greater the sound pressure (subjectively experienced by humans as an increase in loudness), the greater the amplitude of the voltage signal.

The biophysical structure of the calling songs is highly stereotyped and very constant among members of the same species. By contrast, the calling songs of different species may differ considerably. These differences encompass mainly the **temporal structure** of the songs—for example, the number of syllables produced within a chirp and the arrangement of chirps—but to a lesser extent the carrier frequency.

➤ The calling songs of different cricket species may differ considerably. These differences encompass mainly the temporal structure of the songs.

Mechanism of sound production

The male cricket uses a **file-and-scraper-mechanism** to produce his songs. During this so-called **stridulation,** the cricket moves one front wing over the other in a motion resembling the closing of a pair of scissors. In this way, the **scraper**, a protuberance of cuticle on the inner edge of the wing, is drawn over a row of regularly spaced teeth, referred to as the **file**, on the underside of the other wing. Movement of the scraper over

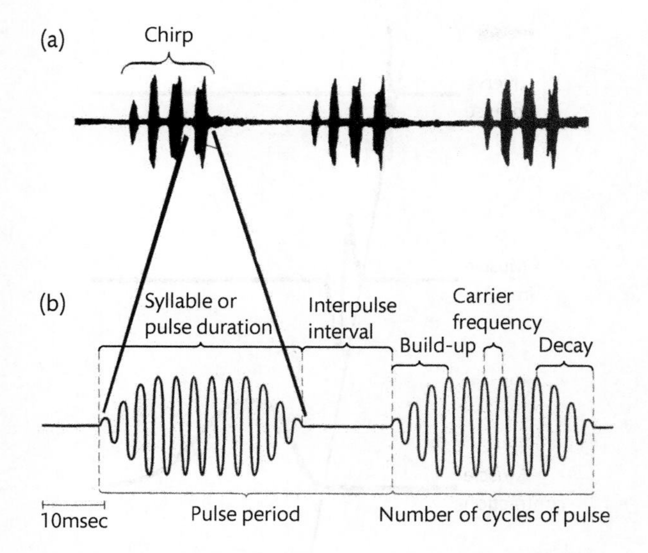

Figure 12.7 Oscillogram of a calling song produced by a male cricket. (a) On a compressed time scale, the gross structure of the song becomes evident. Each song bout consists of chirps, which in turn are made up of syllables. (b) On an expanded time scale, the oscillogram reveals that the individual syllables are comprised of uninterrupted trains of sound waves. While the amplitude of the syllable varies, particularly at the beginning and end of each syllable, the duration of the individual wave periods, and thus the carrier frequency of the song, is rather constant. (After: Huber, F. and Thorson, J. (1985), and Huber, F., Moore, T. E., and Loher, W. (1989).)

one tooth of the file produces a single sound cycle. This mechanism is schematically shown in Fig. 12.8. A complete closing stroke of the wings results in a single syllable. The number of cycles in this syllable is determined by the number of teeth of the file.

➤ Male crickets employ a file-and-scraper stridulation mechanism to produce their songs.

The striking of the scraper over the teeth of the file sets the surface of the wings into oscillation. This vibration is enhanced particularly by the harp frame and a region that includes the file, which **resonate** at the carrier frequency of the sound syllable. At the end of the closing strike, the wings are slightly separated, so that the reopening of the wings is not accompanied by any sound production.

Resonance Oscillation of a system when energy is supplied from an external force at, or close to, the resonant frequency. The resonant frequency is the system's natural oscillation frequency. After supply of external energy at the resonant frequency, the resulting amplitude is particularly large, if the damping of the oscillating system is low.

In two of the most commonly examined crickets, the European field crickets *Gryllus campestris* and *Gryllus bimaculatus*, each syllable lasts for approximately 15–20msec. Most commonly, these species produce trains of four syllables, which follow each other at intervals of about 35msec. This set of four syllables—the chirp—is followed by a brief period of silence. The chirps are characteristically repeated at a rate of 2–4 chirps/sec.

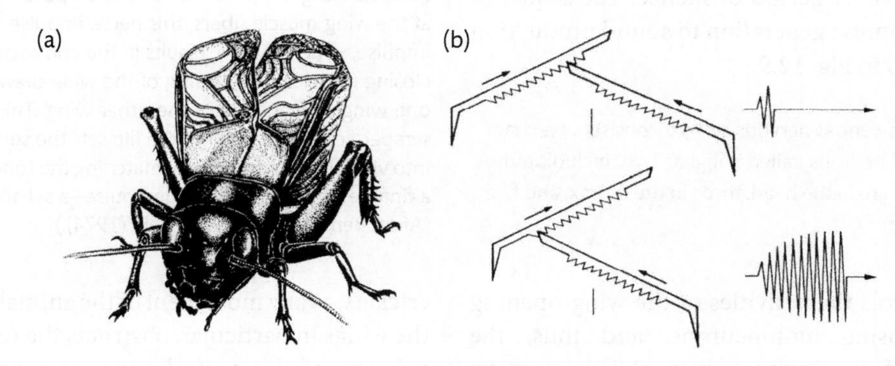

Figure 12.8 Song production by field crickets. (a) Male adopting singing position. (b) Stridulation mechanism. The animal, whose forewings are shown schematically, is oriented as in (a). Both forewings move toward the center, indicated by the vertical line. The scraper of the left (lower) forewing moves across the file of the right (upper) forewing. The movement across one tooth of the file results in the production of a single sound cycle. A complete closing strike of the wings leads to the generation of a syllable. (After: Dambach, M. (1988).)

Neural control of sound production

The closing and opening of the wings during sound production is achieved by the same set of 'twitch' muscles that move the wings during flight, namely **opener** and **closer muscles** of the second thoracic segment. The rhythm of opening and closing, and thus of the production of sound, is controlled by the two thoracic ganglia. Experiments in which these ganglia were deprived of sensory input by cutting the peripheral nerves have shown that the cricket's central nervous system continues to produce the normal motor pattern. This suggests that the neural control of the calling song pattern is independent of sensory input.

Each muscle of the singing and flight muscle system is driven either by a single motoneuron, or by up to five such neurons. This action potential triggers a **muscle impulse** in the bundle of muscle fibers innervated by the respective motoneuron(s). This action potential results in the **contraction** of one of the wing-closing muscles. At the same time, the probability of firing in some neighboring wing-closing motoneurons is enhanced, leading to a powerful contraction of the entire set of muscles.

In contrast, during discharge of the wing-opening motoneurons, the wing-closing neurons are inhibited. The latter neurons fire again after cessation of this inhibition. This alternating activation of wing-closing and wing-opening motoneurons leads to the coordinated alternate closing and opening of the forewing, and thus, to the production of sound, followed by a short period of silence. The sequence from nerve impulse generation to sound production is summarized in Fig. 12.9.

> ➤ The cricket's central nervous system consists of a chain of ten knots of neurons called ganglia. Two, including the brain ganglion, are in the head, three in the thorax, and five in the abdomen.

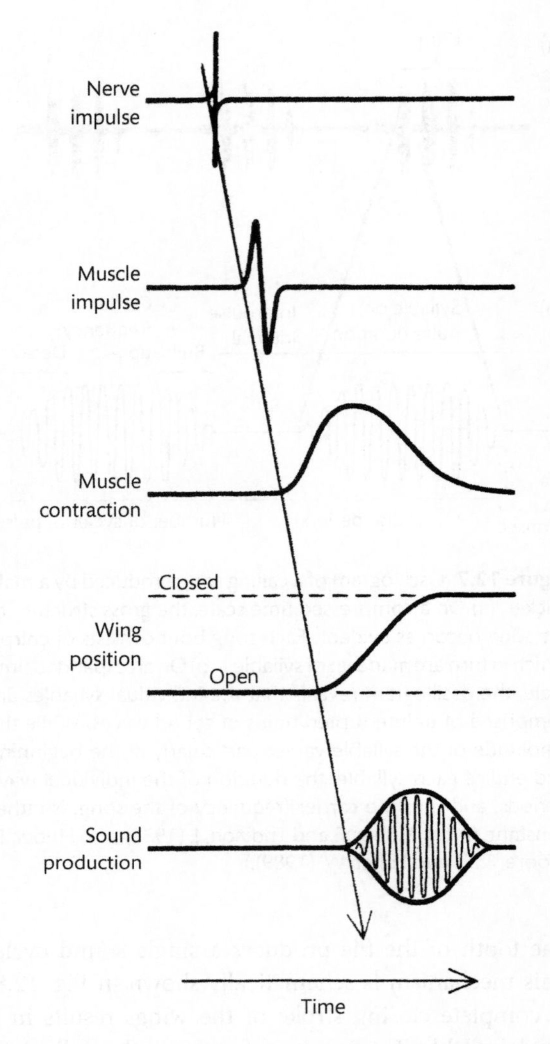

Figure 12.9 Sequence of events leading to a single syllable of the male cricket song. At the beginning, a motoneuron in the thoracic ganglion produces an action potential. Upon arrival at the wing muscle fibers, this nerve impulse causes a muscle impulse, which, in turn, results in the contraction of the wing-closing muscles. The closing of the wing draws the scraper of one wing over the file of the other wing. This rubbing of the scraper across the teeth of the file sets the surface of the wing into vibration at a frequency matching the tooth-strike rate. As a final result, a tone-like sound pulse—a syllable—is produced. (After: Bentley, D. and Hoy, R. R. (1974).)

What controls the activities of the wing-opening and wing-closing motoneurons, and thus, the rhythmicity of the singing behavior? This question was addressed in a series of studies by the laboratory of Berthold Hedwig of the University of Cambridge (United Kingdom). Such investigations are virtually impossible to a carry out in normally behaving crickets, as any movement of the animal in general, or the wings in particular, obstructs the recording from neurons of the central nervous system. However, as an alternative, physiological experiments can be performed by examining **fictive singing**. This elegant approach is based on the analysis of the neural activity of motoneurons, instead of the behavior controlled by these motoneurons. It was discussed, in a different context—escape swimming in *Xenopus* tadpoles—in Chapter 6. In crickets, a preparation that exhibits fictive singing can be obtained by severing all the sensory and motor neurons of the ventral nerve cord. Despite this procedure, the alternating neural activity of the wing-opener and wing-closer motoneurons continues to be generated.

To localize the part of the central nervous system where the rhythmicity of the singing behavior is controlled, singing was induced by microinjection of **eserine** (= physostigmine), which stimulates both nicotinic and muscarinic acetylcholine receptors, into the brain. This is possible because in each half of the brain one neuron, critical for initiating the calling song, is located. The morphology of these two neurons resembles a mirror image of each other, and together they are referred to as the **calling song command neurons**.

In the actual experiments, Hedwig and co-workers progressively truncated the abdominal nerve cord in the fictively singing males by transecting the connectives between adjacent ganglia (Fig. 12.10(a)). This procedure had only limited effect on the singing, as long as only abdominal ganglia posterior to the A3–A4 ganglion connectives were disconnected from the ventral nerve cord (Fig. 12.10(b)). If the A2–A3 connective had been severed, singing stopped completely. However, closer examination also revealed that the more abdominal ganglia were truncated, the more the chirp pattern would become variable (Fig. 12.10(c, d)). During fictive singing with the ventral nerve cord intact, chirps contained 3–5 syllables, and the chirp-rate ranged between 2.0 and 2.6 per second. After severing the A3–A4 connective, 1–8 syllables were generated per chirp, and the chirp rate varied between 1.8 and 4.0 per second.

This and other results led Herwig and coworkers to propose the following neural organization of singing behavior: The abdominal nerve cord houses **two independent timing networks for the control of singing**, one for the chirp activity, and another for the syllable activity. Presumably, the abdominal ganglia project anteriorly, and information related to the chirp pattern from the ganglia A5 and A6 is integrated in A4 with the information related to syllable timing. Further refinement of the syllable pattern takes place in the A3 ganglion. Probably, from the A3 and A4 ganglia the chirp rate and syllable timing information is conveyed to the thoracic ganglion T2, where the wing motoneurons are located. Although less clear, the activity of the calling song command neuron of the brain may also be integrated in the central pattern generating networks distributed along the A3–A6 ganglia. Figure 12.10(e) summarizes the proposed organization of the singing network.

> ➤ Two independent timer networks in the abdominal nerve cord control the singing in crickets—one is responsible for the chirp activity, the other for the syllable activity.

Behavioral analysis of auditory communication

A female cricket that is in the state of copulatory readiness responds to the male calling songs by flying or walking toward the source of the sound, until she reaches the male. This behavioral response is called **positive phonotaxis**. That the female is guided solely by auditory stimuli was demonstrated in an experiment conducted, as early as 1913, by Johann Regen, a high-school teacher in Vienna. Making use of the then newly developed telephone, he transmitted the calling songs of a male to a female cricket. The result speaks for itself: The female approached the receiver as soon as the male songs were broadcast.

> **Positive phonotaxis** Movement of an animal toward the source of a sound.

One approach used to analyze the features used by the female to detect and localize the male's calling songs reliably and with high accuracy was developed by Ernst Kramer and Peter Heinecke at the Max Planck Institute for Behavioral Physiology in Seewiesen, Germany (see Box 12.1). The central device employed in this approach is a polystyrene sphere 50cm in diameter. The cricket walks freely on top of this sphere. Any movement of the animal is tracked by an infrared-sensitive device consisting of a small disk of light-reflecting foil glued to the back of the cricket, an infrared light source, and photodetectors sensible in the infrared range of light. When the cricket attempts to move away from the top of the sphere,

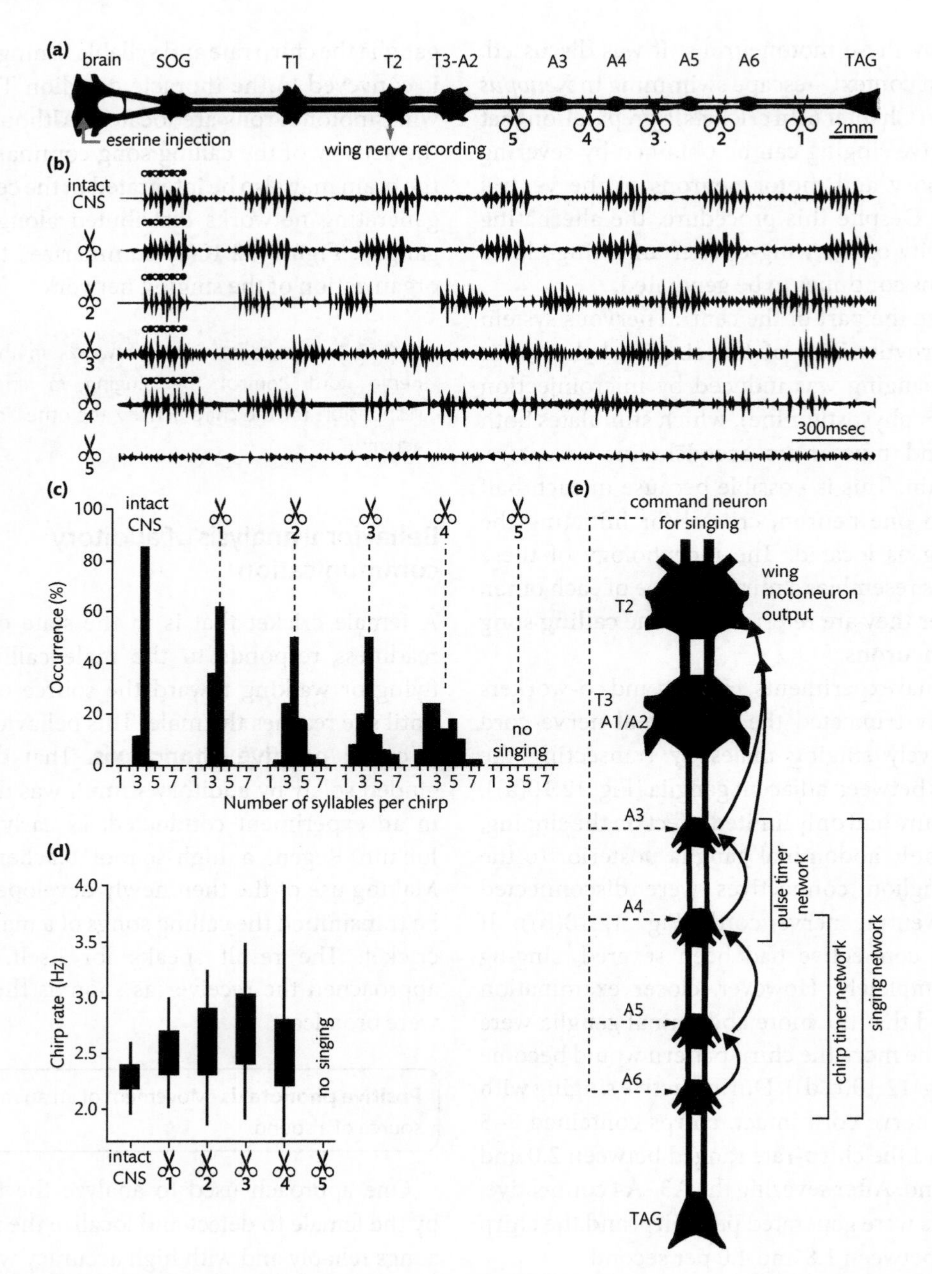

Figure 12.10 Identification of the abdominal ganglia that control the timing of the calling song pattern, in males of the Mediterranean field cricket (*Gryllus bimaculatus*). (a) Sketch of the central nervous system of the cricket, showing the brain with the site of the microinjection of serine; the suboesophageal ganglion (*SOG*); the thoracic ganglia (*T*) and the site from which the fictive singing motor pattern was recorded (*gray arrow*); the abdominal ganglia (*A*), including the connective transection sites (*scissors 1–5*); and the terminal abdominal ganglion (*TAG*). (b) Extracellular recordings of the neural activity of the wing nerve (which reflects the fictive singing pattern) after progressively truncating the abdominal nerve cord, as indicated to the left of each trace. The timing of the wing-opener and wing-closer bursts is indicated by the *open* and *closed circles*, respectively, on top of each of the first chirps. (c) The variability in the number of syllables per chirp increases with progressive truncation of the abdominal nerve cord. (d) Similarly, the chirp rate becomes more variable as more abdominal ganglia are disconnected from the ventral nerve cord (the boxes indicate the so-called interquartile range, and the whiskers the 5th–95th percentile—both are measures of the variability). (e) Model of the organization of the singing neural network in crickets. The dotted lines indicate putative projections of the command neuron of the brain. The proposed locations of both the syllable timer network and the chirp timer network are shown. The arrows indicate the directions of information flow. (After Fig. 1 in Schöneich and Hedwig 2011, © Springer-Verlag; Fig. 10 in Jacob and Hedwig 2016, © Jacob and Hedwig 2016 (CC BY).)

Box 12.1 Principle of Kramer locomotion compensator

The Kramer locomotion compensator is a device for the analysis of the features used by the female cricket to detect and localize the male's calling songs. Part (a) of the figure shows a schematic drawing of the compensator. The cricket walks freely on top of a polystyrene sphere. Sound is broadcast from a loudspeaker placed on the left (L) and right (R) sides, respectively. The cricket's movements are scanned by a camera sensitive to infrared light produced by a circular lamp and reflected by a small disk of foil glued on the back of the animal. To exclude visual cues that could have an effect on the outcome of the experiment, the cricket is surrounded by a fabric cylinder. The camera's signals are fed into a computer-controlled motor system (designated M_x and M_y) that compensates for the movements of the cricket by rotating the sphere in the opposite direction. This compensation mechanism forces the cricket to walk in position on top of the sphere. Part (b) shows time profiles obtained in a particular experiment by monitoring a cricket's movement with the locomotion compensator. The top trace shows the velocity, the middle trace displays the direction of walking of a female during a phonotaxis experiment, and the bottom trace is a polar plot of all the locomotion vectors using the same data obtained during the presentation of the stimuli through the two loudspeakers. Each locomotion vector represents the angle of the walking direction relative to the coordinates of the Kramer locomotion compensator, and a length that is proportional to the walking velocity, both for that 1sec interval. The trial started with 0.5min of silence, followed by stimulation with a synthetic calling song through a loudspeaker placed at an angle of 270°. At 3.5min, this loudspeaker fell silent, and the stimulus was switched to a second speaker placed at a 90° angle. Note the tracking of the female, which meanders around the direction of the active speaker.

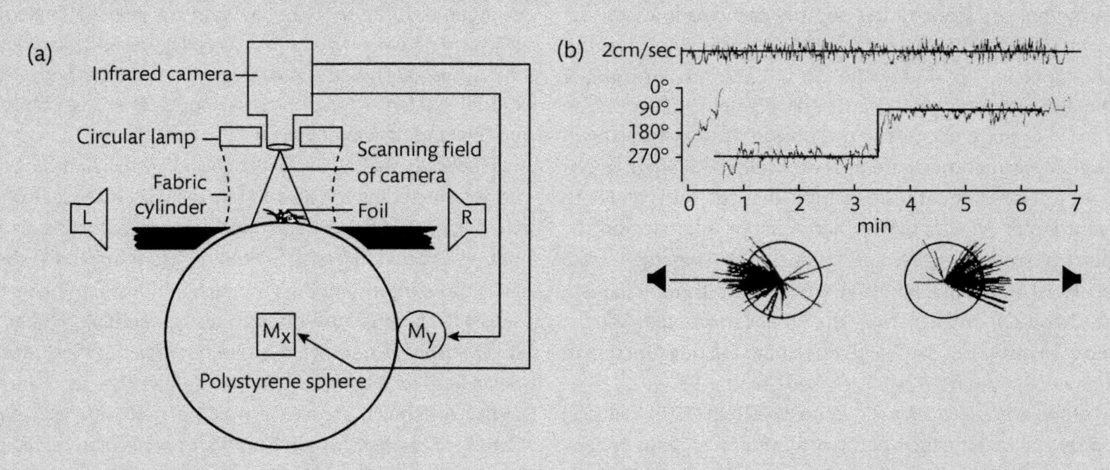

(a) Kramer locomotion compensator. (b) Time profiles obtained by monitoring a cricket's movement with the Kramer locomotion compensator. (After: Schmitz, B., Scharstein, H., and Wendler, G. (1982), and Pires, A. and Hoy, R. R. (1992).)

the photodetectors sense the motion by means of the reflected light. The corresponding signals are fed into a motor system that can compensate the movement by rotating the sphere in the opposite direction, thus forcing the animal to walk near the top of the sphere. The signals encoding the movement of the cricket are stored in a computer, thereby providing a precise record of both direction and speed. In recognition of the merits of one of the inventors, this device is often referred to as the **Kramer locomotion compensator** or Kramer treadmill.

Kramer locomotion compensator A polystyrene sphere on top of which a cricket can walk freely. A control device monitors the movement of the cricket and counter-rotates the sphere to keep the cricket near the top.

Experiments based on the Kramer locomotion compensator, conducted at the Max Planck Institute for Behavioral Physiology in the laboratory of Franz Huber (see Box 12.2), have shown that the following three parameters of the male song are

Box 12.2 Franz Huber

Franz Huber in 2003. (Courtesy: G. K. H. Zupanc.)

Regarded by many as one of the 'fathers of insect neuroethology,' Franz Huber became particularly known for his research on cricket behavior. Huber, born in Nussdorf (Germany) in 1925, grew up on a Bavarian farm where animals were an integral part of his life. In 1947, he began to study biology, chemistry, and physics at the University of Munich. His mentors were the entomologist Werner Jacobs and the discoverer of the bee dance language, Karl von Frisch (see Chapter 3). Others who had a major influence on the young biologist were Konrad Lorenz, Niko Tinbergen, Erich von Holst (for more information on these three scientists, see Chapter 3), and especially the American Kenneth Roeder, whose work on insects paved the road to relating function of the nervous system to behavior (see Chapter 5).

Huber completed his doctoral thesis in 1953 with an anatomical investigation of the orthopteran nervous system. It was also animals from this order that he chose as a suitable model system to study the neural mechanisms of behavior. These animals, crickets in particular, remained the focus of his scientific work throughout most of his life. During his subsequent position at the University of Tübingen, Huber was the first to conduct focal brain stimulation experiments in insects—a then novel approach he had learned from Walter Rudolf Hess in Zurich (see Chapter 3).

It is characteristic of Huber to have kept his enthusiasm to learn new techniques and concepts throughout his life. He was among the first who applied intracellular recording techniques to neuroethological model systems—an approach he learned in the laboratory of Theodore Bullock at the University of California at Los Angeles (see Chapter 3), which he joined as a visiting scientist during 1961–1962. In 1963, Huber took over the chair at the University of Cologne. Finally, from 1973 until his retirement in 1993, he was one of the directors at the Max Planck Institute for Behavioral Physiology in Seewiesen.

By employing an integrative approach, he and his group pioneered the study of behavior at the level of single nerve cells and neural networks in insects. This research culminated in the establishment of one of the first neuroethological model systems for which a comprehensive biological understanding was achieved.

Huber's work has been recognized by the award of many honors, including the Karl-Ritter-von-Frisch Medal from the German Zoological Society and honorary doctorates from distinguished universities, both in Germany and abroad. Even more important than the recognition by his peers is the satisfaction he gained through his research. As Huber puts it: '*One should search for a suitable model system, study its behavioral tactics in the field, select those that can be treated under controlled conditions with no hesitation to adopt a variety of methods to solve riddles at molecular, cellular, and network levels. But behind all is the curiosity for the living world and how it evolved.*'

Franz Huber died in 2017, at the age of 91.

important in determining the female's phonotactic response:

- Syllable rate.
- Carrier frequency.
- Intensity.

In the European field cricket *Gryllus campestris*, the best response is evoked by acoustic stimuli comprised of approximately 30 syllables/sec—a **syllable rate** found in natural chirps produced by the males. Somewhat surprisingly, the number of syllables per chirp is not important, as long as three or more syllables are generated in a row. The female will even track a continuous **trill** consisting of a continuous repetition of synthesized syllables, provided the stimulus is presented at the correct syllable rate. Also, the duration of the syllable relative to the following interval of silence is not crucial in eliciting a response.

In addition to the syllable rate, the **carrier frequency** also plays an important role. In *Gryllus campestris*, the most effective stimulus to trigger a phonotactic response by the female is a 5kHz sound (a frequency found in natural songs) modulated at a syllable rate of approximately 30Hz.

If a song optimized in such a way is simultaneously broadcast through two differently positioned loudspeakers at different **intensities**, the females consistently choose the louder sound. Such an experimental situation might imitate two singing males at different positions in the field. At equal intensity, females walk between the two sound sources.

> ➤ The important features in the male's calling song that trigger a phonotactic response in females are: intensity, carrier frequency of the syllables, and syllable rate.

Perception of auditory signals

When a female cricket detects a calling song, she has to solve the following two problems to successfully approach the singing male:

1. **Localization** of the sound source.
2. **Differentiation** of the calling song produced by conspecific males from sound generated by other animals, or from different types of song, used by conspecific males in different behavioral situations.

These tasks are solved both at the level of the ear and by higher auditory processing stations.

In field crickets, the **ears** are encased in special structures associated with the tibiae of the forelegs (**prothoracic legs**) on each side of the body. Externally, the ear is bordered by an eardrum, the **tympanum** (there is one tympanum on the anterior and one on the posterior surface of the tibia, but only the latter appears to be involved in auditory perception). The ear on one side is coupled to the ear on the other side of the body by the lower branches of an air-filled **tracheal tube**, which is part of the respiratory system. Each of the two upper tracheal branches ends in a **spiracle**, an opening connecting the tracheae with the outside world. Figure 12.11 provides a schematic overview of the tracheal tube system.

The primary auditory organ is the **tympanal organ**. It is attached to the tympanal trachea. To each tympanal trachea, a few dozen auditory receptor cells (**scolopidia**) are attached. Their axons constitute the **auditory nerve.** Its fibers run through the knee joint of the foreleg, up the leg, and terminate, by forming numerous branches, in the prothoracic ganglion of the central nervous system on the same side as the leg from which they arise (Fig. 12.11). This terminal field is known as the **auditory neuropil.** Tracing experiments have shown that the fiber branches of the auditory nerve are entirely restricted to the ipsilateral side; they do not cross the midline. In the region of the auditory neuropil, the terminal branches of the auditory nerve make synaptic contact with interneurons, which relay the information further within the central nervous system.

As the ears are connected to the tracheal tube system, sound pressure reaches the tympanum of each ear directly at its outer surface and indirectly at its inner surface from the contralateral ear as well as the ipsilateral and contralateral spiracles (Fig. 12.12). To examine possible contributions of stimulation of the inner tympanal surface, **laser**

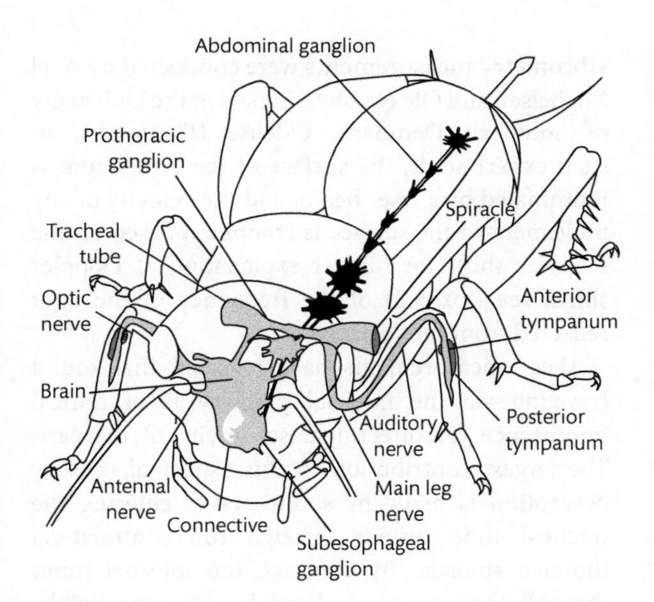

Figure 12.11 The central nervous system and tracheal tube arrangement of the cricket. The ears are situated in the prothoracic legs and bordered by the anterior and posterior tympana. The ears on the two sides of the body are connected via the tracheal tube. Each of the two upper tracheal branches ends in a spiracle. Auditory information is conveyed from the ear to the prothoracic ganglion via the auditory nerve. On top of the hierarchy of central structures processing auditory information is the brain, the frontmost ganglion of the cricket's central nervous system. (After: Huber, F. and Thorson, J. (1985).)

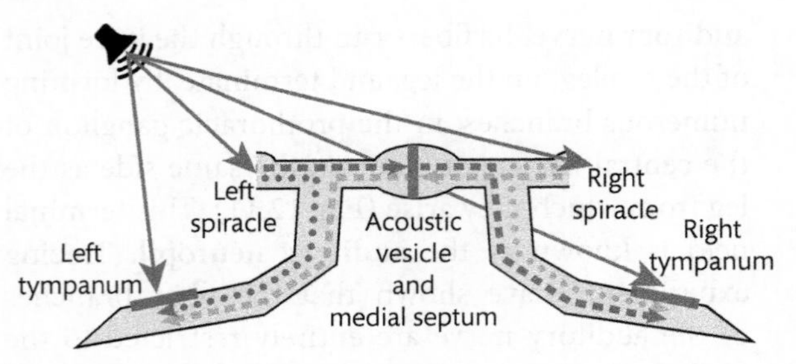

Figure 12.12 Auditory system of crickets. Acoustic input to the tympanum of each ear includes sound from the outside, and sound from the inside that has entered the tracheal tube system via the ipsilateral and contralateral spiracles. The pathways of sound propagation from the spiracles through the air-filled tracheae to the ipsilateral and contralateral ears are indicated by dotted and dashed lines, respectively. Internal input from the contralateral ear is negligible, and hence not shown in the diagram. The left and right tracheae are connected via a tracheal inflation called the acoustic vesicle. At the midline, a thin membrane—the medial septum—separates the left from the right half of the vesicle. (© 2017 Martin J. Lankheet, Uroš Cerkvenik, Ole N. Larsen, Johan L. van Leeuwen)

vibrometry measurements were conducted by Axel Michelsen and Ole Naesbye-Larsen of the University of Southern Denmark, Odense (Denmark). In such experiments, the surface of the tympanum is illuminated by a laser beam, and the velocity of any movement of the surface is computed based on the Doppler shift (for further explanation of Doppler shifts, see Box 5.2) of the frequency of the light reflected from the surface.

These measurements have revealed that sound traveling via the internal pathway is of critical importance for directional sensitivity of the ears. The largest contribution to this aspect of sensory perception is made by sound waves entering the tracheal tube system through the contralateral thoracic spiracle. By contrast, the internal input through the contralateral ear is rather negligible. Furthermore, directionality of the ear depends on the input frequency. The cricket ears are tuned to frequencies near the calling frequency of the male. There is an important consequence of the contribution of internal input at this frequency to the ear's directional sensitivity: When an amplitude peak of a sinusoidal signal at the male's frequency arrives at the outer tympanal surface, then the inner tympanal surface of the same ear experiences a pressure trough. This is due to the length of the internal pathway that sound has to take to arrive at the inner tympanal surface. Measurements have shown that this pathway is approximately half a wavelength longer than the path sound has to travel to reach the outer surface. Thus, the tympanum is simultaneously pushed inward from the outside (due to the pressure peak) and pulled inward from the inside (due to the pressure trough). Correspondingly, half a period later, there is a pressure trough on the outside and a pressure peak on the inside of the tympanal membrane. This causes the tympanum to be simultaneously pulled outward from the outside and pushed outward from the inside. This synergistic effect is particularly pronounced when the male calling song is coming directly from the side of the ear that is closest to the sound source.

On the other hand, if the calling song originates directly from the right side, for example, sound has to travel about the same distance to reach the inner surface of the left tympanum, as it needs to arrive at the outer surface of the left tympanum. As a result, an inward push, caused by a pressure peak on the outside, is counterbalanced by an outward push, caused by the pressure peak on the inside. Now, the overall effect is a largely reduced net movement of the tympanal membrane.

Ears like those of crickets, which are designed in such a way that sound pressure can reach both the inner and the outer surface of the tympanum, are called **pressure gradient ears.** Such a mechanism was first described by the German sensory physiologist Hansjochem Autrum (1907–2003) in 1940. Pressure gradient ears allow the animal to determine the direction from which the sound comes by following a simple behavioral rule: Turn toward the ear that receives the highest sound pressure. This is, indeed, what a female cricket does when approaching a singing male. Characteristically, she follows a zigzag path. This meandering enables her, on a continuous basis, to compare the levels of sound pressure arriving at the two ears.

➤ The cricket's ear functions as a pressure gradient ear: Sound arrives at both the outer and the inner surface of the tympanum to stimulate the primary auditory organ, the tympanal organ.

Song recognition by auditory interneurons

The phonotactic response of female crickets is best elicited by modulation of a 5kHz sound at a syllable rate matching that of natural calling songs produced by the male (see previous section). Based on these behavioral observations, one would expect to find properties either at the level of the ear, or at sensory processing stations within the cricket's central nervous system, that correspond to these features. Indeed, it has been shown that, due to its mechanical resonance properties, the tympanal membrane is tuned to the carrier frequency of the male calling song. A similar frequency tuning is exhibited by the auditory fibers. However, neither the ear nor the primary receptors are sensitive in terms of the temporal parameters of the song. Preference to a particular syllable rate must, therefore, be a property of higher levels of sensory processing.

As mentioned above, the auditory fibers terminate in a restricted region of the prothoracic ganglion called the auditory neuropil. There, the terminal branches make synaptic contact with second-order neurons. A prominent class of these interneurons is formed by the bilateral pair of 'omega-1 neurons,' commonly referred to as **ON-1 cells.** Each of the two ON-1 neurons has its dendritic field on the same side as the cell body, while the axon projects to the other half of the ganglion. The ON-1 cells have received their name because the shape of their axons resembles the capital Greek letter omega (Ω). Axons of the auditory fibers terminate at the numerous, heavily branched dendrites of the ON-1 cells. This input is confined to the side occupied by the cell bodies of the ON-1 cells. In other words, each ON-1 neuron receives input only from the ipsilateral ear. When the ear is stimulated with a train of sound pulses, the ON-1 neurons are rhythmically depolarized such that they closely copy the temporal pattern of the song (Fig. 12.13). This property is shown over a wide range of pulse repetition rates.

The impulses generated by an ON-1 neuron are conducted to the other (contralateral) side of the prothoracic ganglion via its axon, where they cause inhibition. This inhibition can be removed by selectively killing through photoinactivation (for technical details, see Chapter 2) one of the two ON-1 cells—an experimental approach that provided compelling evidence that the inhibition of an ON-1 cell is, indeed, caused by the contralateral ON-1 partner cell.

Taken together, several lines of evidence suggest the following mechanism: Stimulation of the ear causes excitation in the ipsilateral ON-1 cell and inhibition in contralateral ON-1 cell. The inhibition is reciprocal—the ON-1 cell in the left prothoracic ganglion causes inhibition in its partner neuron on the right side. The ON-1 cell in the right ganglion does the same on the left side. This mechanism results in a sharpening of the **directional sensitivity** of the neural circuit to sound.

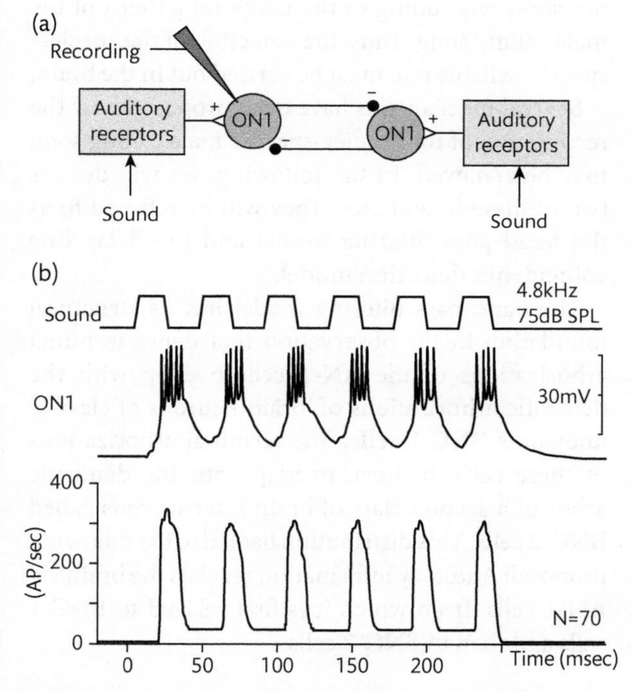

Figure 12.13 Role of ON-1 neurons in song recognition. (a) Connectivity pattern. Each of the two ON-1 cells receives excitatory synaptic input (indicated by open triangles and + signs) from the auditory afferent of the ear ipsilateral to its soma and dendrites, and inhibitory input (indicated by closed circles and – signs) from the contralateral ON-1 cell. (b) Intracellularly recorded activity of one of the two ON-1 cells in response to acoustic stimulation mimicking the species-specific temporal pattern of the calling song. The cell reliably copies the sound pattern (upper panel) by generating a burst of spikes in response to each pulse (middle panel). A plot of the averaged instantaneous spike rate, expressed by the number of action potentials generated per second (AP/sec), reveals a pronounced onset response, with spike rates as high as 300/sec (lower panel). (© 2016 Hedwig (CC-BY).)

Acoustic information encoded by neurons within the prothoracic ganglion is carried to the brain via two pairs of bilateral ascending neurons called 'ascending neurons of type 1' (**AN-1 cells**) and 'ascending neurons of type 2' (**AN-2 cells**). Their dendrites overlap the axonal arborization of the contralateral ON-1 cells, suggesting that the ON-1 neurons transmit information directly to AN-1 cells on the contralateral side of the prothoracic ganglion. Although the various types of ascending neurons differ in terms of their sensitivity to different carrier frequencies of sound, all of them, including the AN-1 cells, still truly copy the temporal pattern of sound signals. Most importantly, however, they do not show any tuning to the temporal pattern of the male calling song. Thus, the detection of the species-specific syllable rate must be carried out in the brain.

Several mechanisms have been proposed how the recognition of the species-specific male calling song may be achieved. In the following, we will discuss two of these hypotheses. They will be referred to as the **band-pass filtering model** and the **delay line coincidence detection model.**

The band-pass filtering model has its structural foundation in the observation that dense terminal arborizations of the AN-1 cells overlap with the dendritic arborizations of 'brain neurons of class 1,' known as **BNC-1 cells.** The terminal arborizations of these cells, in turn, overlap with the dendritic arbor of a second class of brain interneurons called **BNC-2 cells.** This distribution has led to the following proposal: Auditory information reaches the brain via AN-1 cells, from which it is first relayed to BNC-1 cells and then to BNC-2 cells.

➤ A simplified circuit relays auditory information from the ear to the brain: scolopidia → ON-1 cells → AN-1 cells → BNC-1 cells → BNC-2 cells.

The physiological properties of the BNC-1 and BNC-2 cells were examined in detail by Klaus Schildberger in the laboratory of Franz Huber. As his studies have shown, the accuracy of copying of the temporal pattern of the calling songs by the BNC cells is significantly reduced compared with neurons of the prothoracic ganglion. However, they appear to filter out certain features relevant in the phonotactic response of female crickets. This is evident from the following three functional types among BNC-1 and BNC-2 cells (Fig. 12.14):

- One type responds best to syllable rates effective in phonotaxis, plus to higher syllable rates; these cells act like high-pass filters used in electronic circuits, and hence are called **high-pass cells.**
- A second type responds best to syllable rates effective in phonotaxis, plus to lower syllable rates; these cells act like low-pass filters, and hence are called **low-pass cells.**
- A third type, found in a subpopulation of BNC-2 cells, responds only to syllable repetition rates in the range that best elicits phonotaxis in females, in behavioral tests; these cells appear to function as band-pass filters, and hence are termed **band-pass cells.** They can be regarded as **recognition neurons.**

➤ BNC-2 cells appear to act as pattern-recognition neurons: They respond to songs in terms of the number of action potentials elicited by chirps similar to the way the whole animal behaves, in terms of the phonotactic response.

According to the model proposed by Schildberger, the high-pass and low-pass neurons act on the band-pass neurons in a fashion resembling logic AND gates. In other words, only if input is received simultaneously both from the low-pass and the high-pass neurons will the band-pass neurons respond. This model is summarized schematically in Fig. 12.15.

The delay line coincidence detection model was proposed by Stefan Schöneich, Konstantinos Kostarakos, and Berthold Hedwig of the University of Cambridge, U.K. The core element of their model is a coincidence-detection circuit, similar to the one that Lloyd Jeffress has suggested for directional localization of sound (see Chapter 7). However, in contrast to the Lloyd Jeffress model, the delay required in the model by Schöneich, Kostarakos, and Hedwig is not generated by axonal delay lines but instead by an inhibitory mechanism. In their model, the response of brain cells to sound syllables is generated in two parallel pathways. In one pathway, the neural activity is directly transmitted to the coincidence detector, whereas the activity in

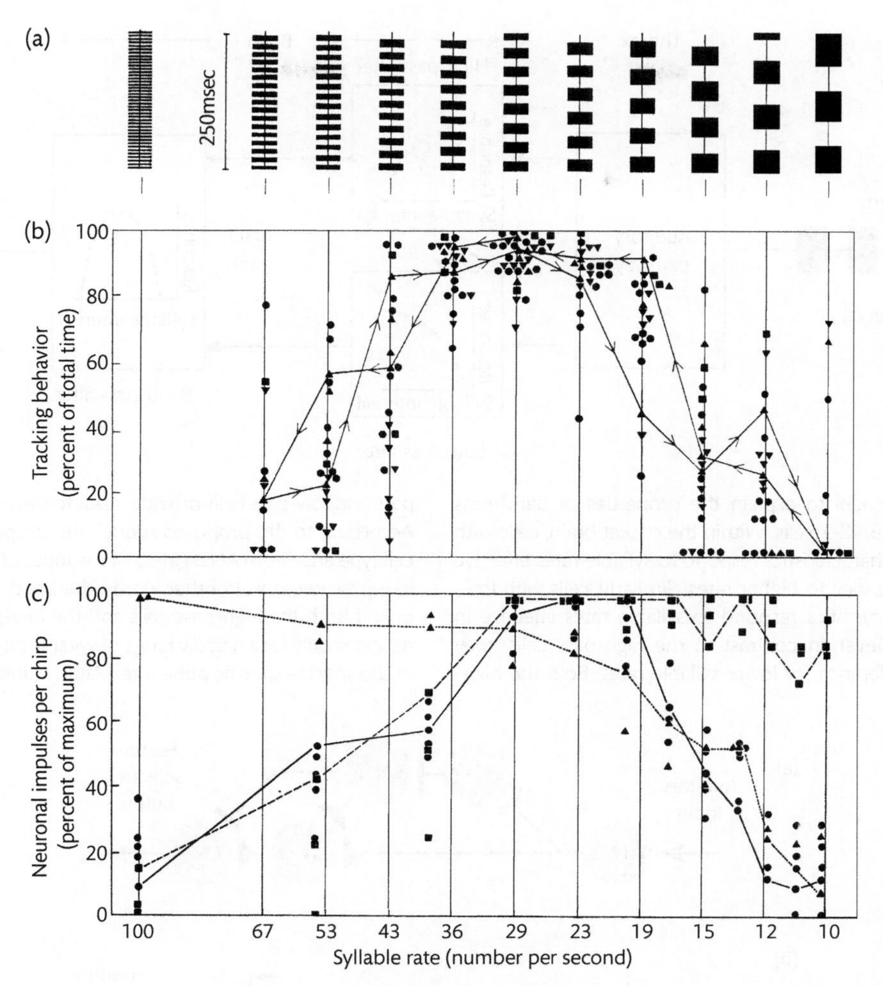

Figure 12.14 Response characteristics of female crickets and various types of BNC cells to synthesized calling songs of varied syllable rates. (a) For stimulation, a series of chirps was used that varied in terms of their syllable rate. (b) Several female crickets (represented by different geometric symbols) were exposed to this series of different chirps. Their behavioral response was monitored using the Kramer locomotion compensator. The data from one of these experiments are connected by arrows to show the sequence of the individual tests. The results of the behavioral experiments indicate a preference for syllable rates near 30/sec, which corresponds to seven to eight syllables over the 250msec period shown. (c) Response of three types of BNC cell, as determined by the number of action potentials elicited by each chirp. Some cells (circles) exhibit a band-pass response; they respond best to syllable rate around 30/sec. Others (squares) have a low-pass response; they respond best to rates at or below 30/sec. A third class of cells (triangles) show a high-pass response; they exhibit the strongest responses to a syllable rate at or above 30/sec. The data of one experiment each are connected by a solid line (band-pass response), a broken line (low-pass response), and a dotted line (high-pass response). Note that, in addition to the syllable rate, other parameters, such as the number of syllables per chirp and the duration of the individual syllables, co-varied in the stimulation experiment shown in (a). However, as has been demonstrated in other investigations, these latter parameters do not affect the response of the neurons. (After: Huber, F. and Thorson, J. (1985).)

the parallel pathway is delayed by the species-specific pulse period of the calling song before it reaches the coincidence detector. In case of the Mediterranean field cricket *Gryllus bimaculatus* (the species used in this research), this delay is approximately 40msec, thus corresponding well to the syllable period of 30–40msec in the male calling song of this species. The circuit constituting these two pathways and some of the physiological properties of its types of neurons are shown in Fig. 12.16.

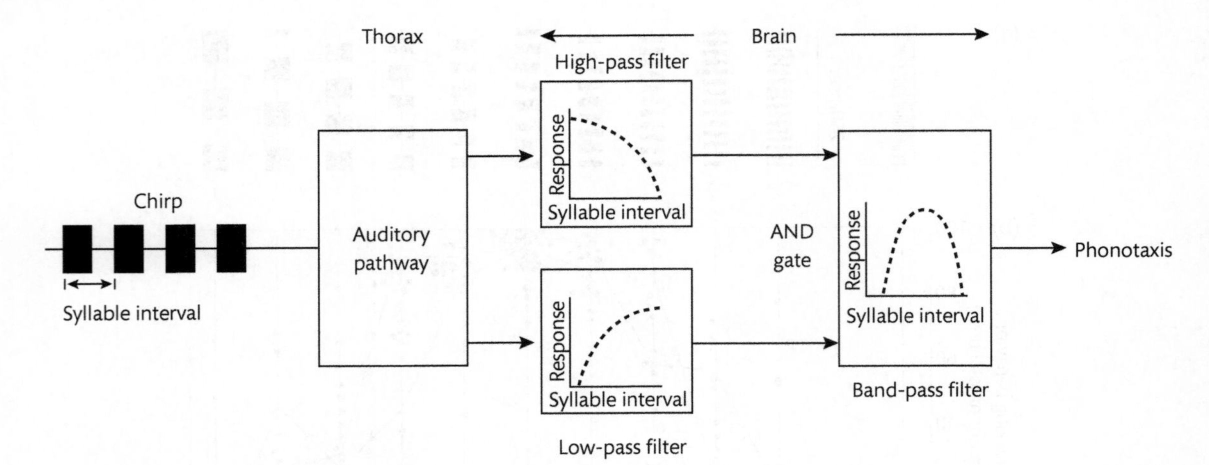

Figure 12.15 Model to explain the properties of band-pass neurons among BNC-2 cells. Within the cricket brain, cells with high-pass filter characteristics respond to syllable rates effective in phonotaxis, as well to higher rates. Similarly, cells with low-pass filter characteristics respond to syllable rates effective in phonotaxis; however, in contrast to the high-pass cells, they also show a preference for lower syllable rates. Both the high-pass and low-pass cells provide input to the band-pass neurons. According to the proposed model, the properties of this latter cell type arise from AND gating of the input of the high-pass and low-pass neurons. In other words, the band-pass cells respond only if both the high-pass cells and the low-pass cells fire. This happens only for a narrow range of syllable rates that correspond to the species-specific pulse rate. (After: Huber, F. (1990).)

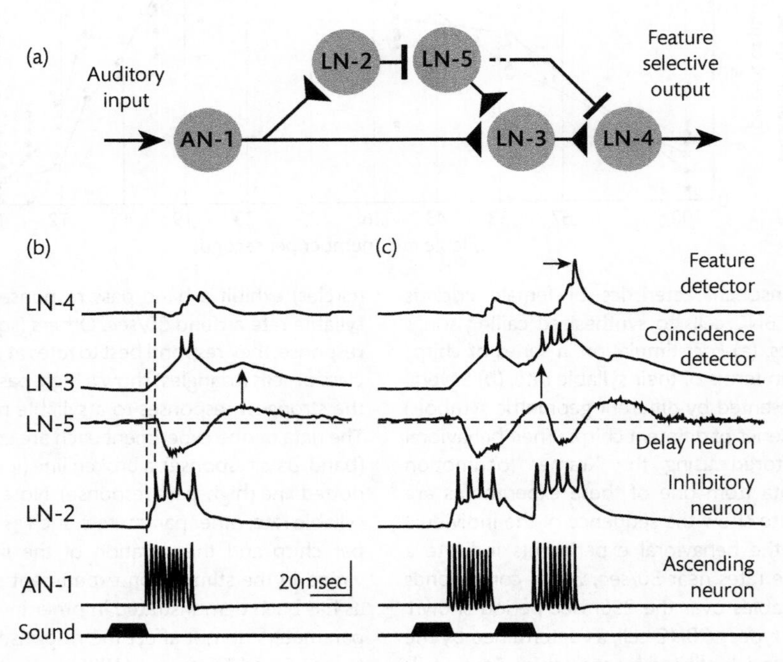

Figure 12.16 Delay line coincidence detector model in the cricket brain. (a) Brain circuitry for detection of pulse periods. Auditory information from AN-1 is split into two parallel pathways—one containing the coincidence detector neuron LN-3, the other LN-2 and LN-5, which delay transmission of the response by a specific delay time using an inhibitory mechanism (for details, see text). In case of two or more pulses with a pulse period corresponding to the species-specific syllable period of the male calling song, the direct input and the delayed input to LN-3 will coincide, making the output of this neuron sufficiently strong to overcome the inhibition of LN-4 from LN-5. LN-4, which serves as the feature detector, will then respond with 1–2 spikes. Response of the feature detecting circuit to a single sound pulse (b) and two sound pulses (c). The activities of AN-1, LN-2, LN-5, LN-3, and LN-4 are shown. The coincidence detector LN-3 integrates the direct response from AN-1 and the delayed response from LN-5. If these two inputs coincide, the output of LN-3 will be strong enough (vertical arrow) to cause the feature detector LN-4 to fire (horizontal arrow). (© 2016 Hedwig (CC-BY).)

According to the model, the AN-1 cell makes excitatory synaptic contact with two local interneurons in the brain, **LN-2** and **LN-3**. The latter serves as the coincidence detector neuron in this circuit. LN-2 provides inhibitory input to **LN-5** for the 20msec duration of a single sound syllable. At the end of the pulse, inhibition is released from LN-2, and LN-5—a non-spiking neuron—generates an excitatory post-inhibitory rebound response (for a detailed description of the rebound phenomenon, see Chapter 6). The rebound response reaches its maximum approximately 40msec after the end of the sound syllable. LN-3 integrates the activity from AN-1 and the delayed activity from LN-5. The integrative action will produce a maximal response in case of coincidence, but a significantly weaker response in case of non-coincidence. Upon maximal response, LN-3 is able to overcome the inhibition exerted by LN-5 on another local interneuron, **LN-4**, causing the latter to fire. Thus, LN-4 functions as the feature neuron in the proposed circuit.

The band-pass characteristics of LN-4 resemble those of the BNC-2 cells in the band-pass filtering model. It is possible that the activity of the latter reflects a step subsequent to the processing by the delay line coincidence detection circuit occurring at an earlier stage.

Temperature coupling

The problem

Communication is possible only if the receiver 'understands' the signal produced by the sender. The fact that the temporal properties of communication signals in **ectothermic animals** are often under the influence of the **ambient temperature** poses a potential challenge to the integrity of the communication system. The following three solutions to this problem have been found:

1. The response criteria of the signal receiver are broadly specified so as to encompass the range of variation of the signal.
2. The response criteria of the signal receiver rely, at least to some extent, on temperature-invariant properties of the signal.
3. The response criteria of the signal receiver change parallel to the temperature.

Of these three solutions, the last, commonly referred to as **temperature coupling**, has been particularly well studied. The subject of a good number of these investigations has been the acoustic communication system of field crickets. Most of this research was carried out by the laboratory of Ron Hoy of Cornell University in Ithaca, New York, using the field cricket *Gryllus firmus*. This North American species lives on the East Coast. In the field, males produce calling songs at ambient temperatures ranging from 12 to 30°C.

Temperature coupling Certain properties of a signal produced by the sender and the response criteria of the receiver change parallel to the ambient temperature changes.

The songs

Figure 12.17 shows an oscillogram of the natural calling song of *Gryllus firmus*. Similar to the European field crickets *Gryllus campestris* and *Gryllus bimaculatus*, the songs are organized into groups of chirps separated by brief periods of silence. Most commonly, each chirp contains four syllables, although occasionally chirps consisting of three or five syllables are also generated. Figure 12.17 also indicates the temporal parameters used to characterize the song:

- The **chirp period** is the time elapsed between the onset of the first syllable of one chirp and the onset of the first syllable of the next chirp; its reciprocal (in Hz) is the chirp rate.
- The **syllable duration** is the time elapsed between the beginning and the end of one syllable.

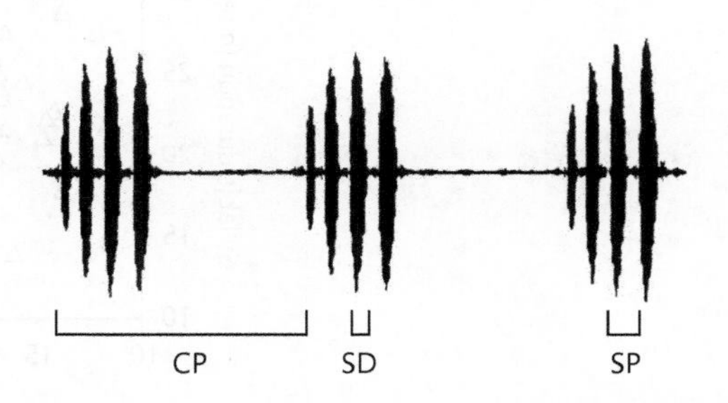

Figure 12.17 Oscillogram of a calling song of the field cricket *Gryllus firmus*. As in any oscillogram, the amplitude of the signal is plotted as a function of time. For further analysis, three temporal parameters have been examined: chirp period (CP), syllable period (SP), and syllable duration (SD). (After: Pires, A. and Hoy, R. R. (1992).)

- The **syllable period** is the time elapsed between the onset of one syllable and the onset of the next syllable; its reciprocal (in Hz) is the syllable rate.

> ➤ Temporal patterns used to characterize cricket song are: chirp rate (reciprocal of chirp period), syllable rate (reciprocal of syllable period), and syllable duration.

Effect of temperature on calling song

In the first set of experiments, Anthony Pires and Ron Hoy made recordings of calling songs in the field. Immediately after each recording, they measured the temperature at the calling site. Back in the laboratory, the calls were played back on an oscilloscope, and the temperature dependence of the three parameters 'chirp period,' 'syllable duration,' and 'syllable period' was analyzed.

Figure 12.18 shows the results of this analysis. Both chirp rate and syllable rate increase, in a linear fashion, with temperature. From 12 to 30°C, chirp rate increases by a factor of 4, from about 0.8 chirps/sec (thus corresponding to a chirp rate of 0.8Hz) to 3.3 chirps/sec (thus corresponding to a chirp rate of 3.3Hz). Over the same temperature range, the syllable rate increases by a factor of 2, from approximately 15 to 29Hz. Conversely, syllable duration is negatively correlated with ambient temperature. From 17 to 30°C, the mean syllable duration decreases by about 40%.

These results might give the impression that each parameter associated with the calling songs changes with temperature. However, this is not the case. In the population of field crickets studied by Pires and Hoy, the carrier frequency of the calling song ranged from 3.6 to 4.6kHz. Despite this variability, a plot of the individual carrier frequencies as a function of the temperature at which the corresponding songs were produced indicated that this parameter is not affected by temperature.

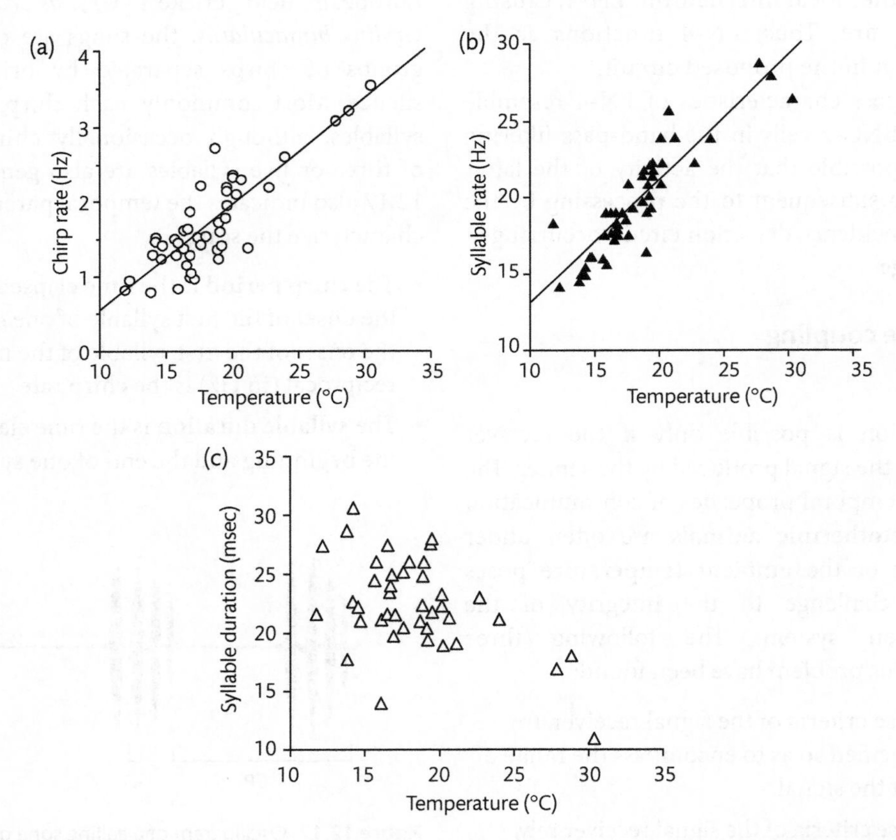

Figure 12.18 Three temporal parameters of the calling song of the field cricket *Gryllus firmus* as a function of temperature. The songs were recorded in the field, and the temperature was measured at the calling site. Each point is the mean of 100 intervals of the respective parameter determined from the song of one individual. (a) Chirp rate. (b) Syllable rate. (c) Syllable duration. (After: Pires, A. and Hoy, R. R. (1992).)

➤ Chirp rate and syllable rate are positively correlated with temperature, whereas syllable duration is negatively correlated with ambient temperature.

Effect of temperature on calling song recognition

As shown earlier in this chapter (see 'Behavioral analysis of auditory communication'), sexually receptive females respond to the calling song produced by the males with positive phonotaxis. However, the experiments by Pires and Hoy have demonstrated that the temporal properties of the male calling songs are strongly affected by temperature. How do the females cope with this environmentally induced variability in the signal?

To answer this question, Pires and Hoy synthesized model songs with temporal patterns corresponding to those recorded in the field at different temperatures. To model a song produced at 15°C, a chirp rate of 1.3Hz and a syllable rate of 14.3Hz was used. Similarly, songs modeled after those generated at 21°C exhibited a chirp rate of 1.8Hz and a syllable rate of 20.0Hz. The values of these parameters were further increased in synthetic songs imitating natural songs produced at 30°C, namely to a chirp rate of 3.3Hz and a syllable rate of 25.0Hz.

➤ Model songs can be synthesized so that their temporal characteristics correspond to the differences found in natural songs under different temperatures.

To avoid a possible effect of previous experience, crickets were reared in the laboratory, and only female nymphs isolated from males were allowed in the experiments. To quantify phonotaxis, the Kramer locomotion compensator was used. The experiments were conducted at three different ambient temperatures, 15, 21, and 30°C. The two loudspeakers were placed at 180° angles from each other. For stimulation, the female crickets were placed on the spherical treadmill and the synthetic male songs modeled on temperatures of 15, 21, and 30°C were presented.

Question: *Using the above framework, how would you design experiments to find out whether or not females respond to stimulation by male calling songs produced under different temperatures with a correct phonotactic response?*

Pires and Hoy conducted, in the laboratory, the following two types of experiments:

1 Single-stimulus sequential experiments.

2 Two-stimuli simultaneous choice experiments.

In the single-stimulus sequential experiments, each trial lasted for 7min. In the first 30sec, there was silence. The female was then stimulated for 3min with song from one speaker, followed immediately by another 3min of stimulation from the opposite speaker, and ending with 30sec of silence. To analyze the female's response, a **locomotion vector** was calculated. This vector, sampled once every second, was defined by an angle describing the cricket's walking direction relative to the coordinates of the Kramer locomotion compensator and a length proportional to her walking velocity. To screen out trivial movements, only the 30 'best' responses in each 3min presentation, of magnitude greater than the mean velocity for the stimulus presentation, were analyzed. In the next step, these 30 locomotion vectors were transformed relative to the direction of the speaker active during the experiment and pooled with the 30 corresponding vectors from the other stimulus presentation. As a result, a mean vector characterizing an individual animal's response to each trial was obtained.

Vector A physical quantity, such as force, that possesses both 'magnitude' and 'direction.' This quantity can be represented by an arrow having appropriate length and direction and emanating from a given reference point. Correspondingly, the 'locomotion vector' in cricket research is defined by the speed and the direction of movement.

In the two-stimuli simultaneous choice experiments, a song was matched to a given temperature at which the experiment was performed. This song was presented through one loudspeaker, while simultaneously one of the other two synthetic songs was played back through the opposite speaker. This paired presentation lasted 8min, with a switch of the songs between the two speakers after the first 4min. Similar to the single-stimulation sequential experiments, the locomotor activity of the female on the Kramer locomotion compensator was sampled and mean vectors were calculated.

> ➤ Two approaches to test a female cricket's response to different songs involve the conduction of single-stimulus sequential experiments and two-stimuli simultaneous choice experiments.

The results of the single-stimulus sequential experiments conducted under the three ambient temperatures—15, 21, and 30°C—are shown in Fig. 12.19. This figure demonstrates that at all three temperatures the strongest responses, characterized by the largest mean speaker components and the highest mean walking velocities, were obtained to the model song matching the ambient temperature. Thus, females tested at 15°C showed the strongest responses to songs modeled after natural songs produced at 15°C (Fig. 12.19a). Females tested at 21 and 30°C responded best to 21 and 30°C songs, respectively (Fig. 12.19b, c). Similar results were obtained in the two-stimuli simultaneous choice experiments.

Taken together with the data obtained in the experiments where the effect of temperature on the calling songs was examined, these findings demonstrate that the signal and response properties change parallel to the changes in temperature. In other words, the cricket's song communication system is temperature coupled.

> ➤ Female crickets exhibit the strongest phonotactic response to model songs matching the ambient temperature.

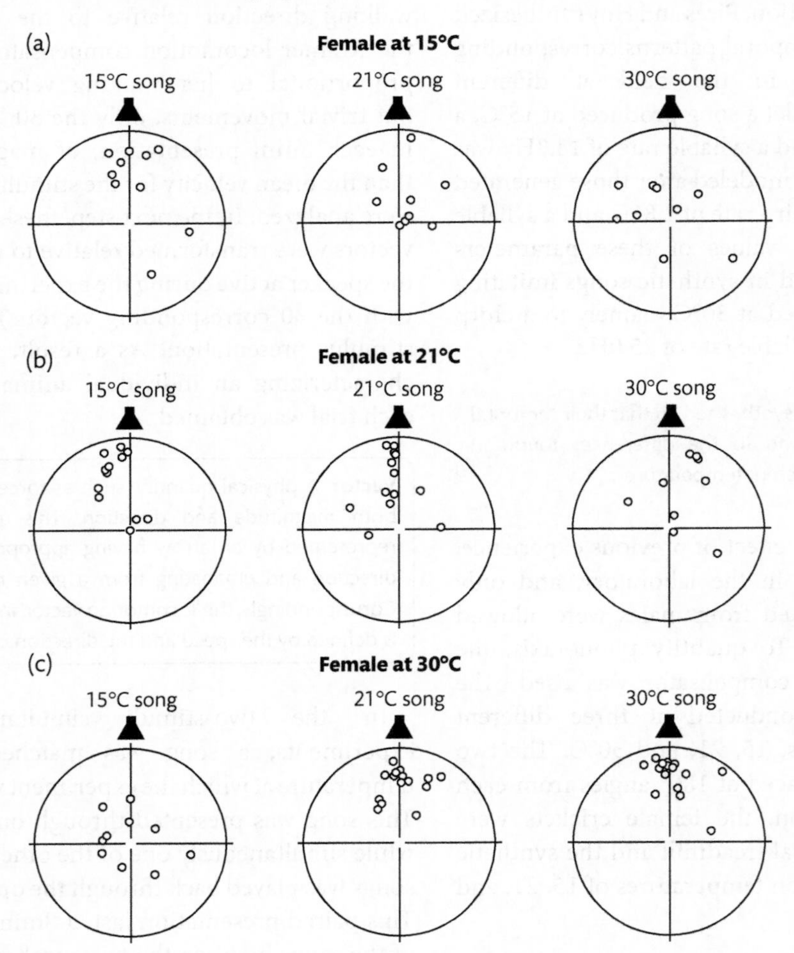

Figure 12.19 Orientation responses of female field crickets, *Gryllus firmus*, to synthetic calling songs modeled after songs produced at three different ambient temperatures, 15, 21, and 30°C. In each of the single-stimulus sequential experiments, the females were tested at three different ambient temperatures. Each point represents the coordinates of the mean vector calculated from a single trial. Results are shown for females kept at 15°C (a), 21°C (b), and 30°C (c) ambient temperature during the experiment. The direction of the loudspeaker relative to the coordinates of the Kramer locomotion compensator is indicated by the loudspeaker icon. (After: Pires, A. and Hoy, R. R. (1992).)

Model system: The development of bird song

Introduction

Songbirds form their own taxonomic suborder, the Passeri (sometimes also referred to as the Oscines), and comprise about 4000 species—nearly half of all birds. Their songs play an important role in **communication**. During the breeding season, the male songs are used to attract females and to defend territories against other males. In some species, females also sing. This may lead, in a few species, to **duetting** (a coordination of vocalization) of the paired birds, which is thought to serve multiple functions, including maintenance of year-round territories and pair bonding.

Although people have been fascinated by bird songs for a long time, its systematic scientific exploration, particularly its **development during ontogeny**, began only in the 1950s. At about that time, two new technologies critical for the advancement of research into bird song became available to ornithologists—the **tape recorder**, which made it possible to record bird song, and the **sound spectrograph**, which enabled scientists to visualize the recorded song as a **sound spectrogram** (or **sonogram**) and to analyze some of its key physical properties. The investigations performed since then have provided fascinating insights into the behavioral aspects and the underlying neural mechanisms of bird song development. In the following sections, we will discuss some of the key accomplishments of this research.

Sound spectrogram Also referred to as a **sonogram**. A visual representation of an acoustic signal. Most commonly, the horizontal axis represents time, and the vertical axis represents frequency. The amplitude at a particular time and particular frequency is encoded either by the gray level (the darker the marking, the higher the sound intensity) or by using a color map (ranging, for example, from light yellow to dark red as sound intensity increases). Sound spectrograms are generated by analog or digital versions of sound spectrographs.

The behavior

The songs of different bird species typically differ greatly. However, variations in song occur also within species. In particular, the songs of many songbirds exhibit geographical variability, resulting in local **dialects**. One of the species in which this phenomenon has been particularly well studied is the white-crowned sparrow (*Zonotrichia leucophrys*) (Fig. 12.20), an abundant North American songbird. Much of the initial work was pioneered by Peter Marler (Box 12.3) and his associates. The spectrographs in Fig. 12.21 show 12 song types of white-crowned sparrows recorded in nine 'island' locations, in the San Francisco Bay area (the term 'islands' includes both four true islands separated by water and five mainland patches of coastal vegetation separated by unsuitable urban or industrial areas). Each population is characterized by one (sometimes two) distinct subtypes of the song. Similar local dialects of the songs have been found in many other species of songbirds.

Figure 12.20 White-crowned sparrow (Courtesy: Wolfgang Warner (CC BY-SA 3.0).)

What causes the differences in the songs between the populations of one species? Are these differences genetically determined, or are they learned during ontogenetic development, or are they the result of a combination of genetic determinants and environmental influences?

Presumably, the first scholar who addressed these questions was Johann Ferdinand Adam Pernauer, who held the baronial title, hence called Baron von Pernau. He was born in 1660 at Steinach in Lower Austria, but because of his family's Lutheran faith they had to leave Austria and settled in Franconia (today part of Bavaria, Germany). He died in 1731. In a book published in 1716, he wrote that the songs of the chaffinch (*Fringilla coelebs*)

Box 12.3 Peter Marler

Fig. Box 12.3 Peter Marler with a hand-reared wood thrush. (Courtesy: Rockefeller Archive Center.)

When great scholars pass away, obituaries will be published in newsletters of professional societies and in scientific journals. Rarely, the general public learns about their deaths. Yet, this is exactly what happened when Peter Marler died, in 2014. Articles in recognition of his life and achievements reached millions of readers by appearing not only in specialized professional journals but also in general science magazines, such as *Nature*, and in major American and British newspapers, including *The New York Times*, the *Los Angeles Times*, *The Telegraph*, and *The Guardian*.

Marler was born to working-class parents in 1928 in Slough, a town near London, England. From his early years, he had been intrigued by birds—a passion that let him write, in an autobiographical essay, that 'if as a child I had believed in reincarnation, I would undoubtedly have chosen to be reborn as a bird'. Nevertheless, doubting that ornithology would enable him to make a living, he studied plant ecology at the University College London, from which he obtained his Ph.D., in 1952. As part of his first job, with the Nature Conservancy, he was assigned the task to survey potential nature reserves in Britain, France, and the Azores. Marler used this opportunity to study, in his spare time, the geographic variation in the song of the chaffinch (*Fringilla coelebs*). He interpreted this variation as local dialects that the birds acquire by listening to the songs of neighboring birds—similar as humans learn to speak. This idea was novel because then, the general view among scientists was that bird vocalization was entirely genetically determined.

Marler's interest in birds came to the attention of William Homan Thorpe, who played a pivotal role in establishing ethology in the English-speaking world. He invited Marler to join him and his colleague Robert Aubrey Hinde at the newly founded Madingley Ornithological Field Station of Cambridge University. Under Thorpe, Peter Marler conducted a comprehensive study of the behavior, including the vocalization, of the chaffinch. This investigation earned him in 1954 a second Ph.D., this time in zoology from the University of Cambridge. Subsequently, he continued his work at Cambridge, focusing on the structure and function of the vocalization of several bird species. Taking advantage of two technological innovations that had become available at that time—the portable tape recorder and the audio spectrograph—he greatly advanced his research, while setting new standards in the use of biophysical analysis of signals for the then largely unexplored area of animal communication.

In 1957, Peter Marler accepted an offer to join the faculty of the University of California, Berkeley. To systematically examine how birds learn songs during development, he chose the white-crowned sparrow (*Zonotrichia leucophrys*), a North American species that exhibits regional song dialects, similar to ones he and Thorpe had described in the chaffinch. While laboratory experiments confirmed that the song dialects are culturally transmitted, they also revealed the importance of genetic factors on the process of learning. These innate learning predispositions became apparent when young birds are raised in isolation and exposed to recorded songs of different species. Under such experimental conditions, they favored songs of their own species. Other species-specific genetic constraints determine, for example, the critical period during which the learning takes place. Marler coined for this interplay of innate influences and environmental effects in the development of learning the term 'instinct to learn'—a concept that had impact far beyond ornithology, as it questioned the traditional dichotomy of nature versus nurture.

As a logic extension of Marler's work on vocal learning in birds, he and his students carried out numerous studies in which they examined the structure of vocal signals and their function for communication in birds and other animals. During a sabbatical year in Uganda, he was the first to study these aspects under field conditions in several primate species. Later, in collaboration with Jane Goodall at the Gombe National Park in Tanzania, he was the first to record the vocal repertoire of chimpanzees, and analyze their sounds by using spectrographs. These studies laid the groundwork for research by numerous primatologists and anthropologists, including his student Thomas Struhsaker and his postdocs Dorothy Cheney and Robert Seyfarth. Struhsaker discovered that the African vervet monkey produces distinct alarm calls indicating different predators, such as leopards, eagles, or snakes. Playback experiments by Cheney and Seyfarth demonstrated that the responses of the vervet troops are specific and appropriate to each of these calls. While previous theories had interpreted such calls merely as manifestations of emotions,

... continued

the observations by Marler's group indicated that, in addition to expressing affection, these signals had a symbolic character and conveyed specific meanings, for example 'leopard,' 'eagle,' or 'snake'. Besides primatology, these findings have borne relevance to a variety of research fields, including the studies of animal cognition and development of human language.

During his tenure at Berkeley, Marler, together with William J. Hamilton III, wrote what turned out to be one of the first attempts of a synthesis of the proximate causes of animal behavior, a nearly 800-pages long book entitled *Mechanisms of Animal Behavior*. At a time when large parts of research on animal behavior in North America were dominated by behavioristic concepts, the book provided a more ethologically oriented perspective, analyzing structure and function of behavior within an evolutionary framework, while at the same time combining field observations with rigorous experimental analysis in the laboratory, and insisting on the application of quantitative methods.

In 1980, Marler moved to Rockefeller University, where he became Director of the newly established field station in Millbrook in upstate New York. In 1989, he switched coasts again to accept a faculty appointment at the University of California, Davis, where he helped to establish the Center for Neurosciences. In recognition of his achievements, he was elected a member of the U.S. National Academy of Sciences and a foreign member of the British Royal Society.

Marler retired in 1994. As Professor Emeritus, he continued to publish research articles, review papers, and books. For his last book, *Nature's Music: The Science of Bird Song*, he and his co-editor Hans Slabbekoorn brought together several of the world's experts on bird song to review the advances in this field—a field that would have been different without him.

Peter Marler died in 2014, at the age of 86, in Winters, California.

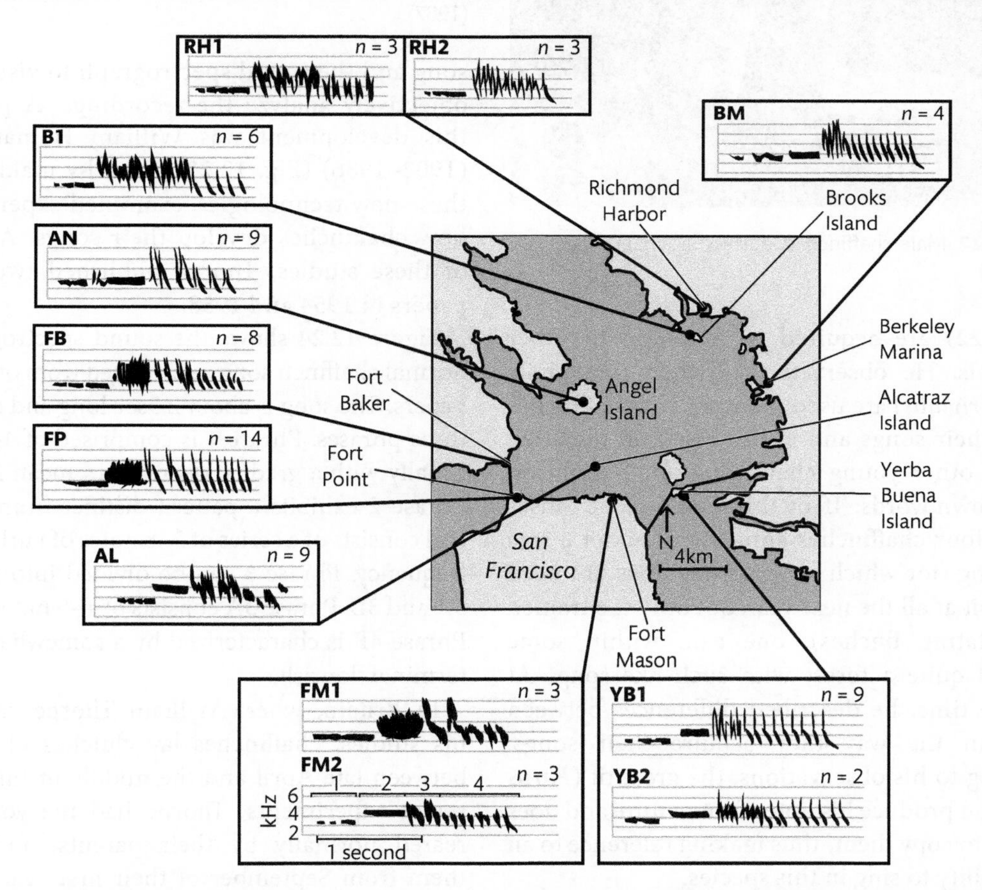

Figure 12.21 Local song dialects of the white-crowned sparrow in the San Francisco Bay area. The map shows the nine locations where songs were recorded. The sound spectrograms reveal a total of twelve song types at the nine locations. The bird populations at three locations (Richmond Harbor, Fort Mason, and Yerba Buena Island) harbored two different song types each. For song type FM2, time scale and frequency range of the sound spectrogram, as well as details of the song pattern are indicated (1: introductory whistle; 2: second phrase; 3: complex syllables; 4: simple syllables). (Courtesy: Slabbekoorn, H., Jesse, A., and Bell, D.A. (2003). Microgeographic song variation in island populations of the white-crowned sparrow (*Zonotrichia leucophrys nuttalli*): innovation through recombination. *Behaviour* **140**:947–963. © Brill)

Figure 12.22 Male chaffinch. (Courtesy: Michael Maggs (CC BY-SA 2.5).)

Figure 12.23 William Homan Thorpe. (From: Hinde, R. A. (1987).)

(Fig. 12.22) are acquired by listening to other individuals. He observed that when tree pipits (*Anthus trivialis*) are used as tutors, the chaffinches learned their songs and even passed on the alien songs to other young chaffinches. In Baron von Pernau's own words: 'If, by this method, one causes three or four chaffinches annually to adopt a tree pipit's song (for which purpose one does not need a tree pipit at all the next year, but only the trained and imitating finches), one can, within some years, fill quite a forest with such-like song.' At the same time, he described differences between species in the way they acquire their songs. According to his observations, the great tit (*Parus major*) 'can produce his notes from nature and does not need to copy them,' thus making reference to an innate ability to sing in this species.

Despite these remarkable observations, it took until the 1950s for significant further advances to be made in the study of song acquisition by birds. This progress was made possible by the introduction of the tape recorder to record bird song and the sound spectrograph to visualize and objectively analyze the recordings. A pioneer in this development was William Homan Thorpe (1902–1986) (Fig. 12.23), who, by making use of these new technologies, examined experimentally how chaffinches develop their songs. As a result of these studies, Thorpe published two seminal papers in 1954 and 1958.

Figure 12.24 shows the sound spectrogram of a normal chaffinch song reproduced from one of these papers. The song is about 2.3sec long and consists of three phrases. Phrase 1 is comprised of 4–14 notes, usually with a gradual decrease in mean frequency. Phrase 2 exhibits a pattern distinct from Phrase 1 and consists of a series of 2–8 notes of fairly constant frequency. Phrase 3 can be divided into two parts, 3A and 3B. Phrase 3A consists of 1–5 notes, whereas Phrase 4B is characterized by a somewhat complex terminal flourish.

In Britain, where William Thorpe carried out his studies, chaffinches lay clutches of 4–5 eggs between late April and the middle of June. In one set of experiments, Thorpe had the young birds reared normally by their parents, and isolated them from September of their first year onwards. He found that, although these young birds did not see any other singing chaffinches during the time of isolation, they developed a normal chaffinch song in the next spring, closely resembling the one shown in Fig. 12.24.

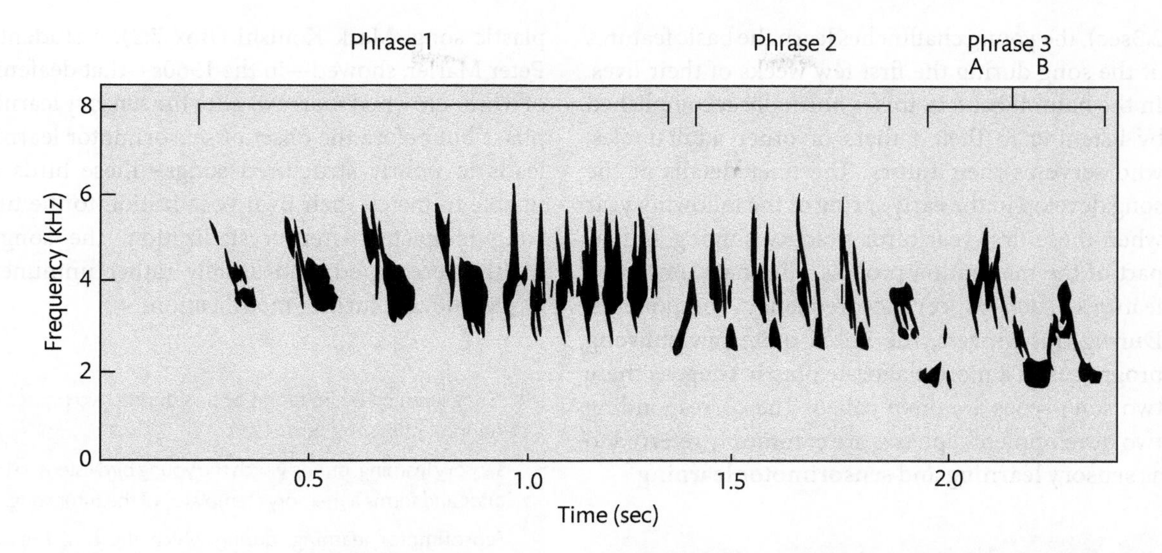

Figure 12.24 Sound spectrogram of the song of a male chaffinch. The song is typical of a chaffinch reared normally by his parents. Three phrases can be distinguished. The third phrase is divided into two sub-phrases, labeled A and B. The physical properties of the phrases are explained in detail in the text. (Thorpe, W.H. (1954). © 1954 Nature Publishing Group)

However, if, in a different set of experiments, the young birds were hand-reared and isolated from experienced birds since the first few days of nestling life, very different song patterns developed—Phrases 1 and 2 were often inseparable, and Phrase 3A was always lacking. Phrase 3B was also often lacking or consisted just of a single 'squeak'. Figure 12.25 shows sound spectrograms of such songs of two hand-reared chaffinches.

Based on these experiments, Thorpe concluded that **the adult song of the chaffinch male is the result of both innate and learned elements.** Within genetically determined limits, which are rather broadly defined (e.g. the length of the song is about

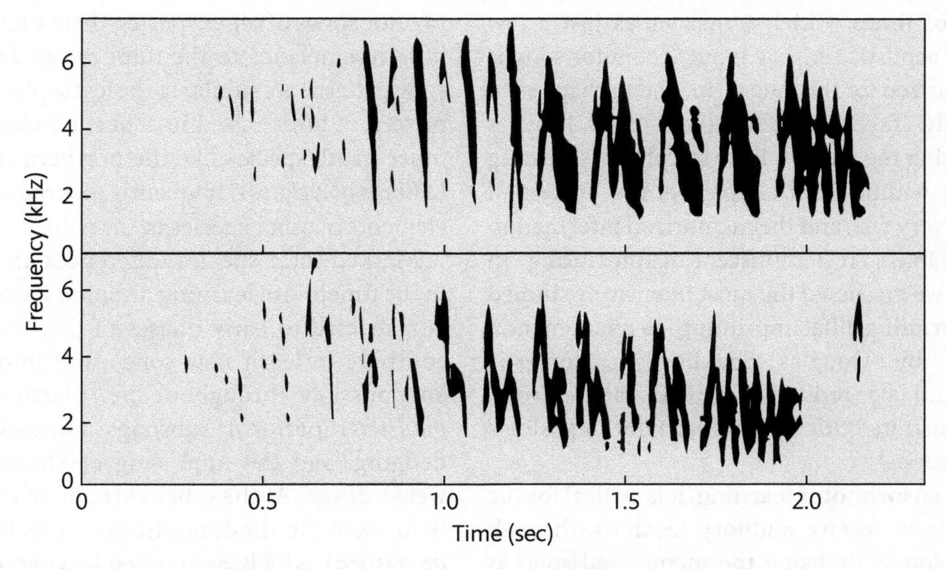

Figure 12.25 Sound spectrograms of two unrelated hand-reared chaffinches. The young birds were isolated from about day 4 of their lives onwards. During the sensory learning phase, they were allowed to hear only each other's songs. Comparison of their song spectrograms with the song spectrogram of a chaffinch reared normally by his parents (see Fig. 12.24) indicates that the hand-reared birds, which lacked the auditory exposure to any adult tutor during the critical phase of sensory learning, developed abnormal and quite simple songs. (Thorpe, W.H. (1954). © 1954 Nature Publishing Group)

2.3sec), the young chaffinches learn the basic features of the song during the first few weeks of their lives. In the natural habitat, this is normally accomplished by listening to their fathers or other adult males, who serve as their **tutors**. The finer details of the song develop in the early spring of the following year when these first-year birds practice singing and, as part of the maturation process, eliminate unwanted features, such as certain frequency components. During this process, the initial quite raw **subsong** progresses to a more elaborate **plastic song**, as these two song types are often called. The corresponding two developmental phases are commonly referred to as **sensory learning** and **sensorimotor learning**.

> ➤ Local song dialects are found in many species of songbirds. These dialects are, within genetically determined limits, learned from adult birds that serve as tutors.

During sensory learning, the young bird listens to its tutor (usually the father) and memorizes the tutor songs. The internal representation of the memorized tutor song is often referred to as the **template**. The sensory learning phase is restricted to a critical period within the first few weeks after hatching. Remarkably, accurate imitation of the tutor song requires listening to this song only a few hundreds of times, which translates as just a few minutes of acoustic sensory input. The tutor song is then memorized for the long term, and perhaps even permanently. Taken together, these three features characterizing the sensory learning phase—learning takes place within a critical period, the process of learning is very fast, and the memorized information is highly stable—are reminiscent of **imprinting**. In Chapter 3, we discussed the most intensively studied type of imprinting, filial imprinting—a phenomenon discovered by Douglas Spalding in domestic chickens, and later rediscovered by Oskar Heinroth, who demonstrated this learned behavior in goslings of graylag geese.

During **sensorimotor learning**, it is critical for the young birds to receive auditory feedback through their own songs. By using the memorized song as a template, they try to match their own songs, with increasing precision over time, to the template song, until the full song 'crystallizes out' of the amorphous plastic song. Mark Konishi (Box 7.2), a student of Peter Marler, showed—in the 1960s—that deafening of white-crowned sparrows after the sensory learning phase but before the onset of sensorimotor learning leads to poorly structured songs—these birds are unable to match their own vocalization to the tutor song template. After crystallization, the song is highly stereotyped, and usually rather immune to any significant further modification.

> ➤ Song learning by imitation occurs during two phases of development in song birds:
>
> Sensory learning, during which the young bird listens to the tutor and forms a memory ('template') of the tutor song.
>
> Sensorimotor learning, during which the bird tries to match, with increasing precision, its own song to the memorized tutor song.

Current comparative evidence suggests that all songbirds acquire their songs—at least to some extent—through learning, although different species may exhibit great differences in the extent of learning and the timeline of learning. White-crowned sparrows, for example, have a clear predisposition to learn songs of their own species. When they are individually isolated at a few days of age and exposed to tutor songs of other species, their adult songs bear little resemblance to the tutor songs. Instead, their song patterns are similar to the less sophisticated song pattern of birds raised in acoustic isolation. On the other hand, species like the northern mockingbird (*Mimus polyglottos*) frequently incorporate also song elements of other species in the wild.

Marked inter-species differences are also found in the timeline of learning. In some species, learning is restricted to early phases of life, whereas other continue to learn new songs well into adulthood and possibly throughout life. Marsh tits (*Poecile palustris*) perform subsongs immediately upon fledging, and the final song crystallizes by 8–10 weeks of age. At the other extreme of the spectrum is, for example, the domestic canary (*Serinus canaria domestica*), which, as an **open learner**, can develop new songs throughout life. Figure 12.26 illustrates the similarities and differences in timeline of song learning of two well-studied species, the zebra

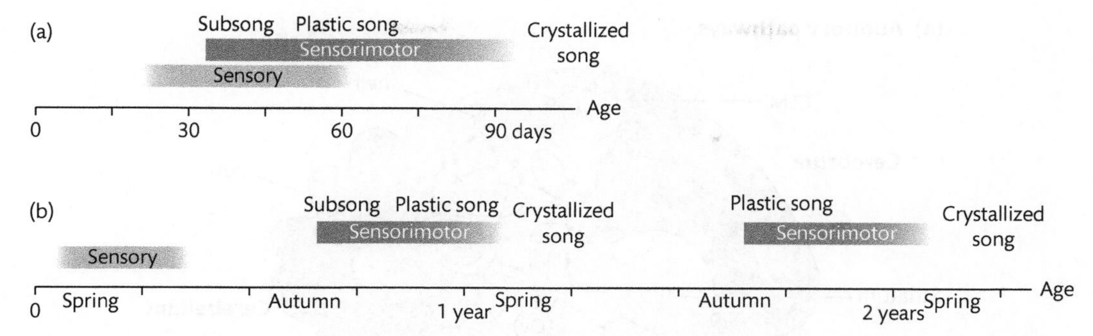

Figure 12.26 Timelines of the development of song learning in two species of songbirds, the zebra finch (a) and the white-crowned sparrow (b). During sensory learning, the young bird listens to and memorizes tutor songs. During sensorimotor learning, the young bird starts to produce its own song. Initially, this so-called subsong resembles a babbling-like vocalization. However, the subsong is continuously modified through auditory feedback to match, with progressing accuracy, the memorized tutor song. The resulting song is more structured but still variable, hence is referred to as plastic song. This maturation process terminates when the song 'crystallizes out' by greatly reducing the variability of the song components. The final product, the crystallized song, is used to attract females, and to establish and defend territories. As comparison of the two timelines of these developmental events shows, zebra finches develop the stage represented by the crystallized song within 90–120 days after hatching, and the sensory and sensorimotor learning phases overlap partially. By contrast, in white-crowned sparrows, as typical of songbirds of temperate zones, the sensory learning and the sensorimotor learning phases are separated by many months. The final stage of this development, song crystallization, is reached only at the end of the bird's first year of life. Notably, sparrows can undergo another cycle of plastic changes and song crystallization during the second year. (After: Mooney, R. (2009). © 2009 by Cold Spring Harbor Laboratory Press)

finch (*Taeniopygia guttata*) and the white-crowned sparrow.

Some intriguing details of the developmental learning of bird song were revealed after a group of scientists from The Rockefeller University, the California Institute of Technology, and the Bell Laboratories developed a fully automated procedure that can measure the similarity between songs. By employing this spectral analysis method, subsequent research by Ofer Tchernichovski of the City University of New York showed that zebra finches, after starting to imitate the tutor songs, display a pronounced daily rhythm: The song structure deteriorates after each night of sleep, but regains after intense morning singing. During the late phase of this morning recovery, the matching of the juvenile's song to the tutor's song increases daily. Although the young bird continues to sing for the rest of the day, little further improvement occurs during this afternoon phase of the daily cycle. Surprisingly, the songs of the birds that exhibit stronger post-sleep deterioration eventually are more similar to the respective tutor songs than the songs of birds that show less deterioration in the morning. One possible function of the increase in the frequency of the less structured songs in the morning is that it provides the birds with a substrate to explore their vocal abilities, and through this strategy to improve imitation of the tutor song.

The neural circuits for song perception, song production, and song learning

In songbirds, three interconnected neural pathways are involved in song perception, song production, and song learning.

The **auditory pathways** (Fig. 12.27(a)) play an essential role in song perception and, as will be discussed in detail below, are likely to also participate in the encoding of tutor song memories.

The **song motor pathway** (Fig. 12.27(b)) is critically involved in song production. At the apex of this pathway is the **HVC** (formerly known as the high vocal center; now the abbreviation is used as a proper name). One of its neuronal populations connects directly with the robust nucleus of the arcopallium. The neurons of the latter project to the tracheosyringeal portion of the hypoglossal

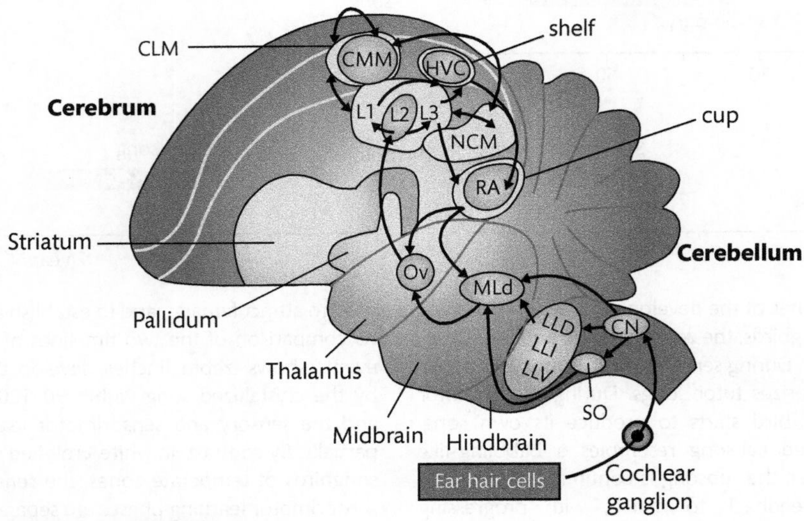

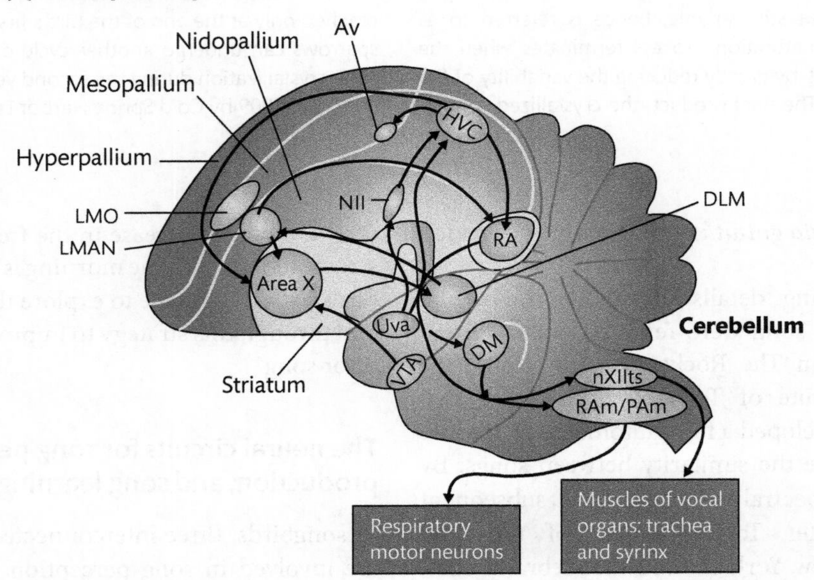

Figure 12.27 Neural pathways in the songbird brain involved in song perception, song production, and song learning. (a) Auditory pathways. (b) Vocal pathways, including the song motor pathway and the anterior forebrain pathway. For a detailed description of these pathways, see text. Area X, Area X of the striatum; Av, avalanche; CLM, caudolateral mesopallium; CN, cochlear nucleus; DLM, medial subdivision of the dorsolateral nucleus of the anterior thalamus; DM, dorsomedial subdivision of nucleus intercollicularis of the mesencephalon; HVC, the acronym is used as a proper name; L1, L2, and L3 are subdivisions of Field L; LLD, lateral lemniscus, dorsal nucleus; LLI, lateral lemniscus, intermediate nucleus; LLV, lateral lemniscus, ventral nucleus; LMAN, lateral magnocellular nucleus of the anterior nidopallium; LMO, lateral oval nucleus of the mesopallium; MLd, dorsal part of the lateral nucleus of the mesencephalon; NIf, interfacial nucleus of the nidopallium; nXIIts, tracheosyringeal portion of the nucleus hypoglossus (nucleus XII); Ov, nucleus ovoidalis; PAm, nucleus parambigualis; RA, robust nucleus of the arcopallium; RAm, nucleus retroambigualis; SO, superior olive; Uva, nucleus uvaeformis; VTA, ventral tegmental area. (After: Moorman, S., Mello, C.V., and Bolhuis, J.J. (2011), © 2011 Wiley Periodicals, Inc.)

motor nucleus (XIIts) and the respiratory premotor neurons of the ventral respiratory group. The hypoglossal motor nucleus innervates the muscles of the trachea and the syrinx, the vocal organs. The ventral respiratory group is composed of two nuclei: the nucleus retroambigualis, which controls expiration; and the nucleus parambigualis, which controls inspiration.

Two lines of evidence indicate that the song motor pathway plays a vital role in song production. First, as demonstrated by Fernando Nottebohm, Tegner M. Stokes, and Christiana M. Lenard of Rockefeller University and City University of New York, lesions of the HVC and the robust nucleus of the arcopallium in adult male canaries cause severe song deficits. Second, James S. McCasland and Mark Konishi of the California Institute of Technology showed by extracellular recordings, using chronically implanted electrodes, that neurons in the HVC greatly increase their neural activity time-locked with certain elements of songs. This activity is not due to sensory perception of these songs because the neural activity *leads*, with a relatively constant latency, the onset of sound production, and the song-related neural discharges in HVC persist even in deafened birds.

Further exploration of the function of the HVC in song production has revealed that those HVC neurons that project to the robust nucleus of the arcopallium fire extremely sparsely during singing. Each of these neurons generates a single brief burst of action potentials at a specific time point of a song motif. (A song motif is a phrase of the song that consists of several syllables.) Different neurons of this HVC population burst at different time points during the song. Taken together, these two observations have led to the hypothesis that the population of HVC neurons that project to the robust nucleus of the arcopallium control temporal features of the song. This notion has received further support from experiments carried out by Michael A. Long and Michale S. Fee of the Massachusetts of Technology. They found that cooling of the HVC slows the tempo of the song. By contrast, local manipulation of the temperature of the robust nucleus of the arcopallium has no observable effect on the timing of the song.

The third pathway is commonly referred to as the **anterior forebrain pathway**. It originates in HVC, passes through Area X of the striatum and the thalamic nucleus dorsolateralis anterior pars medialis, which, in turn, innervates the lateral magnocellular nucleus of the anterior nidopallium, and finally connects to the song motor pathways via the robust nucleus of the arcopallium. The combination of the song motor pathway and the anterior forebrain pathway has become known as the **song system**.

Like neurons in the song motor pathway, neurons in the anterior forebrain pathway display singing-related neural activity. However, as will be discussed in detail in the next section, several lines of evidence suggest that the anterior forebrain pathway is involved in song variability and plasticity, rather than in the production of the crystallized song.

> ➤ The song motor pathway in the brain of songbirds is critically involved in song production, whereas the anterior forebrain pathway plays a major role in song variability and plasticity.

The role of the anterior forebrain pathway in song variability

A critical element of the anterior forebrain pathway is its output nucleus, the **lateral magnocellular nucleus of the anterior nidopallium**. Several studies in juvenile zebra finches have shown that lesions of this pathway disrupt song development, resulting in repetitive, simplified songs. Similarly, inactivation of this nucleus in juvenile males by injection of tetrodotoxin (a sodium channel blocker; see Chapter 2) leads to an immediate, reversible loss of acoustic variability across song renditions. This pharmacologically induced conversion of the variable song of juveniles into a highly stereotyped song is illustrated by Fig. 12.28.

In adults, inactivation of the lateral magnocellular nucleus of the anterior nidopallium through lesions has little effect on the previously learned song. However, such a procedure affects song plasticity, as demonstrated in a study by Mimi H. Kao and Michael S. Brainard of the University of California, San Francisco. This effect became

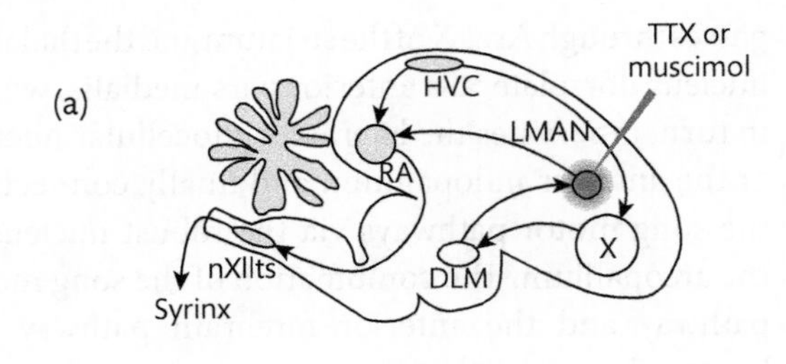

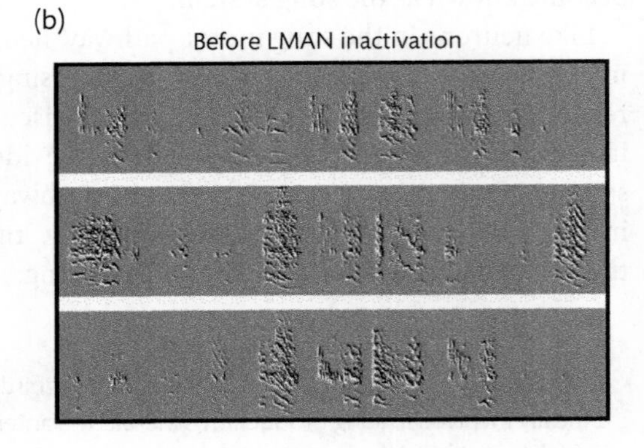

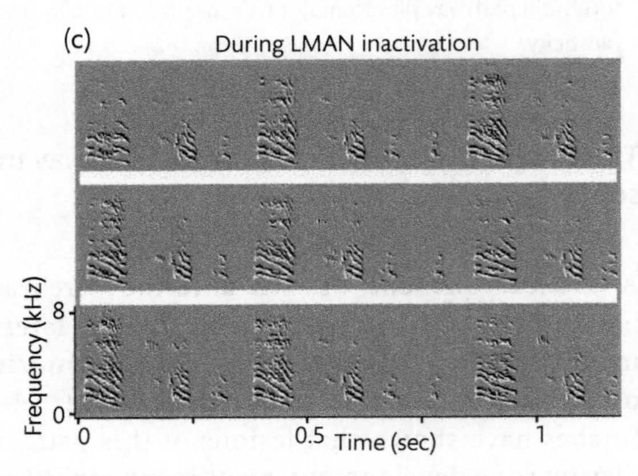

Figure 12.28 Effect of pharmacological inactivation of the lateral magnocellular nucleus of the anterior nidopallium on song variability in juvenile zebra finch males. In this experiment carried out by Bence P. Ölveczky, Aaron S. Andalman, and Michale S. Fee of the Massachusetts Institute of Technology and Harvard University, both in Cambridge, Massachusetts (U.S.A.), the lateral magnocellular nucleus of the anterior nidopallium of juvenile zebra finches was injected with tetrodotoxin (TTX) in both hemispheres (a). The songs were recorded immediately before the inactivation (b) and 1 hour after the inactivation (c). The frequency-time plots show spectral derivatives, a representation that facilitates the identification of song similarity. Visual inspection of these plots demonstrates that the normal song of the juvenile zebra finch male shown here is highly variable in sequence and acoustic structure of song syllables. However, inactivation reduces this variability so that a highly stereotyped song results. (Courtesy: Ölveczky, B.P., Andalman, A.S., and Fee, M.S. (2005). Vocal experimentation in the juvenile songbird requires a basal ganglia circuit. *PLoS Biology* **3:**e153. © 2005 Ölveczky et al.)

apparent when the scientists compared the songs of male zebra finches after lesion under two different behavioral situations. As is known from behavioral observations, when male zebra finches sing alone (thereby producing so-called undirected songs), the variability in syllable structure is much greater than when they sing to a female (thereby producing so-called directed songs). Lesions of the lateral magnocellular nucleus of the anterior nidopallium eliminate this difference between the two song types, thus preventing context-dependent song plasticity.

How does the activity of the lateral magnocellular nucleus of the anterior nidopallium drive song variability?

Axons arising from the lateral magnocellular nucleus of the anterior nidopallium terminate on neurons of the robust nucleus of the arcopallium, adjacent to axonal terminals of HVC neurons. Although both inputs are glutamatergic, their postsynaptic properties differ. Whereas the excitation from the lateral magnocellular nucleus of the anterior nidopallium is mediated almost exclusively by NMDA receptors, the excitatory input from the HVC is mediated predominantly by AMPA receptors and only to a minor extent by NMDA receptors. It is thought that the lateral magnocellular nucleus of the anterior nidopallium drives variability by adding 'noise' to the premotor signals from HVC, and that this effect is based on the binding of glutamate to postsynaptic NMDA receptors on neurons of the robust nucleus of the arcopallium. As expected, blockade of NMDA receptors by injection of the NMDA receptor antagonist 2-amino-5-phosphonovalerate (better known by its abbreviation, AP5) into the robust nucleus of the arcopallium markedly reduces the acoustic variability in song syllables (Fig. 12.29). However, such pharmacological interference does not affect the acoustic structure of the syllables.

Where in the brain is the tutor song template stored?

As detailed above, behavioral experiments have indicated that male songbirds learn to sing from adult tutors during a critical period in juvenile life. An important phase in this process is the formation

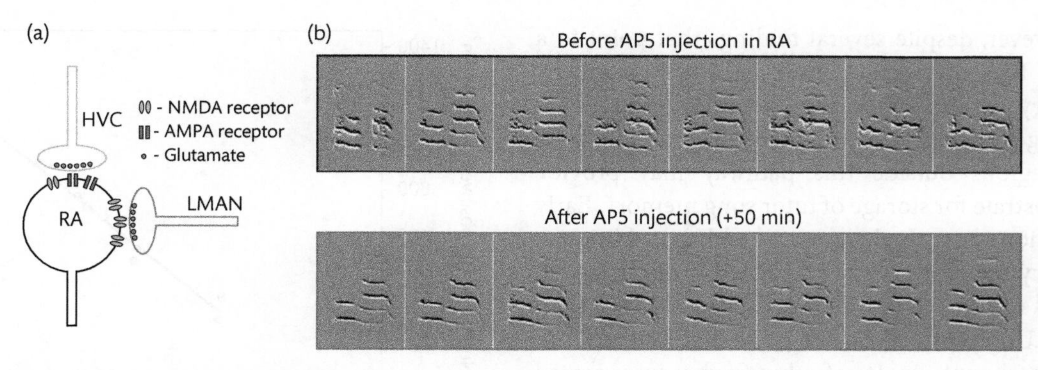

Figure 12.29 Effect on song variability of pharmacological blockade of synaptic input from the lateral magnocellular nucleus of the anterior nidopallium to the robust nucleus of the arcopallium. (a) The inputs to the robust nucleus of the arcopallium from the lateral magnocellular nucleus of the anterior nidopallium are mediated by release of glutamate from the presynaptic terminals, and by binding of this excitatory transmitter to distinct populations of postsynaptic glutamatergic receptors—NMDA receptors associated with synapses formed by neurons of the lateral magnocellular nucleus of the anterior nidopallium, and a mix of NMDA receptors and AMPA receptors formed by HVC neurons. This observation prompts the hypothesis that application of the NMDA receptor antagonist AP5 should block input completely from the lateral magnocellular nucleus of the anterior nidopallium, but lead only to partial inactivation of input from HVC. The following experiment conducted by Bence P. Ölveczky, Aaron S. Andalman, and Michale S. Fee is based on this consideration. (b) Spectral derivatives of eight sequential renditions of one song syllable in a 63-day old zebra finch before (top row) and after (bottom row) AP5 injection into the robust nucleus of the arcopallium. Syllable variability in the song prior to AP5 application is evident from the rapid fluctuations in pitch, the appearance of a noisy acoustic structure, and the variations in syllable duration. (c) The reduction in syllable variability after application of AP5 to the robust nucleus of the arcopallium can be assessed by calculating a variability score. This quantitative measure yields values between 0 (if two syllables are identical) and 1 (if two syllables are unrelated). For the plot shown here, 11 syllables in 4 birds before and after injection of AP5 into the robust nucleus of the arcopallium were analyzed (open circles). In control experiments, saline, instead of AP5, was administered (closed diamonds). While the variability scores of the syllables before and after saline injection are very similar (the line indicates identical pre- and post-injection values), the scores after AP5 injection are markedly reduced compared to the scored obtained before blocking of NMDA receptors. (Courtesy: Ölveczky, B.P., Andalman, A.S., and Fee, M.S. (2005). Vocal experimentation in the juvenile songbird requires a basal ganglia circuit. *PLoS Biology* 3:e153. © 2005 Ölveczky et al.)

of a memory, often referred to as the 'template', of the tutor song. Where in the brain is this memory created?

There is broad consensus among scientists that the song template represents an **auditory memory**—it is sufficient that the juvenile bird *hears* the tutor, but it is not necessary that the bird can also *see* the tutor. Given the importance of the anterior forebrain pathway for sensorimotor learning, it was initially assumed that memory formation takes place within this pathway.

Indeed, electrophysiological recordings in anesthetized zebra finches from neurons in the lateral magnocellular nucleus of the anterior nidopallium, carried out by Michele M. Solis and Allison J. Doupe of the University of California, San Francisco, are in agreement with this hypothesis: In 30-day old birds, these neurons respond equally well to the bird's own song as they do to songs of other birds. However, this situation changes rapidly: At 60 days of age, the responses of the neurons of the lateral magnocellular nucleus of the anterior nidopallium have become selective—they respond now more to the bird's own song than to songs of other zebra finches.

This selectivity of the neurons develops during song learning. Intriguingly, as shown subsequently by Solis and Doupe, tutor song-selective neurons can be identified even in birds that produce abnormal song induced by denervation of the syrinx, before song onset. This experiment suggests that it is not simply the similarity to the bird's own song that evokes a response in these neurons, but that they may encode a neural representation of the tutor song.

However, despite several findings that point to a role of the anterior forebrain pathway in auditory memory formation during the imprinting phase of song learning, other lines of evidence indicate that regions outside this pathway may provide the substrate for storage of tutor song memory. Early indication for such alternate, or perhaps additional, auditory memory sites originated from research carried out by Johan J. Bolhuis and associates, then at the University of Leiden (The Netherlands). In the experiments, males of zebra finches were reared without their father, but exposed, during a critical period of song learning, to a tape-recorded tutor song. When, as adults, these birds were re-exposed to the tutor song, neurons in the **caudal medial nidopallium** exhibited increased expression of the immediate early genes *zenk* and *c-fos*—indications of increased neuronal activity. The level of expression of these immediate early genes correlated with the number of song elements that the experimental birds had copied from the tutor song (Fig. 12.30).

Question: *What would you suggest as controls for this experiment?*

Answer: *As a control experiment, Bolhuis and associates did not re-expose the adult males to the tutor song. In addition, the scientists examined possible changes in anti-Fos immunoreactivity in two control areas of the song system, the HVC and Area X. There was no statistically significant increase in immunolabeling in these two nuclei after exposure of the adult birds to the tutor song.*

The notion that the caudal medial nidopallium is involved in auditory memory formation has received further support from a study carried out by Sarah E. London and David F. Clayton of the University of Illinois, Urbana-Champaign. They transiently suppressed the extracellular signal-regulated kinase signaling pathway (which regulates the expression of *zenk*) in this brain region during controlled tutor experience. Following maturation, male zebra finches that had been subjected, as juveniles, to such treatment produced, as adults, poor copies of the tutor song.

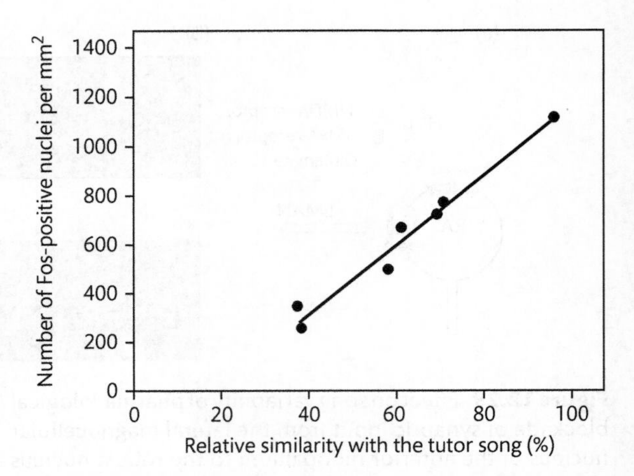

Figure 12.30 Correlation between Fos immunoreactivity in the caudal medial nidopallium and the degree of tutor song learning in males of the zebra finch. The birds were exposed as juveniles to recorded tutor song during a critical period of development, and re-exposed to this song as adults. Expression of the immediate early gene *c-fos* is used as an indicator of neuronal activity. The protein product of this gene is Fos. The degree of tutor song learning was assessed by calculating the percentage of song elements that the birds had copied from the tutor song. (Courtesy: Bolhuis, J.J., Hetebrij, E., Den Boer-Visser, A. M., De Groot, J. H., and Zijlstra, G. G. O. (2001). Localized immediate early gene expression related to the strength of song learning in socially reared zebra finches. *European Journal of Neuroscience* **13:**2165–2170. © 2001 John Wiley & Sons, Inc.)

Summary

- Communication involves the transfer of information from a sender to a receiver. In a more narrowly defined sense, this exchange of information benefits both sender and receiver.
- The vehicle that transfers information is the signal. Depending on the sensory modality involved, different types of communication signals are distinguished, for example visual, acoustic, chemical, tactile, and electric signals.
- The first model system covered in this chapter centers around the neuroethology of cricket song. Male crickets produce different types of sound.

One type, commonly referred to as a calling song, triggers positive phonotaxis in females. This approaching behavior is followed by courtship and mating.

- The calling songs consist of individual pulses called syllables. Each syllable is composed of an uninterrupted train of sound waves. The duration of the individual waves determines the carrier frequency of the song. The interval between the onset of one syllable and the onset of the next syllable defines the syllable rate. Syllables occurring in repeated short sequences are called chirps.
- The calling songs of different species may differ considerably. This is mainly due to differences in the temporal structure of the songs, such as the syllable rate and the arrangement of syllables.
- The songs are produced by a file-and-scraper stridulation mechanism localized in the forewings. The movements of the forewings are achieved through the action of the wing-opening and wing-closing muscles of the second thoracic segment.
- The activity of the wing-opener and wing-closer motoneurons, and thus the rhythmicity of the singing behavior, is controlled by two independent timing networks in the abdominal nerve cord—one that defines the chirp rate, and another that determines the syllable timing pattern.
- An approach commonly used to collect quantitative behavioral data on the phonotaxic response of receptive females to sound broadcast from loudspeakers employs the Kramer locomotion compensator consisting of a spherical treadmill and an infrared monitoring device. This setup forces the cricket to walk on top of the sphere throughout the experiment, while recording the direction and velocity of the cricket's movements.
- Results of such behavioral experiments have shown that the features important for eliciting phonotaxis, in the female, are the carrier frequency and the syllable repetition rate of the songs.
- The cricket ear acts as pressure gradient ears. Sound pressure reaches both the outer and the inner surface of the tympanum. This, as well as the physiological properties of the primary auditory organ, the tympanal organ, enables the cricket to localize sound. The directional sensitivity is sharpened by omega neurons, auditory interneurons in the first thoracic ganglion.
- No preference for species-specific syllable rates is found at either the level of the ear and the primary receptors, or at the level of the neurons in the prothoracic ganglion. Such a preference occurs only in higher-order brain neurons.
- Several mechanisms have been proposed how the brain might recognize the species-specific syllable repetition rate of the male calling song, including band-pass filtering and delay line coincidence detection.
- Comparison—in the communication system—of the biophysical structure of the male calling songs with the behavioral and physiological response of the females has demonstrated a close match between sender and receiver.
- This sender–receiver matching is challenged by the fact that, in the field, male crickets produce calling songs over a wide range of temperatures. The temporal structure of the songs may vary considerably, as chirp rate and syllable rate increase linearly with temperature. However, both sequential and two-choice paradigms have shown that the strongest phonotactic response can be evoked in females to model songs appropriate for the temperature at which they are kept ('temperature coupling' of sender and receiver).
- As a second model system, the development of bird song has been covered in this chapter. Systematic research into this area began in the 1950s after tape recorders and sound spectrographs became available to scientists.
- Studies in a variety of songbirds have suggested that their songs are acquired, within genetically determined limits, through learning during ontogenetic development. The general principles of this learning processes are similar across species, but the exact timelines, and the results

of this learning process, exhibit marked inter-species differences.

- The first step of this development ('sensory learning') is characterized by the young songbird listening to the song of an adult tutor, usually the father.
- As part of the subsequent step ('sensorimotor learning'), the young bird produces its own song, which it tries to match, with progressing accuracy, to the memorized tutor song. In the course of this maturation process, the rather crude subsong advances to the more structured plastic song, until the crystallized adult song has developed. Auditory feedback plays a critical role in this progressive developmental process.
- In the brain of songbirds, song production is controlled by the song motor pathway. At the apex of this pathway, the HVC is situated. An important function of those neurons of this nucleus that project to the robust nucleus of the arcopallium appears to be to control temporal features of the song.
- Song variability and plasticity are controlled by the anterior forebrain pathway. Pharmacological inactivation of its output nucleus, the lateral magnocellular nucleus of the anterior nidopallium, in juvenile zebra finches results in loss of acoustic variability. Lesions of this nucleus in adult zebra finches prevent context-dependent song plasticity.
- Axons arising from the lateral magnocellular nucleus of the anterior nidopallium terminate on neurons of the robust nucleus of the arcopallium, adjacent to axonal terminals of HVC neurons. Presumably, the lateral magnocellular nucleus of the anterior nidopallium drives song variability by adding 'noise' to the premotor signal from HVC.
- The formation of the 'template' of the tutor song appears to take place both within the anterior forebrain pathway (particularly in the lateral magnocellular nucleus of the anterior nidopallium) and in areas outside this pathway (particularly in the caudal medial nidopallium).

The bigger picture

In the first chapter of this book, a number of requirements were listed that a good neuroethological model system should meet. They include the attributes simple, robust, ethologically relevant, and suitable for study in the laboratory. While no one doubts that courtship behaviors of animals are ethologically relevant, they are usually anything but simple. However, neuroethologists as good animal watchers have found ways around that problem. Out of the entire set of natural behavioral patterns displayed during courtship, they typically focus on components that are sufficiently simple and robust to be accessible for ethological and neurobiological analysis. The phonotaxis behavior of crickets, one of the first comprehensively studied model systems of neuroethology, is an example par excellence for illustrating this strategy. The female cricket approaches the singing male by exhibiting positive phonotaxis, a behavior that can be readily quantified by direction and velocity of movement. However, these two parameters are difficult to record in the field. Neuroethologists have, therefore, developed, in collaboration with engineers, a solution—the Kramer locomotion compensator. This sophisticated device has enabled researchers to examine the female's behavioral response to acoustic stimuli under the controlled conditions of the laboratory. The behavioral identification of those features in the male song that are critical for triggering positive phonotaxis subsequently opened up the possibility to search for the neural correlates of this behavior in females. The translation of this strategy into numerous discoveries became one of the first success stories of neuroethology. As such, it has served as a template for the establishment of many other model systems.

Recommended reading

Gerhardt, H. C. and Huber, F. (2002). *Acoustic Communication in Insects and Anurans: Common Problems and Diverse Solutions*. University of Chicago Press, Chicago.

A marvelous synthesis of how frogs and insects produce their calls, what messages are encoded within the sounds, and how the intended recipients receive and decode these signals. The two authors have placed special emphasis on a discussion of the common solutions that the different animal groups have evolved to shared challenges, such as ectothermy, as well as on a presentation of the diversity of solutions that reflect the differences in evolutionary history. A must for any student of animal communication.

Hedwig, B. (2006). Pulses, patterns and paths: Neurobiology of acoustic behavior of crickets. *Journal of Comparative Physiology A* **192**:677–689.

An update of the research on the acoustic behavior of crickets, integrating recent discoveries into the framework established by the classical studies of Franz Huber, Ronald Hoy, and others. It covers both the central nervous system control of the singing behavior of the males, and the neuronal filter mechanisms underlying auditory pattern recognition by the females.

Huber, F., Moore, T. E., and Loher, W. (eds) (1989). *Cricket Behavior and Neurobiology*. Cornell University Press, Ithaca/London.

The 'bible' of the cricket neuroethologists: 565 pages and 15 chapters, mostly on various aspects of auditory communication, written by many of the leading figures in the field.

Mooney, R. (2009). Neural mechanisms for learned birdsong. *Learning & Memory* **16**:655–669.

A valuable review of neural mechanisms underlying imitative learning in songbirds. Emphasis is on the neural circuits for singing and song learning, and on how the individual components of these circuits contribute to song variability and adaptive song plasticity.

Short-answer questions

12.1 Define the term 'communication'. What criterion distinguishes true communication from communication in a broader sense?

12.2 Name three types of song generated by male crickets. Briefly characterize the behavioral situations in which these songs are produced.

12.3 Oscillograms have revealed two major components that constitute the calling song of male crickets. Name and briefly describe these components.

12.4 Which body structure generates the songs of crickets? What is the name of the underlying mechanism?

12.5 In the abdominal nerve cord of male crickets, two independent timing networks for the control of singing have been identified. What aspects of this behavior do they control?

12.6 Describe, in one or two sentences, the positive phonotaxis of female crickets.

12.7 What behavioral aspects of phonotactic behavior of female crickets can be analyzed by use of the Kramer locomotion compensator?

12.8 In which part of the body of field crickets are the ears located? What are their main structural components?

12.9 Sketch the pathway, in the cricket brain, that is involved in processing auditory information.

12.10 How does ambient temperature influence the chirp period, syllable duration, syllable period, and carrier frequency of the calling songs of male crickets?

12.11 What solutions have crickets developed to the problem that some parameters of the calling song of male crickets vary in concert with changes in the ambient temperature?

12.12 Name two technical innovations that greatly advanced the study of bird song in the 1950s?

12.13 What are the two major developmental stages of young songbirds in the acquisition of songs? Characterize them in one sentence each.

12.14 What features of bird song learning support the hypothesis that this process involves imprinting?

12.15 Define, in one sentence, the term 'song crystallization'.

12.16 What characteristic feature distinguishes open learners from other songbirds in terms of song acquisition? Name one songbird species that is an open learner.

12.17 What are the names of the three neural pathways that are involved in song perception, song production, and song learning in songbirds?

12.18 It has been hypothesized that the subpopulation of HVC neurons that project to the robust nucleus of the arcopallium control the tempo of the bird song. Describe, in one sentence, an experiment that has provided support for this notion.

12.19 How does inactivation, by injection of tetrodotoxin, of the lateral magnocellular nucleus of the anterior nidopallium affect song variability in juvenile zebra finches?

12.20 What correlative evidence suggests that the caudal medial nidopallium plays a role in auditory memory formation during the imprinting phase of song learning?

12.21 **Challenge question** Chemical communication through pheromones plays a pivotal role in the context of reproduction and aggregation of many insects. Synthetic versions of such pheromones have been increasingly used over the last half a century to manage pest. One application of these pheromones is their use for detection of harmful insects and for monitoring the populations of these organisms. Explain how the data collected through this approach can aid the use of more specific tools, rather than broad-spectrum insecticides, to manage pest.

Essay questions

12.1 What is the Kramer locomotion compensator? How does it work? What is the advantage of using this device, instead of an arena, for phonotaxis experiments in crickets?

12.2 Male crickets use a file-and-scraper mechanism to produce songs. What are the main features of this mechanism? How does the biophysical structure of a song syllable relate to the mechanical structure of the song-producing system?

12.3 Starting with a biophysical analysis of the calling song, describe how male field crickets produce signals to attract females, and how the female perceives and processes features relevant to elicit phonotaxis.

12.4 How are crickets able to localize sound? Discuss the role of both the ears and the central structures that carry out the analysis of auditory information in this process.

12.5 Changes in ambient temperature lead to alterations in certain parameters of the calling song of field crickets. How do female crickets cope with this potential problem, to ensure a correct phonotactic response?

12.6 Summarize the evidence obtained through field observations and laboratory experiments that has suggested that learning plays a major role in the development of bird song. What phases in this development can be distinguished, and what behavioral events characterize them?

12.7 The lateral magnocellular nucleus of the anterior nidopallium plays a critical role in song variability in juvenile birds and song plasticity in adult birds. How do lesioning and pharmacological inactivation experiments support this hypothesis? Describe the synaptic mechanism through which activity of this nucleus is thought to drive song variability.

12.8 What experiments point to the involvement of the anterior forebrain pathway in the formation of the tutor song template? On the other hand, what evidence supports the notion that the substrate for this auditory memory is provided by regions outside this pathway?

Advanced topic Genetic coupling in the cricket mate-recognition system

Background information

A question that has challenged biologists for a long time is how communication systems evolve. If changes occur, for example in the sender's signaling system, then certain changes may also be required in the receiver to maintain its ability to respond correctly to the signals produced by the sender. This is particularly evident in mate-recognition systems. If signaling and receptive systems diverge too much, mating may fail. Thus, both signal and receptor are likely to be under stabilizing selection. How then can mate-recognition systems change, while maintaining the coordination of the signals and the receptors?

Examining acoustic communication in crickets, Richard D. Alexander of the University of Michigan, Ann Arbor, addressed this issue in a seminal paper published in 1962. He suggested that there is genetic coupling between the song-producing system of males and the acoustic preference system of females. Through such a mechanism, the evolution of a new signaling system would automatically be accompanied by the development of suitable receptor structures. As Alexander put it: *'If there is a linkage—or an identity—here, it would represent an interesting simplification of the process of evolutionary change in a communication system—something of an assurance that the male and the female, or the signaler and the responder, really will evolve together . . . The question has significance in connection with speciation, as well as the evolution of communication . . .'*

Essay topic

Despite numerous attempts to demonstrate genetic coupling of sender and receiver structures, the actual experimental evidence is still a matter of dispute. In an extended essay, address the following aspects: First summarize the ethological evidence that shows behavioral coupling between sender and receiver in crickets. Then present the genetic coupling hypothesis as well as alternative hypotheses that have been proposed to explain the evolution of male signals and female preferences. Finally, discuss the results of studies aimed at verifying these different hypotheses.

Starter references

Alexander, R. D. (1962). Evolutionary change in cricket acoustical communication. *Evolution* **16:**443–467.

Butlin, R. K. and Ritchie, M. G. (1989). Genetic coupling in mate recognition systems: What is the evidence? *Biological Journal of the Linnean Society* **37:**237–246.

Hoy, R. R. and Paul, R. C. (1973). Genetic control of song specificity in crickets. *Science* **180:**82–83.

Hoy, R. R., Hahn, J., and Paul, R. C. (1977). Hybrid cricket auditory behavior: Evidence for genetic coupling in animal communication. *Science* **195:**82–84.

Shaw, K. L. and Lesnick, S. C. (2009). Genomic linkage of male song and female acoustic preference QTL underlying a rapid species radiation. *Proceedings of the National Academy of Sciences U.S.A.* **106:**9737–9742.

Wiley, C., Ellison, C. K., and Shaw, K. L. (2012). Widespread genetic linkage of mating signals and preferences in the Hawaiian cricket *Laupala*. *Proceedings of the Royal Society B* **279:**1203–1209.

To find answers to the short-answer questions and the essay questions, as well as interactive multiple choice questions and an accompanying Journal Club for this chapter, visit www.oup.com/uk/zupanc3e.

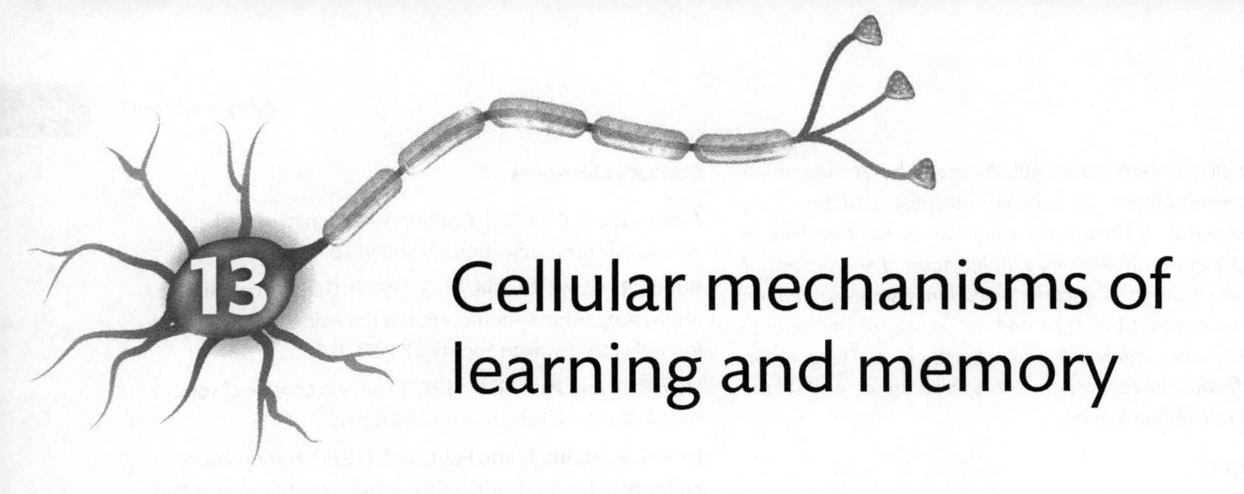

13 Cellular mechanisms of learning and memory

Introduction

Whereas a large part of this book has focused on neural correlates of simple behaviors, the reductionist approach of behavioral neurobiology can equally well be applied to the study of more complex behavioral and cognitive processes. As we will demonstrate in this chapter, one of these processes involves the way memory is stored in the brain. Research in this area has been stimulated in particular by the work of the eminent Canadian psychobiologist Donald O. Hebb, who, in his book *The Organization of Behavior: A Neuropsychological Theory*, published in 1949, made concrete proposals on how the physiological properties of neural networks might contribute to perception and memory formation (see Box 13.1). Since then, an ever-growing number of researchers have tested and expanded the ideas of Hebb using a variety of empirical approaches. Out of the numerous aspects studied over the past half a century, we will cover mainly one—the cellular mechanisms that underlie memory storage. Before we can take a closer look at the major findings in this area, we first need to discuss the evidence supporting the concept of two different memory systems.

Key concepts

In 1957, a seminal discovery was reported by the American neurosurgeon William Scoville, and the Canadian psychologist Brenda Milner who had done her Ph.D. thesis under the supervision of Donald Hebb at McGill University in Montreal. They reported the case of a patient, Henry Gustav Molaison, better known as 'H. M.' (see Journal club at www.oup.com/uk/zupanc3e). As a child, H. M. had a head injury that eventually led to epilepsy. Over the years, his seizures

Box 13.1 Donald O. Hebb

Donald O. Hebb. (Courtesy: McGill University Archives.)

A rather modest and self-critical man during his lifetime, Donald Hebb has become a legend, especially after his death. His theories on the self-assembly of neurons, particularly on the relationship between brain and behavior, revolutionized psychology, then dominated by Freudian views or radical behaviorism, and made Hebb the father of cognitive psychobiology.

Donald Olding Hebb was born in Chester, Nova Scotia (Canada), in 1904 and graduated from Dalhousie University in 1925. He first went into education and became a school principal in the Province of Quebec. In 1936, he obtained his Ph.D. from Harvard with a thesis on the effects of early visual deprivation upon size and brightness perception in rats. Following an invitation by Wilder Penfield, he accepted a fellowship at the Montreal Neurological Institute at McGill University to examine the impact of brain injury and surgery on human behavior and intelligence. In 1942, Hebb joined Karl Lashley (see Chapter 3) at the Yerkes Laboratory of Primate Biology, where he explored fear, anger, and other emotional processes in chimpanzees. Particularly stimulated by the intellectual climate of this laboratory, Donald Hebb made a successful attempt to bridge, in the first half of the twentieth century, the wide gap between neurophysiology and psychology through the publication of *The Organization of Behavior: A Neuropsychological Theory* in 1949. In this influential book, Hebb postulated pivotal ideas, which have laid the path for numerous empirical studies. Among them is what has become famous as the 'Hebb synapse'—the prediction that the efficacy of connections between neurons increases with repeated stimulation of the postsynaptic neuron by the presynaptic neuron. Experimental confirmation of this postulate was only obtained decades later when long-term potentiation and the generation of new synapses as a result of increased neuronal activity was discovered (see 'Long-term potentiation' later in this chapter).

Hebb returned to McGill University as Professor of Psychology and was appointed chair of this institution in 1948. In this position, he attracted many outstanding scientists and made McGill University one of the premier centers of psychobiology in North America. He died in 1985.

worsened to the point where he became severely incapacitated. As a last resort to control the epileptic seizures, in 1953, when H. M. was 27, Scoville removed his medial temporal lobe, including the **hippocampus**, on both sides of the brain. Although the operation indeed relieved H. M. of his seizures, it also left him with a devastating loss of memory.

Similar profound and irreversible deficits in memory have been found in other patients with lesions of temporal lobe structures, especially of the hippocampus. Clinically, their memory deficit is typically manifested as a severe **anterograde amnesia** and a partial **retrograde amnesia** covering a certain time period preceding the lesion. Following the event that led to the lesion, these patients can still perform learning tasks, and they do these with an efficiency similar to that of normal people. However, when these patients carry out learning tasks, they do well only as long as they focus on a given task. As soon as their attention shifts to a new topic, the whole event is forgotten. However, old memories from their childhood appear to be intact, which indicates that the hippocampus is not the sole site of memory storage within the brain.

Amnesia Disturbance in long-term memory, manifested by total or partial inability to recall past experiences.

Anterograde amnesia Deficit in memory in reference to events occurring after the traumatic event or disease that caused the condition.

Retrograde amnesia Deficit in memory in reference to events that occurred before the trauma or disease that caused the condition.

Extensive testing of the learning and memory capabilities has revealed another remarkable phenomenon in patients with temporal lobe lesions. Despite their severe memory impairments, they are able to learn new sensorimotor skills quite well, with stable retention over time. One of these tasks is illustrated in Fig. 13.1(a). A human subject is shown a picture of a double-margined star and asked to draw a line between the two margins. Both a normal subject and a memory-impaired patient can do this very easily. The person is then asked to draw the lines while looking at the hand and the star in a mirror. This task is quite difficult, as one tends to draw the line from the points in the wrong direction. However, one can learn this skill with practice. Interestingly, the memory-impaired patients exhibit learning curves similar to the normal subjects (Fig. 13.1b). However, while they show steady improvement over the training sessions and retain this motor skill beyond the training phase, in the end they are unable to remember that they have carried out this task before.

This observation, made by Brenda Milner, was one of the first pieces of experimental evidence indicating that more than one memory system exists in the brain. Together with similar findings in other amnesic patients, this led to the distinction of the following two types of memory:

- **Explicit memory** (also called **declarative** or **episodic memory**) indicates what is ordinarily associated with the term 'memory.' In humans, it requires conscious recollection of people, places, objects, and events. This type of memory is dependent on the integrity of regions within the medial temporal lobe of the cerebral cortex, including the hippocampus.
- **Implicit memory** (also called **non-declarative** or **procedural memory**) forms the basis for perceptual and motor skills. In humans, it does not involve a conscious recall of the past. Furthermore, it does not require an intact temporal lobe.

Implicit memory includes not only classical conditioning, but also forms of behavioral changes

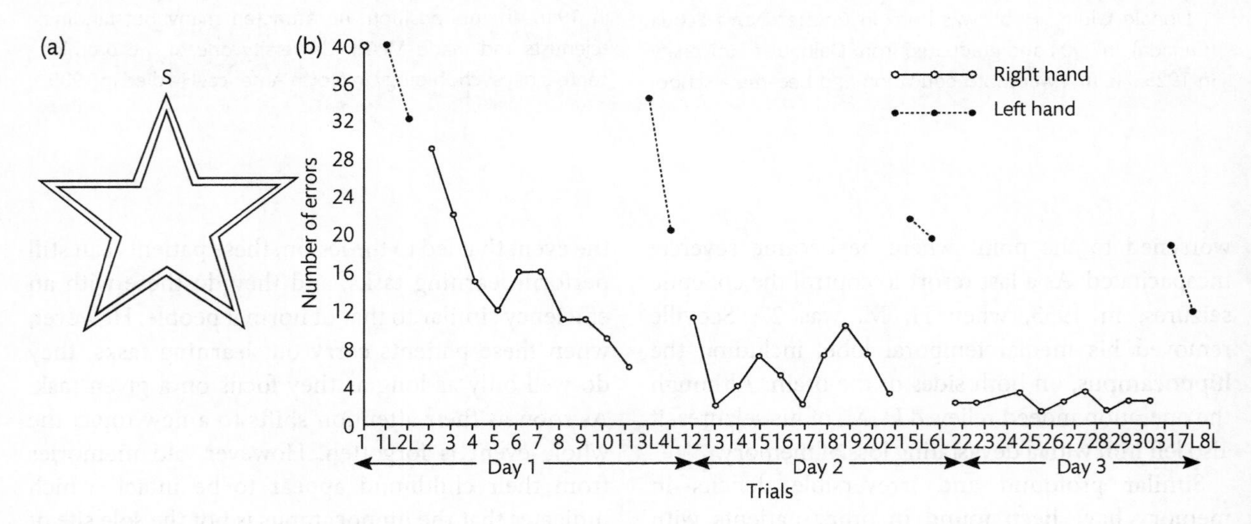

Figure 13.1 Mirror drawing task. (a) The test person is asked to draw a line between the two outlines of the star, starting from point 'S', while viewing the star and the hand in a mirror. (b) Learning curve of a patient who underwent temporal-lobe surgery. Over the three days of testing, he steadily improved in performing the mirror drawing task. The number of errors indicates the number of times in each trial he strayed outside the boundaries of the star while drawing. (After: Milner, B., Squire, L. R., and Kandel, E. R. (1998).)

traditionally not considered to be types of learning by ethologists, such as habituation and sensitization. In both explicit and implicit memory, a short-term form of information storage can be distinguished from a long-term form of memory. Most importantly, and as will be shown in the following, both explicit and implicit memory make use of similar molecular strategies to implement short-term and long-term behavioral changes.

Model system: The cell biology of an implicit memory system—sensitization in *Aplysia*

At about the same time that Brenda Milner's behavioral observations suggested an association of the hippocampus with memory function, scientists made the first attempts to apply a cellular approach to the study of learning and memory. However, it soon became clear that, with the techniques then available, a focus on explicit memory processes was unlikely to succeed. Therefore, scientists turned to implicit memory and searched extensively for simple model systems, particularly among invertebrates. One of these scientists, who pioneered this research like no one else, was the American Eric Kandel. A portrait of his life and work is presented in Box 13.2. Kandel focused on the marine mollusk ***Aplysia***, which has since become one of the major model systems in neurobiology.

Why *Aplysia*?

The sea hare (*Aplysia californica*) is a gastropod mollusk that lives along the coastal waters of southern California.

Box 13.2 Eric R. Kandel

Eric R. Kandel. (Courtesy: The Nobel Foundation.)

Like no other scientist before him, Eric Kandel pioneered a reductionist approach to the study of learning and memory. His merit is based on successfully linking various forms of behavioral changes, including simple types of learning, to specific subcellular processes and synaptic plasticity.

Eric Kandel was born in Vienna, Austria, in 1929. In 1939, he emigrated with his family to the U.S.A. after Nazi Germany's annexation of his country. Following graduation from Harvard College with a degree in history and literature, he studied medicine at New York University School of Medicine. During his postdoctoral training with Wade Marshall in the Laboratory of Neurophysiology of the National Institute of Mental Health in Bethesda from 1957 to 1960, he studied the cellular properties of the hippocampus, the part of the mammalian brain closely associated with complex memory. Contrary to the initial expectation, he found that the intrinsic signaling properties of the hippocampus are not much different from neurons in other areas of the brain. Rather, it appeared to be the pattern of functional interconnections that determined the unique functions of the hippocampus. However, due to the immense number of neurons and interconnections in the hippocampus and the limitations of the techniques available at that time, this aspect was not approachable. Therefore, an intensive search for a simple model system began. Finally, during a postdoctoral fellowship with Ladislav Tauc at the Institut Morey in Paris, he found such a suitable experimental animal—the marine mollusk *Aplysia*.

In the following decades, *Aplysia* formed the basis for Kandel's research. After completing his residency in clinical psychiatry at the Massachusetts Mental Health Center of the Harvard Medical School, Kandel held faculty positions at the

Harvard Medical School and the New York University School of Medicine. In 1974, he joined Columbia University in New York as founding director of the Center for Neurophysiology and Behavior, which was established with the aim of understanding the biological basis of behavior by employing an integrated approach. Since then, he has remained on the faculty of Columbia. Among the numerous honors received for his research are such prestigious awards as the Wolf Prize, the Lasker Award, the National Medal of Science, and the Nobel Prize in Physiology or Medicine, which he shared in the year 2000 with Arvid Carlsson of the University of Gothenburg (Sweden) and Paul Greengard of Rockefeller University in New York.

After three decades of work on *Aplysia*, during which he and his laboratory staff discovered key molecular mechanisms underlying habituation, sensitization, and classical conditioning, Erich Kandel returned, at the age of 60, to the hippocampus. Again, he pioneered research. By using genetically modified mice, the work of his group provided the framework to understanding the cellular and subcellular mechanisms of long-term potentiation and its relationship to spatial memory.

As an experimental animal, *Aplysia* offers several important advantages:

1. It exhibits several simple reflex behaviors that can be modified by different forms of learning.
2. The total number of neurons in the nervous system of *Aplysia* is low (approximately 20 000, compared with a trillion (10^{12}) or so in the central nervous system of mammals), and less than 100 are involved in the reflex pattern studied by Kandel and his group.
3. The neurons are among the largest known, some reaching diameters of up to 1000μm, and thus are visible to the naked eye. This makes it relatively easy to record data from these cells, even over long periods of time. Moreover, thanks to their large size, individual cells can be dissected out of the nervous system and used for biochemical analysis. With the modern techniques available, one can obtain sufficient messenger RNA (mRNA) from a single cell to make a cDNA library!
4. Many of the nerve cells are uniquely identifiable. This makes it possible to record results from the same type of cell in different individuals, or to return to the same cell in a later recording session. Moreover, dyes or molecular constructs can be injected into these identified cells, which provides the researcher with a very powerful tool to perform a molecular analysis.

The gill-withdrawal reflex

The behavior Kandel and his group chose for their investigations consisted of a defensive reflex. When the siphon of the sea hare is touched, both the siphon and the gill are withdrawn into the mantle cavity under the mantle shelf (Fig. 13.2). This response is referred to as the **gill-withdrawal reflex**. In spite of its simplicity, this reflex can be modified by several forms of learning, including habituation, dishabituation, sensitization, classical conditioning, and operant conditioning. Most of the work conducted by Kandel and his group has focused on sensitization. The effect of this simple type of learning can best be illustrated using the response of *Aplysia* to external stimuli. Without sensitization, a weak tactile stimulus applied to the siphon evokes a weak response, and the siphon and the gill withdraw only briefly. However, after an aversive shock has been applied to another part of the body, such as the tail, the same weak touch increases both the size and duration of the reflex response. This effect is called **sensitization**.

Sensitization Enhancement of a behavioral response to a stimulus after application of a different, noxious stimulus.

The duration of this memory depends on the number of repetitions of this noxious stimulation. If a single shock is applied to the tail, then the sensitization is effective for only a few minutes. If, on the other hand, four to five spaced shocks are applied to the tail, then a modulation of the gill-withdrawal reflex can be observed for several days. A typical result of such experiments is shown in Fig. 13.3. The duration of the sensitization effect allows the investigator to distinguish between **short-term memory** and **long-term memory**. These two types of memory are paralleled by two different cellular mechanisms mediating their behavioral outcome. As

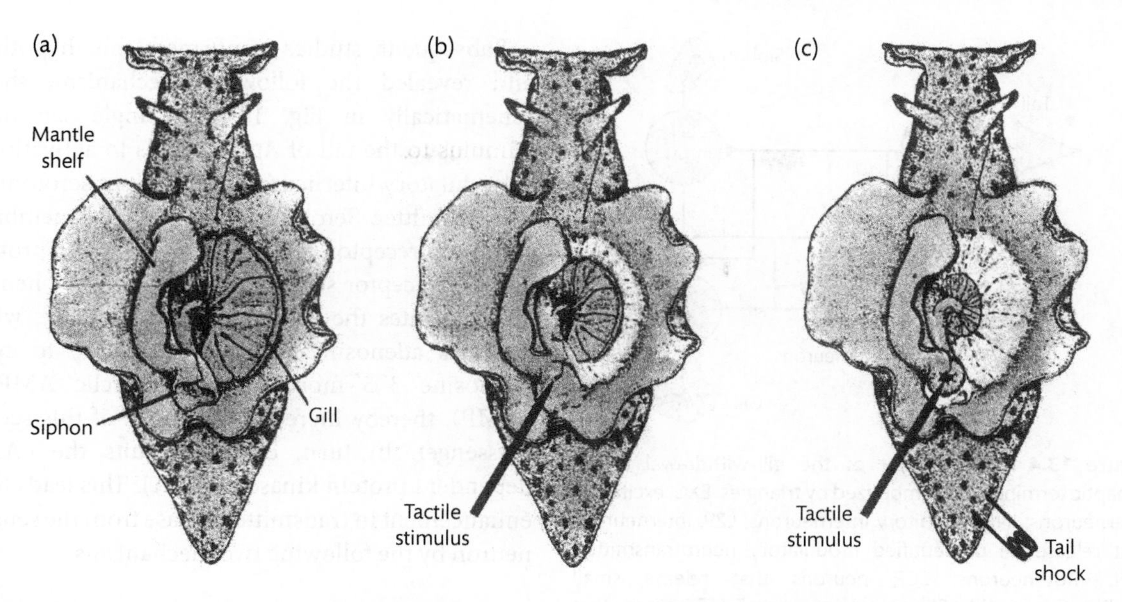

Figure 13.2 The gill-withdrawal reflex. (a) Dorsal view of *Aplysia*. The mantle shelf has been retracted to expose the siphon and the gill. (b) Upon applying a light touch to the siphon with a fine probe, the siphon contracts, and the gill is withdrawn into the mantle cavity for protection. (c) Application of a noxious stimulus, such as an electric shock, to the tail leads to sensitization of the gill-withdrawal reflex. Following such sensitizing training, the same light touch as applied in (b) now causes a much larger siphon- and gill-withdrawal response. In addition, the siphon and gill are withdrawn for much longer periods of time than before sensitization. (After: Kandel, E. R. (2001).)

will become evident later in this chapter, the essential difference between these two mechanisms is that short-term memory does not require the synthesis of new protein, whereas long-term memory does.

Neural circuit of the gill-withdrawal reflex

Before we describe in detail the molecular mechanism of sensitization, we need to take a look at the neural substrate underlying the gill withdrawal reflex. Figure 13.4 sketches the major components of this neural network, which is located in the abdominal ganglion. It consists of six motoneurons that make direct synaptic connections onto the gill. These motoneurons are connected to 24 sensory neurons that innervate the siphon. The sensory neurons also connect to excitatory and inhibitory interneurons that, in turn, are connected to the motoneurons.

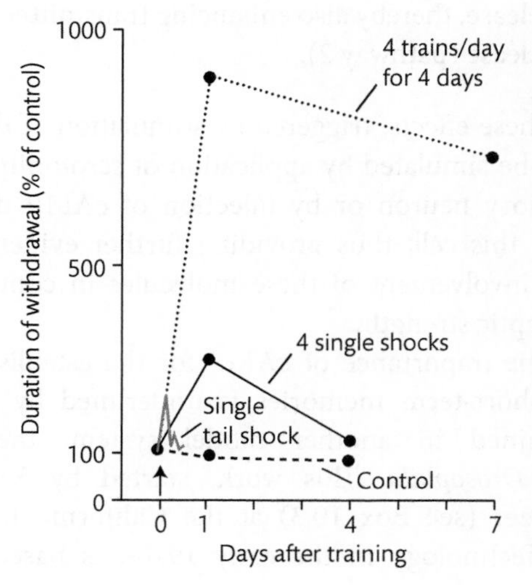

Figure 13.3 Conversion of short-term to long-term memory storage in *Aplysia*. As an indicator of the behavioral response, the duration of the withdrawal of the siphon and gill was measured and expressed relative to unsensitized (control) reflex responses arbitrarily set at 100%. Unsensitized animals respond to a light touch of the siphon with a weak siphon- and gill-withdrawal reflex ('control' curve). Application of a single electric shock to the tail causes the same light tactile stimulus to produce a much larger siphon- and gill-withdrawal reflex; however, this enhancement lasts for only about 1h ('single tail shock' curve). Four spaced shocks applied to the tail result in an enhanced gill- and siphon-withdrawal reflex lasting for several days ('4 single shocks' curve). More intense training, consisting of four brief trains of shocks a day applied for four days in a row, increases the size and duration of the reflex response even more; the underlying memory now lasts for weeks ('4 trains/day for 4 days' curve). (After: Kandel, E. R. (2001).)

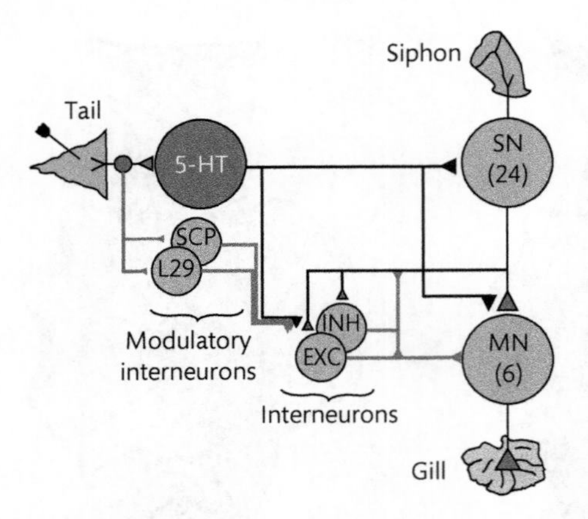

Figure 13.4 Neural circuit of the gill-withdrawal reflex. Synaptic terminals are symbolized by triangles. EXC, excitatory interneurons; INH, inhibitory interneurons; L29, interneurons that release an unidentified modulatory neurotransmitter; MN, motoneurons; SCP, neurons that release small cardioactive peptide; SN, sensory neurons; 5-HT, neurons that release serotonin (5-hydroxytryptamine, or 5-HT). For further explanations, see text. (After: Kandel, E. R. (2001).)

Stimulation of the tail activates three types of modulatory interneuron: the so-called L29 cell, cells that release small cardioactive peptide at their terminal site, and cells that release serotonin. These interneurons connect to the sensory neurons and the excitatory interneurons. The serotonergic pathway is particularly important for modulation of the gill-withdrawal reflex, as blocking of the serotonergic cells abolishes the effect of the sensitizing tail stimulus.

Molecular biology of short-term sensitization

In the initial stage of the molecular analysis of the mechanisms of short-term memory, Kandel and his group focused on short-term sensitization. They found that the associated behavioral changes result from alterations in the strength of the synaptic connections between neurons, a phenomenon referred to as **facilitation**. Furthermore, their investigations provided clear evidence that both the synaptic changes and the short-term behavioral changes are expressed even when protein synthesis is inhibited. This suggested that this type of behavioral and neuronal plasticity is mediated by a second-messenger system.

Facilitation Enhancement of synaptic transmission.

Subsequent studies confirmed this hypothesis and revealed the following mechanism, shown schematically in Fig. 13.5. A single sensitizing stimulus to the tail of *Aplysia* leads to activation of a modulatory interneuron that releases serotonin as its transmitter. Serotonin acts on a transmembrane serotonin receptor, which belongs to the G protein-coupled receptor superfamily, on a sensory neuron. This activates the enzyme adenylyl cyclase, which converts adenosine triphosphate (ATP) to cyclic adenosine 3′,5′-monophosphate (cyclic AMP or cAMP), thereby increasing the level of this second messenger. In turn, cAMP recruits the cAMP-dependent protein kinase A (PKA). This leads to an enhancement in transmitter release from the sensory neuron by the following two mechanisms.

1. Activation of PKA by cAMP causes phosphorylation of specific K^+ channels. This results in closure of these channels, and thus in a reduction of repolarizing K^+ currents. As a consequence, action potentials are broadened and the influx of Ca^{2+}, which occurs as a normal event during the action potential, is prolonged. The increased concentration of Ca^{2+} ions, which are necessary for vesicle exocytosis (see Chapter 2), contributes to a greater transmitter release and thus to an enhanced activation of the motoneuron, as observed during sensitization (pathway 1).
2. Activation of PKA directly affects, through a mechanism not yet understood, one or more steps in vesicle mobilization and exocytotic release, thereby also enhancing transmitter release (pathway 2).

These effects, triggered by stimulation of the tail, can be simulated by application of serotonin to the sensory neuron or by injection of cAMP directly into this cell, thus providing further evidence for the involvement of these molecules in controlling synaptic strength.

The importance of cAMP for the establishment of short-term memories is underlined by results obtained in another model system, the fruit fly, *Drosophila*. This work, started by Seymour Benzer (see Box 10.3) at the California Institute of Technology in the early 1970s, is based on a

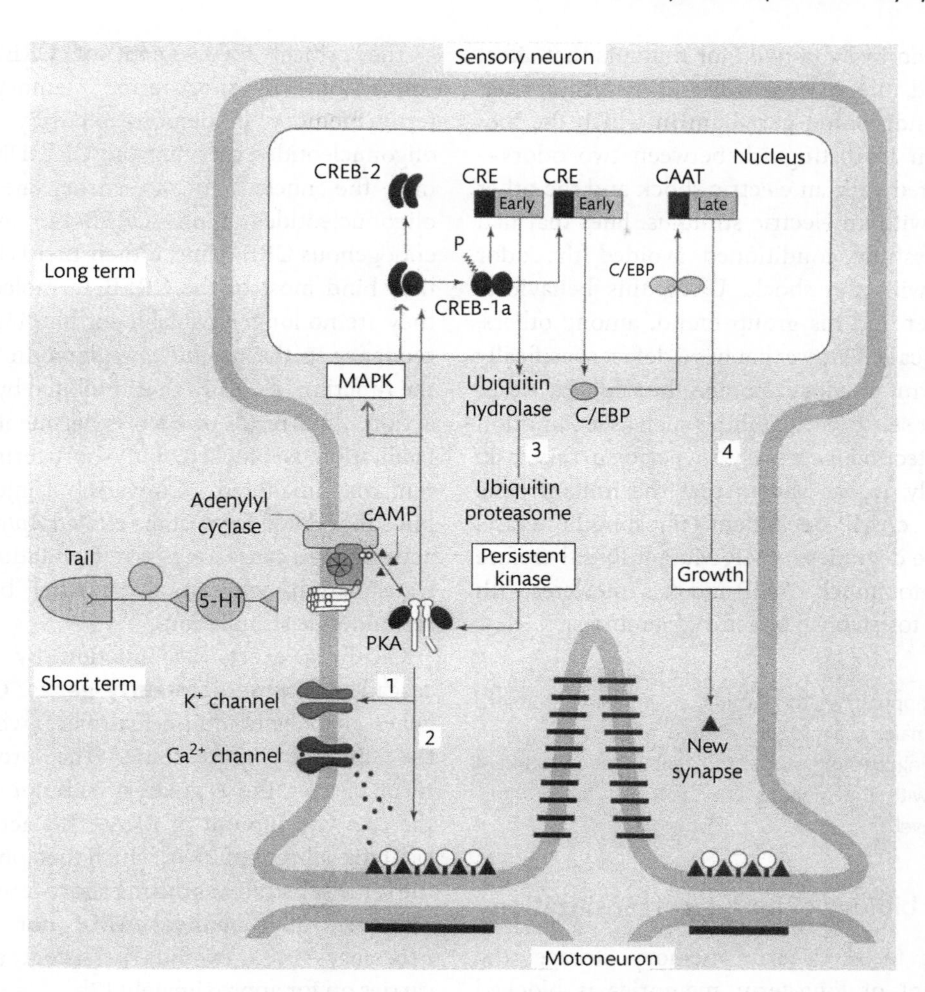

Figure 13.5 Molecular signaling in short- and long-term sensitization of the gill-withdrawal reflex in *Aplysia*. A noxious stimulus, such as an electric shock, applied to the tail activates sensory neurons that excite facilitatory interneurons. These serotonergic (5-HT) interneurons synapse onto the terminals of sensory neurons, which detect tactile stimuli given to the siphon skin and synapse onto motoneurons controlling the withdrawal of the siphon and the gill. At the axoaxonic synapses, the facilitatory interneurons are able to enhance transmitter release from the sensory neurons. This so-called presynaptic facilitation is mediated by serotonin released by the facilitatory interneurons upon stimulation of the tail. Depending on the mode of serotonergic activation, the synapse of the sensory neuron can undergo short-term facilitation, lasting for minutes, or long-term facilitation, lasting for days. In short-term facilitation, presynaptic facilitation induced by a single tail shock (or a single pulse of serotonin) activates pathways (1) and (2). Serotonin acts on a transmembrane serotonin receptor that activates the enzyme adenylyl cyclase to convert ATP to the second messenger cyclic AMP (cAMP). The cAMP, in turn, activates protein kinase A (PKA), which phosphorylates a number of target proteins involved in channel functions (pathway 1) and exocytosis of synaptic vesicles (pathway 2). As a result, transmitter availability and release are enhanced. These molecular modifications last for minutes and thus underlie short-term memory storage. In long-term facilitation, presynaptic facilitation induced by repeated-shock stimulation of the tail (or multiple pulses of serotonin) activates pathways (3) and (4). The switch to these mechanisms is initiated by PKA. This kinase recruits another kinase, mitogen-activated protein kinase (MAPK). Both kinases translocate to the nucleus where they phosphorylate the cAMP response element binding (CREB) protein and remove the repressive action of CREB-2, which inhibits CREB-1a. CREB-1a activates several immediate-response genes. One of these encodes ubiquitin hydrolase, which is a crucial factor in the production of a persistently active PKA. The other immediate-response gene activated by CREB-1a encodes the transcription factor CCAAT/enhancer binding protein (C/EBP), which is involved in the activation of downstream genes. The expression of these genes ultimately leads to the generation of new synapses. (After: Mayford, M. and Kandel, E. R. (1999).)

genetic screen in *Drosophila* for mutants that affect learning and memory. Benzer and his group used a simple conditioning paradigm in which the flies had to learn to distinguish between two odors—one associated with an electric shock and the other not paired with an electric stimulus. Flies that had been successfully conditioned avoided the odor associated with the shock. Using this behavioral assay, Benzer and his group found, among others, one mutant called *dunce* that has a defect specifically in short-term memory storage, but not in other behaviors or sensory capabilities such as locomotion or odor detection necessary to perform this task. Subsequently, it was shown that the mutant gene encodes a cAMP-dependent phosphodiesterase. This enzyme degrades cAMP. Mutant flies therefore accumulate too much cAMP and this interferes with their ability to establish olfactory memories.

➤ Genetic approaches to learning and memory storage frequently make use of mutagenized animals that are screened using learning assays. The genetic modifications associated with the mutants are then analyzed at the molecular level.

Molecular biology of long-term sensitization

In contrast to short-term memory storage, the establishment of long-term memories is blocked when protein synthesis is inhibited. This suggests that the expression of certain genes is induced in the process of transferring information from the short-term memory to the long-term memory.

What are these genes and how is their expression regulated? Studies on long-term sensitization of the gill-withdrawal reflex in *Aplysia* have revealed the following mechanism turned on by repeated tail shocks or by several puffs of serotonin to the terminal region of the sensory neuron (Fig. 13.5). First, as with short-term sensitization, PKA is activated. However, now the PKA recruits another kinase, the **mitogen-activated protein kinase** (MAPK). Both of these kinases translocate into the cell's nucleus where they activate a cascade of transcriptional processes. At the beginning of this cascade, the transcriptional activator CREB-1a (cAMP response element binding protein 1a) is activated. This protein binds to a cAMP response element (CRE) in the promoters of target genes.

The critical involvement of CREB-1a in the conversion of short-term memory to long-term memory is demonstrated by injection of oligonucleotides carrying the CRE DNA sequence into the nucleus of a sensory neuron. These oligonucleotides bind CREB-1a, as does the endogenous CRE. Thus, if their titer is high enough, they bind most of the CREB-1a molecules so that they are no longer available for binding to the CRE sequence in the promoter regions. In other words, the function of CREB-1a is inhibited by competitive action. As a result of such experiments, long-term facilitation is blocked, but short-term facilitation remains unaltered. Conversely, injection of a phosphorylated form of the cloned *Aplysia* CREB-1a activator can cause long-term facilitation of synaptic transmission, without any further behavioral or serotonergic stimulation.

CREB-1a exerts its function by activating a cascade of **immediate-early genes**. One of these genes encodes ubiquitin hydrolase, which activates the ubiquitin proteasome. The proteasome, in turn, cleaves the regulatory subunit that inhibits the catalytic subunit of PKA. This action frees the catalytic subunit of PKA, which then phosphorylates the same substrates as during short-term facilitation. However, now neither cAMP nor serotonin is necessary—PKA exhibits persistent activity. This carries on for approximately 12h.

Immediate-early genes A class of genes whose expression is low or undetectable in quiescent cells, but whose transcription is activated within minutes after proper stimulation.

The memory of this long-term facilitation process is stabilized by the establishment of new synaptic connections. The growth of these synapses is achieved through the activation by CREB-1a of a second immediate-early gene encoding the transcriptional factor CCAAT/enhancer binding protein (C/EBP). This factor acts on downstream genes that give rise to the growth of new synapses. If expression of the C/EBP gene is blocked, then the growth of new synaptic connections is suppressed and the establishment of the long-term memory inhibited.

The results of these molecular biology experiments expanded earlier electron microscopic findings made by Craig Bailey and Mary Chen, both, like

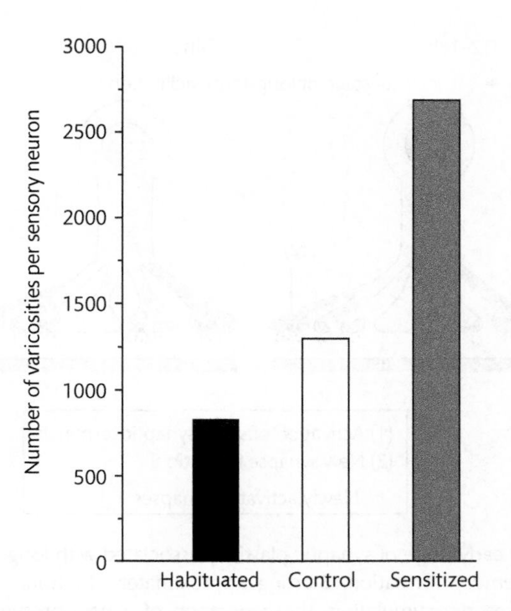

Figure 13.6 Structural changes in synaptic organization involved in long-term habituation and sensitization. The histogram shows the total number of presynaptic terminals per completely reconstructed sensory neuron in *Aplysia*. This number is highest in sensitized animals and lowest in habituated animals. (After: Bailey, C. H. and Chen, M. (1988).)

Kandel, from Columbia University. They combined intracellular labeling techniques (see Chapter 2) with the ultrastructural analysis of completely reconstructed identified sensory neurons of animals from different behavioral groups. Their results demonstrated that the establishment of long-term memory is accompanied by the following two major types of change in synaptic organization:

1 The number, size, and vesicle complement of the active zones of the presynaptic terminals of sensory neurons are larger in sensitized animals than in controls, and are smaller in habituated individuals.
2 The total number of presynaptic terminals per sensory neuron depends on the behavioral status of the animal. Control animals have about 1200 synaptic terminals. Sensory neurons of long-term habituated animals have, on average, 35% fewer presynaptic terminals than controls, whereas in animals that underwent long-term sensitization, the total number of synaptic terminals doubled compared with controls (Fig. 13.6).

The alterations in both the number of presynaptic varicosities and active zones of sensory neurons persist in parallel with the behavioral retention of the memory. This correspondence in the time course of the two events provides additional evidence for the notion that the structural modifications are an important causal factor in the maintenance of long-term sensitization or long-term habituation, respectively.

To learn more about the dynamics of the structural changes associated with long-term sensitization, Kandel, Bailey, and their co-workers took advantage of the possibility of reconstituting the monosynaptic sensory-to-motoneuron connection of *Aplysia* in culture. They monitored these changes by performing live time-lapse **confocal imaging** on the sensory neurons that contained three fluorescent markers: a whole-cell marker; synaptophysin tagged with enhanced green fluorescent protein to label synaptic vesicles in the presynaptic terminals; and synapto-pHluorin, a genetically encoded indicator to monitor changes at the active transmitter-release sites.

> **Confocal microscopy** A microscopy technique widely used in biomedicine. It enables investigators to image either fixed or living tissue labeled with one or several fluorescent probes. Compared to conventional optical microscopy, confocal microscopy offers the advantage of elimination of out-of-focus glare caused by fluorescently labeled specimens thicker than a few micrometers; the ability to control the depth of field; and the capability to collect serial optical sections and to use these data for three-dimensional reconstruction of tissue structures.

Examination of cultured neurons labeled with these three markers showed that approximately 12% of the presynaptic terminals were 'empty'—they were not stained by the synaptophysin-enhanced green fluorescent protein and the synapto-pHluorin. Thus, they are incompetent to release transmitter.

As one of the next steps in their study, Kandel, Bailey, and associates mimicked the sensitizing effects of tail shocks in the intact animal by applying five pulses of serotonin to the cultured neurons. This treatment induced a rapid filling of the empty presynaptic terminals with synaptic vesicles and active-zone material. The un-silencing of these synaptic terminals was complete within 3–6 hours after the serotonin application. This process is

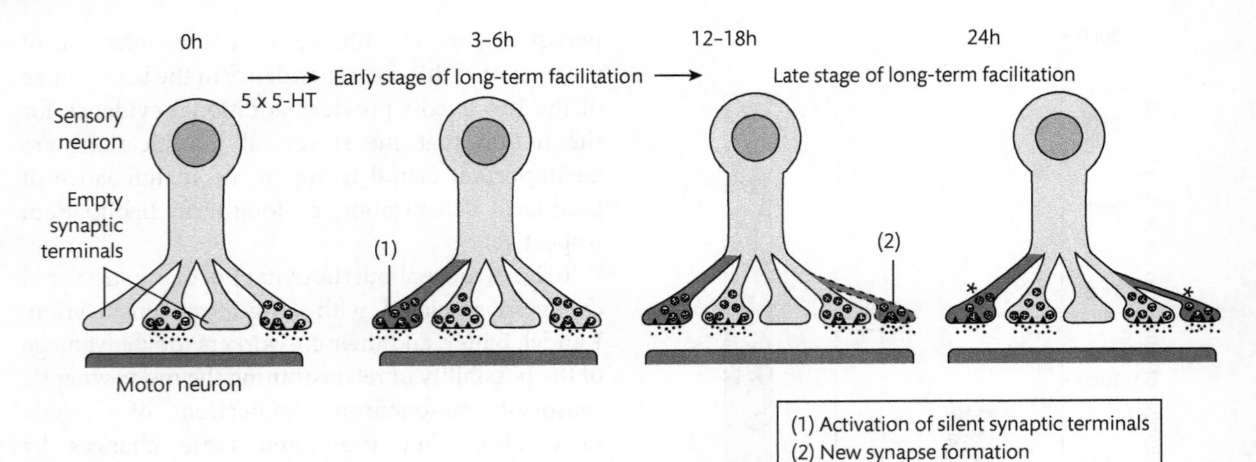

Figure 13.7 Time course of two key changes in the structure of the presynaptic sensory terminals associated with early and late stages of long-term facilitation in *Aplysia*. Shown is a sensory neuron with presynaptic terminals at different time points after application of five pulses of serotonin. Two of these synaptic terminals are initially empty. Within 3–6 hours, one of these terminals is filled with synaptic vesicles (empty circles) and active zone material (filled triangles). It is thought that the modification of existing synaptic terminals represents an early stage of synaptic plasticity associated with long-term memory formation. Starting approximately 12 hours after serotonin stimulation, the generation of a new presynaptic terminal begins by fission of an existing terminal. This process is complete about 24 hours after application of serotonin. Both filled and the newly formed presynaptic terminals are able to release transmitter upon electrical stimulation, thereby demonstrating their functional competence. (After: Bailey, C.H., Kandel, E.R., and Harris, K.M. (2015) © 2003 by Cell Press)

thought to represent an early formative stage of long-term memory storage.

Approximately 12 hours after exposure of the cultured neurons to the serotonin pulses, the generation of new presynaptic terminals begins. This is achieved through division of some of the existing mature presynaptic terminals. The formation of the new synapses is thought to be associated with the establishment of persistent memory storage.

Figure 13.7 summarizes the two processes—the remodeling and activation of existing silent synapses, and the formation of new synapses.

> ➤ The conversion of short-term memory storage into long-term memory storage is characterized by the synthesis of new proteins and the establishment of structural changes at the synaptic level.

Model system: The cell biology of an explicit memory system—the hippocampus of mammals and birds

As suggested by the observations of patients with defects of their temporal lobes mentioned at the beginning of this chapter, explicit memory in mammals requires specialized structures within this part of the cortex, including the hippocampus. This is a feature fundamentally different from implicit memory, which does not involve specialized anatomical systems devoted to memory storage. Rather, as the above detailed molecular analysis of sensitization has shown, implicit memory storage is achieved by modification of the synaptic properties of the neural circuitry concerned with processing of the information relevant to the execution of the respective behavior (e.g. of the gill-withdrawal reflex in *Aplysia*). Despite this difference, research has demonstrated that the molecular mechanisms for converting a labile short-term memory into a stable long-term memory are, at least to a certain degree, remarkably similar among implicit and explicit forms of learning. In the following, we will discuss, in some detail, the biochemical pathways involved in this process in the mammalian hippocampus.

Non-human models to study explicit memory

The knowledge gained through the study of human patients with severe memory loss after brain surgery pointed to the importance of structures within the medial temporal lobe for memory function. The identity of these structures and their precise role in memory formation, however, remained enigmatic due to the low number of such patients available and the frequently poor definition of their brain lesions. A better definition was possible only when the patient's brain was available for autopsy after death. (Today, these difficulties can be overcome by the use of magnetic resonance imaging of the patient's brain during their lifetime.)

As a result of these difficulties in humans, efforts began, in the late 1950s, to establish animal models in which defined lesions could be correlated with specific memory deficits. Such model systems included mice, rats, rabbits, and monkeys, all of which have a medial temporal lobe system, including a hippocampus.

Lesioning studies, combined with behavioral testing, in such mammalian model systems have suggested that the key structures of the medial temporal-lobe system are the hippocampus proper, the dentate gyrus, the subicular complex, and the entorhinal cortex (collectively referred to as the **hippocampal formation**), as well as two adjacent parts of the cortex, the perirhinal and parahippocampal cortices. Lesions of any of these structures impair explicit memory.

> ➤ The hippocampal formation is comprised of the following structures of the neocortex in mammals: hippocampus proper, dentate gyrus, subicular complex, and entorhinal cortex.

The function of the medial temporal-lobe memory system appears to be to direct changes in the organization of cortical areas that represent stored memory. These areas within the cortex are typically geographically separate and need to be bound together in order to fulfill their function in permanent memory storage. In the process of this cortical reorganization, information transiently stored in the medial temporal-lobe system is passed on to the neocortex. As a consequence, the medial temporal-lobe system is needed for memory storage only for a limited amount of time after learning. This explains why patients with temporal lobe ablations have intact memory of events that occurred years before the surgery, but—in addition to the anterograde amnesia—they also exhibit retrograde amnesia covering a period of between several months and a few years prior to the removal of parts of the temporal lobe.

The structure of the hippocampus

A remarkable feature of the hippocampus is its highly regular cellular organization and connectivity. This regularity makes it possible to obtain a number of cross-sections, each of which is very similar to the others, in terms of the morphological and physiological properties. Figure 13.8 shows such a cross-section through the hippocampus on one side of the brain. As this simplified diagram reveals, two principal neuronal fields—the granule cells of the dentate gyrus and the pyramidal cells of areas CA1 and CA3—and the following three principal neural pathways exchange information within the hippocampus:

1. The **perforant pathway**, which originates in the entorhinal cortex and forms excitatory connections with the granule cells of the dentate gyrus.
2. The **mossy fiber pathway**, which is formed by the axons of the granule cells of the dentate gyrus that make synaptic contact with the pyramidal cells in the CA3 area.

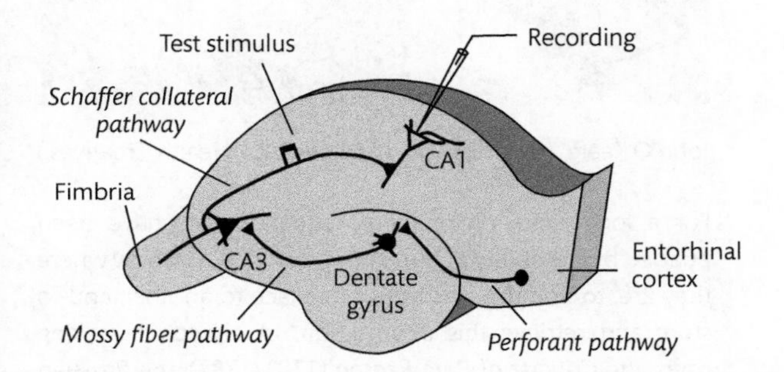

Figure 13.8 The hippocampus and its major neural pathways. The perforant pathway originates in the entorhinal cortex and forms excitatory connections with the granule cells of the dentate gyrus. The mossy fiber pathway is formed by the axons of the granule cells of the dentate gyrus and terminates onto pyramidal cells in the CA3 area of the hippocampus. The Schaffer collateral pathway is formed by the pyramidal cells of the CA3 area projecting to the pyramidal cells of the CA1 area. In addition to these three pathways, the drawing shows a typical experimental arrangement to demonstrate long-term potentiation. A stimulation electrode is placed next to the Schaffer collateral fibers. The postsynaptic potentials evoked by stimulation of this fiber path are recorded from the cell bodies of the pyramidal cells of the CA1 area. (After: Kandel, E. R. (2001).)

3 The **Schaffer collateral pathway**, which connects the pyramidal cells of the CA3 area with the pyramidal cells in the CA1 region.

> ➤ The three major pathways within the hippocampal formation are the perforant pathway, the mossy fiber pathway, and the Schaffer collateral pathway. Together, they form the trisynaptic circuitry.

Taken together, the pattern of connectivity exhibited by these three pathways is often referred to as the **trisynaptic circuitry**.

The regular organization of the hippocampus repeats itself along the longitudinal axis. Therefore, each cross-section taken through the hippocampus and kept alive in artificial cerebrospinal fluid (yielding a so-called **slice preparation**) allows the researcher to study many physiological properties of the principal cell groups and connections *in vitro*.

Place cells in the hippocampus and the formation of cognitive maps

As soon as lesioning studies provided evidence of a role of the hippocampal formation in learning and memory, an extensive search began to attribute more specific functions to this part of the brain. A seminal discovery was published in 1971 when John O'Keefe (Box 13.3) and John Dostrovsky of University College London reported the results of their study in a brief

Box 13.3 John O'Keefe

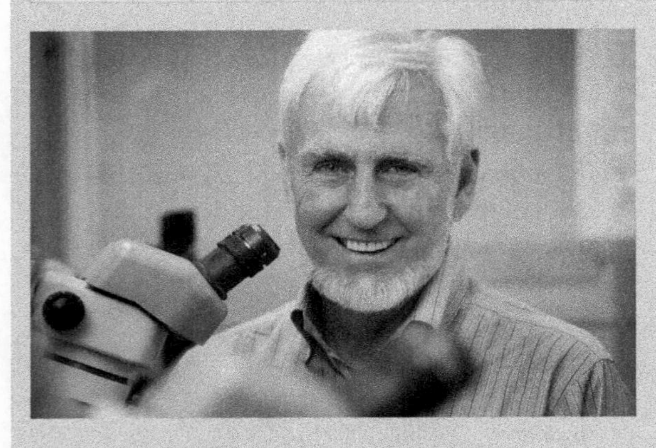

John O'Keefe (Courtesy: David Bishop UCL Health Creatives.)

For a long time, philosophers and scientists have been puzzled by the ability of humans and animals to know where they are, to find the way from one place to another, and to store and retrieve this information. In his epoch-making opus, the *Critique of Pure Reason* (1781/1787), the Prussian philosopher Immanuel Kant argued that the concept of space cannot be derived from outer experience. Rather, sensory input must conform to an a priori representation of space, as part of a cognitive process, in order to generate meaningful spatial information. In the mid-twentieth century, psychologists started to develop approaches to examine, through behavioral experiments, this inner representation of space. Based on labyrinth experiments in rats, the American psychologist Edward Chase Tolman proposed in 1948 that a 'cognitive map' is formed in the brain that enables animals to find their way. However, the neural correlates of this map remained enigmatic until, in the early 1970s, John O'Keefe discovered a key component of the brain's positioning system, the place cells of the hippocampus.

John Michael O'Keefe, born in 1939 in Harlem, New York, to poor Irish immigrant parents, grew up in the South Bronx. Although he won a scholarship to Regis High School, a renowned private Jesuit university-preparatory school on Manhattan's Upper East Side, his low grades prevented him from receiving any financial assistance for college. He, therefore, worked full-time in the daytime to make a living and to pay for evening classes in aeronautical engineering at New York University. After three years of this, as he recalled later, 'grueling schedule,' he enrolled as a full-time student at City College of New York—even back then one of the few tuition-free colleges in the United States. There, he took courses across a wide range of subjects, from filmmaking to philosophy. During this period, he developed a genuine interest in physiological psychology.

After graduating from the City College of New York, he enrolled in the graduate program of the Department of Psychology of McGill University in Montreal, Canada, where influential figures like Donald Hebb and Brenda Milner (see 'Introduction' and 'Key concepts') pioneered this area of research. Supported by his Ph.D. advisor Ronald Melzack, he developed techniques to perform chronic recordings from the amygdala—a brain region within the temporal lobe—of freely moving animals. Using this approach, he discovered cells that responded to complex behavioral stimuli, such as simulated bird calls or a black mouse, thus exhibiting activity

. . . *continued*

strikingly similar to the 'grandmother neurons' intensely debated at that time (cf. Chapter 7).

After receiving his Ph.D. in 1967, O'Keefe went, as a postdoctoral fellow, to University College London to work on somatosensation. One day in 1970, he implanted, by accident, a microelectrode into the hippocampus, instead of the somatosensory thalamus. When he recorded from the first hippocampal neuron, he was immediately struck by the strong correlation of its activity and the animal's motor behavior. Intrigued by this physiological response, John O'Keefe decided to leave the somatosensory system and study the hippocampus. Shortly after this first encounter with a hippocampal cell, he and Jonathan Dostrovsky, an M.Sc. student, discovered a class of pyramidal cells in the hippocampus that are activated when the animal assumes a particular place in the environment. Their research and subsequent studies by other investigators have shown that these 'place cells' are a key component of the cognitive map of the environment proposed more than two decades earlier by Tolman.

O'Keefe and Dostrovsky summarized their findings in the form of a short paper published in the journal *Brain Research* in 1971. In addition, together with his colleague Lynn Nadel of the Anatomy Department at University College London, John O'Keefe wrote a comprehensive book in which the two authors reviewed the evidence that supports the idea that a major function of the hippocampus is its contribution to a cognitive map. After six years of work, the book was published in 1978 by Oxford University Press. Despite this effort, the notion that hippocampal neurons are involved in determining an animal's location in the environment was initially met with skepticism and became only gradually accepted after overwhelming supportive evidence had accumulated. This evidence includes particularly the results of hippocampal lesioning experiments, which produce behavioral deficits in spatial navigation, and the discovery by May-Britt Moser and Edvard I. Moser, in 2005, of grid cells in a nearby part of the brain, the entorhinal cortex. These cells constitute a coordinate system, which, together with other cells, including the hippocampal place cells, form the circuitry of the cognitive map.

John O'Keefe has remained at University College London during his entire academic career. There, he was appointed Professor of Cognitive Neuroscience in 1987, and inaugural Director of the Sainsbury Wellcome Centre in Neural Circuits and Behaviour in 2013. In addition to several other prestigious awards received, he is recipient of the 2014 Nobel Prize in Physiology or Medicine, which he shared with the Mosers.

communication in the journal *Brain Research*. They made extracellular recordings from pyramidal cells of the hippocampus of rats, while the animals could freely move around in the experimental chamber. What O'Keefe and Dostrovsky found was a peculiar firing pattern of some of these cells: A specific cell increased the rate of spike generation when the rat's head was in a particular part of the chamber, regardless of its orientation. When the rat left this area, the discharge rate decreased, but as soon as the rat returned to the area, the cell increased its firing again. Other cells behaved in a very similar way, with the only difference being that they increased firing in a different section of the environment. This remarkable discharge pattern of such cells is shown in Fig. 13.9.

The particular area that evokes firing by the pyramidal cells is referred to as the cell's **place field** or **firing field**. The cells exhibiting this kind of behavior are called **place cells**. Each place cell has a somewhat different spatial preference. This preference is quite stable in a given environment, but may be different in a different environment. Place cells can therefore encode place fields in more than one environment. When the animal is introduced to a new environment, new place fields are formed within minutes, but as soon as the cell's preference is established, the characteristic firing pattern becomes and remains stable over weeks or even months. Collectively, the place fields of the different place cells form an internal representation of the space—a so-called **spatial map**. The establishment of these maps involves a learning process and conversion of the acquired space information from a short-term into a long-term memory. Presumably, storage of the place-cell map is achieved through similar changes in synaptic structure as those that take place during the formation of long-term memory in *Aplysia*.

➤ Place cells are a subpopulation of hippocampal pyramidal cells that encode place fields in an environment. Collectively, the place fields of the different place cells form a spatial map of the space.

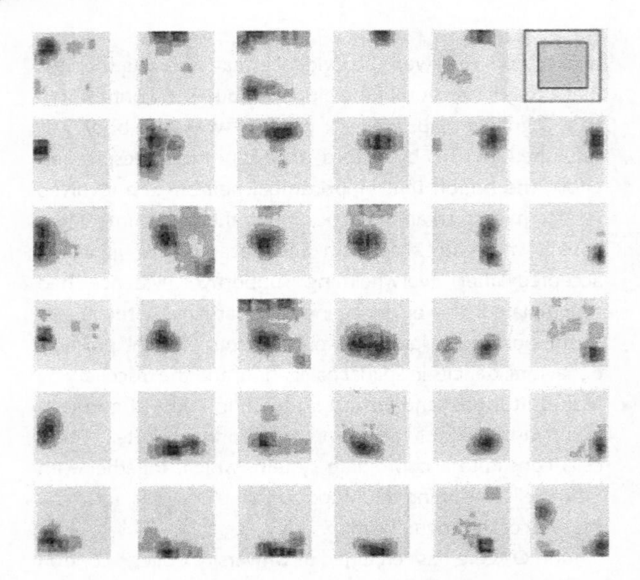

Figure 13.9 Place fields of 35 simultaneously recorded pyramidal place cells in the CA1 region of the rat hippocampus. During the recordings, the freely moving rat searched for grains of rice on a small $40 \times 40cm^2$ open platform. The surface area of the holding box relative to the overall camera view is shown in the upper right-hand corner of the picture. The recordings were obtained by four implanted microelectrodes. During foraging, the rat's location and the pyramidal cells' activities from each electrode were recorded. The firing rates within the place fields are represented by four gray-scale levels, each representing 20% of the peak firing rate. In the picture, the place fields of the individual place cells are arranged according to field location. Fields toward the northwest of the box are shown in the upper left portion of the picture, those toward the southeast in the lower right part, and so on. Note that the fields of the 35 cells collectively cover a considerable area of the total space defined by the box. (After: O'Keefe, J., Burgess, N., Donnett, J. G., Jeffery, K. J., and Maguire, E. A. (1998).)

Since memory has a limited capacity, the following obvious question arises: What factors determine which information is encoded and stored long term? It is well known from psychological experiments (and everyday experience) that **attention** plays an important role in this selection. To study details of this modulatory process, James J. Knierim of Johns Hopkins University in Baltimore, Maryland, tested, in collaboration with researchers from Johns Hopkins and the University of Delaware, the specific hypothesis that attentive behavior drives the formation and potentiation of place fields. For their investigation, the group capitalized on the natural exploratory behavior of rats. When rats explore an environment, they often intersperse periods of forward progression with pauses, during which they perform lateral head-scanning movements. The scans involve active investigation of features of the environment. Knierim and co-workers found that increased neural activity during the head scanning was strongly associated with the formation and potentiation of place fields at the location where the scan had been performed.

What mechanism could mediate the learning of the association of environmental cues and a specific place? May-Britt Moser and Edvard I. Moser, both of the Norwegian University of Science and Technology, addressed this question in a study conducted in collaboration with Kei M. Igarashi, Ki Lu, and Laura L. Colgin. They focused on the coordination of neural activity of the entorhinal cortex and the hippocampus. The connection between these two areas forms part of a circuit in which the entorhinal cortex functions as an interface between the hippocampus and the neocortex. Olfactory input is predominantly received via two subfields, the distal part of the CA1 area of the hippocampus and the lateral part of the entorhinal cortex. By simultaneously recording from these subfields, the investigators showed that, as rats learn to use an odor-cue to guide navigational behavior, neural coupling between these two brain regions develops in the 20–40Hz frequency band. It is thought that the coupling of the 20–40Hz **neural oscillations** plays a key role in synchronizing developing neural representations of olfactory–spatial associations in the entorhinal cortex and the hippocampus.

Neural oscillation Rhythmic neural activity generated by individual neurons or through interactions of neural ensembles. A commonly used technique to monitor the synchronized activity of large numbers of neurons is electroencephalography (EEG). Such analysis has revealed oscillatory activity in distinct frequency bands. These frequency-specific signals have been linked to a variety of behavioral and cognitive states.

Place cells are not the only spatially modulated neurons in the brain. Several other types of such neurons have been discovered. It is likely the collective action of most, if not all, of them that forms the neural basis of a cognitive map acquired through learning and memory consolidation. Two intensively studied types of such cells are head direction cells and grid cells.

Head direction cells were discovered by James B. Ranck Jr. at SUNY Downstate Medical Center in Brooklyn, New York, in 1984 in the dorsal presubiculum of the rat. An individual cell of this type fires maximally when the head of an animal faces a specific direction in the horizontal plane. Different head direction cells show preference for different directions so that the activity of the population of these cells represents essentially all directions of the two-dimensional space.

Grid cells were discovered by May-Britt and Edvard Moser (who both shared for this work the 2014 Nobel Prize in Physiology or Medicine with John O'Keefe; see Box 13.3) and co-workers in 2005 when they recorded from the medial part of the entorhinal cortex. These cells fire in multiple, regularly spaced locations. Collectively, the multiple firing fields of the individual neurons form a triangular or hexagonal grid pattern (Fig. 13.10). The function of grid cells is not well understood, but, based on the constant spacing between firing fields, it has been suggested that they provide information about the distance walked by the animal.

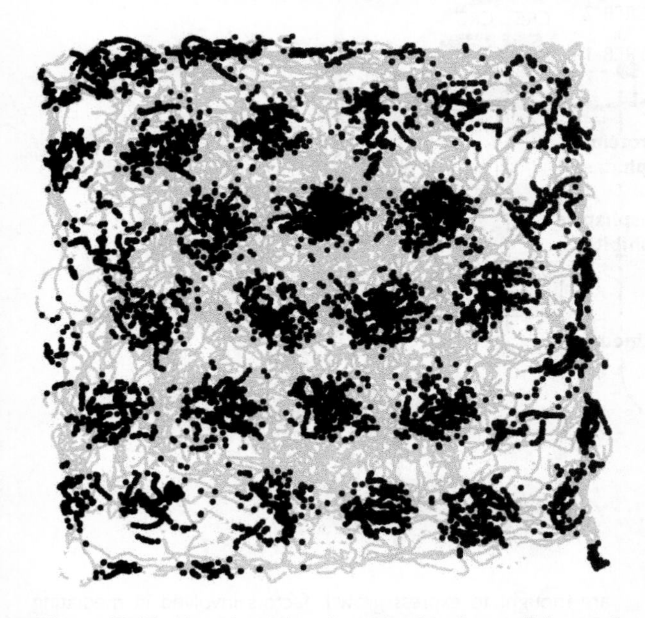

Figure 13.10 Neural activity of a grid cell in the medial part of the entorhinal cortex. The signals were recorded while the rat was freely moving in a squared arena. The rat's path is shown in gray. Each firing of the grid cell is symbolized by a black dot and superimposed onto the rat's path to indicate its position when the cell was active. Viewed together, a regular triangular/hexagonal pattern is visible. (© Stensola, Kavli Institute for Systems Neuroscience.)

Long-term potentiation

In each of the three hippocampal pathways, a type of synaptic plasticity occurs that has become known as **long-term potentiation**, commonly abbreviated to LTP. This phenomenon was discovered in 1973 by Timothy Bliss and Terje Lømo in the Laboratory of Per Andersen at the University of Oslo, Norway. With brief volleys of constant-voltage pulses, lasting for a few seconds and presented at frequencies ranging between 10 and 100Hz, they stimulated the perforant path fibers in rabbits. The stimulations resulted in an increase in the efficiency of synaptic transmission for periods ranging from 30min to 10h after delivering the electric pulses. Since the original discovery made by Bliss and Lømo, LTP has also been found in the other two hippocampal pathways. Moreover, it has been shown that, under proper conditions, LTP can last for days or even weeks. Thus, LTP exhibits a property that makes it suitable for memory storage.

Long-term potentiation (LTP) An abrupt and sustained increase in the efficiency of synaptic transmission after high-frequency stimulation of a monosynaptic excitatory pathway.

As with facilitation of the gill-withdrawal reflex in *Aplysia*, LTP in the hippocampus exhibits short-term and long-term phases. A typical experimental result revealing these two phases is shown in Fig. 13.11. A single-stimulus training produces a short-term phase of LTP, called the **early phase of LTP**. This phase lasts for 1–3h and does not require the synthesis of new protein. By contrast, four or more training sessions of electrical stimuli induce a long-term phase of LTP, called the **late phase of LTP**. This phase lasts for at least one day and requires the synthesis of both new mRNA and new protein.

Molecular biology of long-term potentiation in the mammalian hippocampus

The late-phase LTP has been particularly well studied in the Schaffer collateral pathway. Glutamate released from the presynaptic terminals of the Schaffer

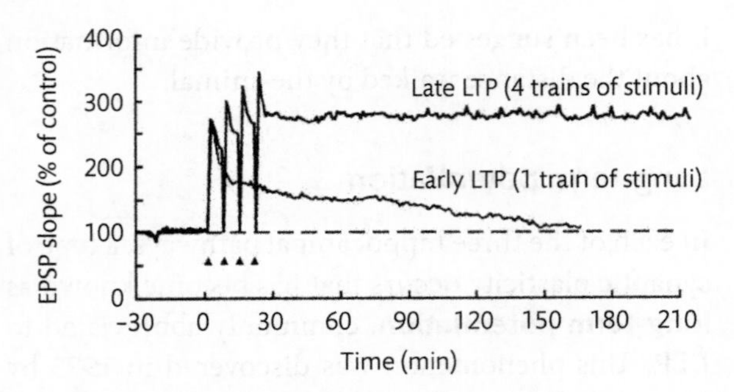

Figure 13.11 Early and late phases of LTP in the Schaffer collateral pathway of the hippocampus. A response was evoked by applying constant test stimuli before and after induction of LTP. For the induction, a single train, or four at 10min intervals, of 100Hz stimuli were delivered for 1sec each (the time points of stimulation are indicated by triangles below the plot). The response is indicated by the plot of the slope of the rising phase of the evoked excitatory postsynaptic potential (EPSP). As revealed by the diagram, one train of this high-frequency stimulation elicits an early LTP, whereas four trains evoke the late phase of LTP. (After: Kandel, E. R. (2001).)

collateral fibers binds to *N*-methyl-D-aspartate (NMDA) receptors on the pyramidal cells of the CA1 region. As shown in Fig. 13.12, this triggers an influx of Ca^{2+} ions into the postsynaptic cell. Upon repeated stimulation, the Ca^{2+} influx recruits an adenylyl cyclase, which, in turn, activates the cAMP-dependent PKA. As in long-term facilitation in *Aplysia*, the activated PKA recruits MAPK, and both translocate to the nucleus, where they activate a transcriptional cascade regulated by CREB proteins. The target genes activated by CREB-1 include those that express growth-promoting effectors, such as brain-derived neurotrophic factor (BDNF). Indeed, there is some evidence that the late-phase LTP eventually leads to the formation of new synapses, which could make the synaptic changes persistent.

To study the role of PKA in the establishment of LTP and long-term memory, Ted Abel, Peter Nguyen, Rusiko Bourtchouladze, and Eric Kandel employed a transgenic approach. They generated transgenic mice

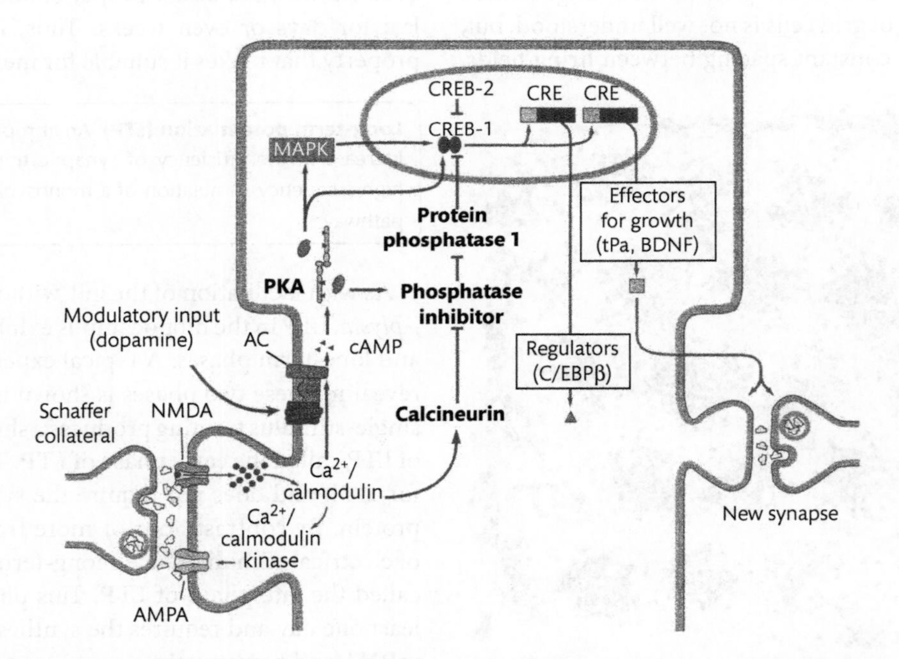

Figure 13.12 Cellular mechanisms of the late phase of LTP in the Schaffer collateral pathway. Repeated stimulation of the Schaffer collateral fibers leads to activation of NMDA receptors and influx of Ca^{2+} ions into the postsynaptic cell. The Ca^{2+} recruits an adenylyl cyclase (AC), which activates a cAMP-dependent PKA. The PKA recruits MAPK, and both kinases translocate into the nucleus, where they phosphorylate CREB-1. The latter protein, which is inhibited by CREB-2, binds to cAMP-response elements (CREs) in the promoter regions of target genes. These genes are thought to express growth factors involved in mediating structural changes in synaptic organization, including the growth of new synapses. This pathway, which results in the establishment of the late phase of LTP, is regulated by constraints (shown in bold) that inhibit long-term memory storage. Abbreviations not explained: AMPA, α-amino-3-hydroxy-5-methyl-4-isoxazole-propionic acid; BDNF, brain-derived neurotrophic factor; C/EBPβ, CCAAT/enhancer binding protein beta; tPa, tissue plasminogen activator. (After: Kandel, E. R. (2001).)

with a reduced activity of PKA in the hippocampus. These mice expressed a dominant-negative form of the regulatory subunit of PKA. This so-called R(AB) protein carries mutations in both cAMP-binding sites and acts as an inhibitor of enzymatic activity. To make the inhibition region-specific, the researchers used a promoter from the gene expressing the α-subunit of the Ca^{2+}/calmodulin-dependent protein kinase II. As expression of the latter gene is specific to somata and dendrites of neurons of the forebrain in adult mice, the expression of the transgene was also limited to this brain region.

As expected, in the R(AB) transgenic mice, the hippocampal PKA activity was significantly reduced. However, this did not affect the early phase of LTP, which was normal. The learning process itself, as well as short-term memory, were also not impaired. However, a marked decrease was found in the late phase of LTP, and this defect was paralleled by deficits in long-term spatial memory.

A more detailed analysis of the R(AB) transgenic mice showed that the diminished PKA activity caused a reduced stability of the place cells in the CA1 region of the hippocampus. When R(AB) mice were placed in a new environment, the place cells were able to establish new place fields, and the resulting map was stable when tested 1h after exposure to the new environment. However, these place fields were not stable when the mice were tested at 24h.

Taken together, these results indicate that the CA1 area of the hippocampus plays an important role in the conversion of short-term memory into long-term memory. In this process of memory consolidation, PKA is thought to induce the transcription of genes encoding proteins that are required to make the synaptic potentiation long-lasting.

New neurons for new memories

Another exciting research direction in the neurobiological study of learning and memory has emerged in recent decades. Based on the discovery that new neurons are continuously produced in the adult hippocampus, the possibility that these cells play a role in memory formation has been intensively examined. The two major model systems used in these investigations are the hippocampus of rodents and, to a lesser extent, the hippocampus of birds.

Adult neurogenesis and memory formation in the hippocampus of rodents

First indications that new neurons are generated in the hippocampus of rodents were obtained by Joseph Altman (1925–2016) and co-workers, then at the Massachusetts Institute of Technology in Cambridge, Massachusetts, in the 1960s. Yet, initially the idea that the adult brain is capable of producing functional neurons from adult stem cells met massive resistance. It was broadly accepted only closer to the end of the twentieth century, after an increasing number of investigations had provided compelling evidence for the existence of this phenomenon, which has become known as **adult neurogenesis**. In rodents, adult-born neurons are primarily found in two brain regions, one of which is the dentate gyrus of the hippocampus. In the following, we will first take a closer look at the morphology and some of the physiological properties of this brain structure. Then, we will describe the site within the dentate gyrus that harbors the adult stem cells, the development of their progeny, and the modulation of some of the underlying developmental processes by environmental factors. Finally, we will discuss experiments conducted to test the hypothesis that adult-born hippocampal neurons play a causal role in learning and memory functions.

Adult neurogenesis The generation of functional neurons in the nervous system from adult stem cells.

The dentate gyrus: Fundamental structural and functional organization

The **dentate gyrus** is composed of three layers. The principal layer is the **granule cell layer**, which is mainly formed by densely packed granule cells. This layer is only 4–8 neurons, or approximately 60μm, thick. The overlying **molecular layer** is largely cell free and occupied by the dendrites of the granule cells. The third layer is the underlying **polymorphic cell layer** (also referred to as the **hilus**), in which a variety of cell types are located.

The granule cells of the dentate gyrus receive their major input from fibers of the **perforant path**, which originate in the **entorhinal cortex**. Their axons, which Ramón y Cajal (see Box 2.1) called **mossy fibers**, travel through the polymorphic cell layer to make synaptic contact with **pyramidal cells** in the hippocampal **CA3 region**—the only projection area of the dentate gyrus granule cells.

A remarkable feature of the dentate gyrus is that its principal neurons outnumber both the input and output regions several-fold. In the adult rat, the total number of dentate gyrus granule cells is approximately 1.2 million, whereas layer II of the entorhinal cortex is composed of 0.11 million, and the CA3 region has approximately 0.25 million principal cells. Even though new granule cells are continuously generated during adulthood, their total number does not vary, indicating that there is a cellular turnover, instead of an addition of new cells to the population of older cells.

Another notable feature is the sparse activity of the granule cells. Using induction of immediate-early genes as a proxy for activation of these cells, investigators have found that only about 2% of the total granule cell population respond to a given behavioral experience.

The subgranular zone of the hippocampal dentate gyrus as a stem-cell niche

The **subgranular zone** of the dentate gyrus of the hippocampus is a narrow zone between the granule cell layer and the hilus. It serves as a **stem-cell niche** in the adult mammalian brain. The **stem cells** are thought to be a specific type of glial cell, so-called **radial glia**. They generate **intermediate progenitor cells** that proliferate to clone themselves, thereby exhibiting transit-amplifying properties, and to produce **neuroblasts**. The latter cells, although still immature, are already committed to the neuronal fate. The neuroblasts migrate a short distance into the granule cell layer, and differentiate, in the course of a few weeks, into mature granule cell neurons of the dentate gyrus. One of the early steps of this differentiation process includes a limited growth of cellular processes, but the cells do not, at this stage, appear to integrate into the neural network by making synaptic contacts with other neurons.

Stem-cell niche A specific microenvironment in tissue, in which stem cells reside. Their behavior is regulated through interaction with cells of the stem-cell niche. As a consequence of this interaction, the stem cells are maintained in a quiescent state, stimulated to proliferate for self-renewal, or induced to commit to differentiate.

During the second week after their birth, the new cells grow dendrites that extend towards the molecular layer, and axons (the mossy fibers) that travel through the hilus towards the CA3 region. They also receive **excitatory GABAergic input**, probably from local interneurons. This input appears to play a critical role in regulating survival and further maturation of the adult-born cells.

Excitatory action of GABA during development GABA (gamma-aminobutyric acid) operates as an inhibitory neurotransmitter in the adult brain in a wide range of species. However, in the immature brain activation of GABAergic synapses in immature neurons produces depolarization, instead of the well-characterized hyperpolarization in mature neurons. The excitatory action is related to a higher intracellular chloride concentration in immature neurons than in mature neurons. Once chloride exporters are expressed during development, and chloride is efficiently pumped from the intracellular milieu to the extracellular milieu, GABA switches from the excitatory to the conventional inhibitory mode.

During the third week, the adult-born cells start to establish connections with the local neuronal network. Their dendrites make synaptic contact with fibers of the perforant path. Subsequently, the young neurons develop synaptic contacts with dendrites of CA3 pyramidal neurons. This stage of development is also characterized by the onset of glutamatergic synaptic input, and the transition of the GABAergic input from excitatory to inhibitory.

During the following few weeks, the young dentate gyrus granule cells are distinguished by their pronounced synaptic plasticity, compared to mature cells. Among others, this is indicated by their low threshold for the induction of long-term potentiation. By eight weeks of age, the adult-born granule cells appear to have fully matured—their physiological properties and their potential for synaptic plasticity are now indistinguishable from those of older cells.

Behavioral modulation of adult neurogenesis in the dentate gyrus

Adult neurogenesis in the hippocampal dentate gyrus is influenced by a number of behavioral factors. Investigations by the laboratory of Fred ('Rusty') Gage at the Salk Institute for Biological Sciences in La Jolla, California, have revealed that, particularly, various environmental conditions play an important role in the behavioral modulation of this phenomenon. As depicted in Fig. 13.13, one group of mice were kept in standard housing cages, which were rather small and did not provide many opportunities for exploratory behavior and social interactions. A second group of mice lived in an **enriched environment**, which consisted of large cages with a variety of objects, such as tunnels, nesting material, and toys. Furthermore, the groups of mice kept under the latter condition were larger (14 mice versus three to four mice under standard conditions). Thus, the enriched environment offered plenty of opportunities for exploration and social interaction among the other individuals compared with the relatively impoverished standard laboratory environment.

Behavioral experiments demonstrated that environmental enrichment enhances memory functions in various learning tasks. To examine the possible involvement of neurogenesis in this behavioral effect, newborn cells were identified by a technique based on the incorporation of 5-bromo-2′-deoxyuridine (commonly known by its acronym **BrdU**) into DNA during the S phase of mitosis. Histological examination of BrdU-labeled tissue sections showed that mice kept under such conditions exhibit significantly more neurons in the dentate gyrus than mice kept under standard laboratory conditions. In principle, the larger number of neurons in the dentate gyrus of mice living in an enriched environment could be due either to a higher proliferative activity in mice exposed to this environmental condition, or to better survival of the new cells. As the studies of Gage and his group have suggested, it is mainly the higher rate of survival of the newborn cells that leads to more neurons in the dentate gyrus of mice kept under enriched conditions (Fig. 13.14).

BrdU labeling technique A method to label cells, based on the administration of 5-bromo-2′-deoxyuridine, commonly known as 'BrdU.' This molecule is an analog of thymidine, and does not exist naturally in the body. Similar to thymidine, BrdU is incorporated into replicating DNA during the S phase of mitosis. As BrdU is foreign to the organism, it can be recognized by antibodies raised against it. After immunohistochemical processing (see Chapter 2), the BrdU label can be detected in those cell nuclei that underwent mitosis at the time of administration of BrdU, or shortly thereafter.

Since an enriched environment offers increased opportunities for exploration and learning, it is possible that learning-associated activation of the

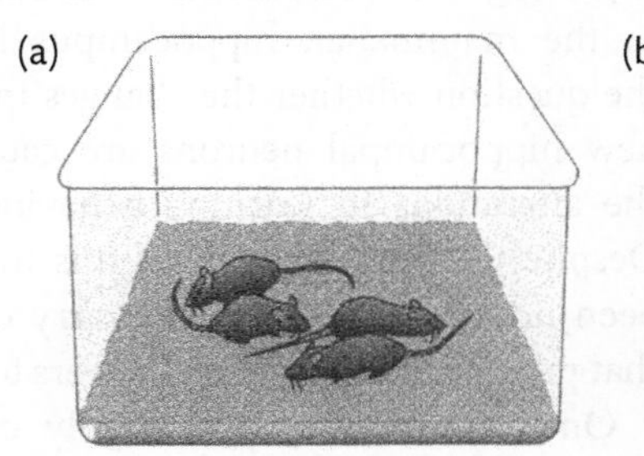

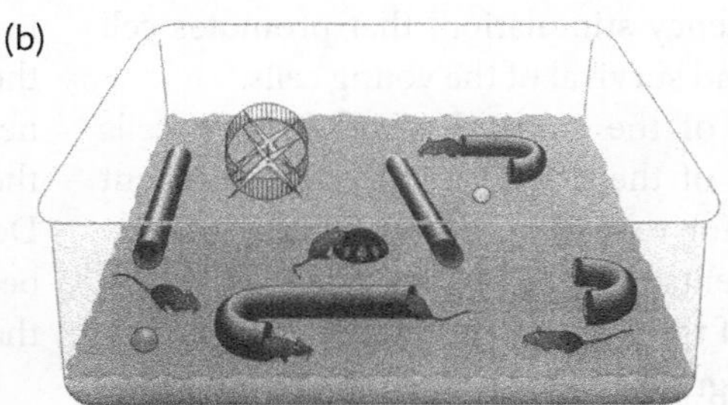

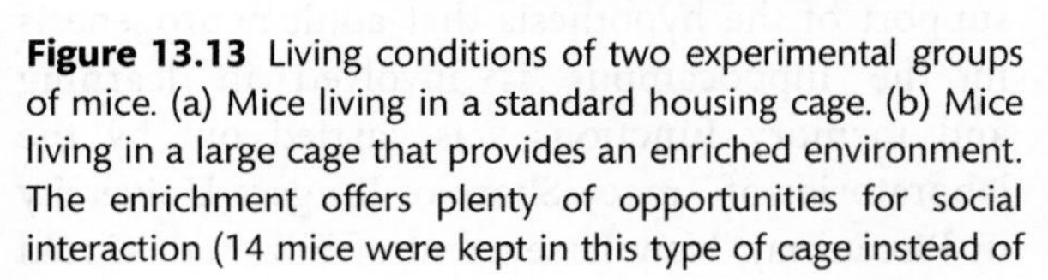

Figure 13.13 Living conditions of two experimental groups of mice. (a) Mice living in a standard housing cage. (b) Mice living in a large cage that provides an enriched environment. The enrichment offers plenty of opportunities for social interaction (14 mice were kept in this type of cage instead of three to four mice kept in the standard cage), exploration of the environment by providing appropriate objects such as toys and a set of tunnels, and physical exercise by being equipped with a running wheel. (After: Van Praag, H., Kempermann, G., and Gage, F. H. (2000).)

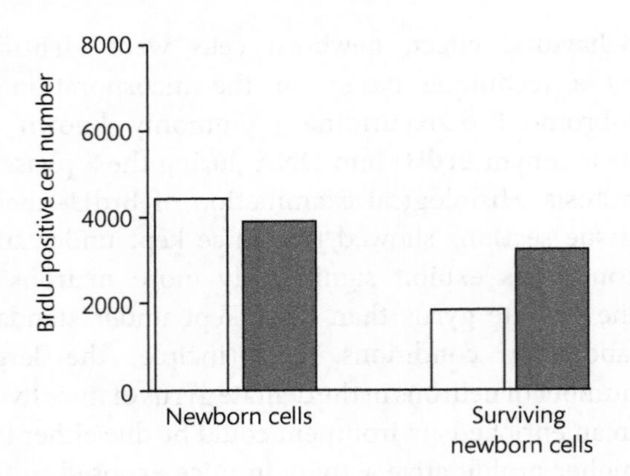

Figure 13.14 Total number of newborn cells and surviving newborn cells per dentate gyrus of mice kept under standard housing conditions (open bars) and in an enriched environment (shaded bars). The number of newborn cells was estimated using the BrdU technique one day after the last administration of a series of BrdU injections. The number of surviving new cells was determined four weeks after the last administration of BrdU. (After: Van Praag, H., Kempermann, G., and Gage, F. H. (1999).)

hippocampus facilitates better survival of the adult-born cells in the dentate gyrus. There is, indeed, experimental evidence to support this hypothesis. Induction of long-term potentiation by high-frequency stimulation at synapses of the perforant path with the granule cell neurons of the dentate gyrus both stimulated proliferation of the progenitor cells in the subgranular zone, and enhanced survival of adult-born dentate gyrus granule cells. However, the enhancement of cell proliferation is abolished when the NMDA receptor antagonist 4-(3-phosphonopropyl)-2-piperazine-carboxylic acid is administered. This finding indicates that it is primarily NMDA receptor activation, rather than the high-frequency stimulation, that promotes cell proliferation and survival of the young cells.

Modulation of the survival of adult-born cells at early stages of their maturation offers excellent opportunities for regulation of the number of new cells in the dentate gyrus. It has been estimated that 80–90% of the adult-born granule cell neurons die before integrating into the neural circuit of the dentate gyrus. During this stage of maturation, the young cells are highly vulnerable to **apoptotic cell death**. Reducing apoptosis is, thus, a highly effective mechanism to increase the number of surviving adult-born cells.

Apoptosis Also referred to as programmed cell death. It is controlled by specific apoptotic molecular pathways. Apoptosis is the predominant type of cell death during development, yet it may also occur in response to disease and injury. Cells undergoing apoptosis are characterized by distinct morphological alterations.

Another behavioral factor that modulates adult neurogenesis in the dentate gyrus is **voluntary physical exercise.** Mice that were provided the opportunity to run in a running wheel in their cage exhibited significantly higher cell proliferation rates in the subgranular zone than controls. Survival of the newborn cells was also increased in runners, although less so than in enriched mice. These results show that, although the effects of running and exposure to enriched environment on adult neurogenesis overlap to a certain degree, they also target different aspects of adult neurogenesis.

➤ Adult stem cells in the subgranular zone give rise to new cells that mature in the course of several weeks to develop into granule cell neurons of the hippocampal dentate gyrus. The generation and development of these cells is modulated by a number of factors, including environmental enrichment and voluntary physical exercise.

Possible functions of adult-born neurons in the hippocampus

Since conditions that improve memory functions, such as environmental enrichment or voluntary physical exercise, frequently have been found to increase the number of new cells, much of the research exploring behavioral functions of adult neurogenesis in the mammalian hippocampus has focused on the question whether the changes in the number of new hippocampal neurons are causally related to the alterations in learning behavior and memory. Despite the attractiveness of this hypothesis, it has been notoriously difficult to carry out experiments that provide unambiguous answers to this question.

One of the early and highly cited studies in support of the hypothesis that adult neurogenesis in the hippocampus is involved in learning and memory functions was carried out by the laboratories of Tracey Shors of Rutgers University in Piscataway, New Jersey, and of Elizabeth Gould of Princeton University, New Jersey. They treated rats with methylazoxymethanol acetate (**MAM**), a DNA-methylating agent that diminishes the number of proliferating cells in the organism, including the hippocampus (Fig. 13.15a). To test the effect of reduction of cell proliferation on memory formation, two conditioning paradigms were used. One involved delay conditioning, while the other employed trace conditioning.

During **delay conditioning**, the conditioned stimulus and the unconditioned stimulus show a temporal overlap. Acquisition of a conditioned response of this type does not require an intact hippocampus. In contrast, during **trace conditioning**, there is a temporal gap between the conditioned stimulus and the unconditioned stimulus. Thus, the animal must resurrect a memory 'trace' of the conditioned stimulus to associate this stimulus with the unconditioned stimulus. Acquisition of this type of conditioning requires the hippocampus.

Rats treated with the cell proliferation-suppressing agent showed a reduction in the number of conditioned responses during trace conditioning, but not during delay conditioning (Fig. 13.15b). This result is consistent with the idea that generation of new cells in the hippocampus during adulthood is causally linked to the formation of hippocampus-dependent memories. However, since MAM targets cell proliferation not only in the hippocampus but also in other parts of the body, researchers have sought more specific ways to experimentally manipulate mitotic activity of progenitor cells in the subgranular layer of the dentate gyrus. One increasingly applied approach utilizes **X-irradiation** to selectively eliminate stem cells in the subgranular zone of the dentate gyrus, while leaving other stem-cell regions in the brain (and the rest of the body) intact.

To test whether adult neurogenesis in the hippocampus is required for the improved learning and memory functions found after environmental enrichment, a research group under the leadership of René Hen of Columbia University, New York, first depleted the neural stem/progenitor cell populations specifically in the dentate gyrus of mice using X-irradiation. Then, they exposed the animals to an enriched environment for six weeks. This enrichment phase was followed by assessment of spatial learning in the **Morris water maze**. This experiment showed that even though adult neurogenesis was blocked specifically in the hippocampus, environmental enrichment still improved spatial learning in the mice tested. This finding raises doubts that the adult-born cells in the dentate gyrus mediate this effect, as suggested by the correlative evidence obtained in earlier studies.

Morris water maze A spatial learning test developed by Richard Morris of the University of St. Andrews in Scotland. The animal, usually a rat or mouse, is placed into a large circular pool of water, from which it can escape by swimming onto a submerged and invisible platform. The test is designed such that no local cues are available to indicate the location of the platform, but the animal can navigate using distal spatial cues.

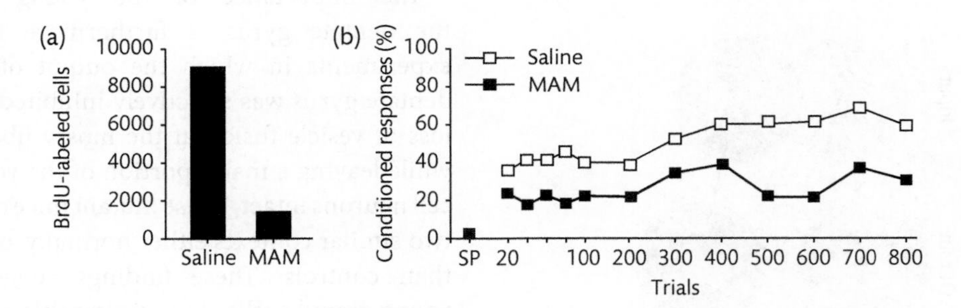

Figure 13.15 Effect of treatment of rats with methylazoxymethanol acetate (MAM) on the number of adult-generated neurons in the dentate gyrus (a) and on trace conditioning (b) in rats. The animals were treated with MAM or saline (as a control) daily for 14 days. On days 10, 12, and 14, they each received one injection of BrdU to label newly generated cells. After being left untreated for two days, they were subjected to an eye-blink conditioning paradigm involving 200 trials per day for four days, for a total of 800 trials. SP indicates the spontaneous blink rate. As the graph shows, treatment with MAM reduced both the number of newly generated cells in the dentate gyrus and the overall number of conditioned responses during trace conditioning. (After: Shors, T. J., Miesegaes, G., Beylin, A., Zhao, M., Rydel, T., and Gould, E. (2001).)

Whereas the role of adult neurogenesis in mediating the effects of environmental enrichment on learning and memory functions remains unclear, research in more recent years has suggested that another phenomenon—**pattern separation**—might critically depend on neurogenesis in the dentate gyrus. Pattern separation relates to the ability of the brain to form distinct memories of similar experiences (Fig. 13.16). Imagine you park your car every morning in a parking structure at your work place. If you are not one of the few privileged employees who have an assigned parking space, you probably end up parking your car in different parking slots on different days. Although each parking space is, in terms of its spatial characteristics, unique, most of its features are shared with other parking spaces. The challenge is, thus, to transform overlapping input patterns into differentiated output patterns by reducing the overlap in their representations, thereby facilitating their recall as distinct entities.

Several lines of evidence support the hypothesis that adult neurogenesis in the hippocampal dentate gyrus plays a critical role in performing pattern separation tasks. A frequently used approach to test this hypothesis involves subjecting a mouse to **contextual fear-discrimination learning**. The mouse is initially transferred from its home cage to the conditioning chamber, which is distinct in terms of its wall patterns, illumination, and odor. As a floor, stainless steel rods are used that are connected to a generator through which brief, mild electric shocks can be delivered. During the training phase, the mice are placed into this chamber, and after a few minutes they receive a single electric shock. The pairing of the previously neutral conditioned stimulus—the chamber—with the aversive unconditioned stimulus—the foot shock—results in the rapid acquisition through classical conditioning of a fear response. Most obviously, the mice will show **freezing**—the absence of any movement except movements related to respiration.

After acquisition of this conditioned response, the exposure of the mice to the chamber environment is varied, in a counterbalanced order, by exposure to a similar, yet slightly different, chamber. The differences include the walls of the chamber, the illumination, and the scent. During the mouse's exposure to this second chamber, no electric shock is delivered.

The freezing response of the mice shows that they rapidly learn to discriminate between the two chambers. At the point when the mice are able to distinguish the two environmental contexts, the effect of the experimental manipulation of adult neurogenesis in the dentate gyrus is examined. Such experiments have demonstrated that reduced neurogenesis impairs the ability of the mice to discriminate the two contexts. In complementary experiments, in which the population of adult-born neurons was expanded through enhancement of their survival, the performance of mice to distinguish two similar contexts was improved.

The importance of the young neurons in the dentate gyrus is furthermore indicated by experiments in which the output of the mature dentate gyrus was selectively inhibited by inducing loss of vesicle fusion at the mossy fiber terminals, while leaving a major portion of the young granule cell neurons intact. These mutant mice distinguished two similar contexts either normally, or even better than controls. These findings suggest that the young granule cells alone are capable of performing pattern separation tasks in the dentate gyrus, and that older granule cell neurons may even negatively interfere with this process. The exact mechanism through which the young dentate gyrus granule cells contribute to behavioral pattern separation is unknown.

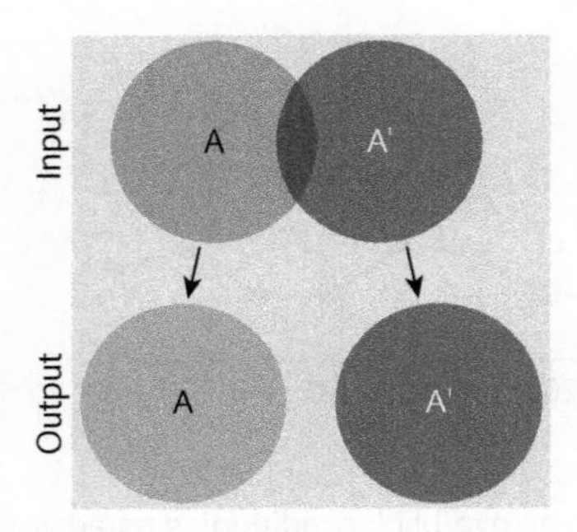

Figure 13.16 Conceptual representation of pattern separation. This process transforms similar patterns of neuronal inputs (indicated by the overlapping A and A′) into more distinct neuronal output representations (indicated by the non-overlapping A and A′). (After: Yassa, M. A. and Stark, C. E. L. (2011), © 2011 Elsevier Ltd.)

Adult neurogenesis and spatial learning in the avian hippocampus

A close correlation between adult neurogenesis in the hippocampus, on the one side, and learning and memory performance, on the other, has also been found by experiments conducted on birds. Most of these studies were carried out in the laboratories of John Krebs at Oxford University and of Fernando Nottebohm at Rockefeller University in New York. Hand-reared juvenile marsh tits (*Poecile palustris*) allowed to store sunflower seeds in storage sites and to retrieve this food showed a significantly higher rate of cell proliferation in the ventricular zone bordering the hippocampus than did controls. The control group consisted of age-matched birds that were treated in an identical way as the test animals, except that they could not store or retrieve food. After several weeks of food storing and retrieval activity, the experienced birds also exhibited a larger number of cells in the hippocampus than the control birds.

Similar results were obtained in free-ranging black-capped chickadees (*Poecile atricapillus*), songbirds that are very common in forests of temperate North America. In the late summer and early fall, their diet shifts from insects to seeds. At that time, they start to hide part of the seeds they find in storing sites, with typically only a few items per site. In the wild, they retrieve these seeds after a period of hours or days. Experiments on captive birds, however, have demonstrated that the memory of cache sites can last much longer, namely for several weeks.

Although determination of neuronal birth dates with radioactively labeled thymidine showed that chickadees form new neurons at all times of the year, these experiments also revealed a marked peak in the fall, as shown in Fig. 13.17. These neurons live for a few months and then appear to die. Moreover, the rate of cell proliferation in the region of the hippocampus is notably higher in juveniles than in adults.

Taken together, the experiments on tits and chickadees suggest that the recruitment of new neurons into the hippocampus may play a role in the acquisition of new spatial memories. In chickadees, the need for such an acquisition is particularly acute in the fall and in juveniles. It is the young birds in particular that encounter a wealth of novel information in the fall. At this time, the birds cease to defend territories and form flocks, change their diet, and experience a rapidly changing environment, such as the trees changing color and losing leaves, and later the arrival of snow.

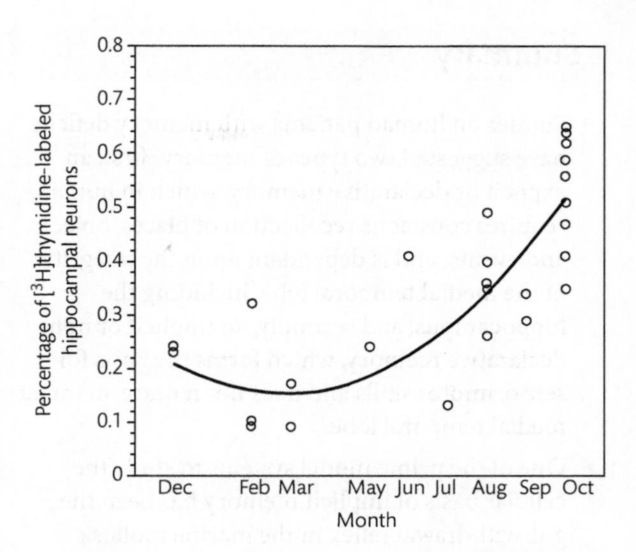

Figure 13.17 Percentage of new neurons in the hippocampus of adult black-capped chickadees at different times of the year. The free-ranging birds were captured and received a single injection of [^{3}H]thymidine (i.e. thymidine radioactively tagged by containing, in one position, tritium instead of non-radioactive hydrogen) to label cells undergoing mitosis at or shortly after the time of administration. The birds were then released. Six weeks later, they were recaptured, and the relative proportion of radioactively labeled neurons in the hippocampus was determined. The graph reveals that, although new neurons are produced throughout the year, their proportion is significantly higher in the fall than at other times of the year. (After: Barnea, A. and Nottebohm, F. (1994).)

➤ Experiments on birds suggest a close relationship between recruitment of new neurons in the hippocampus and acquisition of spatial memories.

On the other hand, sustained learning appears to require the periodical replacement of neurons within the hippocampus. It is thought that, in order to avoid running out of memory space, memory information encoded by older neurons is transferred from the hippocampus to sites elsewhere in the brain. As the older neurons are likely to have lost their plastic properties essential for the formation of memories, these neurons are eventually replaced by newly generated ones.

Summary

- Studies on human patients with memory deficits have suggested two types of memory: first, an explicit or declarative memory, which in humans requires conscious recollection of places, objects, and events, and is dependent upon the integrity of the medial temporal lobe, including the hippocampus; and secondly, an implicit or non-declarative memory, which forms the basis for sensorimotor skills and does not require an intact medial temporal lobe.
- One of the prime model systems to study the cellular basis of implicit memory has been the gill-withdrawal reflex in the marine mollusk *Aplysia*. This defensive reflex involves a withdrawal of the siphon and the gill into the mantle cavity when the siphon of the mollusk is touched.
- The gill-withdrawal reflex can be modified by several forms of learning, including sensitization, which occurs when an aversive shock is applied to another part of the body. A weak tactile stimulus applied to the siphon then evokes a much greater reflex response than it normally would.
- If a single shock is applied to the tail, then the sensitization of the gill-withdrawal reflex lasts for only a few minutes and thus exhibits features of short-term memory. If shocks are applied repeatedly, then the gill-withdrawal reflex remains sensitized for several days, thus displaying properties of long-term memory.
- At the neuronal level, the sensitization effect is mediated by modulatory interneurons in the abdominal ganglion, which make synaptic contact with sensory neurons. The latter neurons innervate the siphon and make direct synaptic contact with motoneurons innervating the gill.
- Cellular analysis of short-term sensitization has revealed an increase in the strength of the synaptic connection between the sensory neuron and the motoneuron. This so-called facilitation is initiated by activation of the modulatory interneuron, which releases serotonin at the synapse making contact with the sensory neuron. The serotonin acts on a transmembrane serotonin receptor on the sensory neuron, which activates adenylyl cyclase, thereby leading to an increase in the intracellular level of cAMP. The cAMP in turn recruits PKA, which results in an enhancement of transmitter release from the sensory neuron.
- Repeated tail shocks, or several puffs of serotonin to the terminal region of the sensory neuron, leading to long-term sensitization, initially activate a similar biochemical pathway as that occurring during short-term sensitization. However, now the PKA recruits MAPK. Both kinases translocate into the cell nucleus, where they activate a cascade of transcriptional processes.
- The activation of several target genes results in two major effects. First, the PKA exhibits persistent activity. Secondly, the synaptic organization of the connection between the sensory and motoneurons is altered; this includes the filling of initially empty presynaptic terminals with synaptic vesicles and active zone material, as well as the growth of new synaptic connections.
- Non-human models to study explicit memory have been established in a number of mammals that have a medial temporal lobe including a hippocampus.
- These studies have revealed place cells among hippocampal pyramidal cells, which collectively form spatial maps of different environments of the animal. The establishment of these maps resembles learning processes and the conversion of short-term spatial memory into long-term memory.
- A second phenomenon discovered in the hippocampus is long-term potentiation (LTP), which is reflected by an abrupt and sustained increase in the efficiency of synaptic transmission after high-frequency stimulation of a monosynaptic excitatory pathway. Due to its long-lasting effect, LTP is thought to be involved in memory storage.
- A detailed molecular analysis of the late phase of LTP, which lasts for at least one day, in the Schaffer collateral pathway of the hippocampus has implicated PKA as a key molecule. The final

step in this process appears to be the formation of new synapses.

- A reduction in the activity of PKA using a transgenic approach causes a marked decrease in the late phase of LTP and a reduced stability of place cells in the CA1 region of the hippocampus. This result underlines the importance of PKA in the process of memory consolidation.
- A phenomenon that contributes to the plasticity of the adult hippocampus is the generation of new granule cells neurons in the dentate gyrus. The function of this so-called adult neurogenesis is unclear, but several lines of evidence indicate involvement of the young granule cell neurons in pattern separation, an ability of the brain to form distinct memories of similar experiences.

The bigger picture

One of the most influential postulates in neurobiology was made by Donald O. Hebb: That the efficacy of connections between neurons increases with repeated stimulation of the postsynaptic neurons by the presynaptic neuron. This prediction is particularly astonishing given that it was published in 1949—several years before the first ultrastructural descriptions of synapses were published. Indeed, it took until 1973 for the synaptic plasticity foreseen by Hebb to be demonstrated. In that year, Timothy Bliss and Terje Lømo reported the discovery of LTP in one of the hippocampal pathways. The finding of LTP in the hippocampus—a brain structure by then known to be involved in learning and memory functions—provided the foundation for a testable hypothesis: That the behavioral processes of learning and memory are causally linked to changes at the synaptic level. Starting in the 1970s, it was especially the work of Eric R. Kandel and associates that showed, initially in *Aplysia* and subsequently in rodents, that short-term memory formation manifests through biochemical changes at existing synapses, whereas long-term memory is primarily associated with the development of new synapses. Evidence in more recent years has implicated even the generation of new neurons in learning and memory processes. From a broader perspective, the discoveries of synaptic plasticity and adult neurogenesis have been key in overturning the erroneous belief of the immutability of the adult brain—a notion that had dominated neurobiology until close to the end of the twentieth century.

Recommended reading

Bailey, C. H., Kandel, E. R., and Harris, K. M. (2015). Structural components of synaptic plasticity and memory consolidation. *Cold Spring Harbor Perspectives in Biology* 7:a021758.

The article reviews the remodeling and growth of pre-existing synapses and the formation of new synapses underlying the consolidation of implicit memory in Aplysia and of explicit memory in the mammalian hippocampus.

Bliss, T. V. P. and Collingridge, G. L. (1993). A synaptic model of memory: Long-term potentiation in the hippocampus. *Nature* **361**:31–39.

A comprehensive review of the physiological and cellular mechanisms of LTP in the hippocampus. Primarily targeted towards readers at a more advanced level.

Drew, L. J., Fusi, S., and Hen, R. (2013) Adult neurogenesis in the mammalian hippocampus: Why the dentate gyrus? *Learning & Memory* **20**:710–729.

This comprehensive review summarizes the major theories of adult hippocampal neurogenesis, highlights the issues and discrepancies associated with these theories, and discusses possible future research directions that might help in elucidating the advantage for the neural circuit of the dentate gyrus of generating new neurons during adulthood.

Gonçalves, J. T., Schafer, S. T., and Gage, F. H. (2016). Adult neurogenesis in the hippocampus: From stem cells to behavior. *Cell* **167**:897–914.

A contribution to the literature from one of the leading laboratories in the field of mammalian adult neurogenesis. This review describes the molecular machinery involved in the generation of new neurons in the dentate gyrus and their integration into functional neural circuits. The authors also discuss potential functions of the adult-born neurons, including their relevance to behavior.

Grieves, R. M. and Jeffrey, K. J. (2017). The representation of space in the brain. *Behavioural Processes* **135:**113–131.

This article summarizes the research that has contributed to the 'cognitive map' concept. It discusses in detail not only place cells, head direction cells, and grid cells, but also other spatially modulated neurons in the brain that are thought to underlie cognitive mapping processes.

Kandel, E. R. (2001). The molecular biology of memory storage: A dialogue between genes and synapses. *Science* 294:1030–1038.

Based on the author's address to the Nobel Foundation, this essay provides an excellent review of the work of Eric Kandel.

Mayford, M. and Kandel, E. R. (1999). Genetic approaches to memory storage. *Trends in Genetics* **15:**463–470.

This article reviews some key advances made through genetic approaches in the study of learning and memory, and discusses possible future directions.

O'Keefe, J., Burgess, N., Donnett, J. G., Jeffery, K. J., and Maguire, E. A. (1998). Place cells, navigational accuracy, and the human hippocampus. *Philosophical Transactions of the Royal Society of London B* **353:**1333–1340.

A useful summary of how the hippocampal formation is involved in spatial navigation, based mainly on the work of the group headed by John O'Keefe and Neil Burgess.

Squire, L. R. and Kandel, E. R. (1999). *Memory: From Mind to Molecules*. Scientific American Library, New York.

A highly stimulating book, written by two of the world leaders in memory research. Easy to read, without oversimplifying the facts. An ideal starting point!

Yuste, R. and Bonhoeffer, T. (2001). Morphological changes in dendritic spines associated with long-term synaptic plasticity. *Annual Review of Neuroscience* **24:**1071–1089.

An authoritative review of the role of morphological changes of dendritic spines in hippocampal LTP in rodents, with special emphasis on a discussion of 'online' studies using time-lapse imaging to visualize the generation of new spines.

Short-answer questions

13.1 Define, in one sentence, the term 'Hebb synapse'.

13.2 Which of the following two types of memory requires the integrity of the medial temporal lobe, including the hippocampus?

a) explicit memory.

b) implicit memory.

13.3 Which of the following two choices is correct? Modification of the gill-withdrawal reflex in *Aplysia* by sensitization is mediated through modulation involving:

a) substance P;

b) serotonin.

13.4 Indicate whether the following statement is true or false: Long-term sensitization of the gill-withdrawal reflex in *Aplysia* leads to the formation of new synapses by sensory neurons.

13.5 Define, in one sentence, the term 'place cells.'

13.6 Define, in one sentence, the term 'long-term potentiation'.

13.7 When mice are exposed to enriched-environment conditions, the dentate gyrus of their hippocampus contains more granule cell neurons than the dentate gyrus of mice kept under standard laboratory conditions. This larger number of hippocampal neurons is the result of:

a) higher proliferative activity of adult stem cells in the dentate gyrus;

b) better survival of the newborn cells.

Indicate the correct statement.

13.8 What is pattern separation? Which cells in which brain region are thought to play a key role in this cognitive ability of the brain?

13.9 Indicate the correct statement. Experiments in black-capped chickadees have revealed a marked increase in adult neurogenesis in the hippocampus in the fall, compared to other times of the year. The recruitment of new neurons into the hippocampus is thought to play a role in the

a) acquisition of new spatial memories;

b) changes in the regulation of body temperature during the summer-winter transition period.

13.10 Challenge question Considering that memory has a limited capacity, why does attention play an important role in memory formation?

13.11 Challenge question It has been hypothesized that the modular organization of grid cells is crucial for the network to be able to generate not only one map of the external environment, but thousands or millions. Explain how this might be achieved.

13.12 Challenge question Consolidation of episodic memories has commonly been explained by gradual transfer over weeks of such memories from the hippocampus, where short-term memory is formed, to neocortical networks for long-term storage. Recent research has, however, indicated that memories are formed simultaneously in the hippocampus and the neocortex. Suggest possible explanations why the formation of memories in the neocortex might have remained undiscovered for so long.

Essay questions

13.1 What is the difference between implicit and explicit memory? What experimental approaches have led to the distinction between these two types of memory?

13.2 How do the behavioral effects of short- and long-term sensitization on the gill-withdrawal reflex of *Aplysia* differ? What are the molecular mechanisms mediating these two effects?

13.3 What role has been proposed for the mammalian hippocampus in the establishment of long-term memory?

13.4 What is long-term potentiation? Describe the molecular mechanisms underlying the late phase of long-term potentiation in the Schaffer collateral pathway of the mammalian hippocampus. How can a transgenic approach assist the researcher in defining the role of molecular factors involved in this process?

13.5 The generation and development of new neurons in the hippocampal dentate gyrus of the adult mammalian brain is well documented. Nevertheless, it has remained notoriously difficult to assign a behavioral function to this phenomenon of adult neurogenesis. Discuss the reasons for this difficulty, and describe how the application of more specific approaches is likely to shed light on this problem.

Advanced topic Structural changes in the adult human brain associated with learning

Background information

In a widely publicized study, Eleanor A. Maguire of University College London and her co-workers compared, in a series of studies, the brains of London taxi drivers with those of control subjects who did not drive taxis. They found that the posterior hippocampus of the taxi drivers was larger, relative to that of control subjects. These results were interpreted as showing increases in volume of the posterior hippocampus during phases of spatial learning, when the taxi drivers were thought to have created a 'mental map' of the city of London.

Essay topic

In an extended essay, describe the principles of magnetic resonance imaging (MRI), the method used by Maguire and co-workers to analyze the volume of brain structures. Then summarize and discuss the main results, considering the following points of criticism raised by other investigators:

1. A generalization of the findings may not be possible because the study was gender-biased, as only male taxi drivers were examined. This point is likely to be important, as an investigation by Georg Grön and associates has pointed to gender differences in neural networks that are activated during human navigation.
2. The fact that only a single time point—after the assumed acquisition of spatial information—was examined leaves open the possibility that the differences between the two groups were not due to hippocampal growth during phases of spatial learning. Instead, such differences could be caused, for example, by genetic factors, resulting in an intrinsic variability in the size of certain hippocampal areas among human populations. According to this interpretation, people with a larger posterior hippocampus have better navigational skills, and thus might more frequently choose the profession of taxi driver.
3. Nothing is known about the nature of the cellular events that mediate the presumptive growth of specific parts of the hippocampus.

Include in your discussion a presentation of the results of other investigations in this area of research that were inspired by the work of Eleanor Maguire and colleagues.

Starter references

Draganski, B., Gaser, C., Kempermann, G., Kuhn, H. G., Winkler, J., Büchel, C., and May, A. (2006). Temporal and spatial dynamics of brain structure changes during extensive learning. *Journal of Neuroscience* **26**:6314–6317.

Draganski, B. and May, A. (2008). Training-induced structural changes in the adult human brain. *Behavioural Brain Research* **192**:137–142.

Grön, G., Wunderlich, A. P., Spitzer, M., Tomczak, R., and Riepe, M. W. (2000). Brain activation during human navigation: Gender-different neural networks as substrate of performance. *Nature Neuroscience* **3**:404–408.

Maguire, E. A., Gadian, D. G., Johnsrude, I. S., Good, C. D., Ashburner, J., Frackowiak, R. S. J., and Frith, C. D. (2000). Navigation-related structural change in the hippocampi of taxi drivers. *Proceedings of the National Academy of Sciences of the U.S.A.* **97**:4398–4403.

Maguire, E. A., Woollett, K., and Spiers, H. J. (2006). London taxi drivers and bus drivers: a structural MRI and neuropsychological analysis. *Hippocampus* **16**:1091–1101.

To find answers to the short-answer questions and the essay questions, as well as interactive multiple choice questions and an accompanying Journal Club for this chapter, visit **www.oup.com/uk/zupanc3e**.

GLOSSARY

Action potential A transient depolarization of the membrane potential generated in an all-or-nothing fashion.

Actogram Plot of the recorded activity of an individual during a given time period.

Adequate stimulus Form of stimulation that elicits an optimal response from a sensory receptor or a sense organ. Most commonly, this stimulus requires minimal energy to cause excitation. For example, the adequate stimulus for visual organs is light of specific wavelengths. Although other sensory modalities may also stimulate eyes or photoreceptors, only light can elicit responses at minimal energies.

Adult neurogenesis Generation of functional neurons in the nervous system from adult stem cells.

Alpha-Bungarotoxin A constituent protein of the venom of the Southeast Asian krait (*Bungarus multicinctus*). By binding to nicotinic postsynaptic receptor sites, this neurotoxin irreversibly blocks cholinergic transmission at the neuromuscular junction, thus producing muscle paralysis.

Amnesia Disturbance in long-term memory, manifested by total or partial inability to recall past experiences. See also **anterograde amnesia, retrograde amnesia**.

AMPA receptor An ionotropic glutamate receptor that can be activated by the glutamate analogue α-amino-3-hydroxy-5-methyl-4-isoxazolepropionic acid, commonly known as AMPA.

Anode Positive pole of a voltage source.

Anterograde amnesia Deficit in memory in reference to events occurring after the traumatic event or disease that caused the condition.

Anthropocentrism The view that places humans as the most important element in the center of the universe, and interprets the world exclusively in terms of human values and experience.

Apoptosis Also referred to as programmed cell death. It is controlled by specific apoptotic molecular pathways. Apoptosis is the predominant type of cell death during development, yet it may also occur in response to disease and injury. Cells undergoing apoptosis are characterized by distinct morphological alterations.

Aposematism The use of a conspicuous signal, such as bright coloration, by an animal to advertise to potential predators that it is toxic or distasteful.

Axoplasm The cytoplasm of an axon.

Azimuth Horizontal plane.

Batesian mimicry A form of superficial biological resemblance in which a noxious species, protected by its toxicity or unpalatability, is mimicked by a harmless, palatable species. The imitated species is known as the model, whereas the imitating species is referred to as the mimic. The mimic benefits because predators mistake it for the model. Batesian mimicry is named after its discoverer, the nineteenth-century English naturalist Henry Walter Bates.

Behaviorism The psychological discipline that studies observable behavior in a select number of model organisms by using objective approaches. Its main emphasis is on the exploration of how environmental stimuli produce and modulate behavior.

BrdU labeling technique A method to label cells, based on the administration of 5-bromo-2′-deoxyuridine, commonly known as 'BrdU.' This molecule is an analog of thymidine, and does not exist naturally in the body. Similar to thymidine, BrdU is incorporated into replicating DNA during the S phase of mitosis. As BrdU is foreign to the organism, it can be recognized by antibodies raised against it. After immunohistochemical processing (see Chapter 2), the BrdU label can be detected in those cell nuclei that underwent mitosis at the time of administration of BrdU, or shortly thereafter.

Cathode Negative pole of a voltage source.

Cell theory The nervous system is composed of individual cells.

Chirps Transient frequency and amplitude modulations of the electric organ discharge of wave-type electric fish of the taxonomic order Gymnotiformes.

Chronobiology The scientific discipline that examines biological rhythms.

Cilia (singular: cilium) Hair-like appendages of many kinds of cells with a bundle of microtubules at their core. The microtubules are arranged such that nine doublet microtubules are located in a ring around a pair of single microtubules. This arrangement gives them the distinctive appearance of a '9 + 2' array in electron micrographs of cross-sections.

Circadian rhythm A rhythm of a biological phenomenon, driven by an endogenous time-keeping system, with a period close, but not equal, to 24 hours.

Confocal microscopy A microscopy technique widely used in biomedicine. It enables investigators to image either fixed or living tissue labeled with one or several fluorescent probes. Compared to conventional optical microscopy, confocal microscopy offers the advantage of elimination of out-of-focus glare caused by fluorescently labeled specimens thicker than a few micrometers; the ability to control the depth of field; and the capability to collect serial optical sections and to use these data for three-dimensional reconstruction of tissue structures.

Connectome Complete description of the structural connectivity of the nervous system of an organism.

Corollary discharge A copy of a motor command that does not produce any motor action. Instead, it is routed to sensory structures where it influences sensory processing.

Curare Generic name for various types of unstandardized extracts derived mainly from the bark of the tropical plants *Strychnos* and *Chondrodendron*. It is prepared for use as an extremely potent arrow poison by Indians in South America. The physiologically active ingredient of curare is the alkaloid tubocurarine, which is employed as a relaxant of skeletal muscles during surgery to control convulsions. The muscle-relaxant effect is caused by interference of the alkaloid with the action of acetylcholine at the neuromuscular junction.

Cytoarchitectonics Sub-discipline of neuroanatomy that uses Nissl staining for parcellation of the brain into distinct areas.

Delay line In electric circuits, a device that introduces a specific delay time in transmission of a signal. Functionally similar structures exist in neural networks.

Duty cycle Proportion of time during which a device is operated. Refrigerators, for example, do not run continuously but are switched on and off for a certain amount of time. If, for example, a refrigerator pump is on for 45min over a 60min period, then its duty cycle is 75%. In bat echolocation research, the duty cycle defines the relative 'on' time of the emitted echolocation calls.

Eclosion Emergence of adult insects from their pupal cases or of larvae from the eggs.

Elevation Vertical plane.

Epigenetics The study of the heritable alterations in a chromosome, without changes in the DNA sequence. Epigenetic alterations are stably transmitted through mitotic cell divisions within an organism, or through meiosis across generations. Several epigenetic processes have been identified, including DNA methylation and chromatin modification.

Ethogram The entire behavioral repertoire of an animal species.

Ethology The biological discipline that studies natural animal behavior in a broad range of organisms by using objective and comparative approaches.

Excitatory action of GABA during development GABA (gamma-aminobutyric acid) operates as an inhibitory neurotransmitter in the adult brain in a wide range of species. However, in the immature brain activation of GABAergic synapses in immature neurons produces depolarization, instead of the well-characterized hyperpolarization in mature neurons. The excitatory action is related to a higher intracellular chloride concentration in immature neurons than in mature neurons. Once chloride exporters are expressed during development, and chloride is efficiently pumped from the intracellular milieu to the extracellular milieu, GABA switches from the excitatory to the conventional inhibitory mode.

Extracellular recording Placement of an electrode in the proximity of a neuron and of the reference electrode at some distance in the extracellular fluid. Through this arrangement, potential changes at the surface of the cell membrane are recorded either from single neurons ('single-unit recording') or, in the form of field potentials, from many neurons.

Facilitation Enhancement of synaptic transmission.

Feature detectors Neurons that respond selectively to specific features of a sensory stimulus.

Fictive behaviors Sequences of motoneuron activity occurring without the production of actual movement or muscular contraction. Such behaviors can be observed in immobilized whole animals or in isolated preparations of the nervous system in which muscles are removed.

Frequency difference (Df) In wave-type electric fish, frequency of the neighbor's signal minus frequency of the fish's signal.

Growth cone Highly motile structure at the tip of a growing axon. Growth cones play a critical role in the guidance of the extending axons.

Gyration Revolution around the longitudinal axis.

Immediate-early genes A class of genes whose expression is low or undetectable in quiescent cells, but whose transcription is activated within minutes after proper stimulation.

Interstitial fluid Fluid in the intercellular spaces.

Intracellular recording Insertion of the tip of a glass microelectrode inside a cell and measurement of the potential difference between the intracellular space and an extracellular reference point.

Kramer locomotion compensator A polystyrene sphere on top of which a cricket can walk freely. A control device monitors the movement of the cricket and counter-rotates the sphere to keep the cricket near the top.

Labyrinth Otolith organ plus semicircular canals plus cochlea.

Lagena The third otolith organ, besides the utriculus and the sacculus, of the inner ear (cf. Chapter 4, 'Model system: Geotaxis in vertebrates'). It is present in elasmobranchs, teleosts, amphibians, reptiles, and birds. In mammals, it is found only in monotremes. The lagena has been implicated in auditory and vestibular functions, but may also be involved in magnetoreception.

Long-term potentiation (LTP) An abrupt and sustained increase in the efficiency of synaptic transmission after high-frequency stimulation of a monosynaptic excitatory pathway.

Lordosis behavior Body posture of females in some mammals (including rodents and felines) during mating, triggered by mounting of the male. Characteristic features of this reflex-like behavior are a ventral arching of the vertebral column, lowering of the forelimbs, raising of the hips, and lateral or dorsal displacement of the tail so that the vagina is presented to the male.

Luciferases Enzymes that catalyze the oxidation of luciferins to generate light in a bioluminescent reaction. The best-known luciferase is that produced by the North American firefly, which emits green light. Luciferase (*luc*) genes can be inserted into the DNA of other organisms so that the light produced acts as a 'reporter' for the activity of regulatory elements that control the expression of *luc*.

Magnetometer An instrument for measuring magnetic fields.

Melatonin A hormone secreted by the pineal gland in the brain and by some other organs, including the retina; its circulating levels exhibit daily cycles.

Meta-analysis Quantitative statistical analysis of several separate, but conceptually similar, studies. The outcome may be a more precise estimate of an effect examined in these previous studies, or the assessment of an effect not done in any of the previous investigations.

Mind-body dualism The philosophical concept that mind and body are fundamentally different kinds of entities. According to this theory, perception, emotion, thoughts, and other 'mental' phenomena are supernatural and not the result of brain action.

Morris water maze A spatial learning test developed by Richard Morris of the University of St. Andrews in Scotland. The animal, usually a rat or mouse, is placed into a large circular pool of water, from which it can escape by swimming onto a submerged and invisible platform. The test is designed such that no local cues are available to indicate the location of the platform, but the animal can navigate using distal spatial cues.

Motivation The physiological state of an animal that defines the frequency and intensity of occurrence of a behavior when elicited by a given endogenous or exogenous stimulus.

Mutagenesis The process by which a mutation occurs in nature, or is induced in the laboratory using mutagens (e.g. X-rays, UV light, specific chemicals).

Mutualism Two organisms of different species interacting, each benefiting from this relationship.

Neural oscillation Rhythmic neural activity generated by individual neurons or through interactions of neural ensembles. A commonly used technique to monitor the synchronized activity of large numbers of neurons is electroencephalography (EEG). Such analysis has revealed oscillatory activity in distinct frequency bands. These frequency-specific signals have been linked to a variety of behavioral and cognitive states.

Neuroethology The biological discipline that attempts to understand how the nervous system controls the natural behavior of animals.

Neurotropic virus Virus with an affinity for nervous tissue.

NMDA receptor An ionotropic glutamate receptor. The agonist *N*-methyl-D-aspartate (NMDA) binds selectively to this type of glutamate receptor but not to other glutamate receptors.

Null mutation Any mutation of a gene that results in a lack of function.

Olfactory imprinting hypothesis The notion that salmon become imprinted to the odor of their home river during the smolt phase; after reaching sexual maturity, they use this information to find their way back home.

One-letter code of amino acids For each of the 20 amino acids, a unique letter of the alphabet is assigned: A, alanine; C, cysteine; D, aspartic acid; E, glutamic acid; F, phenylalanine; G, glycine; H, histidine; I, isoleucine; K, lysine; L, leucine; M, methionine; N, asparagine; P, proline; Q, glutamine; R, arginine; S, serine; T, threonine; V, valine; W, tryptophan; Y, tyrosine.

Operant conditioning (= instrumental conditioning) A learning process by which the consequences of a behavior affect the probability of its occurrence in the future.

Ophthalmic nerve The smallest of the three divisions of the trigeminal nerve. It supplies afferent branches, among other structures, to parts of the eye and the nasal cavity.

Oscillogram Trace on the panel display of an oscilloscope of instantaneous voltage of an electric signal as a function of time. Since a microphone converts the sound pressure of an acoustic signal into voltage, the oscillograms of such a signal represents a plot of instantaneous sound pressure over time. The greater the sound pressure (subjectively experienced by humans as an increase in loudness), the greater the amplitude of the voltage signal.

Parallel plate capacitor Composed of two parallel metal plates isolated against each other. They have equal but opposite charges. By definition, the electric field lines arise from the positively charged plate (anode) and run perpendicular to the plates to the negatively charged plate (cathode).

pH Method of quantitatively expressing the concentration of H_3O^+ ions in a solution. The pH is defined as the negative logarithm to the base 10 of the hydrogen ion concentration. Pure water is neutral and has a pH of 7. A pH of less than 7 indicates an acidic solution and a pH of greater than 7 a basic (alkaline) solution.

Polymorphic networks Anatomically defined networks whose modulation results in multiple functional modes of operation.

Positive phonotaxis Movement of an animal toward the source of a sound.

Pressure and shear forces Contact interactions between different material bodies take place across contact surfaces. Any contact force acting on a surface can, regardless of its direction, be resolved into its normal (= perpendicular to the surface) component and its tangential (= parallel to the surface) component. The normal component is referred to as the pressure force, and the tangential as the shear (or shearing) force.

Projection of a neuron The route of its axon from the site of origin to the target region.

Proprioception The perception of spatial orientation and movement by sensory receptors within the body, which are referred to as proprioceptors. Example: Muscle spindles in limbs are proprioceptors that provide information about changes in muscle length.

Pyramidal neurons Neurons with a pyramidal-shaped cell body and two distinct dendritic trees. They are found in a number of areas within the forebrain. Pyramidal neurons represent the most common excitatory cell type in the mammalian cortex.

Rebound firing The production of action potential(s) by a neuron after it has been hyperpolarized.

Receptive field That part of the sensory space (e.g. skin surface) that can elicit neuronal responses (e.g. change in spike frequency) when stimulated.

Reinforcer A stimulus that increases the probability of a behavioral response in operant conditioning. Reinforcers can be either positive (such as reward) or negative (such as removal of a negative consequence).

Resonance Oscillation of a system when energy is supplied from an external force at, or close to, the resonant frequency. The resonant frequency is the system's natural oscillation frequency. After supply of external energy at the resonant frequency, the resulting amplitude is particularly large, if the damping of the oscillating system is low.

Reticular theory The nervous system forms a continuous network ('reticulum') of fused processes.

Retrograde amnesia Deficit in memory in reference to events that occurred before the trauma or disease that caused the condition.

Sensitization Enhancement of a behavioral response to a stimulus after application of a different, noxious stimulus.

Sequential imprinting hypothesis The notion that salmon learn a series of olfactory intermediary points in the course of their downstream migration.

Sign stimulus The component of the environment that triggers a specific behavior.

Sound pressure level In physics, sound is defined as a longitudinal, mechanical wave, whose amplitude reflects a local pressure deviation from the ambient pressure. Sound pressure is commonly measured with microphones as the force of sound (in Newton) on a surface area (in square meters) perpendicular to the direction of sound. The SI unit of sound pressure is pascal (Pa). Sound pressure level L_p is a logarithmic measure of sound pressure p relative to a reference value p_0, expressed in decibels (dB):

$$L_p = 20 \log (p/p_0) \text{ dB}$$

A logarithmic scale is used because the intensity perceived by humans is proportional to the logarithm of the actual intensity measured. In air, the commonly used reference value is 20μPa, which corresponds roughly to the threshold of human hearing. If sound pressure is doubled, the sound pressure level will increase by 6dB. A sound pressure level of approximately 120dB, relative to 20μPa, is generated by a jet engine at a distance of 100m.

Sound spectrogram Also referred to as a **sonogram**. A visual representation of an acoustic signal. Most commonly, the horizontal axis represents time, and the vertical axis represents frequency. The amplitude at a particular time and particular frequency is encoded either by the gray level (the darker the marking, the higher the sound intensity) or by using a color map (ranging, for example, from light yellow to dark red as sound intensity increases). Sound spectrograms are generated by analog or digital versions of sound spectrographs.

Space-specific neurons Respond only to acoustic stimuli originating from a restricted area in space.

Spinal cord segments The spinal cord is divided into four regions: cervical (which is closest to the head), thoracic, lumbar, and sacral. Each of these regions consists of several segments.

Spinal preparation Animal with its spinal cord isolated from higher regions of the central nervous system by transection of the cord at its upper segments.

Stem-cell niche A specific microenvironment in tissue, in which stem cells reside. Their behavior is regulated through interaction with cells of the stem-cell niche. As a consequence of this interaction, the stem cells are maintained in a quiescent state, stimulated to proliferate for self-renewal, or induced to commit to differentiate.

Suprachiasmatic nucleus A paired neuronal structure in the brain, at the base of the hypothalamus, just above the optic chiasm. It receives photic input from the retina via the optic nerve and plays an important role in the regulation of the body's circadian rhythms.

Sympatric species Two species that exist in the same geographic area.

Synapses Specialized contact zones in the nervous system where a neuron communicates with another cell.

Taxis Orienting reaction or movement in freely moving organisms directed in relation to a stimulus.

Temperature coupling Certain properties of a signal produced by the sender and the response criteria of the receiver change parallel to the ambient temperature changes.

Tesla (T) Unit of magnetic field, also known as magnetic flux density. It is named after the Serbian-American inventor and engineer Nikola Tesla. Commonly, the geomagnetic field on the surface of the Earth is measured in nanotesla ($1 nT=10^{-9}T$).

Transgenesis Process of introducing a foreign gene into the genome of an organism. This organism will exhibit some new properties and transmit these properties to its offspring.

Tropism Involuntary orientation of an organism in response to an environmental stimulus. Such orientation involves a turning movement or differential growth.

True communication Transfer of information from a sender to a receiver benefiting both partners.

Ultrasound Sound in the frequency range above that of human hearing, that is, above approximately 20kHz.

Umwelt That part of the environment which is perceived after sensory and central filtering.

Vector A physical quantity, such as force, that possesses both 'magnitude' and 'direction.' This quantity can be represented by an arrow having appropriate length and direction and emanating from a given reference point. Correspondingly, the 'locomotion vector' in cricket research is defined by the speed and the direction of movement.

Voltage-gated ion channels A class of transmembrane proteins that form ion channels, which open and close in response to changes in the membrane potential.

ZENK An acronym of the initial letters of Zif268, Egr-1, NGFI-A, and Krox-24. ZENK is the avian homolog of these four immediate-early gene products of the zinc finger family. ZENK expression is widely used as a marker of neural activity induced by relevant behavioral stimuli.

BIBLIOGRAPHY

1 Introduction

Baerends, G. P. and Baerends-van Roon, J. M. (1950). An introduction to the study of the ethology of cichlid fishes. *Behaviour* 1 (Suppl.):1–242.

Jørgensen, C. B. (2001). August Krogh and Claude Bernard on basic principles in experimental physiology. *BioScience* 51:59–61.

Wickler, W. (1968). *Das Züchten von Aquarienfischen: Eine Einführung in ihre Fortpflanzungsbiologie.* Franckh'sche Verlagshandlung, Stuttgart.

Zupanc, G. K. H. and Banks, J. R. (1998). Electric fish: Animals with a sixth sense. *Biological Sciences Review* 11:23–27.

2 Fundamentals of neurobiology

Ahrens, M. B., Orger, M. B., Robson, D. N., Li, J. M., and Keller, P. J. (2013). Whole-brain functional imaging at cellular resolution using light-sheet microscopy. *Nature Methods* 10:413–420.

Andres-Barquin, P. J. (2002). Santiago Ramón y Cajal and the Spanish school of neurology. *The Lancet Neurology* 1:445–452.

Byrne, J. H. and Roberts, J. L. (2004). *From Molecules to Networks: An Introduction to Cellular and Molecular Neuroscience.* Elsevier/Academic Press, Amsterdam.

Camhi, J. M. (1984). *Neuroethology: Nerve Cells and the Natural Behavior of Animals.* Sinauer Associates Inc. Publishers, Sunderland, MA.

Furshpan, E. J. and Potter, D. D. (1959). Transmission at the giant motor synapses of the crayfish. *Journal of Physiology* 145:289–325.

Grant, G. (2007). How the 1906 Nobel Prize in Physiology or Medicine was shared between Golgi and Cajal. *Brain Research Reviews* 55:490–498.

Hartline, D. K. and Colman, D. R. (2007). Rapid conduction and the evolution of giant axons and myelinated fibers. *Current Biology* 17: R29–R35.

Jahn, R. and Scheller, R. H. (2006). SNAREs—engines for membrane fusion. *Nature Reviews Molecular Cell Biology* 7:631–643.

Kandel, E. R. and Schwartz, J. H. (1985). *Principles of Neural Science*, 2nd edn. Elsevier, New York/Amsterdam/ Oxford.

Kandel, E. R., Schwartz, J. H., and Jessell, T. M. (2000). *Principles of Neural Science*, 4th edn. McGraw-Hill, New York.

Kole, M. H. P. and Stuart, G. J. (2012). Signal processing in the axon initial segment. *Neuron* 73:235–247.

Ling, C., Hendrickson, M. L., and Kalil, R. E. (2012). Resolving the detailed structure of cortical and thalamic neurons in the adult rat brain with refined biotinylated dextran amine labeling. *PLoS ONE* 7(11):e45886.

Marban, E., Yamagishi, T., and Tomaselli, G. F. (1998). Structure and function of voltage-gated sodium channels. *Journal of Physiology* **508**:647–657.

Marin-Padilla, M. (1987). The Golgi method. In *Encyclopedia of Neuroscience*, Vol. 1 (ed. G. Adelman), pp. 470–471. Birkhäuser, Boston/Basel/Stuttgart.

McKinney, R. A. (2005). Physiological roles of spine motility: Development, plasticity and disorders. *Biochemical Society Transactions* **33**:1299–1302.

Miller, J. P. and Selverston, A. I. (1979). Rapid killing of single neurons by irradiation of intracellularly injected dye. *Science* **206**:702–704.

Numa S. and Noda, M. (1986). Molecular structure of sodium channels. *Annals of the New York Academy of Sciences* **479**:338–355.

Numa S., Noda M., Takahashi H., Tanabe T., Toyosato, M., Furutani Y., and Kikyotani, S. (1983). Molecular structure of the nicotinic acetylcholine receptor. *Cold Spring Harbor Symposia on Quantitative Biology* **48**:57–69.

Penzlin, H. (1980). *Lehrbuch der Tierphysiologie*, 3rd edn. Gustav Fischer Verlag, Stuttgart.

Regehr, W. G., Carey, M. R., and Best, A. R. (2009). Activity-dependent regulation of synapses by retrograde messengers. *Neuron* **63**:154–170.

Seyfarth, E.-A. (2006). Julius Bernstein (1839–1917): Pioneer neurobiologist and biophysicist. *Biological Cybernetics* **94**:2–8.

Shepherd, G. M. (1988). *Neurobiology*, 2nd edn. Oxford University Press, New York/Oxford.

Sherman, D. L. and Brophy, P. J. (2005). Mechanisms of axon ensheathment and myelin growth. *Nature Reviews Neuroscience* **6**:683–690.

Stasheff, S. F. and Masland, R. H. (2002). Functional inhibition in direction-selective retinal ganglion cells: spatiotemporal extent and intralaminar interactions. *Journal of Neurophysiology* **88**:1026–1039.

Stretton, A. O. W. and Kravitz, E. A. (1968). Neuronal geometry: Determination with a technique of intracellular dye injection. *Science* **162**:132–134.

Swanson, L. W. (2004). *Brain Maps III: Structure of the Rat Brain*, 3rd edn. Elsevier/Academic Press, Amsterdam.

Trueta, J. (1952). Ramón y Cajal: The first century of his birth. *The Lancet* **260**:281–283.

Unwin, P. N. T. and Zampighi, G. (1980). Structure of the junction between communicating cells. *Nature* **283**:545–549.

Yang, G. and Masland, R. H. (1992). Direct visualization of the dendritic and receptive fields of directionally selective retinal ganglion cells. *Science* **258**:1949–1952.

Yu, F. H. and Catterall, W. A. (2003). Overview of the voltage-gated sodium channel family. *Genome Biology* **4**:207.

Zigmond, M. J., Bloom, F. E., Landis, S. C., Roberts, J. L., and Squire, L. R. (eds) (1999). *Fundamental Neuroscience*. Academic Press, San Diego/London.

3 The study of animal behavior: a brief history

Bartlett, F. C. (1960). Karl Spencer Lashley: 1890–1958. *Biographical Memoirs of Fellows of the Royal Society* **5**, 107–118.

Becker, J. B., Breedlove, S. M., and Crews, D. (eds) (1993). *Behavioral Endocrinology*. MIT Press, Cambridge, Massachusetts/London.

Bullock, T. H. (1996). Theodore H. Bullock. In *The History of Neuroscience in Autobiography*, Vol. 1 (ed. L. R. Squire), pp. 110–156. Society for Neuroscience, Washington, D.C.

Bullock, T. H. (1999). Neuroethology has pregnant agendas. *Journal of Comparative Physiology A* **185**:291–295.

Bullock, T. H. and Horridge, G. A. (1965). *Structure and Function in the Nervous System of Invertebrates*, Vols I and II. W. H. Freeman & Company, San Francisco/London.

Darwin, C. (1892). *The Expression of the Emotions in Man and Animals*. John Murray, London.

De Bono, M. and Bargmann, C. I. (1998). Natural variation in a neuropeptide Y receptor homolog modifies social behavior and food response in *C. elegans*. *Cell* **94**:679–689.

Dewsbury, D. A. (1989). A brief history of the study of animal behavior in North America. In *Perspectives in Ethology*, Vol. 8: *Whither Ethology?* (eds P. P. G. Bateson and P. H. Klopfer), pp. 85–122. Plenum Press, New York.

Ewert, J.-P. (1980). *Neuroethology: an Introduction to the Neurophysiological Fundamentals of Behavior*. Springer-Verlag, Berlin.

Hassenstein, B. (1964). Erich von Holst. *Zoologischer Anzeiger* 27 (Suppl.):676–682.

Heisenberg, M. (1997). Genetic approach to neuroethology. *BioEssays* **19**:1065–1073.

Hoagland, H. and Mitchell, R. T. (1956). William John Crozier: 1892–1955. *The American Journal of Psychology* **69**:135–138.

Hollard, V. D. and Delius, J. D. (1982). Rotational invariance in visual pattern recognition by pigeons and humans. *Science* **218**:804–806.

Jaynes, J. (1969). The historical origins of 'ethology' and 'comparative psychology'. *Animal Behaviour* **17**:601–606.

Laties, V.G. (2003). Behavior analysis and the growth of behavioral pharmacology. *The Behavior Analyst* **26**:235–252.

Lorenz, K. (1966). *On Aggression*. Methuen & Co. Ltd., London.

Lorenz, K. (1977). *Behind the Mirror: a Search for a Natural History of Human Knowledge*. Methuen & Co. Ltd., London.

Lorenz, K. Z. (1954). *Man Meets Dog*. Methuen & Co. Ltd., London.

Lorenz, K. Z. (1964). *King Solomon's Ring: New Light on Animal Ways*. Methuen & Co. Ltd., London.

Lorenz, K. Z. (1981). *The Foundations of Ethology*. Springer-Verlag, New York/Wien.

McFarland, D. (1993). *Animal Behaviour*, 2nd edn. Addison Wesley Longman, Harlow.

Manning, A. (1972). *An Introduction to Animal Behaviour*, 2nd edn. Edward Arnold Publishers, London.

Morgan, C. L. (1900). *Animal Behaviour*. Edward Arnold, London.

Morris, E. K., Lazo, J. F., and Smith, N. G. (2004). Whether, when, and why Skinner published on biological participation in behavior. *The Behavior Analyst* **27**:153–169.

Nevin, J. A. (1992). Burrhus Frederic Skinner: 1904–1990. *The American Journal of Psychology* **105**:613–619.

Nicolai, G. F. (1907). Die physiologische Methodik zur Erforschung der Tierpsyche, ihre Möglichkeit und ihre Anwendung. *Journal für Psychologie und Neurologie* **10**:1–27.

Osterhout, W. J. V. (1928). Jacques Loeb. *The Journal of General Physiology* **8**:9–92.

Pavlov, I. P. (1960). *Conditioned Reflexes: an Investigation of the Physiological Activity of the Cerebral Cortex*. Dover Publications, Inc., New York.

Pflüger, H.-J. and Menzel, R. (1999). Neuroethology, its roots and future. *Journal of Comparative Physiology A* **185**:389–392.

Rachlin, H. (1995). Burrhus Frederic Skinner 1904–1990. *Biographical Memoirs of the National Academy of Sciences* **67**:363–377.

Skinner, B. F. (1981). Selection by consequences. *Science* **213**:501–504.

Tinbergen, N. (1951). *The Study of Instinct*. Oxford University Press, London.

Tinbergen, N. (1963). On aims and methods of ethology. *Zeitschrift für Tierpsychologie* **20**:410–433.

Tinbergen, N. (1965). *Social Behaviour in Animals: With Special Reference to Vertebrates*. Methuen & Co. Ltd., London.

von Frisch, K. (1950). *Bees: Their Vision, Chemical Senses, and Language*. Cornell University Press, Ithaca, New York.

von Frisch, K. (1966). *The Dancing Bees: An Account of the Life and Senses of the Honey Bee*. Methuen & Co. Ltd., London.

von Frisch, K. (1967). *A Biologist Remembers*. Pergamon Press, Oxford.

von Frisch, K. (1967). *The Dance Language and Orientation of Bees*. The Belknap Press of Harvard University Press/ Oxford University Press, Cambridge, Massachusetts/ London.

von Holst, E. (1973). *The Behavioural Physiology of Animals and Man*, Vol. 1. Methuen & Co. Ltd., London.

Warden, C. J. (1927). The historical development of comparative psychology. *Psychological Review* **34**:57–168.

Watson, J. B. (1912). Instinctive activity in animals: some recent experiments and observations. *Harper's Magazine* 124:376–382.

Wilson, E. O. (1975). *Sociobiology: the New Synthesis*. Belknap Press of Harvard University Press, Cambridge, Massachusetts/London.

Zupanc, G. K. H. (2006). Theodore H. Bullock (1915–2005). *Nature* **439**:280.

Zupanc, G. K. H. and Zupanc, M. M. (2008). Theodore H. Bullock: Pioneer of integrative and comparative neurobiology. *Journal of Comparative Physiology A* **194**:119–134.

4 Orienting movements

Barnes, W. J. P. (1993). Sensory guidance in arthropod behaviour: Common principles and experimental approaches. *Comparative Biochemistry and Physiology* **104A**:625–632.

Corey, D. P. and Hudspeth, A. J. (1979). Ionic basis of the receptor potential in a vertebrate hair cell. *Nature* **281**:675–677.

Eckert, R. and Brehm, P. (1979) Ionic mechanisms of excitation in *Paramecium*. *Annual Review of Biophysics and Bioengineering* **8**:353–383.

Flock, Å. (1965). Transducing mechanisms in the lateral line canal organ receptors. *Cold Spring Harbor Symposia on Quantitative Biology* **30**:133–145.

Fraenkel, G. S. and Gunn, D. L. (1940). *The Orientation of Animals: Kineses, Taxes and Compass Reactions*. Clarendon Press, Oxford.

Howard, J., Roberts, W. M., and Hudspeth, A. J. (1988). Mechanoelectrical transduction by hair cells. *Annual Review in Biophysics and Biophysical Chemistry* **17**:99–124.

Hudspeth, A. J. and Corey, D. P. (1977). Sensitivity, polarity, and conductance change in the response of vertebrate hair cells to controlled mechanical stimuli. *Proceedings of the National Academy of Sciences U.S.A.* **74**:2407–2411.

Jahn, T. L. (1961). The mechanism of ciliary movement. I. Ciliary reversal and activation by electric current; the Ludloff phenomenon in terms of core and volume conductors. *Journal of Protozoology* **8**:369–380.

Jennings, H. S. (1962). *Behavior of the Lower Organisms*. Indiana University Press, Bloomington.

Kandel, E. R. and Schwartz, J. H. (eds) (1985). *Principles of Neural Science*, 2nd edn. Elsevier, New York.

Keeton, W. T. (1980). *Biological Science*, 3rd edn. W. W. Norton & Company, New York.

Kühn, A. (1919). *Die Orientierung der Tiere im Raum*. Fischer Verlag, Jena.

Ludloff, K. (1895), Untersuchungen über den Galvanotropismus. *Archiv für die gesamte Physiologie des Menschen und der Tiere* **59**:525–554.

Machemer, H. (1988a). Galvanotaxis: Grundlagen der elektromechanischen Kopplung und Orientierung bei Paramecium. In *Praktische Verhaltensbiologie* (ed. G. K. H. Zupanc), pp. 60–82. Verlag Paul Parey, Berlin/Hamburg.

Machemer, H. (1988b). Motor control of cilia. In *Paramecium* (ed. H.-D. Görtz), pp. 216–235. Springer-Verlag, Berlin/Heidelberg.

Mogami, Y., Pernberg, J., and Machemer, H. (1990). Messenger role of calcium in ciliary electromotor coupling: A reassessment. *Cell Calcium* **11**:665–673.

Naitoh, Y., Eckert, R., and Friedman, K. (1972). A regenerative calcium response in *Paramecium*. *Journal of Experimental Biology* **56**:667–681.

Oertel, D., Schein, S. J., and Kung, C. (1978). A potassium conductance activated by hyperpolarization in *Paramecium*. *Journal of Membrane Biology* **43**:169–185.

Pickles, J. O. and Corey, D. P. (1992). Mechanical transduction by hair cells. *Trends in Neurosciences* **15**:254–259.

Schöne, H. (1984). *Spatial Orientation: The Spatial Control of Behavior in Animals and Man*. Princeton University Press, Princeton.

von Holst, E. (1950). Die Tätigkeit des Statolithenapparates im Wirbeltierlabyrinth. *Naturwissenschaften* **37**:265–272.

5 Active orientation and localization

Barber, J. R. and Conner, W. E. (2007). Acoustic mimicry in a predator–prey interaction. *Proceedings of the National Academy of Sciences U.S.A.* **104**:9331–9334.

Corcoran, A. J., Conner, W. E., and Barber, J. R. (2010). Anti-bat tiger moth sounds: Form and function. *Current Zoology* **56**:358–369.

Dethier, V. G. (1993). Kenneth David Roeder. In *Biographical Memoirs* Vol. 62 (ed. National Academy of Sciences), pp. 350–366. The National Academies Press, Washington, D.C.

Fullard, J. H. and Heller, B. (1990). Functional organization of the arctiid moth tymbal (Insecta, Lepidoptera). *Journal of Morphology* **204**:57–65.

Fullard, J. H., Simmons, J. A., and Saillant, P. A. (1994). Jamming bat echolocation: The dogbane tiger moth *Cycnia tenera* times its clicks to the terminal attack calls of the big brown bat *Eptesicus fuscus*. *Journal of Experimental Biology* **194**:285–298.

Griffin, D. R. (1953). Bat sound under natural conditions, with evidence for the echolocation of insect prey. *Journal of Experimental Zoology* **123**:435–466.

Griffin, D. R. (1944). Echolocation by blind men, bats and radar. *Science* **100**:589–590.

Griffin, D. R. (1998). Donald R. Griffin. In *The History of Neuroscience in Autobiography*, Vol. 2 (ed. L. R. Squire), pp. 68–93. Academic Press, San Diego.

Griffin, D. R. and Galambos, R. (1941). The sensory basis of obstacle avoidance by flying bats. *Journal of Experimental Zoology* **86**:481–506.

Gross, C. G. (2005). Donald R. Griffin. In *Biographical Memoirs* Vol. 86 (ed. National Academy of Sciences), pp. 188–207. The National Academies Press, Washington, D.C.

Hartridge, H. (1920). The avoidance of objects by bats in their flight. *Journal of Physiology* **54**:54–57.

Kössl, M., Hechavarria, J. C., Voss, C., Macias, S., Mora, E. C., and Vater, M. (2014). Neural maps for target range in the auditory cortex of echolocating bats. *Current Opinion in Neurobiology* **24**:68–75.

Miller, L. A. (1991). Arctiid moth clicks can degrade the accuracy of range difference discrimination in echolocating big brown bats, *Eptesicus fuscus*. *Journal of Comparative Physiology A* **168**:571–579.

Miller, L. A. and Olesen, J. (1979). Avoidance behavior in green lacewings. I. Behavior of free flying green lacewings to hunting bats and ultrasound. *Journal of Comparative Physiology A* **131**:113–120.

Pierce, G. W. and Griffin, D. R. (1938). Experimental determination of supersonic notes emitted by bats. *Journal of Mammalogy* **19**:454–455.

Pollak, G. D., Marsh, D. S., Bodenhamer, R., and Souther, A. (1977). Characteristics of phasic on neurons in inferior colliculus of unanaesthetized bats with observations relating to mechanisms of echo ranging. *Journal of Neurophysiology* **40**:926–942.

Schnitzler, H.-U. and Kalko, E. K. V. (2001). Echolocation by insect-eating bats. *BioScience* **51**:557–569.

Schnitzler, H.-U. and Ostwald, J. (1983). Adaptations for the detection of fluttering insects by echolocation in horseshoe bats. In *Advances in Vertebrate Neuroethology* (eds J.-P. Ewert, R. R. Capranica, and D. J. Ingle), pp. 801–827. Plenum Press, New York.

Simmons, J. A. (2012). Bats use a neuronally implemented computational acoustic model to form sonar images. *Current Opinion in Neurobiology* **22**:311–319.

Simmons, J. A., Fenton, M. B., and O'Farrell, M. J. (1979). Echolocation and pursuit of prey by bats. *Science* **203**:16–21.

Suga, N. (2009). Nobuo Suga. In *The History of Neuroscience in Autobiography*, Vol. 6 (ed. L. R. Squire), pp. 480–512. Oxford University Press, New York.

Suga, N. (1990). Cortical computational maps for auditory imaging. *Neural Networks* 3:3–21.

Suga, N. (1990). Bisonar and neural computation in bats. *Scientific American* **262**:60–68.

Tougaard, J., Casseday, J. H., and Covey, E. (1998). Arctiid moths and bat echolocation: Broad-band clicks interfere with neural responses to auditory stimuli in the nuclei of the lateral lemniscus of the big brown bat. *Journal of Comparative Physiology A* **182**:203–215.

Yager, D. D., May, M. L., and Brock Fenton, M. (1990). Ultrasound-triggered, flight-gated evasive maneuvers in the praying mantis *Parasphendale agrionina*. *Journal of Experimental Biology* **152**:17–39.

6 Neural control of motor output

Adrian, E.D. (1931). Potential changes in the isolated nervous system of *Dytiscus marginalis*. *Journal of Physiology* **72**:132–151.

Adrian, E.D. (1966). Thomas Graham Brown: 1882–1965. *Biographical Memoirs of Fellows of the Royal Society* **12**:22–33.

Arshavsky, Y. I., Orlovsky, G. N., Panchin, Y. V., Roberts, A., and Soffe, S. R. (1993). Neuronal control of swimming locomotion: Analysis of the pteropod mollusc *Clione* and embryos of the amphibian *Xenopus*. *Trends in Neurosciences* **16**:227–233.

Borisyuk, R., al Azad, A. K., Conte, D., Roberts, A., and Soffe, S. R. (2011). Modeling the connectome of a simple spinal cord. *Frontiers in Neuroinformatics* **5**:20.

Brown, T. G. (1911). The intrinsic factors in the act of progression in the mammal. *Proceedings of the Royal Society of London Series B* **84**:308–319.

Brown, T. G. (1914). On the nature of the fundamental activity of the nervous centres; together with an analysis of the conditioning of rhythmic activity in progression and a theory of the evolution of function in the nervous system. *Journal of Physiology (London)* **48**:18–46.

Brown, T.G. (1916). Die Reflexfunktionen des Zentralnervensystems mit besonderer Berücksichtigung der rhythmischen Tätigkeiten beim Säugetier. *Ergebnisse der Physiologie* **15**:480–790.

Buhl, E., Roberts, A., and Soffe, S. R. (2012). The role of a trigeminal sensory nucleus in the initiation of locomotion. *Journal of Physiology* **590**:2453–2469.

Clarke, J. D. W., Hayes, B. P., Hunt, S. P., and Roberts, A. (1984). Sensory physiology, anatomy and immunohistochemistry or Rohon–Beard neurones in embryos of *Xenopus laevis*. *Journal of Physiology* **348**:511–525.

F(ulton), J. F. (1952). Sir Charles Scott Sherrington, O.M. (1857–1952). *Journal of Neurophysiology* **15**:167–190.

Hughes, G. M. and Wiersma, C. A. G. (1960). The co-ordination of swimmeret movements in the crayfish, *Procambarus clarkii* (Girard). *Journal of Experimental Biology* **37**:657–670.

Jankowska, E., Jukes, M. G. M., Lund, S., and Lundbuerg, A. (1965). Reciprocal innervation through interneuronal inhibition. *Nature* **206**:198–199.

Lambert, T. D., Howard, J., Plant, A., Soffe, S., and Roberts, A. (2004). Mechanisms and significance of reduced activity and responsiveness in resting frog tadpoles. *Journal of Experimental Biology* **207**:1113–1125.

Li, W.-C., Perrins, R., Soffe, S. R., Yoshida, M., Walford, A., and Roberts, A. (2001). Defining classes of spinal interneurons and their axonal projections in hatchling *Xenopus laevis* tadpoles. *Journal of Comparative Neurology* **441**:248–265.

Li, W.-C., Soffe, S. R., and Roberts, A. (2004). A direct comparison of whole cell patch and sharp electrodes by simultaneous recording from single spinal neurons in frog tadpoles. *Journal of Neurophysiology* **92**:380–386.

Li, W.-C., Soffe, R., and Roberts, A. (2004). Dorsal spinal interneurons forming a primitive, cutaneous sensory pathway. *Journal of Neurophysiology* **92**:895–904.

Li, W.-C., Soffe, S. R., Wolf, E., and Roberts, A. (2006). Persistent responses to brief stimuli: feedback excitation among brainstem neurons. *Journal of Neuroscience* **26**:4026–4035.

Penfield, W. (1962). Sir Charles Sherrington, O.M., F.R.S. (1857–1952): An appreciation. *Notes and Record of the Royal Society of London* **17**:163–168.

Perrins, R., Walford, A., and Roberts, A. (2002). Sensory activation and role of inhibitory reticulospinal neurons that stop swimming in hatchling frog tadpoles. *Journal of Neuroscience* **22**:4229–4240.

Roberts, A. (1990). How does a nervous system produce behaviour? A case study in neurobiology. *Science Progress* **74**:31–51.

Roberts, A., Conte, D., Hull, M., Merrison-Hort, R., al Azad, A. K., Buhl, E., Borisyuk, R., and Soffe, S. R. (2014). Can simple rules control development of a pioneer vertebrate network generating behavior? *Journal of Neuroscience* **34**:608–621.

Roberts, A., Li, W.-C., and Soffe, S. R. (2010). How neurons generate behavior in a hatchling amphibian tadpole: An outline. *Frontiers in Behavioral Neuroscience* **4**: Article 16.

Sherrington, C. S. (1906). *The Integrative Action of the Nervous System*. Yale University Press, New Haven, CT.

Stuart, D. G. and Hultborn, H. (2008). Thomas Graham Brown (1882–1965), Anders Lundberg (1920–), and the neural control of stepping. *Brain Research Reviews* **59**:74–95.

Wilson, D. M. and Wyman, R. J. (1965) Motor output patterns during random and rhythmic stimulation of locust thoracic ganglia. *Biophysical Journal* **5**:121–143.

7 Neuronal processing of sensory information

Bower, T. G. R. (1966). Heterogeneous summation in human infants. *Animal Behaviour* **14**:395–398.

Carr, C. E. and Konishi, M. (1990). A circuit for detection of interaural time differences in the brain stem of the barn owl. *Journal of Neuroscience* **10**:3227–3246.

DeBello, W. M., Feldman, D. E., and Knudsen, E. I. (2001). Adaptive axonal remodeling in the midbrain auditory space map. *Journal of Neuroscience* **21**:3161–3174.

DeBello, W. M. and Knudsen, E. I. (2004). Multiple sites of adaptive plasticity in the owl's auditory localization pathway. *Journal of Neuroscience* **24**:6853–6861.

Drees, O. (1952). Untersuchungen über die angeborenen Verhaltensweisen bei Springspinnen (Salticidae). *Zeitschrift für Tierpsychologie* **9**:169–207.

Eibl-Eibesfeldt, I. (1974). *Grundriß der vergleichenden Verhaltensforschung*, 4th edn. R. Piper & Co Verlag, München/Zürich.

Ewert, J.-P. (1967). Aktivierung der Verhaltensfolge beim Beutefang der Erdkröte (*Bufo bufo* L.) durch elektrische Mittelhirn-Reizung. *Zeitschrift für Vergleichende Physiologie* **54**:455–481.

Ewert, J.-P. (1968). Der Einfluß von Zwischenhirndefekten auf die Visuomotorik im Beute- und Fluchtverhalten der Erdkröte (*Bufo bufo* L.). *Zeitschrift für Vergleichende Physiologie* **61**:41–70.

Ewert, J.-P. (1969). Quantitative Analyse von Reiz-Reaktionsbeziehungen bei visuellem Auslösen der Beutefang-Wendereaktion der Erdkröte *Bufo bufo* (L.). *Pflügers Archiv* **308**:225–243.

Ewert, J.-P. (1974). The neural basis of visually guided behavior. *Scientific American* **230**:34–42.

Ewert, J.-P. and Borchers, H.-W. (1971). Reaktionscharakteristik von Neuronen aus dem Tectum opticum und Subtectum der Erdkröte (*Bufo bufo* L.). *Zeitschrift für Vergleichende Physiologie* **71**:165–189.

Ewert, J.-P. and Ewert, S. B. (1981). *Warnehmung*. Quelle & Meyer, Heidelberg.

Ewert, J.-P. and Hock, F. (1972). Movement-sensitive neurones in the toad's retina. *Experimental Brain Research* **16**:41–59.

Ewert, J.-P. and Traud, R. (1979). Releasing stimuli for antipredator behaviour in the common toad *Bufo bufo* (L.). *Behaviour* **68**:170–180.

Ewert, J.-P. and Wietersheim, A. v. (1974). Musterauswertung durch Tectum- und Thalamus/Praetectum-Neurone im visuellen System der Kröte *Bufo bufo* (L.). *Journal of Comparative Physiology* **92**:131–148.

Ewert, J.-P. and Wietersheim, A. v. (1974). Einfluß von Thalamus/Praetectum-Defekten auf die Antwort von Tectum-Neuronen gegenüber bewegten visuellen Mustern bei der Kröte *Bufo bufo* (L.). *Journal of Comparative Physiology* **92**:149–160.

Ewert, J.-P., Arend, B., Becker, V., and Borchers, H.-W. (1979). Invariants in configurational prey selection by *Bufo bufo* (L.). *Brain, Behavior and Evolution* **16**:38–51.

Grüsser-Cornehls, U., Grüsser, O.-J., and Bullock, T. H. (1963). Unit responses in the frog's tectum to moving and nonmoving visual stimuli. *Science* **141**:820–822.

Heiligenberg, W. and Kramer, U. (1972). Aggressiveness as a function of external stimulation. *Journal of Comparative Physiology* **77**:332–340.

Jeffress, L. A. (1948). A place theory of sound localization. *Journal of Comparative and Physiological Psychology* **41**:35–39.

Knudsen, E. I. (1981). The hearing of the barn owl. *Scientific American* **245**:82–91.

Knudsen, E. I. (1982). Auditory and visual maps of space in the optic tectum of the owl. *Journal of Neuroscience* **2**:1177–1194.

Knudsen, E. I. (2002). Instructed learning in the auditory localization pathway of the barn owl. *Nature* 417:322-328.

Knudsen, E. I. and Brainard, M. S. (1991). Visual instruction of the neural map of auditory space in the developing optic tectum. *Science* **253**:85–87.

Knudsen, E. I. and Konishi, M. (1979). Mechanisms of sound localization in the barn owl (*Tyto alba*). *Journal of Comparative Physiology A* **133**:13–21.

Knudsen, E. I., Blasdel, G. G., and Konishi, M. (1979). Sound localization by the barn owl (*Tyto alba*) measured with the search coil technique. *Journal of Comparative Physiology A* 133:1–11.

Konishi, M. (1990). Similar algorithms in different sensory systems and animals. *Cold Spring Harbor Symposia on Quantitative Biology* **55**:575–584.

Konishi, M. (1992). The neural algorithm for sound localization in the owl. *The Harvey Lectures* **86**:47–64.

Konishi, M. (1993). Listening with two ears. *Scientific American* **268**:34–41.

Lack, D. (1953). *The Life of the Robin*. Penguin Books, Melbourne/London/Baltimore.

Leong, C.-Y. (1969). The quantitative effect of releasers on the attack readiness of the fish *Haplochromis burtoni* (Cichlidae, Pisces). *Zeitschrift für Vergleichende Physiologie* **65**:29–50.

Lettvin, J. Y., Maturana, H. R., McCulloch, W. S., and Pitts, W. H. (1959). What the frog's eye tells the frog's brain. *Proceedings of the Institute of Radio Engineers* **47**:1940–1951.

Manley, G. A., Köppl, C., and Konishi, M. (1988). A neural map of interaural intensity differences in the brain stem of the barn owl. *Journal of Neuroscience* **8**:2665–2676.

Payne, R. S. (1971). Acoustic location of prey by barn owls (*Tyto alba*). *Journal of Experimental Biology* **54**:535–573.

Staddon, J. E. R. (1975). A note on the evolutionary significance of 'supernormal' stimuli. *The American Naturalist* **109**:541–545.

Tinbergen, N. (1969). *The Study of Instinct*. Oxford University Press, London.

Tinbergen, N. and Perdeck, A. C. (1950). On the stimulus situation releasing the begging response in the newly hatched herring gull chick (*Larus argentatus argentatus* Pont.). *Behaviour* **3**:1–39.

8 Sensorimotor integration

Alexander, R. McN. (1996). Hans Werner Lissmann: 30 April 1909–21 April 1995. *Biographical Memoirs of Fellows of the Royal Society* **42**:234–245.

Bastian, J. and Yuthas, J. (1984). The jamming avoidance response of *Eigenmannia*: properties of a diencephalic link between sensory processing and motor output. *Journal of Comparative Physiology A* **154**:895–908.

Bullock, T. H. and Heiligenberg, W. (eds) (1986). *Electroreception*. John Wiley & Sons, New York/ Chichester/Brisbane/Toronto/Singapore.

Carlson, B. A. and Kawasaki, M. (2004). Nonlinear response properties of combination-sensitive electrosensory neurons in the midbrain of *Gymnarchus niloticus*. *Journal of Neuroscience* **24**:8039–8048.

Heiligenberg, W. (1977). *Principles of Electrolocation and Jamming Avoidance Response in Electric Fish*. Springer-Verlag, Berlin/Heidelberg/New York.

Heiligenberg, W. (1991). The jamming avoidance response of the electric fish, *Eigenmannia*: Computational rules and their neuronal implementation. *Seminars in the Neurosciences* **3**:3–18.

Heiligenberg, W. and Rose, G. (1985). Phase and amplitude computations in the midbrain of an electric fish: intracellular studies of neurons participating in the jamming avoidance response of *Eigenmannia*. *Journal of Neuroscience* **5**:515–531.

Heiligenberg, W., Baker, C., and Matsubara, J. (1978). The jamming avoidance response in *Eigenmannia* revisited: The structure of a neuronal democracy. *Journal of Comparative Physiology A* **127**:267–286.

Heiligenberg, W., Keller, C. H., Metzner, W., and Kawasaki, M. (1991). Structure and function of neurons in the complex of the nucleus electrosensorius of the gymnotiform fish *Eigenmannia*: Detection and processing of electric signals in social communication. *Journal of Comparative Physiology A* **169**:151–164.

Heiligenberg, W., Metzner, W., Wong, C. J. H., and Keller, C. H. (1996). Motor control of the jamming avoidance response of *Apteronotus leptorhynchus*: Evolutionary changes of a behavior and its neuronal substrates. *Journal of Comparative Physiology A* **179**:653–674.

Huston, S. J. and Jayaraman, V. (2011). Studying sensorimotor integration in insects. *Current Opinion in Neurobiology* **21**:527–534.

Kawasaki, M., Maler, L., Rose, G. J., and Heiligenberg, W. (1988). Anatomical and functional organization of the prepacemaker nucleus in gymnotiform electric fish: The accommodation of two behaviors in one nucleus. *Journal of Comparative Neurology* **276**:113–131.

Keller, C. H. (1988). Stimulus discrimination in the diencephalon of *Eigenmannia*: The emergence and sharpening of a sensory filter. *Journal of Comparative Physiology A* **162**:747–757.

Keller, C. H., Maler, L., and Heiligenberg, W. (1990). Structural and functional organization of a diencephalic sensory–motor interface in the gymnotiform fish, *Eigenmannia*. *Journal of Comparative Neurology* **293**:347–376.

Lissmann, H. W. (1958). On the function and evolution of electric organs in fish. *Journal of Experimental Biology* **35**:156–191.

Lissmann, H. W. and Machin, K. E. (1958). The mechanism of object location in *Gymnarchus niloticus* and similar fish. *Journal of Experimental Biology* **35**:451–486.

Matsushita, A. and Kawasaki, M. (2005). Neuronal sensitivity to microsecond time disparities in the electrosensory system of *Gymnarchus niloticus*. *Journal of Neuroscience* 25:11424–11432.

Metzner, W. (1993). The jamming avoidance response in *Eigenmannia* is controlled by two separate motor pathways. *Journal of Neuroscience* **13**:1862–1878.

Metzner, W. (1999). Neural circuitry for communication and jamming avoidance in gymnotiform fish. *Journal of Experimental Biology* **202**:1365–1375.

Rose, G. J. and Heiligenberg, W. (1985). Temporal hyperacuity in the electric sense of fish. *Nature* **318**:178–180.

Rose, G. J. and Heiligenberg, W. (1986). Neural coding of difference frequencies in the midbrain of the electric fish *Eigenmannia*: Reading the sense of rotation in an amplitude–phase plane. *Journal of Comparative Physiology A* **158**:613–624.

Watanabe, A. and Takeda, K. (1963). The change of discharge frequency by A.C. stimulus in a weak electric fish. *Journal of Experimental Biology* **40**:57–66.

Zupanc, G. K. H. and Bullock, T.H. (2006). Walter Heiligenberg: The jamming avoidance response and beyond. *Journal of Comparative Physiology A* 192:561–572.

Zupanc, G. K. H. and Lamprecht, J. (1994). Walter Heiligenberg (1938–1994). *Trends in Neurosciences* **17**:507–508.

9 Neuromodulation: the accommodation of motivational changes in behavior

Dickinson, P. S., Mecsas, C., and Marder, E. (1990). Neuropeptide fusion of two motor-pattern generator circuits. *Nature* **344**:155–157.

Forger, N. G. and Breedlove, S. M. (1987). Seasonal variation in mammalian striated muscle mass and motoneuron morphology. *Journal of Neurobiology* **18**:155–165.

Getting, P. A. and Dekin, M. S. (1985). *Tritonia* swimming: a model system for integration within rhythmic motor systems. In *Model Neural Networks and Behavior* (ed. A. I. Selverston), pp. 3–20. Plenum Press, New York/London.

Hatton, G. I. (1997). Function-related plasticity in hypothalamus. *Annual Review of Neuroscience* 20:375–397.

Herkenham, M. (1987). Mismatches between neurotransmitter and receptor localizations in brain: Observations and implications. *Neuroscience* **23**:1–38.

Huber, R. and Delago, A. (1998). Serotonin alters decisions to withdraw in fighting crayfish, *Astacus astacus*: the motivational concept revisited. *Journal of Comparative Physiology A* **182**:573–583.

Huber, R. and Kravitz, E. A. (1995). A quantitative analysis of agonistic behavior in juvenile American lobsters (*Homarus americanus* L.). *Brain, Behavior and Evolution* **46**:72–83.

Huber, R., Smith, K., Delago, A., Isaksson, K., and Kravitz, E. A. (1997). Serotonin and aggressive motivation in crustaceans: Altering the decision to retreat. *Proceedings of the National Academy of Sciences U.S.A.* **94**:5939–5942.

Kawasaki, M., Maler, L., Rose, G. J., and Heiligenberg, W. (1988). Anatomical and functional organization of the prepacemaker nucleus in gymnotiform electric fish: The accommodation of two behaviors in one nucleus. *Journal of Comparative Neurology* **276**:113–131.

Keller, C. H., Maler, L., and Heiligenberg, W. (1990). Structural and functional organization of a diencephalic sensory–motor interface in the gymnotiform fish, *Eigenmannia*. *Journal of Comparative Neurology* 293:347–376.

Kurz, E. M., Sengelaub, D. R., and Arnold, A. P. (1986). Androgens regulate the dendritic length of mammalian motoneurons in adulthood. *Science* **232**:395–398.

Marder, E. and Richards, K. S. (1999). Development of the peptidergic modulation of a rhythmic pattern generating network. *Brain Research* **848**:35–44.

Swanson, L. W. (1988/89). The neural basis of motivated behavior. *Acta Morphologica Neerlando-Scandinavica* 26:165–176.

VanderHorst, V. G. J. M. and Holstege, G. (1997). Estrogen induces axonal outgrowth in the nucleus retroambiguus-lumbosacral motoneuronal pathway in the adult female cat. *Journal of Neuroscience* **17**:1122–1136.

Yeh, S.-R., Musolf, B. E., and Edwards, D. H. (1997). Neuronal adaptations to changes in the social dominance status of crayfish. *Journal of Neuroscience* **17**:697–708.

Zupanc, G. K. H. (1991). The synaptic organization of the prepacemaker nucleus in weakly electric knifefish, *Eigenmannia*: A quantitative ultrastructural study. *Journal of Neurocytology* **20**:818–833.

Zupanc, G. K. H. (2001). Adult neurogenesis and neuronal regeneration in the central nervous system of teleost fish. *Brain, Behavior and Evolution* **58**:250–275.

Zupanc, G. K. H. and Heiligenberg, W. (1989). Sexual maturity-dependent changes in neuronal morphology in the prepacemaker nucleus of adult weakly electric knifefish, *Eigenmannia*. *Journal of Neuroscience* 9:3816–3827.

Zupanc, G. K. H. and Heiligenberg, W. (1992). The structure of the diencephalic prepacemaker nucleus revisited: Light microscopic and ultrastructural studies. *Journal of Comparative Neurology* **323**:558–569.

Zupanc, G. K. H. and Lamprecht, J. (2000). Towards a cellular understanding of motivation: Structural reorganization and biochemical switching as key mechanisms of behavioral plasticity. *Ethology* 106:467–477.

10 Circadian rhythms and biological clocks

Aschoff, J. (1952). Aktivitätsperiodik von Mäusen im Dauerdunkel. *Pflügers Archiv* **255**:189–196.

Bechtold, D. A., Gibbs, J. E., and Loudon, A. S. I. (2010). Circadian dysfunction in disease. *Trends in Pharmacological Sciences* **31**: 191–198.

Brandes, C., Plautz, J. D., Stanewsky, R., Jamison, C. F., Straume, M., Wood, K. V., Kay, S. A., and Hall, J. C. (1996). Novel features of Drosophila *period* transcription revealed by real-time luciferase reporting. *Neuron* **16**:687–692.

Daan, S. and Gwinner, E. (1998). Jürgen Aschoff (1913–98): pioneer in biological rhythms. *Nature* **396**:418.

Dunlap, J. C., Loros, J. J., and DeCoursey, P. J. (eds) (2004). *Chronobiology: Biological Timekeeping*. Sinauer Associates, Sunderland, Massachusetts.

Hardin, P. E., Hall, J. C., and Rosbash, M. (1990). Feedback of the *Drosophila period* gene product on circadian cycling of its messenger RNA levels. *Nature* **343**:536–540.

Harris, W. A. (2008). Seymour Benzer 1921–2007: The man who took us from genes to behaviour. *PLoS Biology* **6**:e41.

Keenan, K. (1983). Lilian Vaughan Morgan (1870–1952): Her life and work. *American Zoologist* **23**:867–876.

Konopka, R. J. and Benzer, S. (1971). Clock mutants of *Drosophila melanogaster*. *Proceedings of the National Academy of Sciences U.S.A.* **68**:2112–2116.

Menaker, M. (1996). Colin S. Pittendrigh (1918–96). *Nature* **381**:24.

Meyer-Bernstein, E. L. and Sehgal, A. (2001). Molecular regulation of circadian rhythms in *Drosophila* and mammals. *The Neuroscientist* **7**:496–505.

Moore-Ede, M. C., Sulzman, F. M., and Fuller, C. A. (1982). *The Clocks That Time Us*. Harvard University Press, Cambridge, Massachusetts/London.

Morgan, L. V. (1922). Non-criss-cross inheritance in *Drosophila melanogaster*. *Biological Bulletin* **42**:267–274.

Plautz, J. D., Kaneko, M., Hall, J. C., and Kay, S. A. (1997). Independent photoreceptive circadian clocks throughout *Drosophila*. *Science* **278**:1632–1635.

Ralph, M. R., Foster, R. G., Davis, F. C., and Menaker, M. (1990). Transplanted suprachiasmatic nucleus determines circadian period. *Science* **247**:975–978.

Sack, R. L. (2010). Jet lag. *The New England Journal of Medicine* **362**:440–447.

Sehgal, A., Price, J. L., Man, B., and Young, M. W. (1994). Loss of circadian behavioral rhythms and *per* RNA oscillations in the *Drosophila* mutant *timeless*. *Science* **263**:1603–1606.

Silver, R., LeSauter, J., Tresco, P. A., and Lehman, M. N. (1996). A diffusible coupling signal from the transplanted suprachiasmatic nucleus controlling circadian locomotor rhythms. *Nature* **382**:810–813.

Tosini, G. and Menaker, M. (1996). Circadian rhythms in cultured mammalian retina. *Science* **272**:419–421.

11 Large-scale navigation: migration and homing

Beason, R. C. and Nichols, J. E. (1984). Magnetic orientation and magnetically sensitive material in a transequatorial migratory bird. *Nature* **309**:151–153.

Beason, R. C. and Semm, P. (1996). Does the avian ophthalmic nerve carry magnetic navigational information? *Journal of Experimental Biology* **199**:1241–1244.

Berdahl, A., Westley, P. A. H., Levin, S. A., Couzin, I. D., and Quinn, T. P. (2014). A collective navigation hypothesis for homeward migration in anadromous salmonids. *Fish and Fisheries* **17**:525–542.

Blakemore, R. P. (1982). Magnetotactic bacteria. *Annual Review of Microbiology* **36**:217–238.

Brower, L. P. (1996). Monarch butterfly orientation: missing pieces of a magnificent puzzle. *Journal of Experimental Biology* **199**:93–103.

Childerhose, R. J. and Trim, M. (1979). *Pacific Salmon and Steelhead Trout*. Douglas & McIntyre, Vancouver.

Cooper, J. C., Scholz, A. T., Horrall, R. M., Hasler, A. D., and Madison, D. M. (1976). Experimental confirmation of the olfactory hypothesis with artificially imprinted homing coho salmon (*Oncorhynchus kisutch*). *Journal of the Fisheries Research Board of Canada* **33**:703–710.

Dickhoff, W. W., Folmar, L. C., and Gorbman, A. (1978). Changes in plasma thyroxine during the smoltification of coho salmon, *Oncorhynchus kisutch*. *General and Comparative Endocrinology* **36**:229–232.

Dittman, A. W. and Quinn, T. P. (1996). Homing in Pacific salmon: mechanisms and ecological basis. *Journal of Experimental Biology* **199**:83–91.

Emlen, S. T. (1967). Migratory orientation in the Indigo Bunting, *Passerina cyanea*. Part I. Evidence for use of celestial cues. *Auk* **84**:309–342.

Emlen, S. T. (1967). Migratory orientation in the Indigo Bunting, *Passerina cyanea*. Part II. Mechanisms of celestial orientation. *Auk* **84**:463–489.

Emlen, S. T. and Emlen, J. T. (1966). A technique for recording migratory orientation of captive birds. *Auk* **83**:361–367.

Hasler, A. D. (1966). *Underwater Guideposts: Homing of Salmon*. University of Wisconsin Press, Madison/Milwaukee/London.

Hasler, A. D. and Scholz, A. T. (1983). *Olfactory Imprinting and Homing in Salmon: Investigations into the Mechanism of the Imprinting Process*. Springer-Verlag, Berlin/Heidelberg/New York/Tokyo.

Hasler, A. D. and Wisby, W. J. (1951). Discrimination of stream odors by fishes and relation to parent stream behavior. *American Naturalist* **85**:223–238.

Helbig, A. J. (1991). Inheritance of migratory direction in a bird species: a cross-breeding experiment with SE- and SW-migrating blackcaps (*Sylvia atricapilla*). *Behavioral Ecology and Sociobiology* **28**:9–12.

Heyers, D., Zapka, M., Hoffmeister, M., Wild, J. M., and Mouritsen, H. (2010). Magnetic field changes activate the trigeminal brainstem complex in a migratory bird. *Proceedings of the National Academy of Sciences U.S.A.* **107**:9394–9399.

Hoar, W. S. (1976). Smolt transformation: evolution, behavior and physiology. *Journal of the Fisheries Research Board of Canada* **33**:1234–1252.

Hoffmann, K. (1954). Versuche zu der im Richtungsfinden der Vögel enthaltenen Zeiteinschätzung. *Zeitschrift für Tierpsychologie* **11**:453–475.

Hoffmann, K. (1960). Experimental manipulation of the orientational clock in birds. *Cold Spring Harbor Symposia on Quantitative Biology* **25**:379–387.

Keeton, W. T. (1971). Magnets interfere with pigeon homing. *Proceedings of the National Academy of Sciences U.S.A.* **68**:102–106.

Keeton, W. T. (1980). *Biological Sciences*. W. W. Norton & Company, New York/London.

Kramer, G. (1950). Orientierte Zugaktivität gekäfigter Singvögel. *Naturwissenschaften* **37**:188.

Kramer, G. (1950). Weitere Analyse der Faktoren, welche die Zugaktivität des gekäfigten Vogels orientieren. *Naturwissenschaften* **37**:377–378.

Kramer, G. and von Saint Paul, U. (1950). Stare (*Sturnus vulgaris* L.) lassen sich auf Himmelsrichtungen dressieren. *Naturwissenschaften* **37**:526–527.

Kreithen, M. L. (1979). The sensory world of the homing pigeon. In *Neural Mechanisms of Behavior in the Pigeon* (eds A. M. Granda and J. H. Maxwell), pp. 21–33. Plenum Press, New York/London.

Kreithen, M. L. and Keeton, W. T. (1974). Detection of changes in atmospheric pressure by the homing pigeon, *Columba livia*. *Journal of Comparative Physiology* **89**:73–82.

Matthews, G. V. T. (1953). Navigation in the Manx shearwater. *Journal of Experimental Biology* **30**:370–396.

Merkel, F. W. and Fromme, H. G. (1958). Untersuchungen über das Orientierungsvermögen nächtlich ziehender Rotkehlchen, *Erithacus rubecula*. *Naturwissenschaften* **45**:499–500.

Nordeng, H. (1971). Is the local orientation of anadromous fishes determined by pheromones? *Nature* **233**:411–413.

Nordeng, H. (1977). A pheromone hypothesis for homeward migration in anadromous salmonids. *Oikos* **28**:155–159.

Papi, F. (1982). Olfaction and homing in pigeons: Ten years of experiments. In *Avian Navigation* (eds F. Papi and H. G. Wallraff), pp. 149–159. Springer-Verlag, Berlin/Heidelberg/New York.

Putman, N. F., Lohmann, K. J., Putman, E. M., Quinn, T. P., Klimley, A. P., and Noakes, D. L. G. (2013). Evidence for geomagnetic imprinting as a homing mechanism in Pacific salmon. *Current Biology* **23**:312–316.

Putman, N. F., Scanlan, M. M., Billman, E. J., O'Neil, J. P., Couture, R. B., Quinn, T. P., Lohmann, K. J., and Noakes, D. L. G. (2014). An inherited magnetic map guides ocean navigation in juvenile Pacific salmon. *Current Biology* **24**:446–450.

Quinn, T. P. and Dittman, A. H. (1990). Pacific salmon migrations and homing: mechanisms and adaptive significance. *Trends in Ecology and Evolution* **5**:174–177.

Ritz, T., Adem, S., and Schulten, K. (2000). A model for photoreceptor-based magnetoreception in birds. *Biophysical Journal* **78**:707–718.

Ritz, T., Thalau, P., Phillips, J. B., Wiltschko, R., and Wiltschko, W. (2004). Resonance effects indicate a radical-pair mechanism for avian magnetic compass. *Nature* **429**:177–180.

Sallan, L. C. and Coates, M. I. (2014). The long-rostrumed elasmobranch *Bandringa* Zangerl, 1969, and taphonomy within a carboniferous shark nursery. *Journal of Vertebrate Paleontology* **34**:22–33.

Sauer, F. (1957). Die Sternorientierung nächtlich ziehender Grasmücken (*Sylvia atricapilla, borin* und *curruca*). *Zeitschrift für Tierpsychologie* **14**:29–70.

Schmidt-Koenig, K. (1960). Internal clocks and homing. *Cold Spring Harbor Symposia on Quantitative Biology* **25**:389–393.

Scholz, A. T., Cooper, J. C., Madison, D. M., Horrall, R. M., Hasler, A. D., Dizon, A. E., and Poff, R. J. (1973). Olfactory imprinting in coho salmon: behavioral and electrophysiological evidence. *Proceedings of the 16th Conference of Great Lakes Research* **16**:143–153.

Scholz, A. T., Horrall, R. M., Cooper, J. C., and Hasler, A. D. (1976). Imprinting to chemical cues: The basis for homestream selection in salmon. *Science* **196**:1247–1249.

Schulten, K., Swenberg, C., and Weller, A. (1978). A biomagnetic sensory mechanism based on magnetic field modulated coherent electron spin motion. *Zeitschrift für Physikalische Chemie Neue Folge* **111**:1–5.

Semm, P. and Beason, R. C. (1990). Responses to small magnetic variations by the trigeminal system of the bobolink. *Brain Research Bulletin* **25**:735–740.

Teichmann, H. (1959). Über die Leistung des Geruchssinnes beim Aal (*Anguilla anguilla*). *Zeitschrift für Vergleichende Physiologie* **42**:206–254.

Treiber, C. D., Salzer, M. C., Riegler, J., Edelman, N., Sugar, C., Breuss, M., Pichler, P., Cadiou, H., Saunders, M., Lythgoe, M., Shaw, J., Keays, D. A. (2012). Clusters of iron-rich cells in the upper beak of pigeons are macrophages not magnetosensitive neurons. *Nature* **484**:367–370.

von Frisch, K. (1941). Die Bedeutung des Geruchssinnes im Leben der Fische. *Naturwissenschaften* **29**:321–333.

Walcott, C. (1996). Pigeon homing: observations, experiments and confusions. *Journal of Experimental Biology* **199**:21–27.

Walcott, C., Gould, J. L., Kirschvink, J. L. (1979). Pigeons have magnets. *Science* **205**:1027–1029.

Walcott, C. and Green, R. P. (1974). Orientation of homing pigeons altered by a change in the direction of an applied magnetic field. *Science* **184**:180–182.

Walker, M. M., Diebel, C. E., Haugh, C. V., Pankhurst, P. M., Montgomery, J. C., and Green, C. R. (1997). Structure and function of the vertebrate magnetic sense. *Nature* **390**:371–376.

Wiltschko, R. (1996). The function of olfactory input in pigeon orientation: does it provide navigational information or play another role? *Journal of Experimental Biology* **199**:113–119.

Wiltschko, R. and Wiltschko, W. (1978). Evidence for the use of magnetic outward-journey information in homing pigeons. *Naturwissenschaften* **65**:112–113.

Wiltschko, W. (1968). Über den Einfluß statischer Magnetfelder auf die Zugorientierung der Rotkehlchen (*Erithacus rubecula*). *Zeitschrift für Tierpsychologie* **25**:537–558.

Wiltschko, W. and Wiltschko, R. (1996) Magnetic orientation in birds. *Journal of Experimental Biology* **99**: 29–38.

Wu, L.-Q. and Dickman, J. D. (2012). Neural correlates of a magnetic sense. *Science* **336**:1054–1057.

12 Communication

Autrum, H. (1940). Über Lautäusserungen und Schallwahrnehmung bei Arthropoden II: Das Richtungshören von *Locusta* und Versuch einer Hörtheorie für Tympanalorgane vom Lucustidentyp. *Zeitschrift für vergleichende Physiologie* **28**:326–352.

Bennet-Clark, H. C. (2003). Wing resonances in the Australian field cricket *Teleogryllus oceanicus*. *Journal of Experimental Biology* **206**:1479–1496.

Bentley, D. and Hoy, R. R. (1974). The neurobiology of cricket song. *Scientific American* **231**:34–44.

Bolhuis, J. J., Hetebrij, E., Den Boer-Visser, A. M., De Groot, J. H., and Zijlstra, G. G. O. (2001). Localized immediate early gene expression related to the strength of song learning in socially reared zebra finches. *European Journal of Neuroscience* 13:2165–2170.

Bolhuis, J. J. and Moorman, S. (2015). Birdsong memory and the brain: In search of the template. *Neuroscience and Biobehavioral Reviews* **50**:41–55.

Bolhuis, J. J., Zijlstra, G. G. O., den Boer-Visser, A. M., and Van der Zee, E. A. (2000). Localized neuronal activation in the zebra finch brain is related to the strength of song learning. *Proceedings of the National Academy of Sciences U.S.A.* **97**:2282–2285.

Bradbury, J.W. and Vehrencamp, S.L. (2011). *Principles of Animal Communication*, 2nd edn. Sinauer Associates, Sunderland, MA.

Dambach, M. (1988). Sozialverhalten und Lauterzeugung bei der Feldgrille. In *Praktische Verhaltensbiologie* (ed. G. K. H. Zupanc), pp. 101–111. Paul Parey, Berlin/Hamburg.

Derégnaucourt, S., Mitra, P. P., Fehér, O., Pytte, C., and Tchernichovski, O. (2005). How sleep affects the developmental learning of bird song. *Nature* **433**:710–716.

Ewert, J.-P. (1980). *Neuroethology: an Introduction to the Neurophysiological Fundamentals of Behavior*. Springer-Verlag, Berlin.

Falls, B. (2015). Peter Robert Marler FRS (1928–2014). *Ibis* **157**:430–432.

Hahnloser, R. H. R., Kozhevnikov, A. A., and Fee, M. S. (2002). An ultra-sparse code underlies the generation of neural sequences in a songbird. *Nature* **419**:65–70.

Hedwig, B. (1996). A descending brain neuron elicits stridulation in the cricket *Gryllus bimaculatus* (de Geer). *Naturwissenschaften* **83**:428–429.

Hedwig, B. G. (2016). Sequential filtering processes shape feature detection in crickets: A framework for song pattern recognition. *Frontiers in Physiology* **7**:46.

Hinde, R. A. (1987). William Homan Thorpe: 1 April 1902–7 April 1986. *Biographical Memoirs of Fellows of the Royal Society* **33**:620–639.

Horseman, G. and Huber, F. (1994). Sound localisation in crickets: I. Contralateral inhibition of an ascending auditory interneuron (AN1) in the cricket *Gryllus bimaculatus*. *Journal of Comparative Physiology A* **175**:389–398.

Horseman, G. and Huber, F. (1994). Sound localisation in crickets: II. Modelling the role of a simple neural network in the prothoracic ganglion. *Journal of Comparative Physiology A* **175**:399–413.

Huber, F. (1990). Cricket neuroethology: neuronal basis of intraspecific acoustic communication. *Advances in the Study of Behavior* **19**:299–356.

Huber, F. (1990). Nerve cells and insect behavior: studies on crickets. *American Zoologist* **30**:609–627.

Huber, F. and Thorson, J. (1985). Cricket auditory communication. *Scientific American* **253**:46–54.

Huber, F., Moore, T. E., and Loher, W. (eds) (1989). *Cricket Behavior and Neurobiology*. Cornell University Press, Ithaca/London.

Jacob, P. F. and Hedwig, B. (2016) Acoustic signaling for mate attraction in crickets: Abdominal ganglia control the timing of the calling song pattern. *Behavioural Brain Research* **309**:51–66.

Kao, M. H. and Brainard, M. S. (2006). Lesions of an avian basal ganglia circuit prevent context-dependent changes to song variability. *Journal of Neurophysiology* **96**:1441–1455.

Keeton, W. T. (1980). *Biological Sciences*. W. W. Norton & Company, New York/London.

Kleindienst, H.-U., Wohlers, D. W., and Larsen, O. N. (1983). Tympanal membrane motion is necessary for hearing in crickets. *Journal of Comparative Physiology A* **151**:397–400.

Lankheet, M. J., Cerkvenik, U., Larsen, O. N., and van Leeuwen, J. L. (2017). Frequency tuning and directional sensitivity of tympanal vibrations in the field cricket *Gryllus bimaculatus*. *Journal of the Royal Society Interface* **14**:20170035.

Larsen, O. N. and Michelsen, A. (1978). Biophysics of the ensiferan ears: III. The cricket ear as a four-input system. *Journal of Comparative Physiology A* **123**: 217–227.

Loher, W. and Dambach, M. (1989). Reproductive behavior. In *Cricket Behavior and Neurobiology* (eds F. Huber, T. E. Moore, and W. Loher), pp. 43–82. Cornell University Press, Ithaca/London.

London, S. E. and Clayton, D. F. (2008). Functional identification of sensory mechanisms required for developmental song learning. *Nature Neuroscience* **11**:579–586.

Long, M. A. and Fee, M. S. (2008). Using temperature to analyse temporal dynamics in the songbird motor pathway. *Nature* **456**:189–194.

Manning, A. and Stamp Dawkins, M. (1992). *An Introduction to Animal Behaviour*, 4th edn. Cambridge University Press, Cambridge/New York/Melbourne.

Marler, P. (1956). Behaviour of the chaffinch *Fringilla coelebs*. *Behaviour Supplement* 5:1–184.

Marler, P. (1970). A comparative approach to vocal learning: song development in white-crowned sparrows. *Journal of Comparative and Physiological Psychology Monograph* 71:1–25.

Marler, P. and Tamura, M. (1964). Culturally transmitted patterns of vocal behavior in sparrows. *Science* 146:1483–1486.

McCasland, J. S. and Konishi, M. (1981). Interaction between auditory and motor activities in an avian song control nucleus. *Proceedings of the National Academy of Sciences U.S.A.* 78:7815–7819.

Mooney, R. (1992). Synaptic basis for developmental plasticity in a birdsong nucleus. *Journal of Neuroscience* 12:2464–2477.

Mooney, R. (2009). Neural mechanisms for learned birdsong. *Learning & Memory* 16:655–669.

Mooney, R. and Konishi, M. (1991). Two distinct inputs to an avian song nucleus activate different glutamate receptor subtypes on individual neurons. *Proceedings of the National Academy of Sciences U.S.A.* 88:4075–4079.

Moorman, S., Mello, C. V., and Bolhuis, J. J. (2011). From songs to synapses: Molecular mechanisms of birdsong memory. *Bioessays* 33:377–385.

Moynihan, M. (1966). Communication in the Titi monkey, *Callicebus*. *Journal of Zoology (London)* 150:77–127.

Nottebohm, F. (2014). Peter Marler (1928–2014). *Nature* 512:372.

Nottebohm, F., Stokes, T. M., and Leonard, C. M. (1976). Central control of song in the canary, *Serinus canarius*. *Journal of Comparative Neurology* 165:457–486.

Ölveczky, B. P., Andalman, A. S., and Fee, M. S. (2005). Vocal experimentation in the juvenile songbird requires a basal ganglia circuit. *PLoS Biology* 3:e153.

Pires, A. and Hoy, R. R. (1992). Temperature coupling in cricket acoustic communication: I. Field and laboratory studies of temperature effects on calling song production and recognition in *Gryllus firmus*. *Journal of Comparative Physiology A* 171:69–78.

Regen, J. (1914). Über die Anlockung des Weibchens von *Gryllus campestris* L. durch telephonisch übertragene Stridulationslaute des Männchens. *Pflüger's Archiv für die gesamte Physiologie des Menschen und der Tiere* 155:193–200.

Schildberger, K. (1984). Temporal selectivity of identified auditory neurons in the cricket brain. *Journal of Comparative Physiology A* 155:171–185.

Schildberger, K. and Hörner, M. (1988). The function of auditory neurons in cricket phonotaxis. I. Influence of hyperpolarization of identified neurons on sound localization. *Journal of Comparative Physiology A* 163:621–631.

Schmitz, B., Scharstein, H., and Wendler, G. (1982). Phonotaxis in *Gryllus campestris* L. (Orthoptera, Gryllidae): I. Mechanisms of acoustic orientation in intact female crickets. *Journal of Comparative Physiology A* 148:431–444.

Schmitz, B., Scharstein, H., and Wendler, G. (1983). Phonotaxis in *Gryllus campestris* L. (Orthoptera, Gryllidae): II. Acoustic orientation of female crickets after occlusion of single sound entrances. *Journal of Comparative Physiology A* 152:257–264.

Schöneich, S. and Hedwig, B. (2011). Neural basis of singing in crickets: Central pattern generation in abdominal ganglia. *Naturwissenschaften* 98:1069–1073.

Schöneich, S., Kostarakos, K., and Hedwig, B. (2015). An auditory feature detection circuit for sound pattern recognition. *Science Advances* 1:e1500325.

Selverston, A. I., Kleindienst, H.-U., and Huber, F. (1985). Synaptic connectivity between cricket auditory interneurons as studied by selective photoinactivation. *Journal of Neuroscience* 5:1283–1292.

Seyfarth, R. M. and Cheney, D. L. (2015). Obituary: Peter Marler (February 24, 1928—July 5, 2014). *International Journal of Primatology* 36:14–17.

Slabbekoorn, H., Jesse, A., and Bell, D.A. (2003). Microgeographic song variation in island populations of the white-crowned sparrow (*Zonotrichia leucophrys nuttalli*): innovation through recombination. *Behaviour* 140:947–963.

Solis, M. M. and Doupe, A. J. (1997). Anterior forebrain neurons develop selectivity by an intermediate stage of birdsong learning. *Journal of Neuroscience* 17:6447–6462.

Solis, M. M. and Doupe, A. J. (1999). Contributions of tutor and bird's own song experience to neural selectivity in the songbird anterior forebrain. *Journal of Neuroscience* 19:4559–4584.

Stark, L. L. and Perkel, D. J. (1999). Two-stage, input-specific synaptic maturation in a nucleus essential for vocal production in the zebra finch. *Journal of Neuroscience* 19:9107–9116.

Stout, J. F. and Huber, F. (1981). Responses to features of the calling song by ascending auditory interneurons in the cricket *Gryllus campestris*. *Physiological Entomology* 6:199–212.

Stresemann, E. (1947). Baron von Pernau, pioneer student of bird behavior. *The Auk* 64:35–52.

Tchernichovski, O., Nottebohm, F., Ho, C. E., Pesaran, B., and Mitra, P. P. (2000). A procedure for an automated measurement of song similarity. *Animal Behaviour* 59:1167–1176.

Tchernichovski, O., Mitra, P. P. Lints, T., and Nottebohm, F. (2001). Dynamics of the vocal imitation process: How a zebra finch learns its song. *Science* 291:2564–2569.

Thorpe, W. H. (1954). The process of song-learning in the chaffinch as studied by means of the sound spectrograph. *Nature* 173:465–469.

Thorpe, W. H. (1958). The learning of song patterns by birds, with special reference to the song of the chaffinch *Fringilla coelebs*. *Ibis* 100:535–570.

Thorson, J., Weber, T., and Huber, F. (1982). Auditory behavior of the cricket: II. Simplicity of calling-song recognition in *Gryllus*, and anomalous phonotaxis at abnormal carrier frequencies. *Journal of Comparative Physiology A* **146**:361–378.

Weber, T., Thorson, J., and Huber, F. (1981). Auditory behavior of the cricket: I. Dynamics of compensated walking and discrimination paradigms on the Kramer treadmill. *Journal of Comparative Physiology A* **141**:215–232.

Wickler, W. (1978). Das Problem der stammesgeschichtlichen Sackgassen. In *Evolution II: Ein Querschnitt der Forschung* (ed. H. von Ditfurth), pp. 29–47. Hoffmann und Campe, Hamburg.

Wignall, A. E. and Herberstein, M. E. (2013). Male courtship vibrations delay predatory behaviour in female spiders. *Scientific Reports* **3**:3557.

Wohlers, D. W. and Huber, F. (1982). Processing of sound signals by six types of neurons in the prothoracic ganglion of the cricket, *Gryllus campestris* L. *Journal of Comparative Physiology A* **146**:161–173.

Young, D. and Ball, E. (1974). Structure and development of the auditory system in the prothoracic leg of the cricket *Teleogryllus commodus* (Walker): I. Adult structure. *Zeitschrift für Zellforschung und mikroskopische Anatomie* **147**:293–312.

13 Cellular mechanisms of learning and memory

Abel, T., Nguyen, P. V., Barad, M., Deuel, T. A. S., Kandel, E. R., and Bourtchouladze, R. (1997). Genetic demonstration of a role for PKA in the late phase of LTP and in hippocampus-based long-term memory. *Cell* **88**:615–626.

Abel, T., Martin, K. C., Bartsch, D., and Kandel, E. R. (1998). Memory suppressor genes: inhibitory constraints on the storage of long-term memory. *Science* **279**:338–341.

Bailey, C. H. and Chen, M. (1983). Morphological basis of long-term habituation and sensitization in *Aplysia*. *Science* **220**:91–93.

Bailey, C. H. and Chen, M. (1988). Long-term memory in *Aplysia* modulates the total number of varicosities of single identified sensory neurons. *Proceedings of the National Academy of Sciences U.S.A.* **85**:2373–2377.

Bailey, C. H., Kandel, E. R., and Harris, K. M. (2015). Structural components of synaptic plasticity and memory consolidation. *Cold Spring Harbor Perspectives in Biology* **7**:a021758.

Barnea, A. and Nottebohm, F. (1994). Seasonal recruitment of hippocampal neurons in adult free-ranging black-capped chickadees. *Proceedings of the National Academy of Sciences U.S.A.* **91**:11217–11221.

Barnea, A. and Nottebohm, F. (1996). Recruitment and replacement of hippocampal neurons in young and adult chickadees: an addition to the theory of hippocampal learning. *Proceedings of National Academy of Sciences U.S.A.* **93**:714–718.

Bartsch, D., Ghirardi, M., Skehel, P. A., Karl, K. A., Herder, S. P., Chen, M., Bailey, C. H., and Kandel, E. R. (1995). *Aplysia* CREB2 represses long-term facilitation: Relief of repression converts transient facilitation into long-term functional and structural change. *Cell* **83**:979–992.

Bliss, T. V. P. and Lømo, T. (1973). Long-lasting potentiation of synaptic transmission in the dentate area of the anaesthetized rabbit following stimulation of the perforant path. *Journal of Physiology* **232**:331–356.

Byers, D., Davis, R. L., and Kiger Jr., J. A. (1981). Defect in cyclic AMP phosphodiesterase due to the *dunce* mutation of learning in *Drosophila melanogaster*. *Nature* **289**:79–81.

Carew, T. J. and Sahley, C. L. (1986). Invertebrate learning and memory: From behavior to molecules. *Annual Review of Neuroscience* **9**:435–487.

Dudai, Y., Jan, Y.-N., Byers, D., Quinn, W. G., and Benzer, S. (1976). *dunce*, a mutant of *Drosophila* deficient in learning. *Proceedings of the National Academy of Sciences U.S.A.* **73**:1684–1688.

Engert, F. and Bonhoeffer, T. (1999). Dendritic spine changes associated with hippocampal long-term synaptic plasticity. *Nature* **399**:66–70.

Frost, W. N., Castellucci, V. F., Hawkins, R. D., and Kandel, E. R. (1985). Monosynaptic connections made by the sensory neurons of the gill- and siphon-withdrawal reflex in *Aplysia* participate in the storage of long-term memory for sensitization. *Proceedings of the National Academy of Sciences U.S.A.* **82**:8266–8269.

Glanzman, D. L., Mackey, S. L., Hawkins, R. D., Dyke, A. M., Lloyd, P. E., and Kandel, E. R. (1989). Depletion of serotonin in the nervous system of *Aplysia* reduces the behavioral enhancement of gill withdrawal as well as the heterosynaptic facilitation produced by tail shock. *Journal of Neuroscience* **9**:4200–4213.

Hafting, T., Fyhn, M., Bonnevie, T., Moser, M.-B., and Moser, E. I. (2008). Hippocampus-independent phase precession in entorhinal grid cells. *Nature* **453**:1248–1252.

Hebb, D. O. (1949). *The Organization of Behavior: A Neuropsychological Theory*. John Wiley & Sons, New York.

Igarashi, K. M., Lu, L., Colgin, L. L., Moser, M.-B., and Moser, E. I. (2014). Coordination of entorhinal–hippocampal ensemble activity during associative learning. *Nature* **510**:143–147.

Kandel, E. R. (2001). The molecular biology of memory storage: a dialogue between genes and synapses. *Science* **294**:1030–1038.

Kempermann, G., Kuhn, H. G., and Gage, F. H. (1997). More hippocampal neurons in adult mice living in an enriched environment. *Nature* **386**:493–495.

Kempermann, G., Kuhn, H. G., and Gage, F. H. (1998). Experience-induced neurogenesis in the senescent dentate gyrus. *Journal of Neuroscience* **18**:3206–3212.

Kim, J.-H., Udo, H., Li, H.-L., Youn, T. Y., Chen, M., Kandel, E. R., and Bailey, C. H. (2003). Presynaptic activation of silent synapses and growth of new synapses contribute to intermediate and long-term facilitation in *Aplysia*. *Neuron* **40**:151–165.

Lechner, H. A. and Byrne, J. H. (1998). New perspectives on classical conditioning: A synthesis of Hebbian and non-Hebbian mechanisms. *Neuron* 20:355–358.

Markakis, E. A. and Gage, F. H. (1999). Adult-generated neurons in the dentate gyrus send axonal projections to field CA_3 and are surrounded by synaptic vesicles. *Journal of Comparative Neurology* 406:449–460.

Mayford, M. and Kandel, E. R. (1999). Genetic approaches to memory storage. *Trends in Genetics* 15:463–470.

Meshi, D., Drew, M. R., Saxe, M., Ansorge, M. S., David, D., Santarelli, L., Malapani, C., Moore, H., and Hen, R. (2006). Hippocampal neurogenesis is not required for behavioral effects of environmental enrichment. *Nature Neuroscience* 9:729–731.

Milner, B., Squire, L. R., and Kandel, E. R. (1998). Cognitive neuroscience and the study of memory. *Neuron* 20:445–468.

Monaco, J. D., Rao, G., Roth, E. D., and Knierim, J. J. (2014). Attentive scanning behavior drives one-trial potentiation of hippocampal place fields. *Nature Neuroscience* 17:725–731.

Nakashiba, T., Cushman, J. D., Pelkey, K. A., Renaudineau, S., Buhl, D. L., McHugh, T. J., Barrera, V. R., Chittajallu, R., Iwamoto, K. S., McBain, C. J., Fanselow, M. S., and Tonegawa, S. (2012). Young dentate granule cells mediate pattern separation, whereas old granule cells facilitate pattern completion. *Cell* 149:188–201.

O'Keefe, J., Burgess, N., Donnett, J. G., Jeffery, K. J., and Maguire, E. A. (1998). Place cells, navigational accuracy, and the human hippocampus. *Philosophical Transactions of the Royal Society of London B* 353:1333–1340.

O'Keefe, J. and Dostrovsky, J. (1971). The hippocampus as a spatial map. Preliminary evidence from unit activity in the freely-moving rat. *Brain Research* 34:171–175.

O'Keefe, J. and Nadel, L. (1978). *The Hippocampus as a Cognitive Map*. Clarendon Press, Oxford.

Patel, S. N., Clayton, N. S., and Krebs, J. R. (1997). Spatial learning induces neurogenesis in the avian brain. *Behavioural Brain Research* 89:115–128.

Quinn, W. G., Harris, W. A., and Benzer, S. (1974). Conditioned behavior in *Drosophila melanogaster*. *Proceedings of the National Academy of Sciences U.S.A.* 71:708–712.

Rotenberg, A., Abel, T., Hawkins, R. D., Kandel, E. R., and Muller, R. U. (2000). Parallel instabilities of long-term potentiation, place cells, and learning caused by decreased protein kinase A activity. *Journal of Neuroscience* 20:8096–8102.

Sahay, A., Scobie, K. N., Hill, A. S., O'Carroll, C. M., Kheirbek, M. A., Burghardt, N. S., Fenton, A. A., Dranovsky, A., and Hen, R. (2011). Increasing adult hippocampal neurogenesis is sufficient to improve pattern separation. *Nature* 472:466–470.

Shors, T. J., Miesegaes, G., Beylin, A., Zhao, M., Rydel, T., and Gould, E. (2001). Neurogenesis in the adult is involved in the formation of trace memories. *Nature* 410:372–376.

Squire, L. R. and Zola, S. M. (1996). Structure and function of declarative and nondeclarative memory systems. *Proceedings of the National Academy of Sciences U.S.A.* 93:13515–13522.

Squire, L. R. and Zola-Morgan, S. (1991). The medial temporal lobe memory system. *Science* 253:1380–1386.

van Praag, H., Kempermann, G., and Gage, F. H. (1999). Running increases cell proliferation and neurogenesis in the adult mouse dentate gyrus. *Nature Neuroscience* 2:266–270.

van Praag, H., Kempermann, G., and Gage, F. H. (2000). Neural consequences of environmental enrichment. *Nature Reviews* 1:191–198.

Yassa, M. A. and Stark, C. E. L. (2011). Pattern separation in the hippocampus. *Trends in Neurosciences* 34:515–525.

INDEX

A

C

D

E

M

O

P

Q

R

T

U

V

W

X

Y

Z